# Lecture Notes in Physics

## The Lecture Notes in Physics

The series Lecture Notes in Physics (LNP), founded in 1969, reports new developments in physics research and teaching – quickly and informally, but with a high quality and the explicit aim to summarize and communicate current knowledge in an accessible way. Books published in this series are conceived as bridging material between advanced graduate textbooks and the forefront of research and to serve three purposes:

- to be a compact and modern up-to-date source of reference on a well-defined topic

- to serve as an accessible introduction to the field to postgraduate students and nonspecialist researchers from related areas

- to be a source of advanced teaching material for specialized seminars, courses and schools

Both monographs and multi-author volumes will be considered for publication. Edited volumes should, however, consist of a very limited number of contributions only. Proceedings will not be considered for LNP.

Volumes published in LNP are disseminated both in print and in electronic formats, the electronic archive being available at springerlink.com. The series content is indexed, abstracted and referenced by many abstracting and information services, bibliographic networks, subscription agencies, library networks, and consortia.

Proposals should be sent to a member of the Editorial Board, or directly to the managing editor at Springer:

Christian Caron
Springer Heidelberg
Physics Editorial Department I
Tiergartenstrasse 17
69121 Heidelberg / Germany
christian.caron@springer.com

J. Daillant
A. Gibaud (Eds.)

# X-ray and Neutron Reflectivity

## Principles and Applications

 Springer

Jean Daillant
CEA-Saclay
IRAMIS, LIONS
91191 Gif-sur-Yvette Cedex
France
jean.daillant@cea.fr

Alain Gibaud
Université du Maine
Faculté des Sciences
bd. O. Messiaen
72085 Le Man Cedex 9
France
alain.gibaud@univ-lemans.fr

Daillant, J., Gibaud, A. (Eds.), *X-ray and Neutron Reflectivity: Principles and Applications*, Lect. Notes Phys. 770 (Springer, Berlin Heidelberg 2009), DOI 10.1007/978-3-540-88588-7

ISBN: 978-3-642-10017-8                    e-ISBN: 978-3-540-88588-7

DOI 10.1007/978-3-540-88588-7

Lecture Notes in Physics ISSN: 0075-8450          e-ISSN: 1616-6361

© Springer-Verlag Berlin Heidelberg 2009
Softcover reprint of the hardcover 1st edition 2009

*Cover design:* Integra Software Services Pvt. Ltd.

Printed on acid-free paper

9  8  7  6  5  4  3  2  1

springer.com

# Foreword to the First Edition

The reflection of x-rays and neutrons from surfaces has existed as an experimental technique for almost 50 years. Nevertheless, it is only in the last decade that these methods have become enormously popular as probes of surfaces and interfaces. This appears to be due to the convergence of several different circumstances. These include the availability of more intense neutron and x-ray sources (so that reflectivity can be measured over many orders of magnitude and the much weaker surface diffuse scattering can now also be studied in some detail); the growing importance of thin films and multilayers in both technology and basic research; the realization of the important role which roughness plays in the properties of surfaces and interfaces; and finally the development of statistical models to characterize the topology of roughness, its dependence on growth processes and its characterization from surface scattering experiments. The ability of x-rays and neutrons to study surfaces over 4–5 orders of magnitude in length scale regardless of their environment, temperature, pressure, etc., and also their ability to probe buried interfaces often makes these probes the preferred choice for obtaining global statistical information about the microstructure of surfaces, often in a complementary manner to the local imaging microscopy techniques. This is witnessed by the veritable explosion of such studies in the published literature over the last few years. Thus these lectures will provide a useful resource for students and researchers alike, covering as they do in considerable detail most aspects of surface x-ray and neutron scattering from the basic interactions through the formal theories of scattering and finally to specific applications.

It is often assumed that neutrons and x-rays interact weakly with surfaces and in general interact weakly enough so that the simple kinematic theories of scattering are good enough approximations to describe the scattering. As most of us now appreciate, this is not always true, e.g., when the reflection is close to being total, or in the neighborhood of strong Bragg reflections (e.g., from multilayers). This necessitates the need for the full dynamical theory (which for specular reflectivity is fortunately available from the theory of optics), or for higher order approximations, such as the distorted wave Born approximation to describe strong off-specular scattering. All these methods are discussed in detail in these lectures, as are also the ways in which the magnetic interaction between neutrons and magnetic moments can yield information on the magnetization densities of thin films and multilayers. I commend the organizers for having organized a group of expert lecturers to present this subject in a detailed but clear fashion, as the importance of the subject deserves.

Argonne, IL                                                          *S.K. Sinha*

# Contents

# Introduction

In his paper entitled "On a New Kind of Ray, A Preliminary Communication" relating the discovery of x-rays, which was submitted to the Würzburg Physico-Medical Society on December 28, 1895, Röntgen stated the following about the refraction and reflection of the newly discovered rays: "The question as to the reflection of the X-ray may be regarded as settled, by the experiments mentioned in the preceding paragraph, in favor of the view that *no noticeable regular reflection of the rays takes place from any of the substances examined*. Other experiments, which I here omit, lead to the same conclusion.[1]"

This conclusion remained unquestioned until in 1922 Compton [1] pointed out that if the refractive index of a substance for x-rays is less than unity, it ought to be possible, according to the laws of optics, to obtain total external reflection from a smooth surface of it, since the x-rays, on entering the substance from the air, are going into a medium of smaller refractive index. This was the starting point for x-ray (and neutron) reflectivity. The demonstration that the reflection of x-rays on a surface was indeed obeying the laws of electromagnetism was pursued by Prins [2,3] and others who investigated the role of absorption on the sharpness of the limit of total reflection and showed that it was consistent with the Fresnel formulae. This work was continued by Kiessig [4] using nickel films evaporated on glass. Reflection on such thin films gives rise to fringes of equal inclination (the "Kiessig fringes" in the x-ray literature), which allow the measurement of thin film thicknesses, now the most important application of x-ray and neutron reflectivity. It was, however, not until 1954 that Parratt [5] suggested inverting the analysis and interpreting x-ray reflectivity as a function of angle of incidence via models of an *inhomogeneous* surface density distribution. The method was then applied to several cases of solid or liquid [6] interfaces. Whereas Parratt noticed in his 1954 paper that "it is at first surprising that any experimental surface appears smooth to x-rays. One frequently hears that, for good reflection, a mirror surface must be smooth to within about one wavelength of the radiation involved..." it soon appeared that effects of surface roughness were important, the most dramatic of them being the asymmetric

---

[1] A more complete citation of Röntgen's paper is given in an appendix to this introduction.

surface reflection known as Yoneda wings [7]. These Yoneda wings were subsequently interpreted as diffuse scattering of the enhanced surface field for incidence or exit angle equal to the critical angle for total external reflection. The theoretical basis for the analysis of this surface diffuse scattering was established in particular through the pioneering work of Croce et al. [8]. In a context where coatings, thin films and nanostructured materials are playing an increasingly important role for applications, the number of studies using x-ray or neutron reflectivity dramatically increased during the 1990s, addressing virtually all kinds of interfaces: solid or liquid surfaces, buried solid–liquid or liquid–liquid interfaces, interfaces in thin films and multilayers. Apart from the scientific and technological demand for more and more surface characterisation, at least two factors explain this blooming of x-ray and neutron reflectivity. First, the use of second and third generation synchrotron sources has resulted in a sophistication of the technique now such that not only the thicknesses but also the morphologies and correlations within and between rough interfaces can be accurately characterised for in-plane distances ranging from atomic or molecular distances to hundreds of microns. In parallel more and more accurate methods have been developed for data analysis. Second, the development of neutron reflectometers (Chap. 5) has been decisive, in particular for polymer physics owing to partial deuteration and for magnetism.

This book follows summer schools on reflectivity held in Luminy in June 1997, Le Croisic in June 2000 and Giens in May 2008. Since the first edition of the book published in 1999, x-ray and neutron reflectivity have continued to develop and new related techniques like grazing incidence small angle scattering (GISAXS) have become very popular. The first aim of this second edition was therefore to include these important new developments. Moreover, the first edition was organised into two parts, "principles" and "applications" whose aim was to give examples of the use of reflectivity in different fields. Several excellent reviews have been published since then and it was no longer necessary to include a review in the book. We found it more useful to include examples in the different chapters devoted to reflectivity-related methods as tutorials. This is the second main change made to the book.

As strongly suggested by the short historical sketch given above, most of the revolutions in the use of x-rays (not only for interface studies) arise by considering new potentialities related to their nature of electromagnetic waves, which was so controversial in the days of Röntgen. The book therefore starts with a panorama of the interaction of x-rays with matter, giving both a thorough treatment of the basic principles, and an overview of more advanced topics like magnetic or anisotropic scattering, not only to give a firm basis to the following developments but also to stimulate reflection on new experiments.

Then, a rigorous presentation of the statistical aspects of wave scattering at rough surfaces is given. This point, obviously important for understanding the nature of surface scattering experiments, as well as for their interpretation, is generally ignored in the x-ray literature (this chapter has been written mainly by a researcher in optics). The basic statistical properties of surfaces are introduced first. Then an ideal scattering experiment is described, and the limitations of such a description, in particular the fact that the experimental resolution is always finite, are discussed.

The finiteness of the resolution leads to the introduction of ensemble averages for the calculation of the scattered intensities and to a natural distinction between coherent (specular, equal to the average of the scattered field) and incoherent (diffuse, related to the mean-square deviation of the scattered field) scattering. These principles are immediately illustrated within the Born approximation in order to avoid all the mathematical complications resulting from the details of the interaction of an electromagnetic wave with matter.

These more rigorous aspects of the scattering theory are treated in Chaps. 3 and 4 for specular and diffuse scattering. The matricial theory of the reflection of light in a smooth or rough stratified medium and its consequences are treated in Chap. 3. A new section on the inversion of reflectivity data has been added to this chapter. The developments of Chap. 3 are used in Chap. 4 for the treatment of diffuse scattering. The Croce approach to the distorted-wave Born approximation (DWBA) based on the use of Green functions is mainly used. It is currently the most popular for data analysis and is extensively used in particular in Chap. 6. The presentation of the DWBA is complemented by the discussion of a more simple approximation, very useful in particular for thin films. The derivation of the scattered intensity from the scattering cross section is described in detail as well as the implications for reflectivity experiments of a finite resolution. Examples are finally discussed in detail, in particular for liquid surfaces and thin films for which a full calculation of the scattering cross section can be made.

The specific aspects of neutron reflectometry require a separate treatment given in Chap. 5. After an introduction to neutron–matter interactions, neutron reflectivity of non-magnetic and magnetic materials is presented and the characteristics of the neutron spectrometers are given. Special attention is paid to the case of non-perfect layers. The theoretical presentation is followed by examples including biological and magnetic films, off-specular reflectivity and grazing incidence scattering.

Multilayers are discussed in Chap. 6. The experimental set-ups are described and examples of reflectivity studies and non-specular scattering measurements are discussed with the aim of reviewing all the important situations that can be encountered. Examples include rough multilayers, stepped surfaces, interfaces in porous media, the role of roughness in diffraction experiments and multilayer gratings.

GISAXS is discussed in Chap. 7. The emphasis is put on the characterisation of nano-objects on surfaces or buried in a substrate. The application of the DWBA to GISAXS is discussed after an introduction to the GISAXS scattering geometry. Form factors are given for a large number of nano-objects and the effect of their correlations is discussed. Examples in hard and soft condensed matter are finally given.

## Appendix: Röntgen's Report on the Reflection of X-Rays

"With reference to the general conditions here involved on the other hand, and to the importance of the question whether the X-rays can be refracted or not on passing from one medium into another, it is most fortunate that this subject may be investigated in still another way than with the aid of prisms. Finely divided bodies in sufficiently thick layers scatter the incident light and allow only a little of it to pass, owing to reflection and refraction; so that if powders are as transparent to X-rays as the same substances are in mass–equal amounts of material being presupposed– it follows at once that neither refraction nor regular reflection takes place to any sensible degree. Experiments were tried with finely powdered rock salt, with finely electrolytic silver-powder, and with zinc-dust, such as is used in chemical investigations. In all these cases no difference was detected between the transparency of the powder and that of the substance in mass, either by observation with the fluorescent screen or with the photographic plate... The question as to the reflection of the X-ray may be regarded as settled, by the experiments mentioned in the preceding paragraph, in favor of the view that no noticeable regular reflection of the rays takes place from any of the substances examined. Other experiments, which I here omit, lead to the same conclusion.

One observation in this connection should, however, be mentioned, as at first sight it seems to prove the opposite. I exposed to the X-rays a photographic plate which was protected from the light by black paper, and the glass side of which was turned towards the discharge-tube giving the X-rays. The sensitive film was covered, for the most part, with polished plates of platinum, lead, zinc, and aluminum, arranged in the form of a star. On the developing negative it was seen plainly that the darkening under the platinum, the lead and particularly the zinc, was stronger than under the other plates, the aluminum having exerted no action at all. It appears, therefore, that these metals reflect the rays. Since, however, other explanations of a stronger darkening are conceivable, in a second experiment, in order to be sure, I placed between the sensitive film and the metal plates a piece of thin aluminum-foil, which is opaque to ultraviolet rays, but it is very transparent to the X-rays. Since the same result substantially was again obtained, the reflection of the X-rays from the metals above named is proved. If we compare this fact with the observation already mentioned that powders are as transparent as coherent masses, and with the further fact that bodies with rough surfaces behave like polished bodies with reference to the passage of the X-rays, as shown as in the last experiment, we are led to the conclusion already stated that regular reflection does not take place, but that bodies behave toward X-rays as turbid media do towards light."

# References

1. Compton, A.H.: Phil. Mag. **45**, 1121 (1923)
2. Prins, J.A.: Zeit. f. Phys. **47**, 479 (1928)
3. James, R.W.: The Optical Principles of the Diffraction of X-Rays, Bell and sons, London (1948)
4. Kiessig, H.: Ann. der Physik **10**, 715, 769 (1931)
5. Parratt, L.G.: Phys. Rev. **95**, 359 (1954)
6. Lu, B.C., Rice, S.A.: J. Chem. Phys. **68**, 5558 (1978)
7. Yoneda, Y.: Phys. Rev. **131**, 2010 (1963)
8. Croce, P., Névot, L., Pardo, B.: C. R. Acad. Sc. Paris **274**, 803, 855 (1972)

# Chapter 1
# The Interaction of X-Rays (and Neutrons) with Matter

F. de Bergevin

## 1.1 Introduction

The propagation of radiation is generally presented according to an optical formalism in which the properties of a medium are described by a refractive index. A knowledge of the refractive index is sufficient to predict what will happen at an interface, that is to establish the Snell–Descartes' laws and to calculate the Fresnel coefficients for reflection and transmission.

One of the objectives in this introduction will be to link the laws of propagation of radiation, and in particular the refractive index, to the fundamental phenomena involved in the interaction of radiation with matter. The main process of interaction in the visible region of the electromagnetic spectrum is the polarization of the molecules (at least for an insulator). At higher energies as with X-rays, it is generally sufficient to take into account the interactions with the atoms and at the highest X-ray energies only the electrons need be considered in the interaction process. Neutrons interact with the nuclei of the materials, and also have another interaction with the electrons for those atoms which carry a magnetic moment.

The conventions used in this book will be defined in Sect. 1.2. In the same section the basics of wave propagation will be revised. The different physical quantities which characterize the scattering of radiation will be defined, and also the properties of Green functions will be explained. In Sect. 1.3 the link between the atomic scattering and the model of a continuous medium represented by a refractive index will be established. Section 1.4 will be devoted to the interaction of X-ray radiation with matter. That will include the inelastic and elastic scattering, and the absorption. The scattering will be described as split into a non-resonant and a resonant part. Together with the questions of resonance and absorption a discussion of the dispersion relations will be given. In Sect. 1.5, the case when the scattering depends on the anisotropy of the material will be briefly examined with reference to magnetic and to Templeton scattering. Neutron scattering will not be presented in detail in this chapter since it is treated in Chap. 5 of this book but we shall frequently refer to it.

F. de Bergevin (✉)
European Synchrotron Radiation Facility, B.P. 220, 38043 Grenoble Cedex 9, France

de Bergevin, F.: *The Interaction of X-Rays (and Neutrons) with Matter.* Lect. Notes Phys. **770**, 1–57 (2009)
DOI 10.1007/978-3-540-88588-7_1

In the present chapter, the **bold italic** font will be used to **define words or expressions** and the *emphasized sentences* will be in *italic*.

Note to the reader of the first edition. A number of minor changes have been introduced in this chapter for the present edition. On top of these, the main corrected errors are the following.

- Equation (1.65), a wrong sign
- Equation (1.105), a factor $c$
- Equation (1.106), $_{in,\ sc}$ exchanged in the third term
- Section 1.5.5, optical activity in the optical range of the light spectrum was wrongly attributed to a quadrupolar term.

Some part of Sect. 1.3 has been rewritten in a different way, a figure added in Sect. 1.3.3 to clear up the meaning of Eq. (1.87) and a few practical remarks added to Sect. 1.4.8.


## 1.2 Generalities and Definitions

### *1.2.1 Conventions, Basic Formulae*

Two conventions can be found in the literature to describe a propagating wave, because complex quantities are not observed and their imaginary part has an arbitrary sign. In optics and quantum mechanics a monochromatic plane wave is generally written as

$$A \propto e^{-i(\omega t - \mathbf{k}.\mathbf{r})}, \tag{1.1}$$

which is also the notation used in neutron scattering, even when doing crystallography. On the other hand, X-ray crystallographers are used to writing the plane wave as

$$A \propto e^{+i(\omega t - \mathbf{k}.\mathbf{r})}. \tag{1.2}$$

The imaginary parts of all complex quantities are the opposite of one another in these two notations. Since the observed real quantities may be calculated from imaginary numbers, it is very important to keep consistently a unique convention. The imaginary part $f''$ of the atomic scattering factor, for example, used in X-ray crystallography is a positive number. This is correct provided that it is remembered that the complex scattering factor $(f + f' + if'')$ ($f$ is the atomic form factor, also positive) is affected by a common minus sign, usually left as implicit. In optics, the opposite convention is commonly used and the most useful quantity is the refractive index. Its imaginary part which is associated with absorption is always positive. The number of alternative choices is increased with another convention concerning the sign of the scattering wave vector transfer $\mathbf{q}$ or scattering vector, which can be written as

$$\mathbf{q} = \mathbf{k}_{sc} - \mathbf{k}_{in} \tag{1.3}$$

or

$$\mathbf{q} = \mathbf{k}_{in} - \mathbf{k}_{sc}, \tag{1.4}$$

where $\mathbf{k}_{in}$ and $\mathbf{k}_{sc}$ are the incident and scattered wave vectors. In this book, the conventions Eqs. (1.2) and (1.3) as used in crystallography will be adopted. *Only one exception will be made in the chapter devoted to neutrons (Chap. 5), in which convention Eq. (1.1) will be used.* The structure factor which describes the scattered amplitude in the Born approximation will therefore be written in all cases (except with neutrons) as

$$f(\mathbf{q}) = \int \rho(\mathbf{r}) e^{i\mathbf{q}\cdot\mathbf{r}} d^3\mathbf{r}, \tag{1.5}$$

where $\rho(\mathbf{r})$ is the scattering density, which will be discussed below. The real part of the refractive index is generally less than 1 with X-ray radiation and the refractive index is usually written as

$$n = 1 - \delta - i\beta \qquad \text{where } \delta \text{ and } \beta \text{ are positive.} \tag{1.6}$$

Indeed the imaginary part $\beta$ is equal to $\lambda\mu/4\pi$, where $\lambda$ is the wavelength and $\mu$ the attenuation coefficient (see Eq. (1.88) and Sect. 1.4.6). When dealing with visible optics, the opposite convention is usual, with an opposite sign for the imaginary part of $n$.

The waves will be assumed to be monochromatic in most instances, with the temporal dependence $e^{i\omega t}$. To satisfy the international standard of units, or SI units, the electromagnetic equations will be written in the rationalized MKSA system of units. The Coulombian force in vacuum is in this system $qq'/4\pi\varepsilon_0 r^2$ with $\varepsilon_0\mu_0 = c^{-2}$, $\mu_0 = 4\pi 10^{-7}$. Note that the same symbol $\mu$ is used here for the magnetic susceptibility and just above for the coefficient of attenuation or absorption. In subsequent use of $\mu$, any confusion should be avoided by consideration of the context.

### 1.2.1.1 Summary of Some Formulae and Constants

The reader will find at the end of the book a table of the main notations in use. Some of the basic formulae and constants considered in this chapter are given here:

| | |
|---|---|
| $n = 1 - \delta - i\beta$ | optical index |
| $\beta = \lambda\mu/4\pi$ | $\lambda$ wavelength, $\mu$ coefficient of attenuation |
| $\mu = \rho\sigma_{\text{atten}}$ | density $\rho$ of objects of attenuation cross-section $\sigma_{\text{atten}}$ |
| $n = 1 - (2\pi/k_0^2)\rho b$ | opt. index from scattering length $b$, Sect. 1.3 |
| $\sigma_{tot} = 2\lambda \,\mathscr{I}\mathrm{m}\,[b(0)]$ | optical theorem, Sect. 1.3.2 |
| $r_e = e^2/4\pi\varepsilon_0 mc^2$ | electron Lorentz classical radius, Sect. 1.4.2 |
| $r_e = 2.818 \times 10^{-15}$ m | |
| $\lambda_{sc} = \lambda_{in} + \lambda_c(1 - \cos 2\theta)$ | Compton formula, Sect. 1.4.5 |
| $\lambda_c = 2\pi\hbar/mc$ | electron Compton wavelength, Sect. 1.4.5 |
| $\lambda_c = 0.002426$ nm | |

511 keV                          electron rest mass
$\lambda$ nm $\times E$ eV $= 1239.842$      wavelength $\times$ energy of a photon

## 1.2.2 Wave Equation

### 1.2.2.1 Propagation in a Vacuum

The propagation of a radiation, whether neutrons or X-rays, obeys a series of second-order partial differential equations which can be presented in a common form. We will discuss first the case of propagation in a vacuum. Electromagnetic radiation can be represented by the 4-vector potential $A_v(v = 0, 1, 2, 3)$ defined by

$$A_0 = \Phi/c, \qquad (A_1, A_2, A_3) = \mathbf{A}, \qquad (1.7)$$

where $\Phi$ is the scalar electric potential and $\mathbf{A}$ is the 3-vector potential. The 4-vector potential obeys in the Lorentz gauge and away from any charge

$$\Delta A_v = \varepsilon_0 \mu_0 \frac{\partial^2 A_v}{\partial t^2}, \quad \left( \Delta = \sum_{x_i = x, y, z} \frac{\partial^2}{\partial x_i^2}, \ \varepsilon_0 \mu_0 = \frac{1}{c^2} \right). \qquad (1.8)$$

For a neutron of wave function $\Psi$, the equivalent form of Eq. (1.8) is the Schrödinger equation without any potential

$$-\frac{\hbar^2}{2m} \Delta \Psi = i \frac{\hbar \partial \Psi}{\partial t} \qquad (1.9)$$

(using the convention of quantum mechanics for the sign of $i$, as discussed above). We shall consider essentially time-independent problems and only monochromatic radiation which has a frequency $\omega/2\pi$. The time variable then disappears from the equations, through use of the relations

$$\frac{1}{c^2} \frac{\partial^2}{\partial t^2} = -\frac{\omega^2}{c^2} = -k_0^2 \qquad \text{(electromagnetic field)} \qquad (1.10)$$

$$i \frac{\hbar \partial}{\partial t} = \hbar \omega = \frac{\hbar^2}{2m} k_0^2 \qquad \text{(Schrödinger equation)}. \qquad (1.11)$$

$k_0$ is the wave vector in a vacuum and $\hbar \omega$ is the energy.

In both cases, writing the generic field or wave function as $A$ yields the ***Helmholtz equation***,

$$\left( \Delta + k_0^2 \right) A = 0. \qquad (1.12)$$

The solutions to this equation are plane waves with the wave vector $k_0$.

In optics this equation is more usually expressed in terms of the electric and magnetic fields $\mathbf{E}$ and $\mathbf{H}$, or the electric displacement and the magnetic induction field $\mathbf{D}$ and $\mathbf{B}$ rather than the vector potential $A_v$. $\mathbf{E}$ is related to the potential through

$$\mathbf{E} = -\operatorname{\mathbf{grad}}\phi - \frac{\partial \mathbf{A}}{\partial t} = -c\operatorname{\mathbf{grad}}A_0 - \frac{\partial \mathbf{A}}{\partial t}. \qquad (1.13)$$

For a monochromatic plane wave in free space, the gauge can be chosen such that $A_0 = 0$, then

$$\mathbf{E} = -\frac{\partial \mathbf{A}}{\partial t} = -i\omega\mathbf{A}. \qquad (1.14)$$

Therefore, $\mathbf{E}$ and $\mathbf{A}$ being proportional to each other, most of the discussion subsequent to Eq. (1.12) applies to $\mathbf{E}$ as well. Nevertheless, in the presence of electric charges or polarizable objects, all the properties of the electromagnetic field cannot be described with the generic field written as a scalar. These particular vector or tensor properties will be addressed when necessary.

### 1.2.2.2 Propagation in a Medium

Equation (1.12) still applies in a modified form even when the radiation propagates in a homogeneous medium rather than a vacuum. All media are inhomogeneous, at least at the atomic scale, so for the moment the homogeneity will be taken as a provisional assumption whose justification will be discussed in Sect. 1.3. We also assume the isotropy of the medium, which is not the case for all materials.

In the case of the electromagnetic radiation the medium is characterized by permeabilities $\epsilon$ and $\mu$ that replace $\epsilon_0$ and $\mu_0$ in Eq. (1.8), although $\mu$ can usually be kept unchanged. Though the static magnetic susceptibility can take different values in various materials, we are concerned here with its value at the optical frequencies and above which it is not significantly different from $\mu_0$. In a medium Eq. (1.12) can be written as either

$$\left(\Delta + k^2\right)A = 0 \qquad (k = nk_0, \quad n^2 = \epsilon\mu/\epsilon_0\mu_0 \simeq \epsilon/\epsilon_0) \qquad (1.15)$$

or

$$\left(\Delta + k_0^2 - U\right)A = 0 \qquad \left(U = k_0^2\left(1 - n^2\right)\right). \qquad (1.16)$$

The first form shows that the wave vector has changed by a factor $n$, which is the refractive index. The second form is similar to the Schrödinger equation in the presence of a potential. Indeed in the case of the Schrödinger equation, the material can be characterized by a potential $V$ and the equation becomes

$$\left[-\frac{\hbar^2}{2m}\left(\Delta + k_0^2\right) + V\right]\Psi = 0 \qquad (1.17)$$

which is equivalent to the previous equation, with

$$U = \frac{2m}{\hbar^2}V \qquad (1.18)$$

and again we may define a refractive index

$$n^2 = 1 - U/k_0^2 = 1 - V/\hbar\omega. \tag{1.19}$$

It is important to realize that describing the propagation in the medium by a Helmholtz equation, with just a simple change of the wave vector by a factor $n$ or with the input of a potential $U$, is really just a convenience. In reality, each atom or molecule produces its own perturbation to the radiation and the overall result is not just a simple addition of those perturbations. It happens in most cases that the Helmholtz equation can be retained in the form indicated above. How $n$ or $U$ depends on the atomic or molecular scattering has to be established. Before addressing this question we have to give some further definitions for the intensity, current and flux of the radiation, and to introduce the formalism of scattering length, cross-section and Green functions which help to handle the scattering phenomena.

### 1.2.3 Intensity, Current and Flux

The square of the modulus of the amplitude of the field, i.e. $|A|^2$, defines the **intensity** of the radiation, which is used to represent either the probability of finding a quantum of radiation in a given volume or the density of energy transported by the radiation. $|A|^2$ is also used when combined with the wave vector direction to measure the flux density. These definitions are trivially correct in vacuum but need to be revised in a material.

The **flux** across a given surface is the amount of radiation, measured as an energy or a number of particles, which crosses this surface per unit time; this is a scalar quantity. The **flux density** or **current density** that we shall also call the **flow** is a vector. For instance, the electromagnetic energy flux density (flow) is designed by **S**; the energy flux in an elementary surface $d\sigma$ is then **S**.$d\sigma$. The **density of energy** $u$ is connected to the flow by a relation which expresses the energy conservation. In a non-absorbing medium the amount of energy which enters a given closed volume must be equal to the variation of the energy inside that volume:

$$\frac{\partial S_x}{\partial x} + \frac{\partial S_y}{\partial y} + \frac{\partial S_z}{\partial z} + \frac{\partial u}{\partial t} = 0. \tag{1.20}$$

Equation (1.20) can be also written in terms of the number of particles instead of the energy; for instance this is appropriate for the case of neutrons or for electromagnetic radiation if it is quantized. The same formalism stands for the flux, the density of current and the density of particles. The dimension of the density of flux is one of the relevant quantity (energy, number of particles or other) divided by dimension $L^2T$.

In the case of electromagnetic radiation, the quantities **E**, **H**, **D** and **B** can be used instead of **A** as discussed above and the dielectric and magnetic permeabilities, $\varepsilon$ and $\mu$, can be used to characterize the medium. The energy density is then given by

$$u = (\varepsilon \mathbf{E}.\mathbf{E}^* + \mu \mathbf{H}.\mathbf{H}^*)/4. \tag{1.21}$$

For a plane wave defined by the unit vector $\hat{\mathbf{k}}$ along the wave vector,

$$\mathbf{H} = \sqrt{\varepsilon/\mu}\,\hat{\mathbf{k}} \times \mathbf{E} \tag{1.22}$$

and the energy density becomes

$$u = \varepsilon\,|\mathbf{E}|^2\,/2. \tag{1.23}$$

The energy flow is then equal to the Poynting vector

$$\mathbf{S} = \mathbf{E} \times \mathbf{H}^*/2 = c\varepsilon\sqrt{\varepsilon_0\mu_0/\varepsilon\mu}\,\,|\mathbf{E}|^2\hat{\mathbf{k}}/2. \tag{1.24}$$

Note that these formulae giving $u$ and $\mathbf{S}$ are written in terms of complex field quantities whose real part represents the physical field. The complex and the real formulations differ by a factor 1/2 in the expressions of second order in the fields.

The change in the wave vector length in going from a vacuum into a medium has been written above, Eq. (1.15), in terms of the refractive index $n$

$$k = nk_0 \tag{1.25}$$

$$n = \sqrt{\varepsilon\mu/\varepsilon_0\mu_0}, \tag{1.26}$$

so that if $\mu \simeq \mu_0$, we obtain

$$u = n^2\,(\varepsilon_0\mu_0/\mu)\,|\mathbf{E}|^2\,/2 \simeq n^2\varepsilon_0\,|\mathbf{E}|^2\,/2 \tag{1.27}$$

$$\mathbf{S} = nc\,(\varepsilon_0\mu_0/\mu)\,|\mathbf{E}|^2\hat{\mathbf{k}}/2 \simeq nc\varepsilon_0\,|\mathbf{E}|^2\hat{\mathbf{k}}/2. \tag{1.28}$$

This shows that *the flux through a surface depends on both the amplitude* $\mathbf{E}$ *and also on the refractive index of the medium.*

A similar expression stands for neutrons (beware, in what follows, as usual in neutron physics $i$ has a sign opposite to our convention). Here the probability density $\rho$ and the current density $\mathbf{j}$ of particles are considered. The amplitude is the wave function $\Psi$:

$$\rho = |\Psi|^2, \qquad \mathbf{j} = (\hbar i/2m)\,(\Psi\,\mathbf{grad}\Psi^* - \Psi^*\,\mathbf{grad}\Psi). \tag{1.29}$$

For a plane wave, $\Psi_0 e^{i\mathbf{k}.\mathbf{r}}$, $\hat{\mathbf{k}}$ being the unit vector along $\mathbf{k}$,

$$\rho = |\Psi_0|^2, \qquad \mathbf{j} = (\hbar k/m)\,|\Psi_0|^2\,\hat{\mathbf{k}}. \tag{1.30}$$

Here too, the current depends on both $\Psi_0$ and on the medium which is characterized by a potential $V$ and

$$\hbar^2\mathbf{k}^2/2m + V = \hbar\omega. \tag{1.31}$$

As in optics, it is possible to introduce a refractive index, which is (Eq. (1.19))

$$n = \sqrt{\frac{\hbar\omega - V}{\hbar\omega}} \tag{1.32}$$

and which from $k_0$ gives the length of the wave vector **k**. Then

$$\mathbf{j} = (n\hbar k_0/m)\,|\Psi_0|^2\,\hat{\mathbf{k}}. \tag{1.33}$$

The above formulae are valid when the medium is isotropic. When it is anisotropic the flow of energy and the current are affected. In the electromagnetic case the direction of the flow does not always coincide with the direction of the wave vector.

*Exercise 1.2.1.* An X-ray beam impinging on a surface gives rise to a reflected and a transmitted beam. The amplitudes of these beams are given by the Fresnel formulae (see Sect. 3.1.2). As assumed above, the two media are not absorbing. Check the conservation of the flux, at least for the (s) polarization.

*Exercise 1.2.2.* Let us consider a wave function $\Psi$, such as an evanescent wave $\Psi_0 e^{(ik_x x - k_z z)}$. Calculate the current density.

### 1.2.4 Scattering Length and Cross-Sections

Let us consider an isolated scattering object (molecule, atom, electron), fully immersed in the field of an incident wave. The object reemits part of the incident radiation. We start with the assumption that its dimensions are small compared to the wavelength so that the scattered amplitude is the same in all the directions; instead for an extended object, direction-dependent phase shifts would appear between the scattered amplitudes coming from different regions in the sample. When examining the scattered amplitude at large distances $r$ from the object, simple arguments yield the following expression of the scattered amplitude:

$$A_{\mathrm{sc}} = -A_{\mathrm{in}}b\,\frac{e^{-ikr}}{r}. \tag{1.34}$$

Indeed this function which has the spherical symmetry ($k$ and $r$ are scalars) is proportional to the incident amplitude $A_{\mathrm{in}}$ and has locally the right wavelength $2\pi/k$; the decay as the inverse of $r$ guarantees the conservation of the total flux since the related intensity decays as the inverse of the surface of a sphere of radius $r$. The remaining coefficient $b$ has the dimension of a length; this coefficient characterizes the scattering power of the sample and is the so-called **scattering length**. The notation $b$ is rather used in the context of neutron scattering. Here we adopt it for X-rays as well. To be fully consistent with this notation we keep, as a mere convention, the minus sign in the definition of $b$. With this sign, the $b$ value is positive for neutron with most nuclei, and also for X-ray Thomson scattering. Yet $b$ can have an imaginary component but if the atom is not strongly absorbing then it is nearly real. A more rigorous justification of Eq. (1.34) will be given in the next section.

To justify that $b$ has the dimension of a length, we have considered the total scattered flux. The ratio of this flux to the incident one per unit of surface (i.e. to the incident flux density or incident flow) has the dimension of a surface and is equal to

$$\sigma_{scat,tot} = 4\pi \, |b|^2 \, . \tag{1.35}$$

This is the so-called **total scattering cross-section**. The scattered flux in the whole space is then equal to the one received by a surface equal to $\sigma_{scat,tot}$ which would be placed normal to the incident beam.

In general, with an extended object, the scattering measured at a large distance depends on the direction of observation, defined by a unit vector $\hat{\mathbf{u}}$, so that $b$ is written as $b(\hat{\mathbf{u}})$. Therefore it is useful to define a cross-section for this particular direction that is called the **differential scattering cross-section**

$$(d\sigma/d\Omega)\,(\hat{\mathbf{u}}) = |b(\hat{\mathbf{u}})|^2 \tag{1.36}$$

which is equal to the flux in the solid angle $d\Omega$ directed towards $\hat{\mathbf{u}}$, divided by $d\Omega$, for a unit incident flow (Fig. 1.1). In this case the definition Eq. (1.35) is replaced by

$$\sigma_{scat,tot} = \iint |b(\hat{\mathbf{u}})|^2 \, d\Omega, \tag{1.37}$$

where the integration is carried out over all the directions defined by $\hat{\mathbf{u}}$.

Any object (atom, molecule) also absorbs some part of the incident radiation without scattering it. Therefore one has to define the so-called **cross-section of absorption**, $\sigma_{abs}$, equal to the ratio of the absorbed flux to the incident flux density. We have used in Eqs. (1.35) and (1.37) a somewhat awkward notation ($\sigma_{scat,tot}$) to recall that it is a scattering cross-section; indeed the **total cross-section** appellation, $\sigma_{tot}$, is also used to name the sum of cross-sections of all the interaction processes (absorption, elastic and inelastic scattering); it is the whole relative flux picked up by the object:

$$\sigma_{tot} = \sigma_{scat,tot} + \sigma_{abs}. \tag{1.38}$$

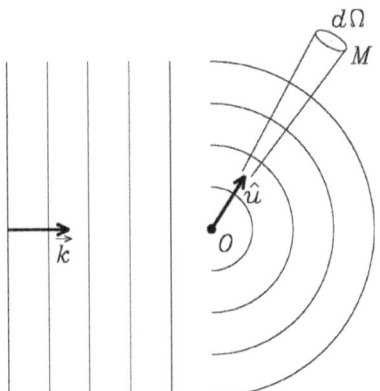

**Fig. 1.1** Definition of the scattering length $b(\hat{\mathbf{u}})$ and of the differential scattering cross-section $(d\sigma/d\Omega)(\hat{\mathbf{u}})$. The incident plane wave is $A_{in}e^{-i\mathbf{k}\cdot\mathbf{r}}$ and the scattered wave $A_{in}(b(\hat{\mathbf{u}})/OM)e^{-ikOM}$. This last expression gives a well-defined flux in the cone $OM$, whatever the distance $OM$. The scattering length and the differential scattering cross-section in the direction $\hat{\mathbf{u}}$ are, respectively, $b(\hat{\mathbf{u}})$ and $|b(\hat{\mathbf{u}})|^2$

## 1.2.5 The Use of Green Functions

The scattering amplitude $b$ in Eq. (1.34) has not been introduced very rigorously and it is possible to define it more formally. The field scattered by a point-like object obeys the wave equation (1.15) everywhere except at the centre of the object, which is both the source and a singular point. The simplest mathematical singularity is the Dirac $\delta$ function. The **Green function** of Eq. (1.15), $G(\mathbf{r})$, is a solution of the equation

$$\left(\Delta + k^2\right) G(\mathbf{r}) = \delta(\mathbf{r}). \tag{1.39}$$

Physically, $G(\mathbf{r})$ represents the field emitted by the source normalized to unity. More generally, any partial derivative equation which is homogeneous in $A$ such as

$$DA(x) = 0 \tag{1.40}$$

(here $D$ represents a sum of differential operators with constant coefficients plus a constant term, and $x$ is a scalar or a vector variable) admits Green functions $G$ which satisfy[1]

$$DG(x) = \delta(x). \tag{1.41}$$

A common application of Green functions is the resolution of non-homogeneous partial derivative equations. For example, if $G(x)$ is a Green function and $A_0(x)$ is any of the solutions of the homogeneous equation, the equation

$$DA(x) = f(x) \tag{1.42}$$

admits the following solutions:

$$A(x) = A_0(x) + \int G(x - x') f(x') dx'. \tag{1.43}$$

This can be shown by substitution into Eq. (1.42) and use of the equation,

$$f(x) = \int \delta(x - x') f(x') dx', \tag{1.44}$$

and by finally applying Eq. (1.41).

Let us now check that the diverging wave Eq. (1.34) (or the converging wave having the opposite sign for $k$) is indeed, to a certain coefficient, a Green function solution of Eq. (1.39). Due to the spherical symmetry, it is worth using the spherical coordinates $r, \hat{\mathbf{u}}$ ($\mathbf{r} = r\hat{\mathbf{u}}$; $\hat{\mathbf{u}}$ is defined by the polar angles $\theta$, $\phi$ which we leave implicit). The differential operators yield

---

[1] A customary definition of Green functions rather sets them as two point functions with
$$DG(x, x') = \delta(x - x').$$
Since we are placed here in free space, translation symmetry allows to centre the $\delta$ function at the origin and to set $x'$ to 0.

$$\mathbf{grad} = \hat{\mathbf{u}}\frac{\partial}{\partial r} + \cdots \tag{1.45}$$

$$\Delta = \frac{\partial^2}{\partial r^2} + \frac{2}{r}\frac{\partial}{\partial r} + \cdots \, ; \tag{1.46}$$

the $\cdots$ are some derivatives with respect to $\theta$, $\phi$ which vanish since we consider a spherical symmetry. We shall also use in the next subsection the same incomplete formulae to calculate waves at large $r$, where the dropped out terms are higher order infinitesimals. We then have

$$\left(\Delta + k^2\right)\frac{e^{\pm ikr}}{r} = 0 \qquad \text{for } r \neq 0. \tag{1.47}$$

At $r = 0$, we must compare the singularity with $\delta(r)$. It is possible to integrate the left-hand side of Eq. (1.47) inside a sphere of radius $r_0$ centred at the origin. Indeed from the definition of $G$, the integral of $(\Delta + k^2)\, G(r)$ must be equal to 1 when performed over the whole volume including the origin. This calculation is proposed in Exercise 1.2.3 and yields $-4\pi$. The Green function of the three dimensional Helmholtz equation is then

$$G_\pm(\mathbf{r}) = -\frac{1}{4\pi r}e^{\pm ikr}. \tag{1.48}$$

Now we retrieve Eq. (1.34). We assume that the amplitude scattered by a point size object sitting in $\mathbf{r} = 0$ satisfies

$$\left(\Delta + k^2\right)A_{sc}(\mathbf{r}) = A_{in}(0)4\pi b\,\delta(\mathbf{r}) \tag{1.49}$$

whose solution is

$$A_{sc}(\mathbf{r}) = A_{in}(0)4\pi b\, G_-(\mathbf{r}). \tag{1.50}$$

It is also useful to express the Green function in one dimension. Indeed some problems related to specular reflectivity can be solved in one dimension. A similar calculation yields

$$G_{1d,\pm}(\mathbf{r}) = -\frac{\pm i}{2k}e^{\pm i|kr|}. \tag{1.51}$$

The Green function of Helmholtz equation in two dimensions can be expressed with the help of Bessel functions. The asymptotic form at large $r$ is yet harmonic, with a $r^{-0.5}$ decay and an additional phase shift equal to $\pm(\pi/4)$.

*Exercise 1.2.3.* Calculate in three dimensions

$$\int_{r<r_0}(\Delta + k^2)(1/r)e^{\pm\,ikr}d^3\mathbf{r}.$$

Hints: The integral of the first term, $\Delta\ldots$, can be transformed into the integral of the gradient over the sphere of radius $r = r_0$; the integral of the second, $k^2\ldots$, can be successively performed over spheres and then over $r$. Note that the independence of the result with respect to $r_0$ yields Eq. (1.47) and is sufficient to prove that the argument after $(\Delta + k^2)$ is a Green function.

## 1.2.6 Green Functions: The Case of the Electromagnetic Field

While the Green functions of the Helmholtz equation are valid for scalar fields, as for instance the neutron wave function, the case of the electromagnetic field is more complicated. Not only is the field a vector (if the potential is used) but also the simplest sources are vibrating dipoles which are represented by vectors and which cannot be described by a simple $\delta$ function.

The 4-vectors $A_v$ and $j_v$ represent respectively the potential and the current-charge density, as follows:

$$A_0 = \Phi/c, \qquad (A_1, A_2, A_3) = \mathbf{A} \qquad (1.52)$$

$$j_0 = c\rho, \qquad (j_1, j_2, j_3) = \mathbf{j}. \qquad (1.53)$$

$\Phi$ is the scalar electric potential as previously defined in Eq. (1.7), $\rho$ the charge density and $\mathbf{j}$ the electric current density. Since $j_v$ describe the charge motion it must fulfil the conservation relationship

$$div\,\mathbf{j} + \frac{1}{c}\frac{\partial j_0}{\partial t} = 0. \qquad (1.54)$$

We shall have to integrate the current density over a volume,

$$\mathbf{J} = \int \mathbf{j}(\mathbf{r})d^3\mathbf{r} \qquad (1.55)$$

to get a 3-vector $\mathbf{J}$. If we consider a conductor wire in which there is a current, $\mathbf{J}$ is the product of the current intensity by the vector identified to the portion of the wire enclosed into the volume of integration. If we consider a moving charge (an electron for instance), $\mathbf{J}$ is the product of the charge by the velocity and if it is a vibrating dipole of amplitude $\mathbf{d}$ such as $\mathbf{d}e^{i\omega t}$, $\mathbf{J}(t)$ yields $i\omega \mathbf{d}e^{i\omega t}$. *Such a vibrating dipole, if infinitesimally small, is the simplest radiating point source.* It is characterized by the following current-charge density which fulfils the conservation law:

$$j_0(\mathbf{r},t) = \frac{ic}{\omega}div\,(\mathbf{J}(t)\delta(\mathbf{r})) = \frac{ic}{\omega}\mathbf{J}(t).\mathbf{grad}\,\delta(\mathbf{r}) \qquad (1.56)$$

$$\mathbf{j}(\mathbf{r},t) = \mathbf{J}(t)\delta(\mathbf{r}). \qquad (1.57)$$

The charge density $j_0/c$ has the form of the derivative in the direction $\mathbf{J}$, of the scalar function $\delta$.[2]

In the presence of the current-charge density $j_v$, the potential $A_v$ that we write with help of the Lorentz gauge verifies, instead of the four homogeneous equations (1.8), the inhomogeneous ones

---

[2] This idealized dipole, isolated in a vacuum, can be used to represent what happens at a microscopic scale in a dielectric material in the range of a few atoms (for X-rays and any material the relevant scale lies inside a single atom). Once the average has been made over a larger volume, these microscopic currents disappear from the equations. They are implicitly accounted for through the dielectric constant and the new fields $\mathbf{D}$ and $\mathbf{H}$, otherwise equal to $\epsilon_0\mathbf{E}$ and $\mathbf{B}/\mu_0$. This is the point of view of Chap. 4.

$$\Delta A_v - \frac{1}{c^2}\frac{\partial^2 A_v}{\partial t^2} = -\frac{j_v}{\varepsilon_0 c^2}. \tag{1.58}$$

We take as $j_v$ the dipole just described. We then keep as the useful solutions $A_v$ those which have the same oscillating time dependence as $j_v$. When $\omega$ is replaced by $ck$, Eq. (1.58) transforms into the inhomogeneous Helmholtz equations

$$\left(\Delta + k^2\right) A_0\left(\mathbf{r},t\right) = -\frac{i}{\varepsilon_0 c^2 k}\,\mathbf{J}(t).\mathbf{grad}\,\delta(\mathbf{r}) \tag{1.59}$$

$$\left(\Delta + k^2\right) \mathbf{A}\left(\mathbf{r},t\right) = -\frac{\mathbf{J}(t)}{\varepsilon_0 c^2}\,\delta(\mathbf{r}). \tag{1.60}$$

The solution of the second equation is a Green function $G_{\pm}(\mathbf{r})$, Eq. (1.48). The first one can be solved by the use of the method proposed in Eq. (1.43). The outgoing solution is

$$A_0\left(\mathbf{r},t\right) = -\frac{i}{\varepsilon_0 c^2 k}\,\mathbf{J}(t).\mathbf{grad}\,G_-(\mathbf{r}) \tag{1.61}$$

$$\mathbf{A}\left(\mathbf{r},t\right) = -\frac{\mathbf{J}(t)}{\varepsilon_0 c^2}\,G_-(\mathbf{r}), \tag{1.62}$$

where $G_-$ is given by Eq. (1.48). Up to a constant factor $|\mathbf{J}|/\varepsilon_0 c^2$, Eqs. (1.61) and (1.62) are *the equivalent for the electromagnetic potential of the Green function for the scalar field*. These particular expressions are due to both the vector character of the field and the electric dipolar character of the source. Other kinds of sources exist that we shall not describe here, as for example magnetic dipoles or multipoles of higher order.

We may need the electric field $\mathbf{E}$. Following Eq. (1.13)

$$\mathbf{E}\left(\mathbf{r},t\right) = \frac{i\omega}{\varepsilon_0 c^2}\left[\mathbf{J}(t)G_-(\mathbf{r}) + \frac{1}{k^2}\mathbf{grad}\,\left(\mathbf{J}(t).\mathbf{grad}\,G_-(\mathbf{r})\right)\right]. \tag{1.63}$$

The second derivative **grad J.grad** can be handled in two different ways. First, since we often consider the radiated field far from the source, we look for an asymptotic value valid when $kr \gg 1$. For this, the expression of the gradient Eq. (1.45) is used, but only the derivative according to $r$ is kept, and in the derivative of $G_- \propto e^{-ikr}/r$, only the derivative of $e^{-ikr}$ is calculated. All the other derivatives are of higher order in $1/kr$. Thus we can write ($\hat{\mathbf{u}}$ is the unit vector along $\mathbf{r}$)

$$\mathbf{grad}\,\left(\mathbf{J}(t).\mathbf{grad}\,G_-(\mathbf{r})\right) \underset{kr\to\infty}{\sim} -\mathbf{grad}\,\left(\mathbf{J}(t).\hat{\mathbf{u}}\,ik\,G_-(\mathbf{r})\right)$$

$$\underset{kr\to\infty}{\sim} -k^2\left(\mathbf{J}(t).\hat{\mathbf{u}}\right)\hat{\mathbf{u}}\,G_-(\mathbf{r}), \tag{1.64}$$

and one can recognize in this expression the projection of $\mathbf{J}$ about the vector $\mathbf{r}$. The asymptotic form of $\mathbf{E}(\mathbf{r},t)$ is

$$\mathbf{E}(\mathbf{r},t) \underset{kr\to\infty}{\sim} -[\mathbf{J}(t) - (\mathbf{J}(t).\hat{\mathbf{u}})\hat{\mathbf{u}}] \frac{i\omega e^{-ikr}}{4\pi\varepsilon_0 c^2 r}, \qquad (1.65)$$

i.e. *the scalar Green function multiplied by the component of the current normal to* $\hat{\mathbf{u}}$ *and by* $i\omega/\epsilon_0 c^2$.

Another way to transform Eq. (1.63), now without any approximation, consists in writing the last term with the alternative form $\mathbf{grad}\,div(\mathbf{J}G_-)$ (this equivalence is seen in Eq. (1.56)). The following equation is also identically valid:

$$\mathbf{grad}\,div \equiv \Delta + \mathbf{curl}\,\mathbf{curl}, \qquad (1.66)$$

and since $G_-$ is the solution of Helmholtz equation away from the origin, $\Delta$ may be replaced by $-k^2$. As a result we have

$$\mathbf{E}(\mathbf{r},t) = -\mathbf{curl}\,\mathbf{curl}\left(\mathbf{J}(t)\frac{ie^{-ikr}}{4\pi\varepsilon_0\omega r}\right) \quad \text{for } r \neq 0. \qquad (1.67)$$

## 1.3 From the Scattering by an Object to the Propagation in a Medium

Though all materials are inhomogeneous at least at the atomic scale, they commonly let radiations to propagate smoothly as in an homogeneous continuum. This was assumed in the previous paragraphs, where the overall character of the material was represented by an optical index. The same assumption is made when representing the response of a material to an electric field through an homogeneous dielectric constant. These representations rely on the feasibility of some procedure which replaces rapidly varying fields by smoothly varying averages and calculates global quantities such as an optical index or a dielectric constant from atomic ones such as a scattering length or an atomic polarizability. The determination and assessment of such an averaging procedure are in general not trivial and have to account for the combination of several parameters: the type of the radiation, usually scalar (neutron) or vector (X-ray), its wavelength, the strength of its interaction with matter and the length scale and other characters of the inhomogeneities. It may sometimes happen that no reasonable average exists and that the radiation does not propagate smoothly through the medium.

The case of X-rays and neutrons is fortunately simple because their interaction with matter happens to be very weak. In other words the density of scattering length is small in all materials. In Sect. 1.3.2 we develop a calculation adapted to this case by adding explicitly to the incident wave the field scattered in the forward direction by all the atoms. It can be viewed as an approximate extension of the optical theorem, briefly described. We then discuss in details the validity of the calculation. We also show (Sect. 1.3.3) that comparing the extinction length, $1/k_0(n-1)$, with other length scales of the problem helps to determine the status of the approximations. Beforehand it is worth to shortly describe in general the problem of field average.

## 1.3.1 The Problem of Defining a Field Average

This problem has been the subject of a large amount of literature, mainly devoted to the electromagnetic field. The case of common dielectric materials was solved for the static electric field through a formula given by Clausius and Mossotti, then Lorentz and Lorenz later showed that a similar one applies to electromagnetic waves (see [12] and [6]). Jackson [12] offers a discussion of the electromagnetic case. A mathematical view with a recent list of references can be found in [7].

We organize the discussion about the case of the propagation of a wave rather than the response to a static field. The subsequent formulae are based on the Helmholtz equation and are written with the generic field $A$. They do not correctly describe the case of the electromagnetic field, but they suffice to a symbolic description of the main issues. Let a medium be comprised of a density $\rho$ of atoms with a scattering length $b$, sitting at positions labelled $i$. It scatters the field $A_{sc}(\mathbf{r})$ when irradiated with $A_{in}(\mathbf{r})$. The wave vector length in vacuum is $k_0$. From Eq. (1.49)

$$(\Delta + k_0^2)A_{sc}(\mathbf{r}) = 4\pi b \sum_i \left(A_{in}(\mathbf{r}_i) + A_{sc}(\mathbf{r}_i) - A_{sc,i}(\mathbf{r}_i)\right)\delta(\mathbf{r} - \mathbf{r}_i). \qquad (1.68)$$

Under the summation is the field exciting the atom $i$. It is the total field at $\mathbf{r}_i$ minus the self-field $A_{sc,i}$, scattered by this atom itself. Let us attempt to find an equation between macroscopic fields—that is field averages—by integrating Eq. (1.68) over a small volume about $\mathbf{r}$. A volume $v$ of fixed size and shape is attached at each value of $\mathbf{r}$, and then called $v(\mathbf{r})$. Writing $\langle\rangle_{v(\mathbf{r})}$ for the average over $v(\mathbf{r})$, the integration yields

$$\langle(\Delta + k_0^2)A_{sc}(\mathbf{r}')\rangle_{v(\mathbf{r})} = \frac{4\pi b}{v} \sum_{i \in v(\mathbf{r})} \left(A_{in}(\mathbf{r}_i) + A_{sc}(\mathbf{r}_i) - A_{sc,i}(\mathbf{r}_i)\right). \qquad (1.69)$$

Before going further, the following questions must be answered:

(i)   Which size of $v$ gives a sound average, if it is ?
(ii)  Can $\langle(\Delta + k_0^2)A_{sc}(\mathbf{r}')\rangle_{v(\mathbf{r})}$ be replaced by $(\Delta + k_0^2)\langle A_{sc}(\mathbf{r}')\rangle_{v(\mathbf{r})}$ ?
(iii) How to manage the self-field exclusion term $-A_{sc,i}(\mathbf{r}_i)$ ?

A suitable answer to (i) is in choosing a volume size larger than the scale of the density fluctuations in the medium and smaller than the wavelength of the radiation. Further analysis shows that the scattering length of one single atom should not be too large. More precisely one single granule of inhomogeneity should give only a small contribution to the scattered field. A favourable answer to (ii) is given at the expense of a more stringent answer to (i). To deal with a derivative of a function we need first to define on some scale the average of that function, then to go to a larger scale to define the derivative. The status of (iii) depends on the type of radiation. If it is scalar the contributions to $A_{sc}(\mathbf{r}_i)$ from scattering by atoms inside the volume $v(\mathbf{r})$ add to each other and the self-field term, being one of many, can be neglected. It is not so for the electromagnetic field because a cancellation of other terms leaves the self-field as an important one.

The calculation in the scalar case may then terminate by neglecting the self-field and adding $A_{in}(\mathbf{r})$ to the left-hand side of Eq. (1.69). The equation between macroscopic fields (overlined) is

$$(\Delta + k_0^2)\left(\overline{A_{in}(\mathbf{r})} + \overline{A_{sc}(\mathbf{r})}\right) = 4\pi\rho b\left(\overline{A_{in}(\mathbf{r})} + \overline{A_{sc}(\mathbf{r})}\right). \tag{1.70}$$

The total macroscopic field obeys the Helmholtz equation with $k = nk_0$ and

$$n^2 = 1 - (4\pi/k_0^2)\rho b. \tag{1.71}$$

The solution in the electromagnetic case relies on two mathematical results. First a spherical distribution of dipoles produces a zero field in its centre. Consequently the field created at the centre of a sphere by a uniform distribution of dipoles in that sphere reduces to the field of the dipole sitting at the centre itself. The second point is that the integral of the field in this sphere, is equal to the total dipolar moment divided by $-3\epsilon_0$. The first point shows that the self-field cannot be neglected. The second result brings a correcting factor $1/(1 + 4\pi\rho b/3k_0^2)$ in the right-hand side of Eq. (1.70). But again we should warn the reader that the equations as written do not actually represent the electromagnetic case; they are merely indicative. See [12] for a rigorous argument in the case of the static field. The optical index satisfies

$$n^2 = 1 - \frac{(2 + n^2)}{3}(4\pi/k_0^2)\rho b. \tag{1.72}$$

This, in our notation, is equivalent to the Lorentz–Lorenz formula.

This approximation applies to visible optics in most of the condensed matter cases. Indeed the wavelength is a thousand times more extended than atomic density fluctuations and a volume of averaging can be chosen between these two ranges. This is not true at X-ray and neutron wavelengths of the same order as the interatomic distances. We can get around this difficulty by weighting the fields with $e^{i\mathbf{k}_0 \cdot \mathbf{r}}$ in order to integrate a smoothly varying quantity, and by using the average form

$$\overline{A(\mathbf{r})} = e^{-i\mathbf{k}_0 \cdot \mathbf{r}}\langle e^{i\mathbf{k}_0 \cdot \mathbf{r}'}A(\mathbf{r}')\rangle_{v(\mathbf{r})}. \tag{1.73}$$

The volume $v$ should be as before large enough to enclose many times the range of atomic density fluctuations. But $e^{i\mathbf{k}_0 \cdot \mathbf{r}}A_{sc}(\mathbf{r})$ can have a reasonable average over $v$, only if the incident and scattered wave vectors are close to each other. This implies $k/k_0 \approx 1$, a condition equivalent to $n \approx 1$ and to $(4\pi/k_0^2)\rho b \ll 1$. It will be shown later in this section that under this condition $A_{sc}(\mathbf{r})$ is negligible everywhere except in the forward direction, at $\mathbf{k} \approx \mathbf{k}_0$. Now, whatever the radiation, scalar or electromagnetic, both Eqs. (1.71) and (1.72) read

$$n - 1 \approx -(2\pi/k_0^2)\rho b. \tag{1.74}$$

The reader may also refer to [16] for the treatment of the X-ray case and look at the footnote in Sect. 2.3.2 of this book. In the next subsection we discuss in detail, through a different route, the X-ray or neutron case, and show how the total field is built up from the waves scattered by individual atoms.

The approximations that we have made may fail in several ways. With X-rays and neutrons the assumption of $A_{sc}$ having a wave vector close to the one of $A_{in}$ is no longer verified when a crystal is at Bragg orientation with respect to the incident beam and diffraction occurs. That case may be treated either in the kinematical theory, that is a Born approximation, considering that the intensity of the diffracted beam is a small fraction of the incident one which propagates normally, or through the more exact, dynamical theory. In the second option, the incident beam may be extincted on a reduced path. Choosing the kinematical or the dynamical theory usually depends on the extension of the perfectly coherent zones in the crystal. The dynamical theory of Bragg diffraction for X-rays [3] is tractable in algebraic form because it neglects the diffraction by all sets of atomic planes but one. This is also true for neutrons. This approximation is again made possible by the weakness of the interaction at the atomic level. An effect similar to Bragg scattering is at the origin of forbidden bands in the electronic energy spectra in crystals; some electronic waves cannot propagate.

On top of the diffraction by periodic structures, disordered inhomogeneities always produce some diffuse scattering. When it is intense the assumptions made above do fail; this is called opalescence. It is well observed at optical wavelengths when the scale of the inhomogeneities is of the same order as the wave. With X-rays, diffuse scattering at wide angles is never intense but some sort of opalescence may occur at scattering angles of the order of millidegrees (see Sect. 1.3.2). Going z wave may be impossible beyond a short path. This is usually the case for electrons in solids, because of the strong interaction between an electron and an atom, and because of the presence of crystalline defects. The electric conduction in metals is an example of diffusive propagation in place of coherent waves. Sometimes, even the diffusive propagation may be stopped and the field is said to be localized [2].

## 1.3.2 The Optical Theorem and Its Extension

### 1.3.2.1 The Optical Theorem

The *optical theorem* exactly relates the total cross-section (absorption plus scattering) $\sigma_{tot}$ of an isolated object with the imaginary part of the amplitude that this object scatters in the forward direction, i.e. $\mathscr{I}m\,[b(0)]$:

$$\sigma_{tot} = 2\lambda\,\mathscr{I}m\,[b(0)]. \tag{1.75}$$

Here $b$ may depend on the scattering direction, but only its value at zero angle is relevant. The proof relies on a balance between absorption and flux going in and out of a surface enclosing the object and the Green theorem is applied (see [12], Sect. 9.14). We stress that no approximation is made. When the field, such as an electromagnetic field, has several components, $b(0)$ represents the scattering in the same polarization as the incident wave.

It is worth to mention this relation because it looks like a partial solution to the problem of finding the optical index of a medium made of such objects. Indeed the total cross-section of the objects (atoms or molecules) which constitute a medium approximately yields the attenuation of that medium, then the attenuation is linked to the imaginary part $\beta$ of the index through $4\pi\beta = \rho\sigma_{tot}\lambda$. What we call attenuation includes the absorption and the loss of radiation due to scattering out of the direction of propagation. Unfortunately the relation between the molecular total cross-section and the attenuation is only approximate because the scattering cross-sections of all the molecules cannot be simply summed, as can be the absorption.

We cannot make use of the optical theorem as it is. First because it gives no access to the real part of the index, and second because of the approximation made when going from the cross-section of the objects to the attenuation in the medium. Its validity can hardly be discussed directly. Instead we shall retrieve Eq. (1.74), which yields the complete refractive index of a medium, with its real and imaginary parts, and explicitly show that it is related to the scattering in the forward direction. That calculation is not exact, but relies on the smallness of $\rho b$. We shall call this condition, "the weak interaction". It is equivalent to say that the index is close to 1. In fact, for X-rays and neutrons, the difference to 1 is of the order of $10^{-5}$, or even less. This is not true for visible light. We write below the formulae for a scalar field. Under the condition that the interaction is weak, those will also be valid for the electromagnetic field. The demonstration follows Jackson [12].

### 1.3.2.2 The Amplitude Scattered by a Planar Assembly of Scattering Objects

Let us consider a population of scattering objects homogeneously located in the surface of a plane $P$ normal to the direction of propagation of the incident plane wave (Fig. 1.2). We shall consider the amplitude of the wave at a point $M$, far enough behind this plane but not at an infinite distance (like in Fresnel diffraction). The field is supposed to be a scalar.

**Fig. 1.2** A plane wave coming from the left encounters a plane $P$ containing an array of scattering objects. The axis $OM$ is normal to the wave planes. The value of the field as modified by the scattering will be calculated at the point $M$

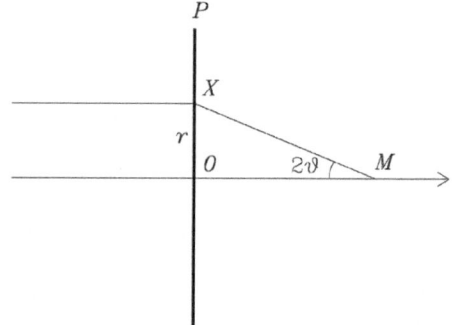

Let $\rho_s$ be the surface density of the scattering objects and $A(O)$ the incident amplitude of the wave at $O$, that is in the plane $P$. The objects located within the surface $ds$ around a point $X$ in the plane $P$ will contribute an amplitude at the point $M$ given by

$$dA_X(M) = -A(O)\rho_s b(\hat{XM}) \frac{e^{-ik_0 XM}}{XM} ds, \tag{1.76}$$

where $b(\hat{XM})$ is the scattering length in the direction $\hat{XM}$. To integrate this expression over the whole plane $P$, one first integrates around a ring of radius $r$ and width $dr$ centred in $O$. The mean value of $b(\hat{XM})$ in this ring is called $b(2\theta)$ ($2\theta$ being the angle $OMX$). The amplitude scattered by this elementary ring is

$$dA_r(M) = -A(O)\rho_s b(2\theta) 2\pi r\, dr \frac{e^{-ik_0 XM}}{XM}. \tag{1.77}$$

This must be integrated over $r$ but since

$$XM^2 = OM^2 + r^2, \quad \text{whence} \quad XM\, dXM = r\, dr, \tag{1.78}$$

it is possible to integrate over $XM$ instead of $r$ and the amplitude $A_{sc}(M)$ scattered by the entire plane $P$ appears as

$$A_{sc}(M) = -A(O)\rho_s 2\pi \int_{OM}^{\infty} b(2\theta) e^{-ik_0 XM} dXM. \tag{1.79}$$

If $b$ is independent of $2\theta$ the calculation is straightforward. In the general case the same result is obtained under the assumption that the point $M$ is far enough. This is shown through an integration by part:

$$A_{sc}(M) = -(2\pi i/k_0) A(O)\rho_s$$
$$\left\{ \left[ b(2\theta) e^{-ik_0 XM} \right]_{XM=OM}^{XM=\infty} - \int_{OM}^{\infty} e^{-ik_0 XM} \frac{db(2\theta)}{dXM} dXM \right\}. \tag{1.80}$$

A treatment of the second term is given in [12], Sect. 9.14. Provided that $b(2\theta)$ presents a non-singular extremum at $\theta = 0$ this term is of the order of $1/(k_0 OM)$. It is then negligible if $OM \gg \lambda$.

As for the first term, it quickly oscillates about zero when $XM$ tends towards infinity in the upper bound so that we make the following approximation

$$b(2\theta) e^{-ik_0 XM} \underset{XM \to \infty}{\sim} 0. \tag{1.81}$$

It is worth noting that to average those oscillations about zero, the upper bound value for the radius $r$ of the ring used in integrating over the plane $P$, should be much larger than some characteristic length. This length is the so-called first Fresnel zone radius which is of the order of $(\lambda OM)^{1/2}$.

Finally the forward scattered amplitude becomes

$$A_{sc}(M) = iA(O)\lambda \rho_s b(0) e^{-ik_0 OM}. \tag{1.82}$$

The forward scattered field adds to the incident field $A(O)e^{-ik_0OM}$ in $M$ and yields a total field $A(M)$

$$A(M) = A(O)e^{-ik_0OM} + A_{sc}(M) = A(O)(1 + i\lambda\rho_s b(0))e^{-ik_0OM}. \qquad (1.83)$$

If we now consider instead of a plane a thin layer of thickness $dx$ the above calculation remains valid provided that the surface density $\rho_s$ is related to the volume density $\rho_v$ by

$$\rho_s = \rho_v dx. \qquad (1.84)$$

The total field for such a layer becomes

$$A(M) = A(O)(1 + i\lambda\rho_v b(0)dx)\ e^{-ik_0OM}. \qquad (1.85)$$

Note that the amplitude scattered by the whole plane in $M$ is outphased by $\pi/2$ relative to the one scattered by an element of this plane; that phase shift results from the summation of amplitudes in the Fresnel diffraction.

### 1.3.2.3 The Propagation of a Wave in a Homogenous Population of Scattering Objects

Let us now consider the plane $P$ as an infinitesimally small layer of thickness $dx$ made of a medium of index $n$. The wave vector in the medium is $nk_0$. If the point $O$ is located at the entrance of the layer, a plane wave which has crossed the thickness $dx$ in the medium of index $n$ has an amplitude at the point $M$ given by

$$A(O)e^{-ink_0dx}e^{-ik_0(OM-dx)} \approx A(O)(1 - i(n-1)k_0dx)e^{-ik_0OM}. \qquad (1.86)$$

The approximation is correct if $|n-1|k_0dx \ll 1$. The comparison of Eqs. (1.86) with (1.85) shows that

$$n = 1 - \lambda^2 \rho_v b(0)/2\pi = 1 - \left(2\pi/k_0^2\right)\rho_v b(0). \qquad (1.87)$$

Equations (1.86) and (1.85) are schematically represented in the complex plane in Fig. 1.3. As shown in this figure, the imaginary part of $b(0)$ or $(n-1)$ modifies the absolute value of the field amplitude in $M$, whereas the real part modifies its phase.

Equation (1.87) links the scattering by elementary objects to the propagation in the medium which is considered to be continuous. It is an extension of the optical theorem. Indeed $\beta$, the imaginary part of $n$, Eq. (1.6), describes the attenuation of the radiation through the medium. The squared modulus of Eq. (1.86) is

$$|A(M)|^2 = |A(O)|^2 \left(1 - 2\beta k_0 dx + o(k_0^2 dx^2)\right), \qquad (1.88)$$

where $o(k_0^2 dx^2)$ is for higher order terms. The attenuation through the medium is described as the sum of attenuation by all objects with an individual cross-section $\sigma_{atten}$:

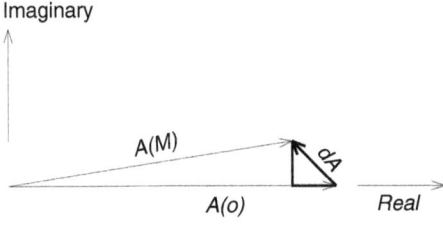

**Fig. 1.3** Representation of the amplitude of the field in the complex plane. Up to the common factor $e^{-ik_0 \overline{OM}}$, the total field $A(M)$ is the sum of the incident field $A(O)$ and of an infinitesimal field $dA = i\lambda \rho_s b(0) A(O)$ in Eq. (1.85) and $dA = -i(n-1)k_0 dx A_0$ in Eq. (1.86). The component of the field $dA$ associated with the real part of $b(0)$ or of $(n-1)$ is turned by $\pi/2$ from the incident field. This produces a phase shift of the total field. On the other hand, the imaginary part of $b(0)$ or of $(n-1)$ decreases the amplitude of the total field

$$|A(M)|^2 = |A(O)|^2 \left(1 - \sigma_{\text{atten}} \rho_v \, dx\right). \tag{1.89}$$

Since from Eq. (1.87), $\beta = (2\pi/k_0^2)\rho_v \, \mathscr{I}\mathrm{m}\,[b(0)]$

$$\sigma_{\text{atten}} = 2\lambda \mathscr{I}\mathrm{m}\,[b(0)]. \tag{1.90}$$

It is, obtained through some approximations, almost the optical theorem, Eq. (1.75) with the total cross-section $\sigma_{tot}$ replaced by the attenuation cross-section in this particular medium, $\sigma_{\text{atten}}$.

All the derivations above consider the field as a scalar. Under the approximations made here ($XM \to \infty$) they can be extended to the electromagnetic field. Indeed under these approximations, only the forward scattering, which is usually independent of polarization and conserving it, is retained. Beyond the above approximations, one must take into account all the scattering directions, and the scalar and vector fields display different behaviours.

### 1.3.2.4  About the Approximations

We must now discuss the approximations which are made. The argument relies on the equality of the amplitudes calculated from the index Equation (1.86) and from the scattering equation (1.85). On the one hand Eq. (1.86) is valid if

$$|n-1|k_0 \, dx \ll 1 \tag{1.91}$$

and on the other hand Eq. (1.85) holds if

$$OM \gg \lambda. \tag{1.92}$$

We are going to show that Eq. (1.87), giving the optical index $n$ for a thin slice $dx$, also applies to a material of finite thickness $x$. That thick material may be divided into layers of thickness $dx$. Let $O$ and $M$ be the points taken at the entrance and at

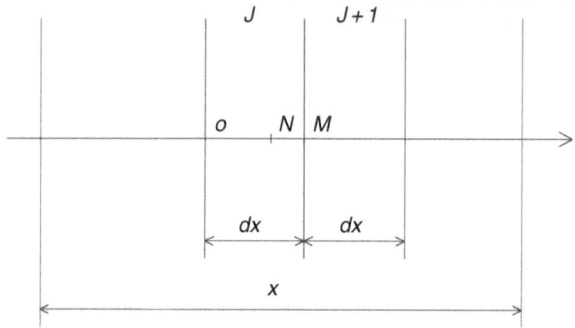

**Fig. 1.4** The point $M$ is located at the border of the two layers (j, j+1) of the material. We assume that the condition $L_e \gg OM \gg \lambda$ (see text) is satisfied. Note that if $OM \gg \lambda$ then $NM \gg \lambda$ for nearly all $N$. Then the amplitude at $M$ only comes from the layer $j$ and is given by Eq. (1.85) (with $\rho_s$ the surface density of the layer). Since $L_e \gg OM$, $(n-1)k_0\,dx$ is infinitesimally small, the approximation Eq. (1.86) does apply, and the material has an index given by Eq. (1.87)

the exit of layer $j$, such as $dx = OM$ (Fig. 1.4). We assume that $dx$ is such that both inequalities Eqs. (1.91) and (1.92) are satisfied.

Having $OM = dx$ does not affect the validity of Eq. (1.86) since $OM$ appears just in a global phase factor. The validity of Eq. (1.85) is affected for the points $N$ located between $O$ and $M$, such that $NM \gg \lambda$ does not hold. But if $OM \gg \lambda$ those points are a minority which contributes only to a negligible part of the scattering amplitude observed at $M$.

In order for the amplitude at $M$ to be given by Eq. (1.85), it is also necessary for the back scattering coming from the layers $j+1$ located behind the point $M$ to be negligible. The different points of that layer scatter towards $M$ with different phase shifts. It is possible to show that the ratio of the amplitude scattered by one layer in the backward direction to the one scattered in the forward direction is of the order of $\lambda/dx$. Therefore the condition $OM \gg \lambda$ is sufficient for Eq. (1.85) to be valid.

Satisfying both inequalities Eqs. (1.91) and (1.92) with $OM = dx$ means

$$\lambda \ll OM \ll \frac{\lambda}{2\pi\,|n-1|}, \tag{1.93}$$

and this in turn implies $|n-1| \ll 1$. *This is the condition for our approximations to be valid.* Actually $|n-1|$ is $10^{-5}$ or less for X-rays and neutrons. These inequalities also suggest that a length $L_e$ defined as

$$L_e = \frac{\lambda}{2\pi\,|n-1|} = \frac{V_a}{\lambda\,|b(0)|} \qquad \text{with } V_a = 1/\rho_v, \tag{1.94}$$

must play an important role in the optical properties of the medium. That length appears in any scattering process. With reference to the dynamical theory of X-ray diffraction we shall call this length the **extinction length**. We should remind ourselves that the theory as previously exposed applies when this length is much larger than the wavelength.

Another condition should also be discussed. We have replaced the sum over the atoms or molecules by integrals to calculate the scattered amplitudes. This is allowed only if the intermolecular distances and more generally the dimensions of heterogeneities are smaller than the range of integration. In the longitudinal direction that range is $OM$ and in the transverse direction the characteristic length for the integration is the radius of the first Fresnel zone, which is of the order of $(OM\lambda)^{1/2}$. The volume $V_{aver}$ which is large enough to represent on the average the material ($V_{aver}$ is defined as a volume larger than the heterogeneity) must be less than $OM^2\lambda$. As the inequality $OM \ll L_e$ must stand, $V_{aver}$ *must be very small compared to* $L_e^2\lambda = V_a^2/\lambda\,|\,b(0)\,|^2$, where $V_a$ is the volume of the unit (namely the atom) of scattering length $b$.

In condensed matter and for X-rays of energy 10 keV or thermal neutrons, $V_a$ is of the order of a few $\lambda^3$. For X-rays and for $Z \simeq 15$, we have $L_e/\lambda \simeq 10^4$. For neutrons this ratio is about ten times larger. The condition $L_e \gg \lambda$ is thus well satisfied. Since $V_a$ is of the order of $\lambda^3$, the volume $L_e^2\lambda$ is of the order of $10^8 V_a$ ($10^{10} V_a$ for neutrons). In reasonably homogeneous materials $V_{aver}$ can easily be chosen between $10^3 V_a$ and $L_e^2\lambda$. The wave propagation according to the continuous medium field equations with the index given by Eq. (1.87) is consequently valid.

If the condition $V_{aver} \ll L_e^2\lambda$ is not fulfilled, the approximation fails. Though the case $L_e \simeq \lambda$ is out of our discussion, let us mention that when it happens together with a size of density fluctuations of the same order, $V_{aver} > L_e^2\lambda$ and then opalescence may occur (see Sect. 1.3.1). This can be observed with visible light. With X-rays a similar situation may occur when the size of fluctuations is for example of the order of $L_e$, but this is much larger than $\lambda$ and the diffuse scattering is visible only at small angle.

Finally we have to humbly set some limitation to the method followed in the present section. When a beam goes through an interface between a vacuum and a medium, our discussion applies quite well to the case of normal incidence. Instead, at grazing incidence approaching the critical angle for total external reflection, it is seen in Exercises 1.3.1 and 1.3.2 that it does not give a correct result. This is because we used a first-order perturbation calculation, like a Born approximation. The grazing incidence conditions lie beyond the domain of validity of this approximation (see next subsection). What we can do is to accept the existence and value of the optical index found by our method and assume without any formal proof of validity that, in those conditions, the medium can still be treated as homogeneous with the same index as in bulk. The method exposed in Chap. 4 offers a basis to complete the discussion.

### *1.3.3 The Extinction Length and the Born Approximation*

The condition Eq. (1.93) ($L_e \gg \lambda$) shows that the extinction length plays a major role in the evaluation of the strength of the interaction of a radiation with matter. When the radiation has travelled a distance $L_e$ in the material, it begins to undergo a measurable phase shift of exactly one radian relative to a propagation in vacuum; also it scatters a substantial amount of radiation. The results may be qualitatively different when the thickness of the material which has been crossed is smaller or bigger than $L_e$. For X-rays of energy 10 keV, the extinction length is of the order of a micron ($|n-1| < 10^{-5}$), and it is one order of magnitude larger for neutrons.

The approximation which has been made to relate $n$ with $b$ is connected to the first Born approximation. We have used a single scattering to produce the plane wave propagating in the medium. The extinction length allows us to decide whether the Born approximation is valid for a given situation. When $L_e \gg \lambda$, the criterion is that the path travelled in the volume of the material giving rise to a coherent scattering must be less than $L_e$. This was the criterion Eq. (1.91) in our approximation. The kinematical theory of diffraction by crystals (equivalent to the Born approximation) is commonly used because the volume of the perfect crystal (coherently scattering) is often smaller than one micron cube, that is $L_e^3$. The property expressed in Eq. (1.93), which tells that the extinction length is much larger than the wavelength, also presents beneficial effects for the physics of X-rays and neutrons. It is associated with the fact that even if the kinematical theory is no longer valid, as in perfect crystals, the dynamical theory remains calculable. In visible optics, where this condition is not valid, the diffraction equations are most often not exactly solvable.

In the study of reflection in grazing incidence on a surface, the extinction length plays a major role. First it is related to the critical angle of total external reflection, discussed in Chap. 3. Indeed the following relation stands:

$$\frac{1}{|\mathbf{q}_c|} = \frac{L_e \sin \theta_c}{2(1+n)} \approx \frac{L_e \sin \theta_c}{4}, \tag{1.95}$$

where $|\mathbf{q}_c|$ is the scattering wave vector transfer corresponding to the specular reflection at the critical angle $\theta_c$ (Fig. 1.5). The left-hand side term represents (up to a factor $1/4\pi$) a sort of wavelength perpendicular to the surface, and the right-hand side term (up to a factor $1/4$) the extinction length projected on the perpendicular

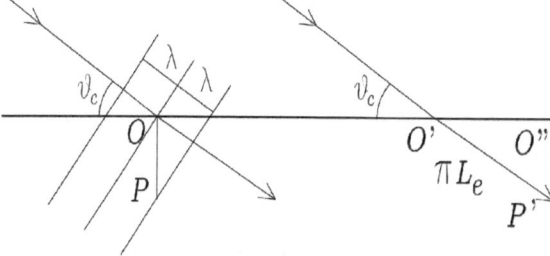

**Fig. 1.5** Relation (Eq. (1.95)) between extinction length and critical angle. A beam coming from vacuum crosses at a glancing incidence angle $\theta$ an horizontal interface limiting a medium of optical index $n$. The extinction length at that interface is $L_e$. The depth $OP$ is the vertical vacuum wavelength $\lambda / \sin \theta$. The depth $O''P'$ is $\pi$ times the extinction depth $L_e \sin \theta$. At the critical condition $\theta = \theta_c$, those lengths are equal, $OP = O''P'$

axis. The quasi equality of these two lengths is the sign showing that at the critical angle the Born approximation is no longer valid. For less shallow angles, the perpendicular wavelength becomes smaller than the perpendicular extinction length and therefore the reflectivity becomes weak and calculable in this approximation. In the case of a rough surface, one must also compare the extinction length to the characteristic lengths of its waviness. If the waviness is longer or shorter than the extinction length, the losses in reflectivity and the scattering are different.

*Exercise 1.3.1.* A plane scalar wave, with the wave vector $\mathbf{k_0}$, enters a medium through a planar interface making the angle $\theta$ with $\mathbf{k_0}$. By dividing the medium in layers parallel to the interface, calculate the scattered amplitude at any point in the medium, as shown in Eq. (1.82). Find the direction of equiphase planes of the total amplitude and compare to Snell–Descartes's law. The validity of the approximation depends on $\theta$.

Hint. One can show that the scattered amplitude at a point located downstream an angled layer is given by Eq. (1.82) divided by $\sin \theta$.

*Exercise 1.3.2.* In the same configuration as the one in the previous exercise, and assuming $b(2\theta)$ constant, find with the same method the amplitude reflected by the interface. Compare with the exact Fresnel expression given in Chap. 3, Sect. 3.1. In Sect. 3.3 in the same chapter the Born approximation is discussed as in this exercise. Note that the amplitude calculated here is the backscattered one, considered as negligible in the discussion of the approximations at the end of Sect. 1.3.2.

Hint. The expressions for the scattered amplitudes at two symmetrical points with respect to an infinitesimal layer are the same. Only $b$ may change from $b(0)$ in one case to $b(2\theta)$ in the other ($2\theta$ is the angle between the reflected and incident wave vectors).

Notice. If the $\theta$ angle is large enough to allow $b(2\theta) \neq b(0)$, the reflectivity, obtained from $b(2\theta)$, differs from the one obtained with the Fresnel formula and an optical index associated with $b(0)$. This is because the shape of the molecules produces both a decrease of $b$ with $\theta$ and a roughness of the interface (Chap. 3, Appendix 3.A).

*Exercise 1.3.3.* In a one-dimension space, the $dx$ element located at $x'$, which receives the scalar field $A$, scatters in the two opposite directions, Eq. (1.51)

$$A \, 2k_0 \rho b_{1d} dx G_{1d,-}(x-x') = A \, i\rho b_{1d} dx e^{-i|k_0(x-x')|}.$$

$G_{1d,-}$ is the one-dimension outgoing Green function and $\rho b_{1d}$ the density of scattering power which in general is real.

Find the relation between $\rho b_{1d}$ and the refractive index in this medium at one dimension.

Hints. Since $\rho b_{1d}$ is not assumed to be weak, the method described above in the present section should not be applied. One may consider an interface at $x = 0$ between the vacuum at negative $x$ and the medium at positive $x$. A wave

$A(x) = A_0 e^{-ik_0 x}$ comes from the vacuum and becomes $A'(x) = A' e^{-ink_0 x}$ in the medium. The field in the medium can be written in two ways:

– By the integral equation of the scattering (see Chap. 4, Sect. 4.1.2)

$$A'(x) = A(x) + \int_0^\infty A'(x') 2k_0 \rho b_{1d} G_{1d,-}(x - x') \, dx';$$

– by the transmission at the interface $A' = tA_0$, where $t$ is the Fresnel transmission coefficient, which here is $2/(n+1)$ (Chap. 3, Sect. 3.1, Eqs. (3.30) and (3.31)).

It may be noticed that the scattering is composed of two terms. The one in $-e^{-ik_0 x}$ is at the origin of the disappearance (so-called **extinction**) of the incident wave (see the extinction theorem [6]).

## 1.4 X-Rays

### 1.4.1 General Considerations

The electromagnetic radiation interacts principally with the electrons, and very weakly with atomic nuclei (the ratio of the amplitudes is as the inverse of masses). The interaction is essentially between the electric field and the charge, but a much weaker interaction is also manifest between the electromagnetic field and the spin, or its associated magnetic moment.

A photon which meets an atom can undergo one of the three following events:

– **elastic scattering**, with no change in energy;
– **inelastic scattering**: part of the energy is transferred to the atom, most frequently with the ejection of an electron (the so-called **Compton effect**); however, it may happen that the lost energy brings the atom in an excited state, without any ionization (**Raman effect**);
– **absorption**: all the energy is transferred to the atom and the photon vanishes. Another photon can be subsequently re-emitted, with a lower energy: this is the so-called *fluorescence*.

These mechanisms are described in many text books; the one of R. W. James [13] is particularly complete (except for the Raman effect which can be found in [17]).

To give an intuitive image, we shall begin with the classical mechanics theory which provides in simple terms a correct result for the scattering by a free electron (Thomson scattering). When the electron is bound, this theory is still convenient enough. However, the Compton scattering cannot be described by this classical theory. Also this theory does not describe correctly the motion of the electrons in the atom. Therefore we shall also review all the following processes in the frame of the quantum theory, i.e.;

- the elastic and inelastic scattering (mainly Compton), for a free or bounded to an atom electron, when the radiation energy is well above the atomic resonance;
- the photoelectric absorption by an atom;
- the dispersion correction brought to the elastic scattering by the atomic resonance.

Finally we shall discuss the general properties of dispersion which are independent of a particular interaction or radiation. One can show that the real and imaginary parts of the scattering are linked by the Kramers–Kronig relations which are extremely general and probe the response of nearly every system to some kind of excitation. The origin of these properties lies in the thermodynamical irreversibility that can be introduced through the principle of causality.

## 1.4.2 Classical Description: Thomson Scattering by a Free Electron

The scattering by a free electron is simple and presents the main characters of the scattering by an atom. We shall start with this case.

The electron undergoes an acceleration, which is due to the force exerted by the incident electric field

$$\mathbf{E}_{\mathrm{in}}(t) = \mathbf{E}_0 e^{i\omega t}. \tag{1.96}$$

Let $\mathbf{z}$ be the electron position and $(-e)$ its charge, then

$$m\ddot{\mathbf{z}} = (-e)\,\mathbf{E}_0 e^{i\omega t}. \tag{1.97}$$

The electron exhibits oscillations of small amplitude, producing a localized current

$$\mathbf{j}(\mathbf{r},t) = (-e)\,\dot{\mathbf{z}}\,\delta(\mathbf{r}) \tag{1.98}$$

$$= \frac{(-e)^2\,\mathbf{E}_{\mathrm{in}}(t)}{i\omega m}\,\delta(\mathbf{r}).$$

The radiation of that vibrating current, similar to a dipole antenna, has been discussed in Sect. 1.2.6. From the Eqs. (1.57) and (1.65) we have at large distances $(kr \gg 1)$,

$$\mathbf{E}_{\mathrm{sc}} \underset{kr \to \infty}{\sim} - \left[ \mathbf{E}_{\mathrm{in}} - (\mathbf{E}_{in}.\mathbf{r})\frac{\mathbf{r}}{\mathbf{r}^2} \right] \frac{(-e)^2 e^{-ikr}}{4\pi\varepsilon_0 mc^2 r}. \tag{1.99}$$

What is measured is the projection of the field on some polarization direction given by the unit vector $\widehat{\mathbf{e}}_{\mathrm{sc}}$, and $\widehat{\mathbf{e}}_{\mathrm{in}}$ is the unit vector which describes the incident polarization. These vectors are chosen so that $\widehat{\mathbf{e}}_{\mathrm{in}}$ is parallel or antiparallel to $\mathbf{E}_0$ and $\widehat{\mathbf{e}}_{\mathrm{sc}}$ normal to $\mathbf{r}$ (see Fig. 1.6):

$$\mathbf{E}_{\mathrm{in}} = (\mathbf{E}_{in}.\widehat{\mathbf{e}}_{in})\,\widehat{\mathbf{e}}_{in} \qquad \text{and} \quad \mathbf{r}.\widehat{\mathbf{e}}_{\mathrm{sc}} = 0. \tag{1.100}$$

**Fig. 1.6** Directions of
incident and scattered
polarizations for (a) the
(s)–(s) or ($\sigma$)–($\sigma$) mode and
(b) the (p)–(p) or ($\pi$)–($\pi$)
mode. The associated
amplitude polarization factor
is, respectively, 1 and $\cos 2\theta$

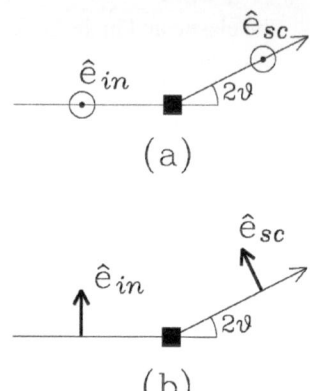

To take account of polarizations the definition of the scattering length $b$ in Eq. (1.34) can be adapted as follows:

$$\mathbf{E}_{sc}.\widehat{\mathbf{e}}_{sc} = -\mathbf{E}_{in}.\widehat{\mathbf{e}}_{in}\, b(\widehat{\mathbf{e}}_{sc},\widehat{\mathbf{e}}_{in})\frac{e^{-ikr}}{r}, \tag{1.101}$$

then Eq. (1.99) yields

$$b(\widehat{\mathbf{e}}_{sc},\widehat{\mathbf{e}}_{in}) = r_e \widehat{\mathbf{e}}_{sc}.\widehat{\mathbf{e}}_{in}, \tag{1.102}$$

where $r_e$ is the **Lorentz classical radius of the electron** with charge $e$ and mass $m$ ($r_e = e^2/4\pi\varepsilon_0 mc^2 = 2.818 \times 10^{-15}$ m).[3] The charge of the electron appears twice, first in the movement and then for the emission of the radiation. Thus it appears as a square and $b$ does not depend on its sign. *The scattered field is however opposite to the incident one* because of its relation with the current (by convention, a positive value of $b$ corresponds to such a sign reversal). If the ingoing polarization is normal or parallel to the plane of scattering, the outgoing one has the same orientation. These polarization modes are called (s) or ($\sigma$) when perpendicular to the plane of scattering and (p) or ($\pi$) when parallel. The polarization factor of the scattering length is 1 in the former case and $\cos 2\theta$ (Fig. 1.6) in the latter. The process that we have described is the so-called **Thomson scattering**.

### 1.4.3 Classical Description: Thomson Scattering by the Electrons of an Atom, Rayleigh Scattering

The simple result of the Thomson scattering is exact, even for the bound electrons of an atom, as far as the frequency of the X-rays is large compared to the characteristic atomic frequencies. Nevertheless it is necessary to take into account both the number of electrons and their position in the electronic cloud when calculating

---

[3] The microscopic phenomena are often described in the Gauss system of units, in which $r_e = e^2/mc^2$.

the scattering from an atom. Every point of the electronic cloud is considered to scatter independently from the others and the scattered amplitudes add coherently. As in any interference calculation within the Born approximation (see Appendix 1.A to this chapter), justified whenever the scattering is weak, one obtains the total scattering length by the Fourier transform of the electron density $\rho(r)$

$$b = r_e \widehat{\mathbf{e}}_{sc}.\widehat{\mathbf{e}}_{in} f(\mathbf{q}), \qquad f(\mathbf{q}) = \int \rho(\mathbf{r}) e^{i\mathbf{q}\cdot\mathbf{r}} dV(\mathbf{r}) \qquad (1.103)$$

($\mathbf{q} = \mathbf{k}_{sc} - \mathbf{k}_{in}$, Sect. 1.2.1). The quantity $f(\mathbf{q})$ is called the **atomic scattering factor** or the **atomic form factor**. The integral of $\rho(\mathbf{r})$ over all $\mathbf{r}$ values must be equal to the number of electrons in the atom:

$$f(0) = Z. \qquad (1.104)$$

There is no safe explanation to support the validity of this interference calculation. The justification comes from the alternative quantum calculation which gives the same result.

The assumption that the frequency of the radiation is greater than the atomic frequency may not be valid especially for the inner electronic shells. The model can be improved by introducing the binding of the electron to the atom which is modelled by a restoring force of stiffness $\kappa$ and a damping coefficient $\gamma$. The damping is the result of the radiation which is emitted by the electron, or of the energy transferred to other electrons. The equation of motion Eq. (1.97), written for a single electron, now becomes

$$m\ddot{\mathbf{z}} + \gamma\dot{\mathbf{z}} + \kappa\mathbf{z} = (-e)\mathbf{E}_0 e^{i\omega t}. \qquad (1.105)$$

One looks for a solution of the kind ($e^{i\omega t}$) which must satisfy

$$(-m\omega^2 + i\gamma\omega + m\omega_0^2)\mathbf{z} = (-e)\mathbf{E}_0 e^{i\omega t}, \qquad (1.106)$$

where $\kappa/m = \omega_0^2$. The current $(-e)\dot{\mathbf{z}}$ is then

$$\mathbf{j}(\mathbf{r},t) = -\frac{i\omega(-e)^2 \mathbf{E}_{in}(t)\delta(\mathbf{r})}{m(\omega^2 - \omega_0^2) - i\gamma\omega}. \qquad (1.107)$$

As shown for the Thomson scattering above, this yields the following scattering length:

$$b = r_e \frac{\omega^2}{\omega^2 - \omega_0^2 - i\gamma\omega/m} \widehat{\mathbf{e}}_{sc}.\widehat{\mathbf{e}}_{in}. \qquad (1.108)$$

We shall now discuss how this expression is modified for different energies when only one electron and one resonance are considered although this discussion could have been more general. Actually it happens that $\omega, \omega_0 \gg \gamma/m$ and we just have to compare $\omega$ with $\omega_0$. For high-energy X-rays and not too heavy atoms we have $\omega > \omega_0$ or even $\omega \gg \omega_0$. Within these approximations Eq. (1.108) is just reduced to Thomson's expression. If on the other hand $\omega \ll \omega_0$, then $b$ becomes

$$b = -r_e \frac{\omega^2}{\omega_0^2} \widehat{\mathbf{e}}_{sc} . \widehat{\mathbf{e}}_{in}. \tag{1.109}$$

This is the so-called **Rayleigh scattering**, originally proposed to explain the scattering of visible light produced by gases or small particles.

Three important features of this kind of scattering should be noticed:

- the polarization factor is the same as for the X-ray Thomson scattering;
- the scattered amplitude is proportional to the square of the frequency and the cross-section to the fourth power;
- the sign of the scattering length is opposite to the one of the Thomson scattering.

The second point explains the blue colour of the sky (the highest frequency in the visible spectrum), which from the first point may appear to be highly polarized. The change of sign noted in the third point is important, since it corresponds to a sign change of $(n-1)$. We shall comment this further when we will dispose of a more quantitative theory (Sect. 1.4.7).

Again for X-rays, the scattering length Eq. (1.108) when summed over all the atomic electrons becomes similar to the one of Thomson equation (1.103) but with real and imaginary corrections:

$$b = r_e \left( f + f' + i f'' \right) \widehat{\mathbf{e}}_{sc} . \widehat{\mathbf{e}}_{in}, \tag{1.110}$$

where $f$ is the Thomson scattering, whereas $f'$ and $f''$, which are real, give the correction due to resonance. This correction is the so-called **dispersion correction** or **anomalous scattering**. [4] One must take into account as in the pure Thomson scattering the sum over all the electrons and their spatial distribution, but this discussion is difficult and uncertain in the classical theory. We shall see that in the quantum theory $f'$ and $f''$ only slightly depend on $\mathbf{q}$ and have an energy dependence that we shall discuss.

The classical model allows the calculation of the absorption as proposed in Exercise 1.4.1. In fact, one rather gets the total cross-section, including absorption and scattering. This result is very realistic, since it agrees with the prediction of the optical theorem discussed in Sect. 1.3.2.

To summarize, the classical model although simple describes most of the phenomena and provides exact values for a certain number of physical quantities. Nevertheless the values of the resonance frequencies and of the damping coefficients are not calculable within this framework and are left arbitrary. In addition, it does not give much indications about the $\mathbf{q}$ dependence of the scattering factor at

---

[4] Originally, it was in optics that the anomalous dispersion was introduced. In the vicinity of resonances, the dispersion is opposite to the usual behaviour for which it is observed that the index of refraction varies in the same sense as the energy. By extension one refers to "anomalous scattering". In French the two adjectives "anormale" and "anomale" are used. "Anormale" means that it does not follow the rule and "anomale" means different from other individuals from the same species. Since the normal behaviour of the dispersion does not constitute a law in itself but only a usual behaviour, the second expression seems to be more appropriate. We acknowledge B. Pardo for his comments.

resonance but more important it does not describe the scattering when an electron is ejected (Compton effect). Although it is possible to give classical description of such an effect by considering the reaction on the scattering of a vibrating electron, only the quantum approach is correct. Therefore the only coherent and completely exact description is given by the quantum theory of the interaction between the radiation and atoms.

*Exercise 1.4.1.* Calculate the total cross-section of an atom which exhibits only one resonance characterized by $\omega_0$ and $\gamma$. We assume that the power taken by an atom from the radiation is the same as the one dissipated by the damping force $\gamma\dot{\mathbf{u}}$ (do not forget that when complex numbers are used to describe the oscillation of real variables, a quantity such as the energy is twice the one obtained with real numbers). The initial power of the radiation is given in Sect. 1.2.3. Check that the optical theorem (Sect. 1.3.2) is satisfied. This gives a simplified view of the radiation damping; see [12], Chap. 17, for a more complete treatment in classical electrodynamics.

## 1.4.4 Quantum Description: A General Expression for Scattering and Absorption

In this description we shall assume that the radiation is quantized as photons. The scattering and absorption probabilities are then the squared modulus of the probability amplitudes. The amplitudes are transformed into scattering lengths and the probabilities into the scattering cross-section. The amplitudes themselves are derived from a perturbative calculation based on the interaction Hamiltonian between the radiation and the electrons.

The expression of the Hamiltonian of one electron in the radiation field contains the following term (we leave aside some other terms such as the potential of the atom)

$$(1/2m)\,(\mathbf{p}-e\mathbf{A})^2 = \mathbf{p}^2/2m + \left(e^2/2mc\right)\mathbf{A}^2 - (e/m)\mathbf{A}\cdot\mathbf{p}. \qquad (1.111)$$

The $\mathbf{p}$ and $\mathbf{A}$ operators are the momentum of the electron and the vector potential of the radiation. The first term of the right-hand side gives the kinetic energy of the electron and the two others the energy of interaction. In this expression, the spin has been neglected which is permitted when the energy of the radiation is weak compared to the rest mass energy of the electron which is 511 keV. A perturbation calculation made at the lowest order on the two interaction terms yields the scattered amplitude. The perturbation terms are sketched in Fig. 1.7. The smallest order of the perturbation is the first order for the term in $\mathbf{A}^2$ and the second order for the term in $\mathbf{A}\cdot\mathbf{p}$. These two terms give rise respectively to one and two terms in the scattering length (with our convention for the sign of imaginaries, unusual in quantum mechanics):

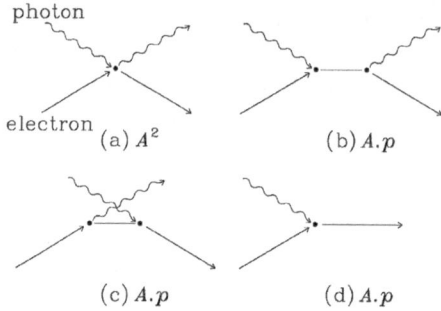

photon

electron
(a) $A^2$      (b) $A.p$

(c) $A.p$      (d) $A.p$

**Fig. 1.7** These diagrams are the symbol of the amplitudes in Eqs. (1.112) and (1.113). A point represents a matrix element and a line the electron or the photon in the initial or final state of the matrix element. For instance in (b) where two matrix elements are represented, the initial and final states display one electron and one photon, and the intermediate state only one electron. In the formulae as written in the text, the photonic states are not made explicit, but their contribution $(\hat{\mathbf{e}}, \mathbf{k})$ is present through the $(\hat{\mathbf{e}} e^{-i\mathbf{k}\cdot\mathbf{r}})$ terms. For any of the four amplitudes, (a) $b_{Th}$, (b) $b_{disp1}$, (c) $b_{disp2}$ and (d) the absorption, the Hamiltonian term is indicated

$$b = r_e \langle s | \hat{\mathbf{e}}_{sc}^* e^{+i\mathbf{k}_{sc}\cdot\mathbf{r}} \hat{\mathbf{e}}_{in} e^{-i\mathbf{k}_{in}\cdot\mathbf{r}} | i \rangle$$

$$- r_e \sum_c \frac{\langle s | \hat{\mathbf{e}}_{sc}^* \cdot \mathbf{p}\, e^{+i\mathbf{k}_{sc}\cdot\mathbf{r}} | c \rangle \langle c | \hat{\mathbf{e}}_{in} \cdot \mathbf{p}\, e^{-i\mathbf{k}_{in}\cdot\mathbf{r}} | i \rangle}{m(E_c - E_i - \hbar\omega_{in} + i\Gamma_c/2)}$$

$$- r_e \sum_c \frac{\langle s | \hat{\mathbf{e}}_{in} \cdot \mathbf{p}\, e^{-i\mathbf{k}_{in}\cdot\mathbf{r}} | c \rangle \langle c | \hat{\mathbf{e}}_{sc}^* \cdot \mathbf{p}\, e^{+i\mathbf{k}_{sc}\cdot\mathbf{r}} | i \rangle}{m(E_c - E_i + \hbar\omega_{sc})}$$

$$= b_{Th} + b_{disp1} + b_{disp2}. \tag{1.112}$$

Here $|i\rangle$ (or $|s\rangle$) stands for the initial (or after scattering) electron states. These two states are identical for elastic scattering and different for inelastic scattering. $\mathbf{r}$ is the position operator of the electron. In the last two rows a sum is made over all the excited states $|c\rangle$ of this electron (bound or continuum states). $E_c - E_i$ represents the energy of excitation and $\hbar\omega_{in}$ ($\hbar\omega_{sc}$) is the energy of the incident (scattered) photon. In elastic scattering $\omega_{sc} = \omega_{in}$. $\Gamma_c$ is the width of the excited level $|c\rangle$ and $2\pi\hbar/\Gamma_c$ its life time. It is also the quantum counterpart of the damping factor $\gamma$, Eq. (1.105), in the classical theory. The polarization vectors may be complex so they can represent elliptical polarization states.[5] The following discussion will show that the first term represents the Thomson scattering found in the classical theory. The two last terms, $b_{disp1}$ and $b_{disp2}$, define the dispersive part of the scattering.

The absorption cross-section is also derived from the interaction Hamiltonian, once again at the lowest order of perturbation [4], Sect. 44,

$$\sigma_{abs}(\hbar\omega_{in}) = \frac{2\pi\hbar c r_e}{m} \sum_c \frac{\hbar\omega_{in}\Gamma_c}{(E_c - E_i)^2} \frac{|\langle c | \hat{\mathbf{e}}_{in} \cdot \mathbf{p}\, e^{-i\mathbf{k}_{in}\mathbf{r}} | i \rangle|^2}{(E_c - E_i - \hbar\omega_{in})^2 + \Gamma_c^2/4}. \tag{1.113}$$

---

[5] In most instances in this book, only linear polarizations are considered and no complex conjugate is indicated. In the case of anisotropic scattering, Sect. 1.5, the circular polarization may be required.

In this process the photon completely disappears. The $\mathbf{A}^2$ term in the Hamiltonian does not contribute and therefore only the $\mathbf{A}.\mathbf{p}$ term is used. Every term of the sum corresponds to the excitation towards a $|c\rangle$ state. The numerator suggests that the electric field transfers some momentum to the electron and changes the $|i\rangle$ level into the $|c\rangle$ level. In a similar way, in Eq. (1.112), one can say that the scattering $b_{\mathrm{disp1}}$ is obtained by excitation $|i\rangle \rightarrow |c\rangle$, then deexcitation $|c\rangle \rightarrow |s\rangle$. This order is reversed in $b_{\mathrm{disp2}}$, since the $|c\rangle$ state, which is virtual, is destroyed before being created (Fig. 1.7).

To calculate the scattering as well as the absorption, one generally uses the *dipolar approximation*, that is to say one replaces the factors $e^{i\mathbf{k}.\mathbf{r}}$ by one, supposing the wavelength much bigger than the atomic dimensions. This approximation, which is excellent in the visible spectrum, is still good for X-rays because the inner electronic levels which are excited are usually very much localized. Under certain conditions, however, this approximation is not sufficient and the next term in the expansion of the exponential (*i*$\mathbf{k}.\mathbf{r}$, the *quadrupolar term*) must be included.

When the energy of a photon is sensibly larger than all the excitation thresholds of the atom ($\hbar\omega \gg E_c - E_i$), only the first term in Eq. (1.112), which represents the Thomson scattering, is significant. In the extreme case of light atom and very high energies, the scattering cross-section given by this first term is even greater than the absorption cross-section Eq. (1.113): see Sect. 1.4.6. We shall start the discussion of the Thomson scattering $b_{Th}$ to show that it can be separated into elastic and inelastic (Compton) scattering. Then we shall describe the absorption spectrum which comes from Eq. (1.113). Finally we shall discuss the dispersive $b_{\mathrm{disp1}} + b_{\mathrm{disp2}}$ scattering, in relation with absorption.

### *1.4.5 Quantum Description: Elastic and Compton Scattering*

For a free electron and in the classical Thomson scattering, the backward move of the electron is ignored. Compton performed a kinematical calculation which took into account the momentum and the energy carried by the radiation quantized as photons. For an electron initially at rest, the conservation of these two quantities implies that the photon releases an energy such that the wavelength after the scattering process $\lambda_{\mathrm{sc}}$ becomes larger than the initial one $\lambda_{\mathrm{in}}$, and satisfies the equation

$$\lambda_{\mathrm{sc}} = \lambda_{\mathrm{in}} + \lambda_c \left(1 - \cos 2\theta\right) \qquad \lambda_c = 2\pi\hbar/mc = 0.002426 \text{ nm}, \qquad (1.114)$$

where $2\theta$ is the angle between the incident and scattered beams and $\lambda_c$ is the *Compton wavelength* of the electron.

In the present calculation, we are doing non-relativistic approximations which are not valid if the photon energy becomes close to the rest energy of the electron. Neglected relativistic effects are the influence of the spin and a factor which diminishes the Compton scattering cross-section.

When the electron is bound to an atom two processes are possible: the radiation may be elastically scattered with the conservation of the electron state (the momentum being transferred to the atom which is assumed to have an infinite mass) or inelastically with the ejection of the electron. One must determine the respective probabilities of these two processes. We start first with the case of an atom which has only one electron.

Let us evaluate the elastic, then the total scattering. The inelastic scattering will be obtained by subtraction. Keeping only $b_{Th}$ from Eq. (1.112) we have

$$b = r_e \widehat{\mathbf{e}}_{sc}^* . \widehat{\mathbf{e}}_{in} f_{si}, \qquad f_{si} = \langle s | e^{i\mathbf{q}.\mathbf{r}} | i \rangle, \qquad (1.115)$$

where $\mathbf{q}$ is equal to $\mathbf{k}_{sc} - \mathbf{k}_{in}$ (Eq. (1.3)). For elastic scattering, $|i\rangle = |s\rangle$. Let $\psi(\mathbf{r})$ and $\rho(\mathbf{r})$ be the wave function and the electron density then

$$f_{si} = f_{ii} = \int \psi^*(\mathbf{r}) \psi(\mathbf{r}) e^{i\mathbf{q}.\mathbf{r}} dV(\mathbf{r}) = \int \rho(\mathbf{r}) e^{i\mathbf{q}.\mathbf{r}} dV(\mathbf{r}). \qquad (1.116)$$

We have derived here more rigorously the form factor which we previously determined by the classical theory (for the atom having one electron). The calculation is completed by the evaluation of the total scattering cross-section, elastic plus inelastic. This total is obtained by summing the modulus square of the scattering factor over all the final states of the electron,

$$\sum_{|s\rangle} \left| \langle s | e^{i\mathbf{q}.\mathbf{r}} | i \rangle \right|^2 = \sum_{|s\rangle} \langle i | e^{-i\mathbf{q}.\mathbf{r}} | s \rangle \langle s | e^{i\mathbf{q}.\mathbf{r}} | i \rangle. \qquad (1.117)$$

Since the sum over the final states is made over all the possible states, these ones form a complete set and satisfy the closure relation

$$\sum_{|s\rangle} |s\rangle \langle s| = \text{unit operator.} \qquad (1.118)$$

The final states disappear from Eq. (1.117) which becomes equal to unity. The inelastic cross-section is obtained by subtraction and finally we have

$$(d\sigma/d\Omega)_{\text{elas+inel}} = |r_e \widehat{\mathbf{e}}_{sc}^* . \widehat{\mathbf{e}}_{in}|^2 \qquad (1.119)$$

$$(d\sigma/d\Omega)_{\text{elas}} = |r_e \widehat{\mathbf{e}}_{sc}^* . \widehat{\mathbf{e}}_{in}|^2 |f_{ii}|^2 \qquad (1.120)$$

$$(d\sigma/d\Omega)_{\text{inel}} = |r_e \widehat{\mathbf{e}}_{sc}^* . \widehat{\mathbf{e}}_{in}|^2 \left(1 - |f_{ii}|^2\right). \qquad (1.121)$$

This calculation prompts two remarks. We first observe that in the sum Eq. (1.117), the terms which do not conserve the momentum seem to play no part: they cancel the matrix element $\langle s | e^{i\mathbf{q}.\mathbf{r}} | i \rangle$. However, these terms must be included in the sum to enable the use of the closure relation Eq. (1.118). Next, some information about the conditions in which this sum is performed should be given. To be correct we must sum over all the final states of the radiation, with the scattering direction $\widehat{\mathbf{u}}$ kept fixed (this is a result of the definition of the differential cross-section) and with the energy conservation obeyed. Instead of Eq. (1.117), the exact expression is ($E_s$ and $E_i$ being the energies of the electron states)

$$\sum_{|s\rangle} \int_{\mathbf{k}_{sc}/|\mathbf{k}_{sc}|=\hat{\mathbf{u}}} \left| \langle s| e^{i\mathbf{q}\cdot\mathbf{r}} |i\rangle \right|^2 \delta(E_s - E_i - \hbar c\,|\mathbf{k}_{in}| + \hbar c\,|\mathbf{k}_{sc}|)d\mathbf{k}_{sc}. \qquad (1.122)$$

Let us note that this expression imposes the two conditions used by Compton which are the energy conservation as shown by the $\delta$ function and the conservation of momentum in the matrix element. Performing the integral over $\mathbf{k}_{sc}$ one gets back the sum Eq. (1.117) over the final states of the electron, with the condition $\mathbf{k}_{sc}/|\mathbf{k}_{sc}| = \hat{\mathbf{u}}$. One can see that $\mathbf{k}_{sc}$ being in the $\delta$ function depends on $\langle s|$, and consequently $\mathbf{q}$ has also that dependence. For the consistency of the above discussion we neglected this dependence otherwise the closure relation Eq. (1.118) could not have been applied to Eq. (1.117). The approximation is very good but the small dependence of $|\mathbf{k}_{sc}|$ on the final state of the electron, which, through the momentum conservation, is also a dependence on the electron initial momentum, can be used to measure the momentum distribution inside the atom or inside the solid. Such an application of Compton scattering will not be developed further in this book.

The inelastic scattering for which we have calculated the cross-section is frequently considered as the Compton scattering. This is not completely correct since the total scattering cross-section also includes the **Raman scattering**. In such a case the final state $\langle s|$ of the electron is not a free plane wave but a bound excited state [17]. To be fully complete we must also consider another inelastic scattering process, the so-called **resonant Raman scattering**. This process does not appear in the above calculation, but rather in the development of the second term in Eq. (1.112), $b_{disp1}$, when one assumes $\langle s| \neq \langle i|$; it is obvious at energies close to an excitation edge. It is thus more associated with absorption and fluorescence than with Compton scattering. However, far from resonances, the dominating inelastic process is usually the Compton scattering.

The calculation that we have just carried out has to be changed for an atom having more than one electron. The electronic states are multi-electron states and each interaction operator is replaced by the sum of operators acting each on one electron. For an atom having two electrons,

$$|i\rangle \quad \rightarrow \quad (1/\sqrt{2})\,|\Psi_1(\mathbf{r}_1)\Psi_2(\mathbf{r}_2) - \Psi_1(\mathbf{r}_2)\Psi_2(\mathbf{r}_1)\rangle \qquad (1.123)$$

$$e^{i\mathbf{q}\cdot\mathbf{r}} \quad \rightarrow \quad e^{i\mathbf{q}\cdot\mathbf{r}_1} + e^{i\mathbf{q}\cdot\mathbf{r}_2}. \qquad (1.124)$$

Expression (1.123) is the Slater's determinant which represents the antisymmetric state with respect to the permutation of the electrons. The elastic scattering factor becomes

$$f = \langle i| e^{i\mathbf{q}\cdot\mathbf{r}_1} + e^{i\mathbf{q}\cdot\mathbf{r}_2} |i\rangle. \qquad (1.125)$$

With $|i\rangle$ given by Eq. (1.123) and using the orthogonality between $\Psi_1$ and $\Psi_2$, this yields

$$f = f_{11} + f_{22} \quad \text{where} \quad f_{jl} = \int \Psi_j^*(\mathbf{r})\Psi_l(\mathbf{r})e^{i\mathbf{q}\cdot\mathbf{r}}dV(\mathbf{r}). \qquad (1.126)$$

To obtain the total cross-section one must sum the amplitude squares over all the final states as in Eq. (1.117). Although it is not necessary to know them, we explicitly write them for more complete view

$$|s\rangle = (1/\sqrt{2})\,|\Psi_1(\mathbf{r}_1)\Psi_2(\mathbf{r}_2) - \Psi_1(\mathbf{r}_2)\Psi_2(\mathbf{r}_1)\rangle \tag{1.127}$$

$$|s\rangle = (1/\sqrt{2})\,|\Psi_x(\mathbf{r}_1)\Psi_2(\mathbf{r}_2) - \Psi_x(\mathbf{r}_2)\Psi_2(\mathbf{r}_1)\rangle \qquad x \neq 1,2 \tag{1.128}$$

$$|s\rangle = (1/\sqrt{2})\,|\Psi_1(\mathbf{r}_1)\Psi_x(\mathbf{r}_2) - \Psi_1(\mathbf{r}_2)\Psi_x(\mathbf{r}_1)\rangle \qquad x \neq 1,2 \tag{1.129}$$

$$|s\rangle = (1/\sqrt{2})\,|\Psi_x(\mathbf{r}_1)\Psi_y(\mathbf{r}_2) - \Psi_x(\mathbf{r}_2)\Psi_y(\mathbf{r}_1)\rangle \qquad x,y \neq 1,2. \tag{1.130}$$

The last state corresponds to a two-electron excitation and gives rise to a zero amplitude, but once again, it must be included to use the closure relation. With this latter, the total cross-section is proportional to

$$\langle i|\,(e^{-i\mathbf{q}\cdot\mathbf{r}_1} + e^{-i\mathbf{q}\cdot\mathbf{r}_2})(e^{i\mathbf{q}\cdot\mathbf{r}_1} + e^{i\mathbf{q}\cdot\mathbf{r}_2})\,|i\rangle. \tag{1.131}$$

Substituting $|i\rangle$ by Eq. (1.123) and using the definition Eq. (1.126), this expression becomes

$$2 + f_{11}f_{22}^* + f_{11}^*f_{22} - |f_{12}|^2 - |f_{21}|^2. \tag{1.132}$$

It is easy to extend this calculation to any number of electrons $Z$. The elastic and inelastic scattering cross-sections become

$$(d\sigma/d\Omega)_{\text{elas+inel}} = (r_e\widehat{\mathbf{e}}_{\text{sc}}^*\cdot\widehat{\mathbf{e}}_{\text{in}})^2 \tag{1.133}$$

$$\left( Z + \sum_{1\leq j\neq l\leq Z} f_{jj}^*f_{ll} - \sum_{1\leq j\neq l\leq Z} |f_{jl}|^2 \right)$$

$$(d\sigma/d\Omega)_{\text{elas}} = (r_e\widehat{\mathbf{e}}_{\text{sc}}^*\cdot\widehat{\mathbf{e}}_{\text{in}})^2 \left| \sum_{1\leq j\leq Z} f_{jj} \right|^2 \tag{1.134}$$

$$(d\sigma/d\Omega)_{\text{inel}} = (r_e\widehat{\mathbf{e}}_{\text{sc}}^*\cdot\widehat{\mathbf{e}}_{\text{in}})^2 \tag{1.135}$$

$$\left( Z - \sum_{1\leq j\leq Z} |f_{jj}|^2 - \sum_{1\leq j\neq l\leq Z} |f_{jl}|^2 \right).$$

*One can see that the elastic scattering factor is written as the Fourier transform of the electron density* (which is the sum of the densities of all the wave functions), c.f. Eq. (1.116). When the terms $|f_{jl}|^2$ are ignored, *the case of the many-electrons atom is naively deduced from the one having one electron: on one hand the elastic scattering lengths and on the other hand the inelastic cross-sections of all the electrons are added.* The $|f_{jl}|^2$ terms constitute in fact a modest correction. They are called the exchange terms since they come from matrix elements in which electrons have been interchanged as for instance in the case of two electrons,

$$\langle \Psi_1(\mathbf{r}_1)\Psi_2(\mathbf{r}_2)|\ldots|\Psi_1(\mathbf{r}_2)\Psi_2(\mathbf{r}_1)\rangle. \tag{1.136}$$

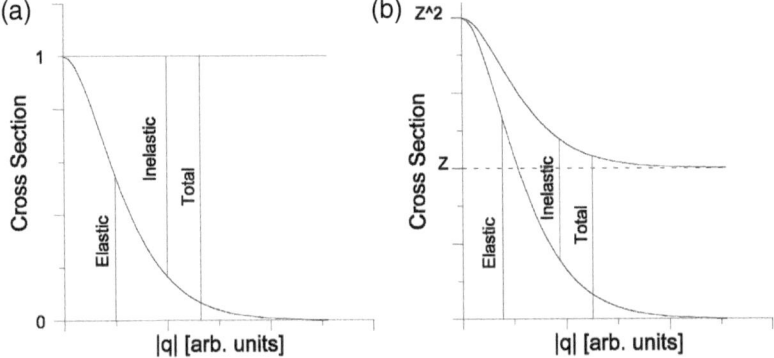

**Fig. 1.8** Schematic representation of elastic, inelastic (Compton) and total cross-sections as a function of $|\mathbf{q}|$, (**a**) for an atom having one electron, (**b**) for an atom having any number of electrons. The unit of cross-section is $r_e^2$

They are subtracted because electrons are fermions.

The general evolution of the cross-sections is presented as a function of $|\mathbf{q}|$ in Fig. 1.8.

To end this section we now discuss how easily the Compton scattering is observed. It is suitable to rewrite the change of wavelength, Eq. (1.114), in the following way as a function of the scattering vector transfer $\mathbf{q}$

$$|\mathbf{k}_{in}| - |\mathbf{k}_{sc}| = (\lambda_c/4\pi)\,\mathbf{q}^2 + (higher\ order\ in\ \mathbf{q}). \qquad (1.137)$$

A radiation of wavelength $\lambda_c$ has an energy of 511 keV, i.e. the mass energy of the electron at rest. For a radiation of energy 10 keV scattered at an angle of say one degree (typical of a grazing incidence surface experiment), the wavelength change is very weak since it depends on the square of $\mathbf{q}$. Things are different in the range of medium and large values of scattering angles where this change is easily measurable, for instance by using an analyser crystal. At energies greater than about 100 keV and medium angles of scattering this change is appreciable.

The Compton cross-section varies in a similar way as shown in Eqs. (1.121) and (1.135) and in Fig. 1.8. For radiation of 10 keV, it is negligible at small angles, but this is not true at wider angles. At higher energies, some tens of keV, the Compton scattering achieves its highest value already at medium angles. At those energies and for light elements it dominates the other processes. Indeed its proportion is larger for small Z scatterers. Table 1.1 gives some values of the total scattering cross-sections (integrated over the whole angular space);[6] the elastic cross-section is condensed in the forward direction in a cone which becomes narrower when the energy increases.

---

[6] More cross-section values can be found in the International Tables for X-ray Crystallography [11], vols. III and IV.

**Table 1.1** Atomic cross-sections of some elements as a function of energy (scattering cross-sections are integrated over a solid angle $4\pi$). In each case are displayed the Compton scattering cross-section/the elastic scattering cross-section/the photoelectric absorption cross-section, in barns (i.e. $10^{-28}$ m$^2$). The irregularities in the evolution of the absorption are due to the presence of an edge close to the chosen energy. After [20]; see also the International Tables for X-ray Crystallography [11]

| Element (Z) | 5 keV | 10 keV | 30 keV | 100 keV |
|---|---|---|---|---|
| C(6) | 2.1/5.8/371 | 2.7/3.2/39 | 3.3/0.67/1.1 | 2.9/0.07/0.02 |
| Cu(29) | 4.65/307/19500 | 8.2/153/22600 | 13/35.6/1090 | 13.3/4.5/30 |
| Ag(47) | 6.5/820/132000 | 11.5/459/20600 | 19/117/6420 | 20.9/15.8/224 |
| Au(79) | 8.22/2630/212000 | 15.3/1580/36100 | 27.8/432/8420 | 33.2/60.8/1590 |

## *1.4.6 Resonances: Absorption, Photoelectric Effect*

In the interaction process, part of the radiation disappears instead of being scattered. As shown in Eq. (1.113), the energy is transferred to an electron which is excited to an empty upper state $|c\rangle$. Most frequently it is expelled from the atom; this is the so-called *photoelectric absorption*. After a delay of about $h/\Gamma_c$, the atom de-excites, according to various processes which can be radiative or not. The most obvious process in a diffraction or scattering experiment is the emission of *fluorescence radiation*. It corresponds to the fall of a second electron of the atom into the level vacated by the first one. Its energy is necessarily lower than the energy of excitation. The *fluorescence yield*, i.e. the fraction of excited atoms which are de-excited in this way, depends on the elements and on the levels; for the K level of copper, the fluorescence yield is 0.5.

Let us note that for a given excited level $|c\rangle$, the cross-section varies with the energy and exhibits a Lorentzian behaviour with a FWHM $\Gamma_c$. In the X-ray domain, $\Gamma_c$ lies between about a bit less than 0.5 eV and a bit more than 5 eV.

The important transitions for X-rays are those of the inner electrons which belong to the K, L, ... shells. The transitions may bring the excited electron towards the continuum of the free states; their spectral signature is then characterized by an *absorption edge*, located at the excitation energy, since any levels above the edge are equally accessible (Fig. 1.9). They can also arise towards the first free bound levels. These states may have a large enough density to give rise to one (or several) well-defined absorption peaks superimposed to the edge, the so-called *white line*(s). The white line(s) spectrum is not an exact image of the density of free states of the atom, molecules or condensed system. The observed spectrum corresponds to a system which has lost a core electron and is deformed by the electric charge of the core hole. Peaks can paradoxically then appear below the edge. In condensed matter, the absorption above the edge exhibits oscillations, the so-called EXAFS (extended X-ray absorption fine structure), which are interpreted as arising from interference effects in the wave function of the ejected electron. These interference are due to the scattering of the ejected electron by the neighbouring atoms.

**Fig. 1.9** Schematics of an absorption edge, with a white line ($\Gamma = 2$ eV). The *solid line* shows the variations of both the absorption cross-section and of the imaginary part of dispersion correction (see next section) and the *dotted line*, the real part of this correction in arbitrary units. The origin of energy, $E_0$, is taken at the edge. This schematic figure is not intended to show the real details of these curves in the vicinity and above the edge. On this short interval of energy, the $E^{-3}$ decay has been neglected

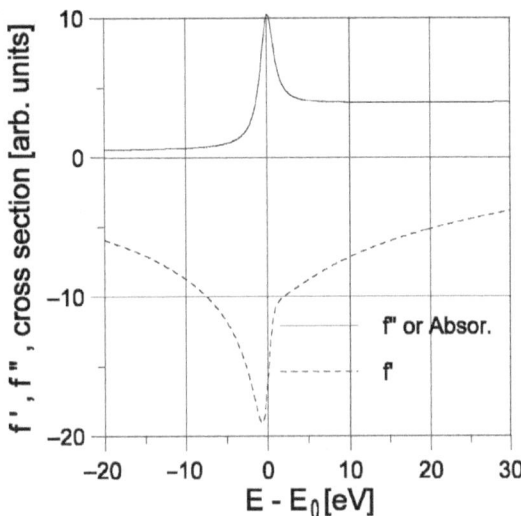

In short, one can say that the absorption varies as $E^{-3}$ and as $Z^4$. This does not take into account the discontinuities at the edges. The K edges produce a discontinuity of the absorption by a factor of about 5–10. Figure 1.10 shows for copper a discontinuity of factor 7 at the K edge and of about the same amount for the three L edges together; one can see that the decay in between the edges is slightly slower than $E^{-3}$. Table 1.1 gives some values of the absorption and scattering cross-sections.

In practice one frequently needs the absorption coefficient $\mu$ rather than the cross-section; this coefficient is defined by the fact that the transmission through a thickness $t$ is given by $e^{-\mu t}$. It is also equal to $4\pi\beta/\lambda$ ($\beta$ the imaginary part of the

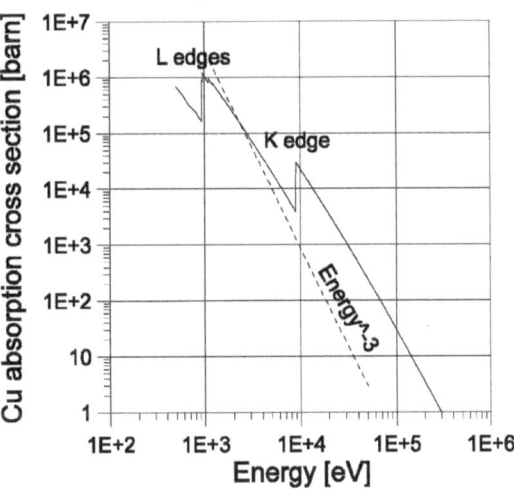

**Fig. 1.10** Absorption cross-sections for the atom of copper in barns ($10^{-22}$ mm$^2$), after Cromer–Libermann (above 10 keV) and Henke (below 10 keV). The slope of the $E^{-3}$ power law is presented by a *dotted line*

refractive index). For a homogeneous material made of a single element, $\mu$ depends on the cross-section $\sigma$ and on the atomic volume $V$ through

$$\mu = \sigma/V. \tag{1.138}$$

To calculate the absorption coefficient of a material, it is sometimes useful to introduce the mass absorption coefficient, given by $\mu/\rho$ ($\rho$ is the density) and commonly tabulated [11]. This coefficient is characteristic of the element and independent of its density. If $A$ is the molar mass and $N$ the Avogadro's number

$$\mu/\rho = N\sigma/A. \tag{1.139}$$

The absorption coefficient of a material composed of several elements $i$, each of them present with the partial density $\rho_i$, is simply given by

$$\mu = \sum_i \rho_i (\mu/\rho)_i. \tag{1.140}$$

### 1.4.7 Resonances: Dispersion and Anomalous Scattering

We return now to the case of the elastic scattering in which we had neglected the dispersive part $b_{\text{disp1}} + b_{\text{disp2}}$ in Eq. (1.112). Actually we shall take into account only $b_{\text{disp1}}$, which represents the second line of this expression. For a term of the sum over $c$ to be appreciable it is necessary for its denominator to be small which never occurs in $b_{\text{disp2}}$.

We reproduce here the formula Eq. (1.110), which gives the separated Thomson and dispersive contributions:

$$b = r_e(f + f' + if'')\widehat{\mathbf{e}}_{\text{sc}}^* . \widehat{\mathbf{e}}_{\text{in}}. \tag{1.141}$$

This separation could appear artificial in the classical expression for $b$, Eq. (1.108), but arises perfectly naturally in the quantum mechanical one, Eq. (1.112). We have assumed that the polarization contributes in $b_{\text{disp1}}$ that is $f' + if''$, through the same polarization factor as in the Thomson term. This is not true in every case, as discussed in Sect. 1.5.

Each of the terms $|c\rangle$ of the sum in Eq. (1.112), from which the dispersion correction $f' + if''$ arises, corresponds to an excitation energy (or commonly a resonance) $E_c - E_i$. The associated correction is

$$r_e(f'_c + if''_c) \propto \frac{1}{x - i} = \frac{x}{1 + x^2} + \frac{i}{1 + x^2} \tag{1.142}$$
$$\text{with } x = [\hbar\omega - (E_c - E_i)]/(\Gamma_c/2).$$

The real and imaginary parts are presented in Fig. 1.11.

**Fig. 1.11** Schematic
representation of the
dispersion correction for a
single resonance at energy
$E_c - E_i = E_0$. $f'$ and $f''$ are
given by $b = r_e(f + f' + if'')$

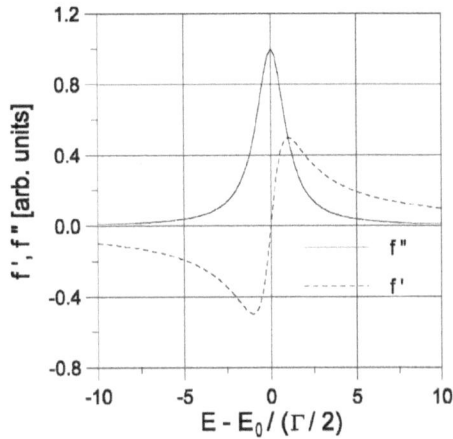

Equations (1.112) and (1.113) show a correspondence between the dispersion
correction in terms of the scattering length and the absorption cross-section. Exactly
at the resonance energy, we check the optical theorem (Im is the imaginary part),

$$\sigma_{abs} = 2\lambda \, \mathscr{I}\text{m}[b(q=0)], \qquad (1.143)$$

discussed in Sect. 1.3.2.[7]

The distribution of the resonance energies $E_c - E_i$, with the edges as main features, has been previously discussed about the absorption. Figure 1.9 shows the
comparison between the variations of the absorption cross-section close to an edge
and the variations of the anomalous scattering.

Let us now look how the real part of the scattering factor, $f + f'$, varies when the
energy changes from X-rays to near infrared, that is to say from several tenths of
keV to 1 eV. The highest energies are far above the edges of most elements and the
Thomson scattering factor $f$ is dominant. For lower energies, a negative contribution
$f'$ appears at every edge and is more important below the edge than above because
of the white lines. Low-energy edges produce the most intense dispersion effects.
Going to low energies, some edges for which $f + f'$ is negative are observed, and
then a transition occurs about 10–100 eV where $f + f'$ definitely changes its sign.
In this range, the very intense absorption lines enormously reduce the propagation
of light in matter, which makes it called **vacuum ultraviolet** radiation, because it
propagates only in vacuum. When the sign of $f + f'$, which is also the sign of $b$,

---

[7] We obtain here an expression for the *absorption* cross-section, while the optical theorem yields
the same expression for the *total* cross-section. The error comes from our calculation of the scattering length, made in the first-order Born approximation. The next order is required to obtain an
imaginary part which expresses also the intensity loss due to scattering (see the end of Appendix
1.A to this chapter). The calculation at that order is made intricate because of some difficulties
of the quantum theory of radiation (the divergences of the field theory). That error is negligible
inasmuch as the absorption is the largest part of the cross-section, which is true up to moderate
energies, but not at the highest.

changes from positive to negative, the refractive index $n$ goes from below to above the unit value (the link between the scattering and the index is discussed in Sect. 1.3).

### 1.4.8 Resonances: Dispersion Relations

The absorption cross-section is easily obtained directly by experiments, as for example, the measurement of the transmission through a known thickness of a material. The imaginary part of the scattering length $b$ is found at the same time. The real part of $b$, however, is more difficult to obtain accurately. Among the different methods, not only diffraction experiments but also reflectivity measurements have been used to extract the scattering length [19]. Having recourse to such methods can be avoided because it is possible to rebuild the real part of $b$ if the imaginary part is known over the entire spectral range. Conversely the imaginary part can be deduced from the real one. This is possible through the so-called dispersion or Kramers–Kronig relations.

   Before showing the theory we give a summary of the practical procedures for retrieving the spectrum of the real part of the dispersion $f'$, when the absorption spectrum has been measured about an edge:

(1) Convert the measured absorption coefficient $\mu$ into the atomic cross-section $\sigma_{abs}$, then apply the optical theorem to obtain the imaginary dispersion correction $f''$. In general the material is composed of several elements, whose atomic cross-sections should be separated out from the global $\mu$.
(2) The aim is to apply the dispersion relation Eq. (1.148) shown below, which needs an integration over a wide range of energy. Instead the absorption is usually measured over a narrow range only. The spectrum of $f''$ should therefore extended with tabulated values. Any step at the junctions should be reduced as far as possible, for example by some correction to the normalization.
(3) Apply the Kramers–Kronig integral Eq. (1.148) to $f''$ in order to obtain $f + f'$. Some care is needed with the algorithm of integration for getting a smooth principal part despite the divergence.
(4) Since the range of integration is not infinite, the result in $f + f'$ is given up to some additive constant, which is retrieved by comparison with tabulated values.

   We now show how the dispersion relations are obtained from first principles.
   Let us start from the classical model, namely the expression Eq. (1.108) for the scattering length,

$$b = r_e \frac{\omega^2}{\omega^2 - \omega_0^2 - i\gamma\omega/m} \widehat{\mathbf{e}}_{sc} \cdot \widehat{\mathbf{e}}_{in}. \tag{1.144}$$

The general case can be represented by summing many expressions of this kind corresponding to each different resonance $\omega_0$ with a different damping constant $\gamma$. Since this model is defined by two independent functions of $\omega_0$, a distribution of the resonance densities and a distribution of the damping constants, one could expect the real and imaginary parts of the scattering length to also constitute two independent functions. However, some constraint is imposed because the damping constants $\gamma$ are necessarily positive. Although this constraint seems to be weak, it is remarkable that it is sufficient to lead to a relation between the real and imaginary parts of $b$. We shall see that such a relation does not come from a particular scattering model such as Eq. (1.144); it is more general and applies to the response of any system to an excitation. For the proof, we return to the model with only one resonance Eq. (1.144) but this could be easily extended to the general case.

To prove the existence of a relation between the real and imaginary parts of $b$, it is necessary to make use of a mathematical trick, the analytical continuation of function $b$ in the complex plane. The trick allows one to express some basic properties of complex functions. Indeed $b$ is a complex function of the real variable $\omega$. If such a function can be represented by a series expansion which converges for any real value of the variable, then it can also be defined for complex values of this variable. Hence, the series still converges in a domain of the complex plane. Inside this domain, the function that we shall call here $\phi(z)$ is analytic and follows Cauchy's theorem. This theorem ensures that for any closed contour $C$ inside the domain of analyticity and for any point $z$ inside the contour,

$$\phi(z) = \frac{1}{2\pi i} \int_C \frac{\phi(z')}{z'-z} dz' \qquad (z, z' \text{ complex}, \tag{1.145}$$

$$C \text{ taken in the positive sense}).$$

For this relation to be useful the integral must be taken only over the region where the function is known, i.e. the real axis. Let $C$ be the real axis plus a curve which continuously approaches infinity in the lower half-plane for instance a semi-circle with a radius approaching infinity (Fig. 1.12, drawn with the variable $\omega$ replacing $z$). We obtain the desired relation provided that (a) the function is analytic in all this half-plane and (b) it approaches zero when the modulus of the variable approaches infinity so that the integral taken over the semi-circle is zero. Under such conditions, Eq. (1.145) is expressed as an integral over $z'$ real. These conditions, however, impose $z$ to be inside the contour and therefore to have an imaginary part strictly negative though one would wish to have only real quantities. Nevertheless $z$ can be on the real axis, but then the expression on the left-hand side is divided by two since

**Fig. 1.12** Integration over a contour $C$ defined by the real axis and a *semi-circle* having its radius approaching infinity. If a function does not have any pole inside the contour it satisfies Eq. (1.145) ($\omega$ has the same role as the $z$ variable). The poles of the scattering length $b(\omega)$, Eq. (1.144), have been represented. They are outside the contour

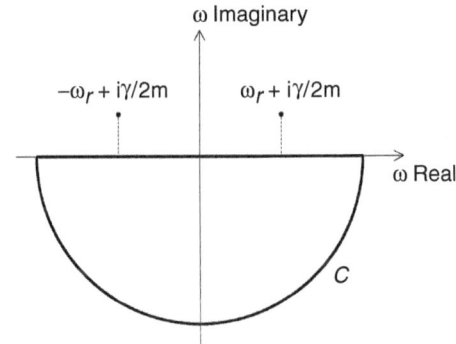

$z$ is at the border (a rigorous proof is available). Finally if $P$ represents the principal part of the integral at the singularity $x' = x$, then

$$\phi(x) = -\frac{1}{\pi i} P \int_{-\infty}^{+\infty} \frac{\phi(x')}{x' - x} dx' \qquad (x, x' \text{ real}). \qquad (1.146)$$

The real and imaginary parts of this relation can be written separately. This shows that *if $\phi(x)$ satisfies the above conditions (a) and (b), i.e. it is analytic in the lower half-plane and tends to zero when $|x|$ goes to infinity, some integral relations exist on the real domain of $x$ between the imaginary and real parts of $\phi$.*

Let us apply this trick to the scattering length written in Eq. (1.144), which is a polynomial fraction of the variable $\omega$. It is analytic over any domain which does not contain its poles, i.e. the zeros of its denominator. These zeros, indicated in Fig. 1.12, are $i\gamma/2 \pm \omega_r$ ($\omega_r$ depends on $\omega_0$ and $\gamma$). The scattering length $b$ satisfies condition (a) because the damping constant $\gamma$ which is necessarily positive yields poles which are in the upper half-plane. To satisfy condition (b) one could divide $b$ by $\omega$. The relations would then be valid in $b(\omega)/\omega$ (after replacing $\phi(x)$), but it is better to divide by $\omega^2$ since we then get more general relations as we are going to comment. Let us notice that the division of $b$ by $\omega^2$ does not add any pole and does not change the domain of analyticity.

A relation such as Eq. (1.146) is not yet completely convenient because the physical domain does not extend over the entire real axis but only over its positive side (the variable is the radiation frequency). Integrating from 0 to $\infty$ is, however, sufficient since $b$ verifies

$$b(-\omega) = b^*(\omega). \qquad (1.147)$$

We should look if such a symmetry of the scattering length is attached to a particular model, or more general. A Fourier transform which transforms the above expression from $\omega$ to time space shows that this is simply the expression of a symmetry by time reversal. It is thus a general property which has, however, a limitation: this symmetry does not hold for magnetic moments so *the following expressions do not hold for magnetic scattering.* In that case, Eq. (1.147) is written with a minus sign and different relations are obtained.[8] With this equality and a bit of algebra, one can rewrite the real and imaginary parts of Eq. (1.146). Replacing $\phi$ by $b(\omega)/\omega^2$ and $x, x'$ by $\omega, \omega'$ yields

$$\mathrm{Re}[b(\omega)/\omega^2] = -\frac{2}{\pi} P \int_0^\infty \frac{\omega' \,\mathscr{I}\mathrm{m}[b(\omega')/\omega'^2]}{\omega'^2 - \omega^2} d\omega' \qquad (1.148)$$

$$\mathscr{I}\mathrm{m}[b(\omega)/\omega^2] = \frac{2\omega}{\pi} P \int_0^\infty \frac{\mathrm{Re}[b(\omega')/\omega'^2]}{\omega'^2 - \omega^2} d\omega'. \qquad (1.149)$$

These relations are the so-called ***Kramers and Kronig or dispersion relations*** for the scattering length.

---

[8] In practice the same dispersion relations can be written for magnetic and non-magnetic scattering lengths provided that the magnetic part is affected by a factor $i$.

In this model, the proof we have given assumes all the poles of $b(\omega)/\omega^2$ to be above the real axis. We have inferred this from the positive value of the damping constant but *it can also be inferred from the principle of causality of very general extent*. To understand the equivalence of these two hypotheses, positive value of the damping constant and principle of causality, it is worth returning back to the resolution of the differential equation (1.105), which describes the movement of the electron in the incident field. We rewrite this equation with noting the displacement $u$ instead of $z$ to avoid any confusion with the variable $z$ in the present section; for simplicity, $u$ will be a scalar. The properties of $u$ that we are going to discuss now are also the ones of the radiated field which is proportional to $u$:

$$m\ddot{u} + \gamma\dot{u} + m\omega_0^2 u = (-e)E_0\, e^{i\omega t}. \qquad (1.150)$$

A systematic method to solve such a differential equation with a right-hand side $f(t)$ consists in using the Green function of the equation. This method has been described in Sect. 1.2.5. Let us recall that a solution of this kind of equation is given by

$$\begin{aligned}
u(t) &= u_0(t) + \int_{-\infty}^{+\infty} G(t - t')f(t')\,dt' \\
&= \int_{-\infty}^{+\infty} G(t - t')f(t')\,dt',
\end{aligned} \qquad (1.151)$$

where the Green function, $G(t)$, is the solution of the equation with $\delta(t)$ instead of $f(t)$ in the right-hand side; the solution $u_0(t)$ of the homogeneous equation (without the right-hand side) becomes nearly zero after a certain amount of time due to damping. Writing the electron displacement $u(t)$ through Eq. (1.151) allows the following physical interpretation to be given. The displacement $u$ at a given time $t$ is the result by linear superposition of the excitation action $f$ at any time $t'$ ; since the laws are invariant by time translation, the coefficient $G$ only depends on the difference $t - t'$. $G(t)$ is obtained through its Fourier transform $g(\omega)$. We replace the right-hand side of Eq. (1.150) by $\delta(t)$, whose Fourier transform is one. The derivatives in the left-hand side transform into powers of $\omega$, so we get

$$g(\omega) = -\frac{1}{m}\frac{1}{\omega^2 - \omega_0^2 - i\gamma\omega/m}, \qquad (1.152)$$

which yields $G(t)$

$$G(t) = -\frac{1}{2\pi m}\int_{-\infty}^{+\infty}\frac{e^{i\omega t}}{\omega^2 - \omega_0^2 - i\gamma\omega/m}\,d\omega. \qquad (1.153)$$

With Eqs. (1.151) and (1.153) we could find the motion such as given by Eq. (1.107), solution of Eq. (1.150), but this is not our purpose; we focus only on $G(t)$. To calculate this integral, it is possible to integrate along a closed path in the complex plane: if the function does not have any poles inside the path of integration, its integral over it is zero. The poles of $g(\omega)$ are those of the scattering length that we have just discussed; the integral taken over the path of integration $C$ (Fig. 1.12) is then zero.

For $t < 0$ the integral over half the circle is also zero since the numerator is bounded and the integral of $d\omega /|\omega|^2$ goes to zero when $|\omega|$ goes to infinite. Then $G(t) = 0$ for $t < 0$. It is important to mention that the proof depends on the position of the poles of $g(\omega)$ and on the positive sign of the damping constant. The condition $G(t)$ equal to zero at negative times constitutes the expression of *a causality principle, according to which an excitation given at a certain instant cannot produce any effect before this instant.* Quite important is the reciprocal, which is true though we do not show the proof here. Because of the principle of causality, $G(t) = 0$ for $t < 0$, then it can be shown that $g(\omega)$ does not have any pole below the real axis and the Kramers–Kronig relations apply to $g(\omega)$ and $u(\omega)$. Making the dispersion relations to depend on this principle gives them a very general extent, beyond the cases where it is possible to clearly define some damping.

In our world most phenomena are irreversible and time is therefore asymmetric. The positive character of the damping and the principle of causality as discussed here are two manifestations of the irreversibility. We showed that one or the other yields the dispersion relations Eqs. (1.148) and (1.149). But this raises a question. Microscopic laws in physics are mainly symmetric with respect to time reversal, while irreversibility is manifest in macroscopic systems with many parameters, when they are drawn out of equilibrium. The scattering of one single photon by one single atom looks like a microscopic elementary phenomenon. It is worth to see how and why it is irreversible. A plane wave travelling in vacuum, such as the incident wave before scattering, constitutes a very unlikely state that can be considered as out of equilibrium. The equilibrium state of a radiation comprises instead many random waves in thermal equilibrium with neighbouring objects. The scattering of a plane wave by an atom is irreversible, a bit like the dilution of an alcohol droplet in a glass of water. The final state is the spherical wave moving away from the atom, superimposed to the incident wave which has a reduced amplitude. Therefore the unique incident plane wave has been changed into a superposition of plane waves travelling in all the directions. In addition, any of the plane components of the diverging wave can be associated to a particular movement of the atom since the momentum must be conserved. This is reminiscent of the dilution effect. If the scattering were reversible, one could produce the reverse operation: starting from a spherical wave converging towards an atom and from a plane wave, one could see the plane wave coming out with an increased amplitude. This would be difficult to realize and may be impossible. For this one should correlate the different plane components of the converging wave to some particular movements of the atom. The difficulty is similar to the one which would be faced in an attempt to invert the dilution of the alcohol droplet by imposing on the molecules of the water–alcohol mixture some initial conditions such as the mixture demixing into two phases after a few instants. When absorption and fluorescence occur the process is still more irreversible since multiple photons can be re-emitted for only one absorbed. As a matter of fact it appears that absorption and resonant scattering contribute much more to the dispersion than pure elastic scattering. From the above arguments one can be convinced that *even though the scattering looks like an elementary phenomenon, it is actually something irreversible, which has to obey the dispersion relations associated with irreversibility.*

# 1.5 X-Rays: Anisotropic Scattering

## 1.5.1 Introduction

In this section we present briefly some other types of X-ray scattering, observed essentially in crystalline materials. These are the magnetic scattering, which depends on the magnetic moment of the atom, and the Templeton anisotropic scattering, which depends on the neighbourhood of the atom in the crystal. A common feature to these two scattering effects is their anisotropy. The usual scattering amplitude which is described in the previous sections can be said isotropic because it depends on the incident and scattered polarization directions through a unique factor, $\widehat{\mathbf{e}}_{\text{sc}} \cdot \widehat{\mathbf{e}}_{\text{in}}$, independent of the orientation of the scattering object. The atomic scattering amplitudes which we discuss now is said to be anisotropic because they depend on the orientation of the characteristic axes of the atom with respect to the incident and scattered polarizations. The characteristic axes may represent the magnetic moment direction of the atom if it exists or the directions of the crystal field which eventually perturbs the state of that atom.

These scattering effects can take their origin from two different mechanisms. The first one is the interaction between the electromagnetic radiation and the spin of the electron. It produces some scattering, the so-called ***non-resonant magnetic scattering***. This one is essentially independent of the binding of the electron in the atom, as is the Thomson scattering. A second type of anisotropic scattering arises as a part of the anomalous or resonant scattering, presented earlier in Sect. 1.4.7. The atomic states $|i\rangle$ and $|c\rangle$ in Eq. (1.112), that is the initial state and the one in which the electron is promoted, may be anisotropic. If that anisotropy originates from a magnetic moment, the resulting scattering is called ***resonant magnetic scattering***. If the anisotropy is some asphericity of the atom kept oriented by the particular symmetry of the material, it is the ***Templeton anisotropic scattering***.

X-ray magnetic scattering is a useful complement to neutron scattering. It can be used with some elements whose common isotopes strongly absorb thermal neutrons. The very good resolution (in all respects, position, angle and wavelength) of X-ray beams is an advantage for some studies. Since X-ray scattering depends on some characters of the magnetic moment in a way different from neutrons, it may raise some ambiguities left by neutron scattering experiments. One of these ambiguities is the ratio of the orbital to spin moment of the atom, because they contribute to neutron scattering exactly in the same way and cannot be discriminated from each other. X-ray amplitudes given by spin and orbital moment depend differently on the geometry of the experiment and they can be separated out. The resonant X-ray magnetic scattering is element dependent and eventually site dependent, which may give some useful information. It is also a spectroscopic method which probes the electronic state of the atom. The availability of small and brilliant X-ray beams compensates for the smallness of magnetic amplitudes in the study of thin films and multilayers. When the magnetic element has a very intense resonant magnetic scattering, even a single atomic layer can be probed.

Applications of Templeton anisotropic scattering have been developed during recent years as a valuable tool for exploring the electronic states of atoms in crystals, inconnected with the specific symmetry of their environment.

In the present section we give a short description of the non-resonant magnetic scattering, the resonant magnetic scattering and the Templeton anisotropic scattering. We also discuss the anisotropy of the optical index. The case of the magnetic neutron scattering is described in the Chap. 5 of this book.

## *1.5.2 Non-resonant Magnetic Scattering*

Similarly to the Thomson scattering, the non-resonant magnetic scattering can be found either in the classical or quantum theory. The quantum calculation can be found in [5]. The spin of the electron is associated with a magnetic moment which, in a classical description, interacts with the magnetic component of the radiation. Figure 1.13 shows schematically how the interaction between the electromagnetic field and the electron, comprised of an electric charge and a magnetic moment, can produce a magnetic-dependent scattering. Having some interplay between spin and motion in space, and having some magnetic properties attached to an electrically charged particle are relativistic effects. That relativistic character introduces the scale factor

$$|\hbar\mathbf{q}| / 2\pi mc = 2\left(\lambda_c/\lambda\right)\sin\theta \tag{1.154}$$

between the magnetic and Thomson scattering amplitudes of an electron. In a diffraction experiment that scale factor is typically of the order of $10^{-2}$. Since only unpaired electrons, which are at most one- or two-tenths of all electrons of a magnetized atom, contribute to the magnetic scattering, the magnetic amplitude is in favourable cases $10^{-3} - 10^{-4}$ of the Thomson amplitude. The intensity of magnetic Bragg peaks of antiferromagnets is then affected by a factor of the order of $10^{-7}$. The orbital moment contributes to the elastic scattering as well as the spin moment and with the same order of magnitude, but with a different dependence on wave vectors and polarizations. We write below the scattering length of an electron of spin $\mathbf{S}$ and orbital moment $\mathbf{L}$

$$b_{mag} = -ir_e\left(\lambda_c/\lambda\right)\left[\left(\widehat{\mathbf{e}}_{sc}^*.\overline{\overline{\mathbf{T}}}_S.\widehat{\mathbf{e}}_{in}\right).\mathbf{S} + \left(\widehat{\mathbf{e}}_{sc}^*.\overline{\overline{\mathbf{T}}}_L.\widehat{\mathbf{e}}_{in}\right).\mathbf{L}\right]. \tag{1.155}$$

The tensors $\overline{\overline{\mathbf{T}}}_S, \overline{\overline{\mathbf{T}}}_L$ simply help to write these bilinear functions of the polarizations. Their elements are vectors. In their expression below, $2\theta$ is the angle between $\widehat{\mathbf{k}}_{in}$ and $\widehat{\mathbf{k}}_{sc}$:

$$\overline{\overline{\mathbf{T}}}_S = \begin{matrix} (s) \\ (p) \end{matrix} \begin{pmatrix} \overset{(s)}{\widehat{\mathbf{k}}_{sc} \times \widehat{\mathbf{k}}_{in}} & \overset{(p)}{2\widehat{\mathbf{k}}_{sc}\sin^2\theta} \\ -2\widehat{\mathbf{k}}_{in}\sin^2\theta & \widehat{\mathbf{k}}_{sc} \times \widehat{\mathbf{k}}_{in} \end{pmatrix} \tag{1.156}$$

**Fig. 1.13** The electron can scatter the electromagnetic radiation through a variety of processes. In each of them, the incident field moves the electron itself or its spin through a driving force at *left*. The back and forth motion is indicated by a pair of *thin opposite arrows*. In this motion, the electron reradiates through a mode indicated at right. The first process is the well-known Thomson scattering. Processes 2–4 describe the scattering by the spin, drawn as a *double arrow*. The process in the fifth line is a correction to Thomson scattering when the electron has a translation motion, indicated by the momentum **p**. When integrated over the orbit of the electron in the atom, it gives rise to a scattering by the orbital moment

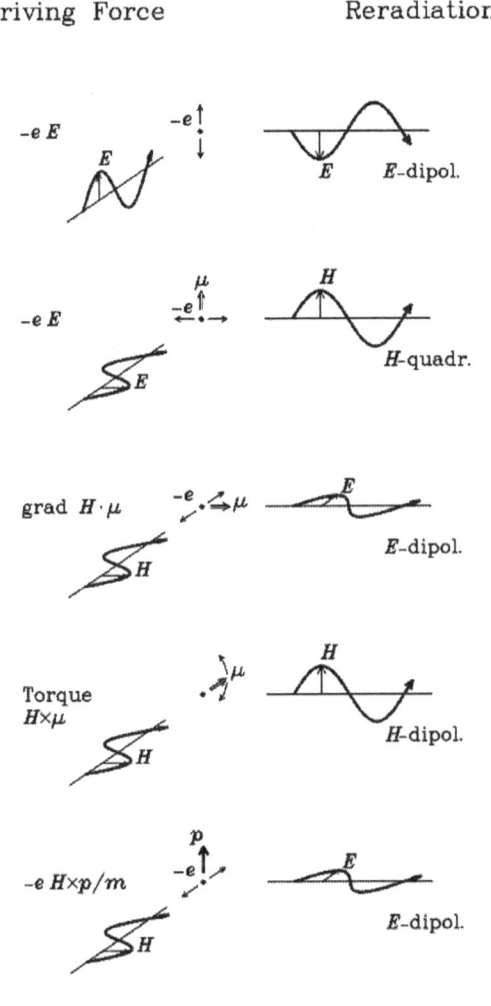

Driving Force          Reradiation

$$
\overline{\overline{\mathbf{T}}}_L = 
\begin{matrix} (s) \\ (p) \end{matrix}
\begin{pmatrix}
\overset{(s)}{0} & \overset{(p)}{\left(\widehat{\mathbf{k}}_{sc} + \widehat{\mathbf{k}}_{in}\right)\sin^2\theta} \\
-\left(\widehat{\mathbf{k}}_{sc} + \widehat{\mathbf{k}}_{in}\right)\sin^2\theta & 2\widehat{\mathbf{k}}_{sc} \times \widehat{\mathbf{k}}_{in}\sin^2\theta
\end{pmatrix}.
\qquad (1.157)
$$

Remember that we use for $i$ a sign opposite to the one used in quantum theory in the frame of which these equations are usually written.

Magnetic Compton scattering is also present but results only from the spin. In Compton scattering $\hbar q / 2\pi mc$ can approach 1 so that the cross-section in magnetic Compton scattering can be significantly larger than in the elastic channel.

## 1.5.3 Resonant Magnetic Scattering

As explained in Sect. 1.4.4 the resonant, or dispersive, part of the scattering is based on the virtual excitation of an electron from a core level to an empty state, which can be just above the Fermi level. In the subsequent discussion in Sect. 1.4.7, we have assumed that the polarization factor was the same as for Thomson scattering, $\widehat{\mathbf{e}}_{sc}.\widehat{\mathbf{e}}_{in}$. This assumption in fact may be wrong. Let us write the numerator of a particular term in $b_{disp1}$, Eq. (1.112), while making the dipolar approximation (the exponentials are reduced to 1)

$$\langle s|\widehat{\mathbf{e}}_{sc}^*.\mathbf{p}|c\rangle\langle c|\widehat{\mathbf{e}}_{in}.\mathbf{p}|i\rangle = \widehat{\mathbf{e}}_{sc}^*.\overline{\overline{T}}_{res}.\widehat{\mathbf{e}}_{in}. \tag{1.158}$$

Again we express this bilinear function of $\widehat{\mathbf{e}}_{sc}^*$, $\widehat{\mathbf{e}}_{in}$ with a tensor $\overline{\overline{T}}_{res}$. Instead of writing this tensor with the (s) and (p) polarizations as a basis, we may use a reference frame $(x,y,z)$ attached to the medium, generally a crystal:

$$\overline{\overline{T}}_{res} = \begin{matrix} & \begin{matrix} (x) & \quad (y) & \quad (z) \end{matrix} \\ \begin{matrix} (x) \\ (y) \\ (z) \end{matrix} & \begin{pmatrix} a_1 & b_3+ic_3 & b_2+ic_2 \\ b_3-ic_3 & a_2 & b_1+ic_1 \\ b_2-ic_2 & b_1-ic_1 & a_3 \end{pmatrix} \end{matrix}. \tag{1.159}$$

With $a_i$, $b_i$, $c_i$ being nine real coefficients, this is the most general tensor in Eq. (1.158). The actual structure of that tensor is determined by the symmetry of the scattering atom. The spherical symmetry is frequently a good approximation, though never completely exact in a crystal; then $\overline{\overline{T}}_{res}$ reduces to the unit matrix [21] and we recover the usual factor $\widehat{\mathbf{e}}_{sc}^*.\widehat{\mathbf{e}}_{in}$.

A case of lowering of the symmetry is the presence of a magnetic moment. We may observe that a time inversion, which should not change the scattering amplitude, exchanges the incident and scattering beams and reverses the magnetization. This shows that the antisymmetrical part of $\overline{\overline{T}}_{res}$, that is the $ic_i$s, is of odd order in the magnetization. That part has the form

$$\propto i\left(\widehat{\mathbf{e}}_{sc}^* \times \widehat{\mathbf{e}}_{in}\right).\widehat{\mathbf{z}}, \tag{1.160}$$

where $\widehat{\mathbf{z}}$ is just the direction of magnetization, at least in the simplest cases.

These symmetry arguments should be completed by an explicit discussion of the physical process. The mechanism is described in [10] and shortly explained in Fig. 1.14 and caption. In the absence of any crystal field,

$$\widehat{\mathbf{e}}_{sc}^*.\overline{\overline{T}}_{res}.\widehat{\mathbf{e}}_{in} \propto \widehat{\mathbf{e}}_{sc}^*.\widehat{\mathbf{e}}_{in}\left(F_{11}+F_{1-1}\right)+i\left(\widehat{\mathbf{e}}_{sc}^* \times \widehat{\mathbf{e}}_{in}\right).\widehat{\mathbf{z}}\left(F_{11}-F_{1-1}\right)$$
$$+\left(\widehat{\mathbf{e}}_{sc}^*.\widehat{\mathbf{z}}\right)\left(\widehat{\mathbf{e}}_{in}.\widehat{\mathbf{z}}\right)\left(2F_{10}-F_{11}-F_{1-1}\right), \tag{1.161}$$

where $F_{1-1}$, $F_{10}$ and $F_{11}$ contain some transition probabilities. These transitions are described by two indices, the first one standing for the change in the orbital moment $\Delta L$ (1 in the dipolar term) and the second one for $\Delta L_z$. The first term is the isotropic

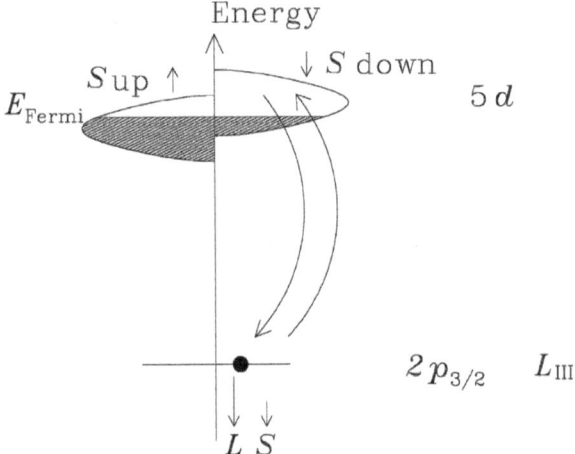

**Fig. 1.14** Mechanism of the resonant magnetic scattering in the case of the $L_{III}$ resonance of a third row transition element, such as the platinum. Due to the magnetic moment the resonance occurs preferentially in the spin down ($S = -1/2$) half valence band, on the right. Because of the strong spin–orbit coupling in the core shell $2p$, the $2p_{3/2}$ level is completely separated out from the $2p_{1/2}$ and contributes alone to the resonance. Therefore the ($S = -1/2$) state involved in the resonance is coupled with a rather defined value of the $\hat{z}$ component ($\hat{z}$ the direction of magnetization) of the orbital moment in the initial state of the electron. For a given polarisation of the radiation, this makes the amplitude to depend on the atom magnetization direction, here the up direction

anomalous scattering, discussed in Sect. 1.4.7. The second term is the just discussed antisymmetrical part. The third one depends on the axis along which the magnetization is lying, but not on its sign; it is responsible for the magnetic linear dichroism. One should not forget that the above expression is to be multiplied by a resonance function of the energy showed in Eq. (1.142) and Fig. 1.11.

*The spin–orbit coupling is a key feature of this mechanism.* In the example displayed in Fig. 1.14, the spin–orbit coupling interaction is very large in the core state, but in some cases it may be present only in the excited state. In addition to the dipolar term written in the above formula, a quadrupolar one also exists. Though smaller, it cannot be neglected if it corresponds to a transition to a strongly magnetized atomic shell such as the $3d$ shell of transition elements or $4f$ shell of lanthanides. It shows a quadrilinear dependence on $\hat{e}_{sc}^*, \hat{e}_{in}, \hat{k}_{sc}, \hat{k}_{in}$.

The order of magnitude of the resonant magnetic scattering may vary on a wide range. The $K$ resonances, accessible for example in the $3d$ transition elements, have amplitudes which are comparable to the non-resonant. Indeed the $K$ shell has no orbital moment so that the effect relies on the spin–orbit coupling in the valence shell, which is much less efficient. Furthermore the dipolar transition occurs to a weakly magnetized $p$ valence shell, while the strongly magnetized $d$ shell can give only a quadrupolar transition. The latter drawback limits also the $L_{II,III}$ resonances of lanthanides, but then the $L$ shell is completely split by the spin–orbit interaction. In that case the amplitude is typically ten times larger than the non-resonant. The $L_{II,III}$ resonances of transition elements are favoured in all respects and enhance the amplitude by several orders of magnitude compared to non-resonant. In the case of $3d$ elements, the long wavelength (of the order of 1.5 nm) can only fit long periodicities and give diffraction on multilayer or be used in reflectivity experiments. The $L$ resonances of $5d$ transition elements arise in the 0.1 nm range but among those only the platinum group elements, and mainly the platinum itself, can take a magnetic moment. The $M_{IV,V}$ resonances of actinides offer the same favourable characters and are also quite effective, with amplitude enhancement by a factor of the order of one thousand. The wavelength near 0.35 nm for the uranium allows for Bragg diffraction experiments.

## 1.5.4 Templeton Anisotropic Scattering

Even without any magnetic moment, the atom may show a low symmetry. Most often, this arises from the crystal field. A spontaneous orbital order (that is an orbital arrangement resulting mainly from the electrostatic interaction between orbitals of neighbouring atoms) is also expected in some materials. As a consequence the symmetrical part of Eq. (1.159), $a_i, b_i$, differs from the unit matrix. Again a quadrupolar term may exist. The Templeton scattering produces some change in the intensity of Bragg peaks at the absorption edges and this can be used to get more structural information. A striking feature is the occurrence of otherwise forbidden reflections [9, 15, 22]. When a reflection is forbidden because of a screw axis or glide plane, the amplitude cancels only if it is a scalar (that is independent of the orientation of the atom). For example with a screw axis, the atom rotates from one site to the next, so that the tensor amplitude in Eqs. (1.158) and (1.159) may not cancel. This breakdown of a crystallographic extinction rule should not be confused with the appearance of, e.g., the 2 2 2 reflection in the diamond structure. In that case the structure factor is a scalar and the broken extinction rule is not a general rule for the space group; it applies only to a special position in the cell.

## 1.5.5 The Effect of an Anisotropy in the Index of Refraction

If the scattering is anisotropic, so may be the index of refraction and the optical properties. The non-resonant magnetic scattering amplitude is zero in the forward direction. From the discussion in Sect. 1.3 it cannot contribute to the refractive index. We shall therefore discuss only the consequences of resonant magnetic and of Templeton scattering. We give only some brief information on this question which could deserve quite a long development. The propagation of neutrons in magnetized materials and the associated reflectivity is examined in the Chap. 4 of this book. It is different from the propagation of the electromagnetic radiation in an anisotropic medium, especially when the interaction is strong. The thermal neutron has a non-relativistic motion which allows for a complete separation of the space and spin variables. The electromagnetic radiation instead is fully relativistic, which intermixes the propagation and polarization properties. In that case, the direction of propagation depends on the polarization. Several unusual effects are consequently observed. For example the direction of a light ray may differ from the normal to the wave planes, or a refracted ray may lie outside of the plane of incidence [6].

Starting from the Helmholtz equation (1.15) the anisotropy of the medium modifies the dielectric constant ($\epsilon/\epsilon_0 \simeq n^2$) which becomes a tensor. Though we wrote the Helmholtz equation for the 4-vector $A$ and the tensor should be of fourth order, all the useful coefficients are contained in a third-order tensor acting in space, similar to Eq. (1.159). In the absence of magnetization, we have the case of crystal optics, described in several textbooks, e.g. [6]. If the medium is magnetized, the antisymmetrical part of the tensor is non-zero and some phenomena occur, such as the Faraday rotation of the polarization plane or the magnetooptic Kerr effect (that is polarization-dependent and polarization rotating reflectivity). The basic theory can be found in [18]. The theory as exposed in the textbooks is drawn in the dipolar approximation, which is legitimate in the range of the visible or near visible optics. I am not aware of a complete description of anisotropic optics including the quadrupolar terms. It seems reasonable in practice to use instead of such a full theory some perturbative corrections since the quadrupolar resonance terms are always small. The incidence of some term beyond the dipolar electric approximation is clear in an effect well observed with visible light and discovered nearly two centuries ago: it is the optical activity, that is the rotation of the polarization plane in substances which lack a centre of symmetry. Indeed the combination of two terms of different multipole orders is required to produce such an effect. In the optical range the second term is magnetic dipolar while in the X-ray range the electric quadrupolar term gives rise to a similar effect. It is a small effect, even in the optical range, since it corresponds to differences of the order of $10^{-4}$ between the indices of the two opposite circular polarizations. Yet it can be easily observed because the absorption of the visible light is still smaller and samples more than $10^5$ wavelengths thick can be probed.

In the X-ray range, at energies of several keV and above, the refractive index differs from one by a small value and its anisotropic part is still smaller. Only a limited list of anomalies are observed and they are interpreted in simple terms. One of the most studied of those is the magnetic circular dichroism. In a ferro or ferrimagnet, where a net magnetization is present, the refractive index changes, according to the helicity of circularly polarized X-rays being parallel or antiparallel to the magnetization. Indeed the optical theorem Eq. (1.75) or its extension Eq. (1.87) yields the absorption atomic cross section or the dispersive part of the index of the medium, from the part of the scattering length written in Eqs. (1.158) and (1.161). For that we make $\widehat{\mathbf{k}}_{sc}$ equal to $\widehat{\mathbf{k}}_{in}$ and $\widehat{\mathbf{e}}_{sc}$ equal to $\widehat{\mathbf{e}}_{in}$. For a circular polarization, the term in Eq. (1.160) is real and reads

$$\pm \widehat{\mathbf{k}}_{in}.\widehat{\mathbf{z}}. \tag{1.162}$$

The sign is switched from $-$ for the right-handed helicity to $+$ for the left handed. This is to be multiplied by the complex resonance factor which we have left out from the formula. The difference in the real, $\delta$, component of the index between both helicities gives rise to the ***Faraday rotation*** of the polarization plane. Similarly the difference in the imaginary, $\beta$, component gives rise to a difference in the absorption, called the ***magnetic circular dichroism***. Once a circularly polarized radiation is available, it is relatively easy to measure that change of absorption, usually by switching the magnetization parallel or antiparallel to the beam. Similar to the resonant scattering the dichroism shows a spectrum in the region of the absorption edge.

At the $L$ edges of the $3d$ elements, the resonances and their magnetic parts are quite large and the full optical theory, in the dipolar approximation, should be considered. Some reflectivity measurements have been done, e.g. [14].

**Acknowledgments** The author of this chapter has been employed at the Laboratoire de Cristallographie of the Centre National de la Recherche Scientifique (CNRS) in Grenoble (France), now a part of the Institut Néel. He is indebted to the CNRS and to the former colleagues in that laboratory for the knowledge acquired and exposed here.

## 1.A Appendix: the Born Approximation

Anne Sentenac, François de Bergevin, Jean Daillant, Alain Gibaud and Guillaume Vignaud

In this appendix we give the Born development for the field scattered by a deterministic object.

In absence of any object, the field (scalar for simplicity) is a solution of the homogeneous Helmholtz equation,

$$\left(\Delta + k_0^2\right) A_{\text{in}}(\mathbf{r}) = 0. \tag{1.A1}$$

The object introduces a perturbation $V$ on the differential operator, see Eqs. (1.16) and (1.17). In this case the field is the solution of

$$\left(\Delta + k_0^2 - V(\mathbf{r})\right) A(\mathbf{r}) = 0. \tag{1.A2}$$

The total field $A$ can be written as the sum of an incident field $A_{\text{in}}(\mathbf{r}) = A_{\text{in}} e^{-i\mathbf{k}_{\text{in}}\cdot\mathbf{r}}$ (plane wave solution of the homogeneous equation) and a scattered field $A_{\text{sc}}$ which satisfies the outgoing wave boundary condition. Following Sect. 1.2.5, we transform Eq. (1.A2) into an integral equation by introducing the Green function

$$G_-(\mathbf{r}) = -\frac{1}{4\pi}\frac{e^{-ik_0 r}}{r} \tag{1.A3}$$

that satisfies outgoing wave boundary condition. We obtain

$$A(\mathbf{r}) = A_{\text{in}}(\mathbf{r}) + \int G_-(\mathbf{r}-\mathbf{r}')V(\mathbf{r}')A(\mathbf{r}')d\mathbf{r}'. \tag{1.A4}$$

Formally, one can write the solution of this integral equation in terms of a series in power of the convolution operator $[G_-V]$,[9]

$$A = A^{(0)} + A_{\text{sc}}^{(1)} + A_{\text{sc}}^{(2)} + \cdots \tag{1.A5}$$

with, $A^{(0)}(\mathbf{r}) = A_{\text{in}}(\mathbf{r})$,

$$A_{\text{sc}}^{(1)}(\mathbf{r}) = \int d^3 r' G_-(\mathbf{r} - \mathbf{r}') V(\mathbf{r}') A_{\text{in}}(\mathbf{r}'),$$

$$A_{\text{sc}}^{(2)}(\mathbf{r}) = \int d^3 r' \int d^3 r'' G_-(\mathbf{r} - \mathbf{r}') V(\mathbf{r}') G_-(\mathbf{r}' - \mathbf{r}'') V(\mathbf{r}'') A_{\text{in}}(\mathbf{r}'').$$

When this series is convergent, one gets the exact value of the field. The main issue of such an expansion lies in its radius of convergence which is not easy to determine. Physically, the potential $V(\mathbf{r}')$ combined with the propagation operator $G_-$ represents the action of the particule (or polarization density) at $\mathbf{r}'$ on the incident wave, i.e. a scattering event. When the potential appears once (in the first-order term), the incident wave is singly scattered by the particules of the object. When it appears twice (in the second-order term), one accounts for the double scattering events, etc. Equation (1.A5) can also be viewed as a perturbative development in which the scattering event $[GV]$ is taken as a small parameter. The first Born approximation consists in stopping the development in Eq. (1.A5) to the first order in $V$ (thus assuming the predominance of single scattering).

We now proceed by evaluating the scattered far-field and the scattering cross-section. We assume that the observation point $\mathbf{r}$ is far from all the points $\mathbf{r}'$ constituting the object (with respect to an arbitrary origin situated inside the object). In this case, one has

$$|\mathbf{r} - \mathbf{r}'| \approx r - \widehat{\mathbf{u}}.\mathbf{r}', \tag{1.A6}$$

so that

$$G_-(\mathbf{r} - \mathbf{r}') = -\frac{1}{4\pi} \frac{e^{-ik_0|\mathbf{r}-\mathbf{r}'|}}{|\mathbf{r}-\mathbf{r}'|} \approx -\frac{1}{4\pi} \frac{e^{-ik_0 r}}{r} e^{ik_0 \widehat{\mathbf{u}}.\mathbf{r}'}. \tag{1.A7}$$

Using this far-field approximation in Eq. (1.A4), one retrieves the expression given in Sect. 1.2.4,

$$A_{\text{sc}}(\mathbf{r}) = -A_{\text{in}} b(\widehat{\mathbf{u}}) \frac{e^{-ik_0 r}}{r}. \tag{1.A8}$$

Bearing in mind the Born development for the field, one can write the scattering length $b$ in the form, $b(\widehat{\mathbf{u}}) = b^{(1)}(\widehat{\mathbf{u}}) + b^{(2)}(\widehat{\mathbf{u}}) + \cdots$, with, for example,

---

[9] In operator notation one can make an analogy with the Taylor expansion of $1/1-x = 1+x+x^2+\cdots$. Indeed, the field can be written as $A = A_{\text{in}}/1 - [G_-V]$ which yields the series

$$A = A_{\text{in}} + [G_-V] A_{\text{in}} + [G_-V][G_-V] A_{\text{in}} + \cdots + [G_-V]^n A_{\text{in}} + \cdots,$$

with $[G_-V]f = \int G_-(\mathbf{r} - \mathbf{r}') V(\mathbf{r}') f(\mathbf{r}') d\mathbf{r}'$.

$$b^{(1)}(\widehat{\mathbf{u}}) = \frac{1}{4\pi} \int d^3 r' V(\mathbf{r}') e^{i(k_0\widehat{\mathbf{u}} - \mathbf{k}_{in}) \cdot \mathbf{r}'}. \tag{1.A9}$$

The calculation of the differential scattering cross-section, Eq. (1.36),

$$\frac{d\sigma}{d\Omega} = |b(\widehat{\mathbf{u}})|^2$$

is then straightforward.

It is worth noting that the perturbative development of the energy (which is proportional to the square of the field) starts at second order in $V$. Hence, to be consistent in our calculation, we should always develop the field up to the second order to account for all the possible terms (of order two) in the energy. A striking illustration of this remark is that the first Born approximation does not satisfy energy conservation. This can be readily shown by injecting the perturbative development of $b$ in the optical theorem which is a direct consequence of the energy conservation. The optical theorem relates the total cross-section to the imaginary part of the forward scattered amplitude. One has, see Sect. 1.3.2, $\sigma_{tot} = 2\lambda \mathscr{I}m[b(\mathbf{k}_{in})]$. If one disregards lossy media, the total cross-section is equal to the scattering cross-section,

$$\sigma_{sc} = \int |b(\widehat{\mathbf{u}})|^2 d\Omega = 2\lambda \mathscr{I}m[b(\mathbf{k}_{in})]. \tag{1.A10}$$

The expansion, Eq. (1.A5), being a formally *exact* representation of the field, the optical theorem, written as a series, is verified at each order of the perturbative development. One gets, to the lowest order,

$$\int |b^{(1)}(\widehat{\mathbf{u}})|^2 d\Omega = 2\lambda \mathscr{I}m[b^{(1)}(\mathbf{k}_{in}) + b^{(2)}(\mathbf{k}_{in})]. \tag{1.A11}$$

This last equation shows clearly that if one wants the Born approximation to conserve energy, one should calculate the scattered amplitude up to second order *in the forward direction*.

# References

1. Als-Nielsen, J., McMorrow, D.: Elements of Modern X-ray Physics. John Wiley & Sons (2001).
2. Anderson, P.W.: Phys. Rev. **109**, 1492–1505 (1958).
3. Batterman, B.W., Cole, H.: Rev. Mod. Phys. **36**, 681–717 (1964).
4. Berestetskii, V.B., Lifshithz, E.M., Pitaevskii, L.P.: Course of theoretical physics. Quantum Electrodynamics, vol. 4. Pergamon Press, Oxford, (1982). E. Lifchitz & L. Pitayevski (L. Landau & E. Lifchitz). Physique théorique. Tome IV, Théorie quantique relativiste, Première partie. Ed. Mir, Moscou (1972).
5. Blume, M.: J. Appl. Phys. **57**, 3615–3618, (1985).
6. Born, M., Wolf, E.: Principles of Optics. 4th edn. Oxford, New York (1980).
7. Brouder, C., Rossano, S.: Eur. Phys. J. B **45**, 19–31 (2005).
8. Cohen-Tannoudji, C., Diu, B., Laloë, F.: Mécanique Quantique. Hermann (1994).

9.  Dmitrienko, V.M.: Acta Cryst. A **39**, 29–35 (1983).
10. Hannon, J.P., Trammel, G.T., Blume, M., Gibbs, D.: Phys. Rev. Lett. **61**, 1245–1248 (1988).
11. International Tables for X-ray Crystallography; vol. III Physical and Chemical Tables; vol. IV Revised and Supplementary Tables to vol. II and III, The Kynoch Press, Birmingham, 1968.
12. Jackson, J.D.: Classical Electrodynamics. 2nd edn. John Wiley & Sons, New York (1975).
13. James, R.W.: The optical principles of the diffraction of X-rays. In: Bragg, S.L. (ed.) The Cristalline State, vol II. G. Bell & Sons ltd, London (1962).
14. Kao, C.C., Chen, C.T., Johnson, E.D., Hastings, J.B., Lin, H.J., Ho, G.H., Meigs, G., Brot, J.-M., Hulbert, S.L., Idzerda, Y.U., Vettier, C.: Phys. Rev. B **50**, 9599–9602 (1994).
15. Kirfel, A., Petcov, A.: Acta Cryst. A **47**, 180 (1991).
16. Landau, L., Lifshitz, E.M.: Course of theoretical physics. Electrodynamics of continuous media, vol. 8. Pergamon Press, Oxford, 1960. L. Landau & E. Lifchitz. Physique théorique. Tome VIII, Electrodynamique des milieux continus. Ed. Mir, Moscou (1969).
17. Schülke, W.: Inelastic scattering by electronic excitations. In: Brown, G., Moncton, D.E. (eds.) Handbook of Synchrotron Radiation, vol. 3. Elsevier Science Publisher, New York (1991).
18. Sokolov, A.V.: Optical properties of metals. Blackie and Son Ltd, London (1967).
19. Stanglmeier, F., Lengeler, B., Weber, W., Göbel, H., Schuster, M.: Acta Cryst. A **48**, 626 (1992).
20. Storm, E., Israel, H.I.: Photon cross sections from 1 keV to 100 MeV for elements Z = 1 to Z = 100. Nuclear Data Tables **A7**, 565–681, (1970).
21. Templeton, D.H., Templeton, L.K.: Acta Cryst. A **36**, 237 (1980).
22. Templeton, D.H., Templeton, L.K.: Acta Cryst. A **42**, 478 (1986).

# Chapter 2
# Statistical Aspects of Wave Scattering at Rough Surfaces

A. Sentenac and J. Daillant

## 2.1 Introduction

The surface state of objects in any scattering experiment is, of necessity, rough. Irregularities are of the most varied nature and length scales, ranging from the atomic scale, where they are caused by the inner structure of the material, to the mesoscopic and macroscopic scale where they can be related to the defects in processing in the case of solid bodies or to fluctuations in the case of liquid surfaces (ocean waves, for example).

The problem of wave scattering at rough surfaces has thus been a subject of study in many research areas, such as medical ultrasonic, radar imaging, optics or solid state physics [1–4]. The main differences stem from the nature of the wavefield and the wavelength of the incident radiation (which determines the scales of roughness that have to be accounted for in the models). When tackling the issue of modelling a scattering experiment, the first difficulty is to describe the geometrical aspect of the surface. In this chapter, we are interested solely in surface states that are not well controlled so that the precise defining equation of the surface, $z = z(x,y)$, is unknown or of little interest. One has (or needs) only information on certain statistical properties of the surface, such as the height repartition or height to height correlations. In this probabilistic approach, the shape of the rough surface is described by a random function of space coordinates (and possibly time as well). The wave scattering problem is then viewed as a statistical problem consisting in finding the statistical characteristics of the scattered field (such as the mean value or field correlation functions), the statistical properties of the surface being given.

In the first section of this contribution we present the statistical techniques used to characterise rough surfaces. The second section is devoted to the description of a surface scattering experiment from a conceptual point of view. In the third section, we investigate to what extent the knowledge of the field statistics such as the mean field or field autocorrelation is relevant for interpreting the data of a scattering experiment which deals necessarily with deterministic rough samples. Finally, we

A. Sentenac (✉)
Institut Fresnel, Campus de Saint Jérôme, av. Escadrille Normandie
13397 Marseille Cedex 20, France

Sentenac, A., Daillant, J.: *Statistical Aspects of Wave Scattering at Rough Surfaces.* Lect. Notes Phys. **770**, 59–84 (2009)
DOI 10.1007/978-3-540-88588-7_2

derive in the fourth section a simple expression of the scattered field and scattered intensity from random rough surfaces under the Born approximation.

## 2.2 Description of Randomly Rough Surfaces

### 2.2.1 Introduction

Let us first consider the example of a liquid surface. The exact morphology of the surface is rapidly fluctuating with time and is not accessible inasmuch as the detector will integrate over many different surface shapes. However, statistical information can be obtained and it provides an useful insight on the physical processes. Indeed, these fluctuations obey Boltzmann statistics and are characterised by a small number of relevant parameters such as the density of the liquid or its surface tension (see Sect. 4.5).

We now consider a set of surfaces of artificial origin (such as metallic optical mirrors) that have undergone similar technological treatments (like polishing and cleaning). Since it is impossible to reproduce all the microscopic factors affecting the surface state, these surfaces have complex and completely different defining equations $z = z(x,y)$. However, if the surface processing is well enough controlled, they will present some similarities, of statistical nature, that will distinguish them from surfaces that have received a totally different treatment.

In these two examples, we are faced with the issue of describing a set of real surfaces which present similar statistical properties and whose defining equations $z(x,y)$ are unknown or of small interest (see Fig. 2.1). It appears convenient [2] to approximate this set of surfaces by a statistical ensemble of surfaces that are realisations of a random continuous process of the plane coordinates $\mathbf{r}_{\parallel} = (x,y)$, whose statistical properties depend on some relevant parameters of the physical processes affecting the surface state (like the grain size of the polishing abrasive in the case of surfaces of artificial origin). It is likely that the characteristic functions $z(\mathbf{r}_{\parallel})$ of the surfaces generated by the random process will be different from that of the real surfaces under study, but the statistical properties of both ensembles should be the same.

### 2.2.2 Height Probability Distributions

Generally speaking, a random rough surface is completely described statistically by the assignment of the n-point $(n \rightarrow \infty)$ height probability distribution $p_n(\mathbf{r}_{1\parallel}, z_1 \ldots \mathbf{r}_{n\parallel}, z_n)$ where $p_n(\mathbf{r}_{1\parallel}, z_1 \ldots \mathbf{r}_{n\parallel}, z_n)dz_1 \ldots dz_n$ is the probability for the surface points of plane coordinates $\mathbf{r}_{1\parallel}, \ldots, \mathbf{r}_{n\parallel}$ of being at the height between $(z_1 \ldots z_n)$ and $(z_1 + dz_1 \cdots z_n + dz_n)$. However, in most cases, we restrict the description of the randomly rough surface to the assignment of the one- and two-point distribution functions

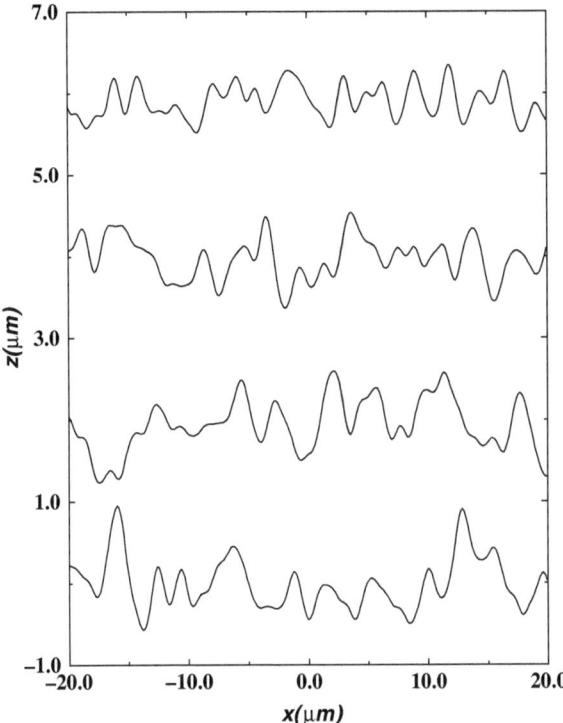

**Fig. 2.1** Examples of various rough surfaces that present the same Gaussian statistical properties

$p_1(\mathbf{r}_\parallel, z)$ and $p_2(\mathbf{r}_{1\parallel}, z_1; \mathbf{r}_{2\parallel}, z_2)$. Indeed, most scattering theories need solely this information.

From these probability functions, one can calculate the ensemble average of any functional of the random variables $(z_1 \dots z_n)$ where $z_i = z(\mathbf{r}_{i\parallel \mathbf{r}_{i\parallel}})$, through the integral,

$$\langle F \rangle (\mathbf{r}_{1\parallel} \dots \mathbf{r}_{n\parallel}) = \int_{-\infty}^{\infty} F(z_1 \dots z_n) p_n(\mathbf{r}_{1\parallel}, z_1 \dots \mathbf{r}_{n\parallel}, z_n) dz_1 \dots dz_n. \qquad (2.1)$$

The domain of integration covers all the possible values for $(z_1 \dots z_n)$. This quantity is equivalent to an average of $F$ calculated over an ensemble of surface realisations $S_p$,

$$\langle F \rangle (\mathbf{r}_{1\parallel} \dots \mathbf{r}_{n\parallel}) = \lim_{N \to \infty} \frac{1}{N} \sum_{p=1}^{N} F(z_1^p \dots z_n^p), \qquad (2.2)$$

where $z_j^p$ is the altitude of the pth surface realisation at plane coordinates $\mathbf{r}_{j\parallel}$.

With this definition, one obtains in particular the mean height of the surface through

$$\langle z \rangle (\mathbf{r}_\parallel) = \int_{-\infty}^{\infty} z(\mathbf{r}_\parallel) p_1(\mathbf{r}_\parallel, z) dz. \qquad (2.3)$$

The mean square height of the surface is given by

$$\langle z^2 \rangle(\mathbf{r}_\parallel) = \int_{-\infty}^{\infty} z^2(\mathbf{r}_\parallel) p_1(\mathbf{r}_\parallel, z) dz. \qquad (2.4)$$

The height–height correlation function $C_{zz}$ is defined by

$$C_{zz}(\mathbf{r}_{1\parallel}, \mathbf{r}_{2\parallel}) = \langle z_1 z_2 \rangle = \int_{-\infty}^{\infty} z_1 z_2 p_2(\mathbf{r}_{1\parallel}, z_1, \mathbf{r}_{2\parallel}, z_2) dz_1 dz_2, \qquad (2.5)$$

where $z_j = z(\mathbf{r}_{j\parallel})$. It is also usual to introduce the pair-correlation function $g(\mathbf{r}_{1\parallel}, \mathbf{r}_{2\parallel})$ which averages the square of the difference in height between two points of the surface,

$$g(\mathbf{r}_{1\parallel}, \mathbf{r}_{2\parallel}) = \langle (z_1 - z_2)^2 \rangle = \int_{-\infty}^{\infty} (z_1 - z_2)^2 p_2(\mathbf{r}_{1\parallel}, z_1, \mathbf{r}_{2\parallel}, z_2) dz_1 dz_2. \qquad (2.6)$$

Note that $g(\mathbf{r}_{1\parallel}, \mathbf{r}_{2\parallel}) = 2\langle z^2 \rangle(\mathbf{r}_\parallel) - 2C_{zz}(\mathbf{r}_{1\parallel}, \mathbf{r}_{2\parallel})$.

### 2.2.3 Homogeneity and Ergodicity

Randomly rough surfaces have frequently the property that the character of the height fluctuations $z$ does not change with the location on the surface. More precisely, if all the probability distribution functions $p_i$ are invariant under any arbitrary translation of the spatial origin, the random process is called homogeneous. As a consequence, the ensemble average of the functional $F(z_1 \ldots z_n)$ will depend only on the vector difference, $\mathbf{r}_{j\parallel} - \mathbf{r}_{1\parallel}$, between one of the $n$ space argument $\mathbf{r}_{1\parallel}$ and the $(n-1)$ remaining others $\mathbf{r}_{j\parallel}$, $j = 2 \ldots n$.

$$\langle F \rangle(\mathbf{r}_{1\parallel}, \ldots, \mathbf{r}_{n\parallel}) = \langle F \rangle(\mathbf{0}_\parallel \ldots \mathbf{r}_{n\parallel} - \mathbf{r}_{1\parallel}). \qquad (2.7)$$

When the random process is isotropic (i.e. has the same characteristics along any direction) the dependencies reduce to the distance $|\mathbf{r}_{j\parallel} - \mathbf{r}_{1\parallel}|$ between one of the space argument and the others. Hereafter we will only consider homogeneous isotropic random processes and we propose a simplified notation for the various functions already introduced.

The mean altitude $\langle z \rangle(\mathbf{r}_\parallel)$ does not depend on the $\mathbf{r}_\parallel$ position and one can find a reference plane surface such as $\langle z \rangle = 0$. The mean square deviation of the surface is also a constant and we define the root mean square (rms) height $\sigma$ as

$$\sigma^2 = \langle z^2 \rangle = \int_{-\infty}^{\infty} z^2 p_1(z) dz. \qquad (2.8)$$

The rms height is often used to give an indication of the "degree of roughness", the larger the $\sigma$ the rougher the surface. Note that the arguments of the probability distribution are much simpler.

Similarly, the height–height correlation function can be written as

$$C_{zz}(\mathbf{r}_{1\|}, \mathbf{r}_{2\|}) = \langle z(\mathbf{0}_{\|})z(\mathbf{r}_{\|})\rangle = C_{zz}(r_{\|}) = \int z_1 z_2 p_2(z_1, z_2, r_{\|}) dz_1 dz_2, \qquad (2.9)$$

where $r_{\|} = |\mathbf{r}_{\|}|$. We also introduce, with these simpler notations, the one-point and two-point characteristic functions,

$$\chi_1(s) = \int_{-\infty}^{\infty} p_1(z) e^{isz} dz, \qquad (2.10)$$

$$\chi_2(s, s', r_{\|}) = \int_{-\infty}^{\infty} p_2(z, z', r_{\|}) e^{isz + is'z'} dz dz'. \qquad (2.11)$$

One of the most important attributes of a homogeneous random process is its power spectrum, $P(\mathbf{q}_{\|})$, that gives an indication of the strength of the surface fluctuations associated with a particular wavelength. Roughly speaking, the rough surface is regarded as a superposition of gratings with different periods and heights. The power spectrum is a tool that relates the height to the period. We introduce the Fourier transform of the random variable $z$,

$$\tilde{z}(\mathbf{q}_{\|}) = \int z(\mathbf{r}_{\|}) e^{i\mathbf{q}_{\|} \cdot \mathbf{r}_{\|}} d\mathbf{r}_{\|}, \qquad (2.12)$$

where $\mathbf{q}_{\|} = (q_x, q_y)$ is the in-plane wave-vector transfer. We define the spectrum as

$$P(\mathbf{q}_{\|}) = \langle |\tilde{z}(\mathbf{q}_{\|})|^2 \rangle = \langle \tilde{z}(\mathbf{q}_{\|})\tilde{z}(-\mathbf{q}_{\|})\rangle. \qquad (2.13)$$

The Wiener–Khintchine theorem [5] states that the power spectrum is the Fourier transform of the correlation function:

$$P(\mathbf{q}_{\|}) = \int d\mathbf{r}_{\|} e^{i\mathbf{q}_{\|} \cdot \mathbf{r}_{\|}} \langle z(\mathbf{0}_{\|})z(\mathbf{r}_{\|})\rangle = 4\pi^2 \tilde{C}_{zz}(\mathbf{q}_{\|}). \qquad (2.14)$$

More precisely, one shows that

$$\langle \tilde{z}^*(\mathbf{q}_{\|})\tilde{z}(\mathbf{q}'_{\|})\rangle = \langle \tilde{z}(-\mathbf{q}_{\|})\tilde{z}(\mathbf{q}'_{\|})\rangle = 4\pi^2 \tilde{C}_{zz}(\mathbf{q}_{\|})\delta(\mathbf{q}_{\|} - \mathbf{q}'_{\|}). \qquad (2.15)$$

The Fourier components of a homogeneous random variable are independent random variables, whose mean square dispersion is given by the Fourier transform of the correlation function. If the power spectrum decreases slowly with increasing $q_{\|}$, the roughness associated to small periods will remain important. Thus, whatever the length scale, the surface will present irregularities. In the real space, it implies that the correlation between the heights of two points on the surface will be small, whatever their separation. As a result, the correlation function will exhibit a singular behaviour about 0 (discontinuity of the derivative for example). An illustration of the influence of the correlation function (or power spectrum) on the roughness aspect of the surface is presented in Fig. 2.2 and detailed in Sect. 2.2.4 in the special case of a Gaussian distribution of heights.

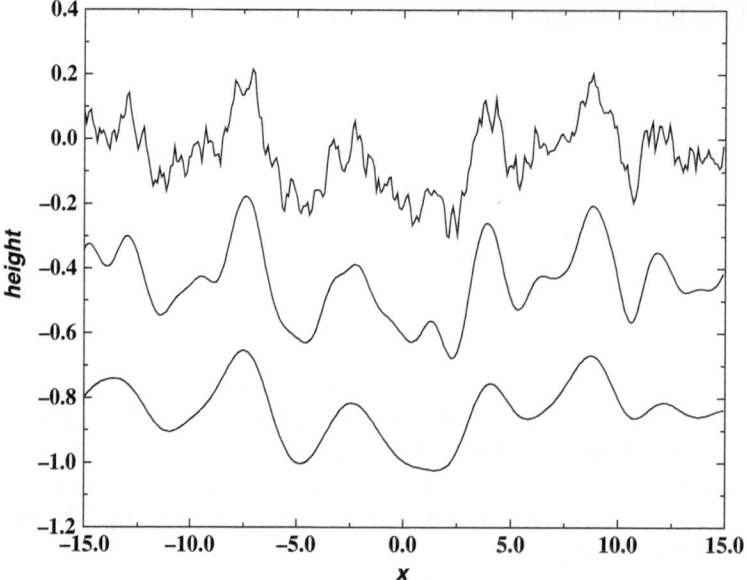

**Fig. 2.2** Various rough surfaces with Gaussian height distribution but various correlation functions. From bottom to top, $C_{zz}(R) = \sigma^2 \xi^4 / (\xi^2 + R^2)^2$, $C_{zz}(R) = \sigma^2 \exp(-\frac{R^2}{\xi^2})$, $C_{zz}(R) = \sigma^2 \exp(-\frac{R}{\xi})$

Until now we have been interested solely in ensemble average, which necessitates the knowledge of the complete set of rough surfaces generated by the homogeneous random process (or the probability distributions). However, sometimes only a single realisation $S_p$ (with dimension $L_x, L_y$ along $Ox$ and $Oy$) of the random process is available and one defines the spatial average of any functional $F(z_1, \ldots, z_n)$ for this surface by

$$\bar{F}_p(\mathbf{0}_{\|}, \ldots, \mathbf{r}_{n\|}) = \lim_{L_x \times L_y \to \infty} \frac{1}{L_x L_y} \int_{L_x \times L_y} d\mathbf{r}'_{\|} F[z(\mathbf{r}'_{\|}) \ldots z(\mathbf{r}'_{\|} + \mathbf{r}_{n\|})]. \qquad (2.16)$$

It happens frequently that each realisation of the ensemble carries the same statistical information about the homogeneous random process as every other realisation. The spatial averages calculated for any realisation are then all equal and coincide with the ensemble average. The homogeneous random process is then said to be an ergodic process. In this case, the following particular relations hold:

$$\sigma^2 = \langle z^2 \rangle = \lim_{L_x, L_y \to \infty} \frac{1}{L_x L_y} \int_{L_x \times L_y} z^2(\mathbf{r}_{\|}) d\mathbf{r}_{\|}, \qquad (2.17)$$

$$C_{zz}(r_{\|}) = \langle z(\mathbf{0}_{\|}) z(\mathbf{r}_{\|}) \rangle = \lim_{L_x, L_y \to \infty} \frac{1}{L_x L_y} \int_{L_x \times L_y} z(\mathbf{r}'_{\|}) z(\mathbf{r}'_{\|} + \mathbf{r}_{\|}) d\mathbf{r}'_{\|}. \qquad (2.18)$$

One can show that Eqs. (2.17) and (2.18) will be satisfied if the correlation function $C_{zz}(r_{\|})$ dies out sufficiently rapidly with increasing $r_{\|}$ (see for demonstration [5]).

Indeed, this property implies that one realisation of the rough surface can be divided up into subsurfaces of smaller area that are uncorrelated so that an ensemble of surfaces can be constructed from a single realisation. Spatial averaging amounts then to ensemble averaging. If the random process is homogeneous and ergodic, all the realisations will look similar while differing in detail. This is exactly what we expect in order to describe liquid surfaces varying with time or set of surfaces of artificial origin. The fact that spatial averaging is equivalent to ensemble averaging when the surface contains enough correlation lengths to recover all the information about the random process is of crucial importance in statistical wave scattering theory.

## 2.2.4 The Gaussian Probability Distribution and Various Correlation Functions

In most theories, the height probability distribution is taken to be Gaussian. The Gaussian distribution plays a central role because it has an especially simple structure and, because of the central limit theorem, it is a probability distribution that is encountered under a great variety of different conditions. If the height $z$ of a surface is due to a large number of local independent events whose effects are cumulative (like the passage of grain abrasive), the resulting altitude will obey nearly Gaussian statistics. This result is a manifestation of the central limit theorem which states that if a random variable $X$ is the sum of $N$ independent random variables $x_i$, it will have a Gaussian probability distribution in the limit of large $N$. Hereafter, we suppose that the average value of the Gaussian variate $z(\mathbf{r}_\parallel)$ is null, $\langle z \rangle = 0$. The Gaussian height distribution function is written as

$$p_1(z) = \frac{1}{\sigma\sqrt{2\pi}} \exp\left(-\frac{z^2}{2\sigma^2}\right). \tag{2.19}$$

Gaussian variates have the remarkable property that the random process is entirely determined by the height probability distribution and the height–height correlation function $C_{zz}$. All higher order correlations are expressible in terms of second-order correlation [5]. The two-point distribution function is given in this case by

$$p_2(z, z', r_\parallel) = \frac{1}{2\pi\sqrt{\sigma^4 - C_{zz}^2(r_\parallel)}} \exp-\left[\frac{\sigma^2(z^2 + z'^2) - 2zz'C_{zz}(r_\parallel)}{2\sigma^4 - 2C_{zz}^2(r_\parallel)}\right]. \tag{2.20}$$

Other useful results on the Gaussian variates are

$$\chi_1(s) = \langle e^{isz} \rangle = e^{-s^2\sigma^2/2}, \tag{2.21}$$

$$\chi_2(s, s', r_\parallel) = \langle e^{i(sz - s'z')} \rangle = e^{-\sigma^2(s^2 + s'^2)/2} e^{ss'C_{zz}(r_\parallel)}. \tag{2.22}$$

The correlation function plays a fundamental role in the surface aspect. It provides an indication of the length scales over which height changes along the surface. It gives in particular the distance beyond which two points of the surface can be considered independent. If the surface is truly random, $C_{zz}(r_\parallel)$ decays to zero with increasing $r_\parallel$. The simplest and often used form for the correlation function is also Gaussian,

$$C_{zz}(r_\parallel) = \sigma^2 \exp(-r_\parallel^2/\xi^2). \tag{2.23}$$

The correlation length $\xi$ is the typical distance between two different irregularities (or bumps) on the surface. Beyond this distance, the heights are not correlated.

In certain scattering experiments, one can retrieve the behaviour of the correlation function for $r_\parallel$ close to zero. We have thus access to the small scale properties of the surface. We have seen that the regularity of the correlation function at zero mirrors the asymptotic behaviour of the power spectrum: the faster the high-frequency components of the surface decay to zero, the smoother the correlation function about zero. The Gaussian scheme whose variations about zero have the quadratic form $\sigma^2(1 - (r_\parallel/\xi)^2)$ is thus indicated solely for surfaces that present only one typical lateral length scale [6].

For surfaces with structures down to arbitrary small scales, one expects the correlation function to be more singular at zero. An example is the self-affine rough surface for which

$$g(r_\parallel) = A_0 r_\parallel^{2h}, \tag{2.24}$$

where $A_0$ is a constant, or

$$C_{zz}(r_\parallel) = \sigma^2 \left(1 - \frac{r_\parallel^{2h}}{\xi^{2h}}\right), \tag{2.25}$$

with $0 < h < 1$. The roughness exponent or Hurst exponent $h$ is the key parameter which describes the height fluctuations at the surface: small $h$ values produce very rough surfaces while if $h$ is close to 1 the surface is more regular. This exponent is associated to fractal surfaces with dimension $D = 3 - h$ as reported by Mandelbrodt [7]. The pair-correlation function given in Eq. (2.24) diverges for $r_\parallel \to \infty$. Hence, all the length scales along the vertical axis are represented and the roughness of the surface cannot be defined. We will see below that in that case, there is no specular reflection. However, very often, some physical processes limit the divergence of the correlation function, i.e. the roughness saturates at some in-plane cut-off $\xi$. Such surfaces are well described by the following correlation function,

$$C_{zz}(R) = \sigma^2 \exp\left(-\frac{R^{2h}}{\xi^{2h}}\right). \tag{2.26}$$

For liquid surfaces other functional forms described in Sect. 4.5 are used.

## 2.2.5 More Complicated Geometries: Multilayers and Volume Inhomogeneities

Up to now we have considered solely the statistical description of a rough surface separating two homogeneous media. The mathematical notions that have been introduced can be generalised to more complicated problems such as stacks of rough surfaces in multilayer components. In this case, one must also consider the correlation function between the different interfaces, $\langle z_i(\mathbf{0}_{\|})z_j(\mathbf{r}_{\|})\rangle$, where $z_i$ represents the height of the $i$th surface. A detailed description of the statistics of a rough multilayer is given in Sect. 6.2. One can also describe in a similar fashion the random fluctuations of the refractive index (or electronic density) $\rho$. In this case $\rho$ is a random continuous variable of the three-dimensional space coordinates $(\mathbf{r}_{\|}, z)$. It will be introduced in Sects. 4.3.3 and 7.3.

## 2.3 Description of a Surface Scattering Experiment, Coherence Domains

We have seen how to characterise, with statistical tools, the rough surface geometry. The next issue is to relate these statistics to the intensity scattered by the sample in a scattering experiment. In this section, we introduce the main theoretical results that describe the interaction between electromagnetic waves and surfaces. Attention is drawn on the notion of "coherence domains" which takes on particular importance in the modelling of scattering from random media. In this foreword, we present briefly the basic mechanisms that subtend this concept.

It can be shown (bear in mind the Huygens–Fresnel principle or see Sect. 4.1.4) that a rough surface illuminated by an electromagnetic incident field acts as a collection of radiating secondary point sources. The superposition of the radiation of those sources yields the total diffracted field. If the secondary sources are coherently illuminated, the total diffracted field is the sum of the complex amplitudes of each secondary diffracted beam. In other words, one has to account for the phase difference in this superposition. As a result, an interference pattern is created. The coherence domain is the surface region in which all the radiating secondary sources interfere. It depends trivially on the nature of the illuminating beam (which can be partially coherent), but more importantly, it depends on the angular resolution of the detector. To illustrate this assertion, we consider the Young's holes experiment [8]. Light from a monochromatic point source (or a coherent beam) falls on two pinholes located in the sample plane (see Fig. 2.3). We study the transmitted radiation pattern on a screen parallel to the sample plane at a distance $D$. In this region, an interference pattern is formed. The periodicity $\Lambda$ of the fringes, which is the signature of the coherence between the two secondary sources, depends on the separation $d$ between the two pinholes, $\Lambda = \lambda D/d$. Suppose now that a detector is moved on the screen to record the diffracted intensity. As long as the detector width $l$ is close

**Fig. 2.3** Scattering geometry
for interpreting surface
scattering

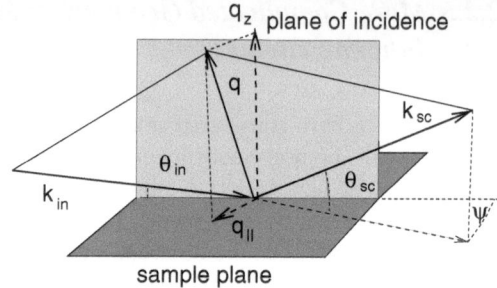

to $\Lambda$, the modulation of the interference pattern will be detected. On the contrary, if $l > 10\Lambda$ the intensity measured by the detector is the average of the fringe intensities. We obtain a constant equal to the sum of the intensities scattered by each secondary source. In this case, one may consider that from the detector point of view, the sources radiate in an incoherent way. We see with this simple experiment that the coherence length is directly linked to the finite extent of the detector (equivalent to a finite angular resolution).[1]

We now turn to a more accurate description of a surface scattering experiment.

### 2.3.1 Scattering Geometry

We consider an ideal scattering experiment consisting in illuminating a rough sample with a (perfectly coherent monochromatic) beam directed along $\mathbf{k}_{in}$ and detecting the flux of Poynting vector in an arbitrary small solid angle in the direction $\mathbf{k}_{sc}$ with a point-like detector located in the far-field region.

The interaction of the beam with the material results in a wave-vector transfer,

$$\mathbf{q} = \mathbf{k}_{sc} - \mathbf{k}_{in}. \tag{2.27}$$

Figure 2.3 shows the scattering geometry in the general case of a surface experiment. The plane of incidence contains the incident wave vector $\mathbf{k}_{in}$ and the normal to the surface $Oz$. In a reflectivity experiment, it is usual to work in the plane of incidence and thus to have $\psi = 0$. Yet the case $\psi \neq 0$ is of special interest for surface diffraction experiments in grazing incidence geometry. When working in the plane of incidence it is also useful to distinguish the symmetric specular geometry for which $\theta_{in} = \theta_{sc}$ and the off-specular geometry for which $\theta_{in} \neq \theta_{sc}$. The following set of Eq. (2.28) gives the components of the wave-vector transfer with the notations introduced in Fig. 2.3:

---

[1] It is also obviously linked to the degree of coherence fixed by, for example, the incidence slit opening. However, for x-ray or neutron experiments the resolution is actually generally limited by the detector slits opening.

$$\begin{cases} q_x = k_0 \left( \cos \theta_{sc} \cos \psi - \cos \theta_{in} \right) \\ q_y = k_0 \left( \cos \theta_{sc} \sin \psi \right) \\ q_z = k_0 \left( \sin \theta_{sc} + \sin \theta_{in} \right) \end{cases} . \tag{2.28}$$

## 2.3.2 Scattering Cross-Section

In the ideal experimental setup presented in the previous section, one exactly measures the differential scattering cross-section as described in Fig. 1.1 (the isolated scattering object is the rough sample in this case). The vectorial electric field $\mathbf{E}$ is written as the sum,

$$\mathbf{E} = \mathbf{E}_{in} + \mathbf{E}_{sc}, \tag{2.29}$$

of the incident plus scattered field. We are interested by the flux of the Poynting vector $\mathbf{S}$ through a surface $dS$ located at the position $\mathbf{R}$ of the detector for a unit incident flux. The precise calculations of the differential scattering cross-section are detailed in Sect. 4.1.4. In this paragraph, we simply introduce the main steps of the derivation.

One assumes that the detector located at $\mathbf{R}$ is placed far from the sample (far-field approximation). We define the scattering direction by the vector $\mathbf{k}_{sc}$ (see Fig. 2.3),

$$\mathbf{k}_{sc} = k_0 \widehat{\mathbf{u}} = k_0 \mathbf{R}/R. \tag{2.30}$$

It is shown in Sect. 4.1.4 that the scattered field can be viewed as the sum of the wavelets radiated by the electric dipoles induced in the material by the incident field (these radiating electric dipoles are the coherent secondary sources presented in the introduction). The strength of the induced dipole located at $\mathbf{r}'$ in the sample is given by the total field times the permittivity contrast at this point, $[k^2(\mathbf{r}') - k_0^2]\mathbf{E}(\mathbf{r}')$. Let us recall that for x-rays,

$$(k^2(\mathbf{r}') - k_0^2) = k_0^2[n^2(\mathbf{r}') - 1] = -4\pi r_e \rho_{el}(\mathbf{r}'), \tag{2.31}$$

where $\rho_{el}$ is the local electron density and $r_e$ the classical electron radius.[2] In the far-field region, the scattered field can be written as, see Eq. (4.19) (the far-field approximation and its validity domain are discussed in more detail in Chap. 4),

---

[2] If one is only interested in materials with low atomic numbers for which the x-ray frequency is much larger than all atomic frequencies, the electrons can be considered as free electrons plunged into an electric field $\mathbf{E}$. In this case, the movement of the electron is governed by $m_e d\mathbf{v}/dt = -e\mathbf{E}$, where $m_e$, $\mathbf{v}$, $-e$, are the mass, the velocity and the charge of the electron, respectively. We find $\mathbf{v} = (ie/m_e \omega)\mathbf{E}$ for a $e^{i\omega t}$ time dependence of the electric field. Thus, the current density is $\mathbf{j} = -e\rho_{el}\mathbf{v} = -(ie^2 \rho_{el}/m_e \omega)\mathbf{E}$ where $\rho_{el}$ is the local electron density. Writing the Maxwell's equations in the form $curl \mathbf{H} = \mathbf{j} + \epsilon_0 \partial \mathbf{E}/\partial t = \partial \mathbf{D}/\partial t = n^2 \epsilon_0 \partial \mathbf{E}/\partial t$ (depending on whether the system is viewed as a set of electrons in a vacuum or as a material of refractive index $n$), one obtains by identification that $n = 1 - (e^2/2m_e \epsilon_0 \omega^2)\rho_{el} = 1 - (\lambda^2/2\pi)r_e \rho_{el} \approx 1 - 10^{-6}$, with $r_e = (e^2/4\pi\epsilon_0 m_e c^2)$ the "classical electron radius". A complete and rigorous demonstration is given in [9].

$$\mathbf{E}_{sc}(\mathbf{R}) = \frac{\exp(-ik_0 R)}{4\pi R} \int d\mathbf{r}'(k^2(\mathbf{r}') - k_0^2)\mathbf{E}_\perp(\mathbf{r}')e^{i\mathbf{k}_{sc}\cdot\mathbf{r}'}, \qquad (2.32)$$

where

$$\mathbf{E}_\perp(\mathbf{r}') = \mathbf{E}(\mathbf{r}') - \hat{\mathbf{u}}.\mathbf{E}(\mathbf{r}')\hat{\mathbf{u}} \qquad (2.33)$$

represents the component of the electric field that is orthogonal to the direction of propagation given by $\hat{\mathbf{u}}$. Expression (2.32) shows that the scattered electric field $\mathbf{E}_{sc}(\mathbf{R})$ can be approximated by a plane wave [8] with wave vector $\mathbf{k}_{sc} = k_0\mathbf{R}/R = k_0\hat{\mathbf{u}}$ and amplitude,

$$\mathbf{E}_{sc}(\mathbf{k}_{sc}) = \mathbf{E}_{sc}(\mathbf{R}). \qquad (2.34)$$

The Poynting vector is then readily obtained,

$$\mathbf{S} = \frac{1}{2\mu_0 c}|\mathbf{E}_{sc}(\mathbf{R})|^2\hat{\mathbf{u}}. \qquad (2.35)$$

The flux of the Poynting vector for a unit incident flux (or normalised by the incident flux through a unit surface normal to the propagation direction) yields the differential scattering cross-section in the direction given by $\mathbf{k}_{sc}$,

$$\frac{d\sigma}{d\Omega} = \frac{1}{16\pi^2|\mathbf{E}_{in}|^2}\left|\int [k^2(\mathbf{r}') - k_0^2]\mathbf{E}_\perp(\mathbf{r}')e^{i\mathbf{k}_{sc}\cdot\mathbf{r}'}d\mathbf{r}'\right|^2. \qquad (2.36)$$

Note that $d\sigma/d\Omega$ involves a double integration, which can be cast in the form,

$$\frac{d\sigma}{d\Omega} = \frac{1}{16\pi^2|\mathbf{E}_{in}|^2}\int d\mathbf{r}\int d\mathbf{r}'(k^2(\mathbf{r}) - k_0^2)(k^2(\mathbf{r}+\mathbf{r}') - k_0^2)$$
$$\mathbf{E}_\perp(\mathbf{r}).\mathbf{E}_\perp^*(\mathbf{r}+\mathbf{r}')e^{i\mathbf{k}_{sc}\cdot\mathbf{r}'}, \qquad (2.37)$$

where $u^*$ stands for the conjugate of $u$. By integrating formally Eq. (2.32) over the vertical axis, one obtains a surface integral,

$$\mathbf{E}_{sc}(\mathbf{R}) = \frac{\exp(-ik_0 R)}{4\pi R}\int \mathscr{E}_\perp(\mathbf{r}'_\parallel, k_{scz})e^{i\mathbf{k}_{sc\parallel}\cdot\mathbf{r}'_\parallel}d\mathbf{r}'_\parallel, \qquad (2.38)$$

with

$$\mathscr{E}_\perp(\mathbf{r}'_\parallel, k_{scz}) = \int [k^2(\mathbf{r}') - k_0^2]e^{ik_{scz}z'}\mathbf{E}_\perp(\mathbf{r}')dz'. \qquad (2.39)$$

We see that Eq. (2.39) is a one-dimensional Fourier transform, thus the variations of $\mathscr{E}_\perp$ with $k_{scz}$ are directly linked to the thickness of the sample. On the other hand, the variations of $\mathbf{E}_{sc}$ with $\mathbf{k}_{sc\parallel}$ are related to the width of the illuminated area (i.e. the region for which $[k^2(\mathbf{r}') - k_0^2]\mathbf{E}$ is non-zero).

### 2.3.3 Coherence Domains

Up to now, we have considered an ideal experiment with a point-like detector. In reality, the detector has a finite size and one must integrate the differential scattering cross-section over the detector solid angle, $\Delta\Omega_{det}$. Since the cross-section is defined as a function of wave vectors, it is more convenient to transform the integration over the solid angle $\Delta\Omega_{det}$ centred about the direction $\mathbf{k}_{sc}$ into an integration in the $(k_x, k_y)$ plane. The measured intensity (scattering cross-section convoluted with the resolution function) is then given by

$$
I = \frac{1}{16\pi^2} \frac{1}{|\mathbf{E}_{in}|} \int d\mathbf{k}_\| \mathscr{R}(\mathbf{k}_\|)
$$
$$
\times \int d\mathbf{r}_\| \int d\mathbf{r}'_\| \mathscr{E}_\perp^*(\mathbf{r}_\| + \mathbf{r}'_\|, k_z) . \mathscr{E}_\perp(\mathbf{r}_\|, k_z) e^{i\mathbf{k}_\| \cdot \mathbf{r}'_\|}, \qquad (2.40)
$$

where $\mathscr{R}(\mathbf{k}_\|)$ is the detector acceptance in the $(k_x, k_y)$ plane. The expression of $\mathscr{R}$ in the wave-vector space is not easily obtained. In an x-ray experiment, it depends on the parameters (height, width) of the collecting slits. The reader is referred to Sect. 4.4 for a detailed expression of $\mathscr{R}$ as a function of the detector shape. In this introductory chapter it is sufficient to take for $\mathscr{R}$ a Gaussian function centred about $\mathbf{k}_{sc\|}$,

$$
\mathscr{R}(k_{scx}, k_{scy}) = C \exp\left[ -\frac{(k_x - k_{scx})^2}{2\Delta k_x^2} - \frac{(k_y - k_{scy})^2}{2\Delta k_y^2} \right]. \qquad (2.41)
$$

The variables $\Delta k_x, \Delta k_y$ govern the angular aperture of the detector. If one assumes that the integrand does not vary significantly along $\mathbf{k}_z$ inside $\Delta k_x \Delta k_y$,[3] the resulting intensity is given by

$$
I = \frac{1}{16\pi^2} \frac{1}{|\mathbf{E}_{in}|} \int\int d\mathbf{r}_\| d\mathbf{r}'_\| \mathscr{E}_\perp^*(\mathbf{r}_\| + \mathbf{r}'_\|, k_{scz}) . \mathscr{E}_\perp(\mathbf{r}_\|, k_{scz}) e^{i\mathbf{k}_{sc\|} \cdot \mathbf{r}'_\|} \tilde{\mathscr{R}}(\mathbf{r}'_\|), \quad (2.42)
$$

where

$$
\tilde{\mathscr{R}}(\mathbf{r}_\|) = 2\pi C \Delta k_x \Delta k_y e^{-\frac{1}{2}\Delta k_x^2 x^2 - \frac{1}{2}\Delta k_y^2 y^2}. \qquad (2.43)
$$

We now examine Eq. (2.38) that gives the scattered field as the sum of the fields radiated by all the induced dipoles in the sample. We see that the electric field radiated in the direction $\mathbf{k}_{sc}$ by the "effective" dipole placed at point $\mathbf{r}_\|$ is added coherently to the field radiated by another dipole placed at $\mathbf{r}_\| + \mathbf{r}'_\|$ whatever the distance between the points. The intensity, measured by an ideal experiment (coherent source and point-like detector), is given by a double integration of infinite extent which contains the incoherent term $|\mathscr{E}_\perp(\mathbf{r}_\|, k_{scz})|^2$ and the cross-product (namely the interference term) $\mathscr{E}_\perp(\mathbf{r}_\|, k_{scz}) . \mathscr{E}_\perp^*(\mathbf{r}_\| + \mathbf{r}'_\|, k_{scz})$. When the detector has a finite size, the double integration is modified by the introduction of the resolution function $\tilde{\mathscr{R}}$

---

[3] This assumption is not straightforward. It is seen in Eq. (2.39) that the thicker the sample, the faster the variations of $\mathscr{E}_\perp$ with $k_z$. In an x-ray experiment, the sample under study is generally a thin film (a couple of microns) and we are interested by the structure along $z$ of the material (multilayers). Hence, the size of the detector is chosen so that its angular resolution permits to resolve the interference pattern caused by the stack of layers. This amounts to saying that the $k_z$ modulation of $\mathscr{E}_\perp^*(\mathbf{r}_\| + \mathbf{r}'_\|, k_z) . \mathscr{E}_\perp(\mathbf{r}_\|, k_z)$ is not averaged in the detector.

which is the Fourier transform of the angular characteristic function of the detector. In our example, $\tilde{\mathcal{R}}$ is a Gaussian whose support in the $(x, y)$ plane is roughly $1/[\Delta k_x \times \Delta k_y]$. This function limits the domain over which the contribution of the cross term to the total intensity is significant. This domain can be called the coherence domain $S_{coh}$ due to the detector. The fields radiated by two points that belong to this domain will add coherently in the detector (the cross term value is important), while the fields coming from two points outside this domain will add incoherently (the cross term contribution is damped to zero). The resulting intensity can be seen as the incoherent sum of intensities that are scattered from various regions of the sample whose sizes coincide with the coherent domain given by the detector. This can be readily understood by rewriting Eq. (2.42) in the form [10],

$$
I \propto \sum_{i=1,N} \int_{S_{coh}} d\mathbf{r}_{\parallel} \int_{S_{coh}} d\mathbf{r}'_{\parallel}
$$

$$
\mathcal{E}_{\perp}^{*}(\mathbf{r}_{i\parallel} + \mathbf{r}_{\parallel} + \mathbf{r}'_{\parallel}, k_{scz}) . \mathcal{E}_{\perp}(\mathbf{r}_{i\parallel} + \mathbf{r}_{\parallel}, k_{scz}) e^{i\mathbf{k}_{sc\parallel} \cdot \mathbf{r}'_{\parallel}} \tilde{\mathcal{R}}(\mathbf{r}'_{\parallel}), \qquad (2.44)
$$

where $\mathbf{r}_i$ is the centre of the different coherent regions $S_{coh}$. Hence, integrating the intensity over a certain solid angle is equivalent to summing the intensities (i.e. incoherent process) from various regions of the illuminated sample. This is the main result of this paragraph. *The finite angular resolution of the detector introduces coherence lengths beyond which two radiating sources can be considered incoherent (even though the incident beam is perfectly coherent).* Note that the plural is not fortuitous, indeed, the angular resolution of the detector can be different in the $xOy$ and $xOz$ plane, thus the coherent lengths vary along $Ox$, $Oz$ and $Oy$. In a typical x-ray experiment (see Sect. 4.4), the sample is illuminated coherently over 5 mm$^2$ but the angular resolution of the detector yields coherence domains of solely a couple of square microns. More precisely, a detection slit with height 100 μm, width 1 cm placed at 1 m of the sample with $\theta_{sc} = 10$ mrad limits the coherent length along $Oz$ to 1 μm, along $Ox$ to 100 μm and that along $Oy$ to 10 nm. Finally, in this introductory section, we have restricted our analysis solely to a detector of finite extent. In general, the incident source has also a finite angular resolution. However, coherence domains induced by the incident angular resolution is usually much bigger than that given by the detector angular resolution so that we do not consider it here. (The calculation scheme would be very similar.) A more complete description of the resolution function of the experiment is given in Sect. 4.4.2.

## 2.4 Statistical Formulation of the Diffraction Problem

In this section, we point out, through various numerical simulations, the pertinence of a statistical description of the surface and of the scattered power for modelling a scattering experiment in which *the rough sample is necessarily deterministic*. The main steps of our analysis are as follows: Within the coherence domain, the field radiated by the induced dipoles (or secondary sources) of the sample interfere. We call *speckle* the complicated intensity pattern stemming from these interferences. The angular resolution of the detector yields an incoherent averaging of the speckle structures (the intensities are added over a certain angular domain). This angular integration can be performed with an ensemble average by invoking

1. The ergodicity property of the rough surface (i.e. we assume that the sample is one particular realisation of an ergodic random process)
2. The equivalence between finite angular resolution and limited coherence domains

It appears finally that the diffused intensity measured by the detector is adequately modelled by the mean square of the electric field viewed as a function of the random variable $z$. Throughout this section, the numerical examples are given in the optical domain. The wavelength is about $1\,\mu m$ and the perfectly coherent incident beam is directed along the $Oz$ axis.

## 2.4.1 To What Extent Is a Statistical Formulation of the Diffraction Problem Relevant?

In Sect. 2.3 it has been shown how to calculate formally the electromagnetic power measured by the detector in a scattering experiment. To obtain the differential scattering cross-section, one needs to know the permittivity contrast at each point of the sample and the electric field at those points, Eq. (2.37). If the geometry of the sample is perfectly well known (i.e. deterministic like gratings), various techniques (such as the integral boundary method [11,12]) permit to obtain without any approximation the field inside the sample. It is thus possible to simulate with accuracy the experimental results. In the case of scattering by gratings (i.e. periodic surfaces) the good agreement between experimental results and calculations confirms the validity of the numerical simulations [12].

We study the scattered intensity from different rough deterministic surfaces $s_n$ (e.g. those presented in Fig. 2.1) illuminated by a perfectly coherent beam. In this experiment, we suppose that the size of the coherence domains induced by the finite resolution of the detector is close to that of the illuminated area $A$. In other words, all the fringes of the interference pattern stemming from the coherent sum of the fields radiated by every illuminated point of the surface are resolved by the detector. We observe in Fig. 2.4 that the angular distribution of the intensity scattered by each surface presents a chaotic behaviour. This phenomenon can be explained by recalling that the scattered field consists of many coherent wavelets, each arising from a different microscopic element of the rough surface, see Eq. (2.38). The random height position of these elements yields a random dephasing of the various coherent wavelets which results in a granular intensity pattern. This seemingly random angular intensity behaviour, known as speckle effect, is obtained when the coherence domains include many correlation lengths of the surface, when the roughness is not negligible as compared to the wavelength (so that the random dephasing

**Fig. 2.4** Simulations of the differential scattering cross-section for the surfaces presented in Fig. 2.1. The illuminated area covers 40 μm which explains the large angular width of the speckle. The incident wavelength is 1 μm, the refractive index is $n = 1.5$. Normal incidence. The calculations are performed with a rigorous integral boundary method (no approximation in solving Eq. (2.37) other than the numerical discretisations) [13]

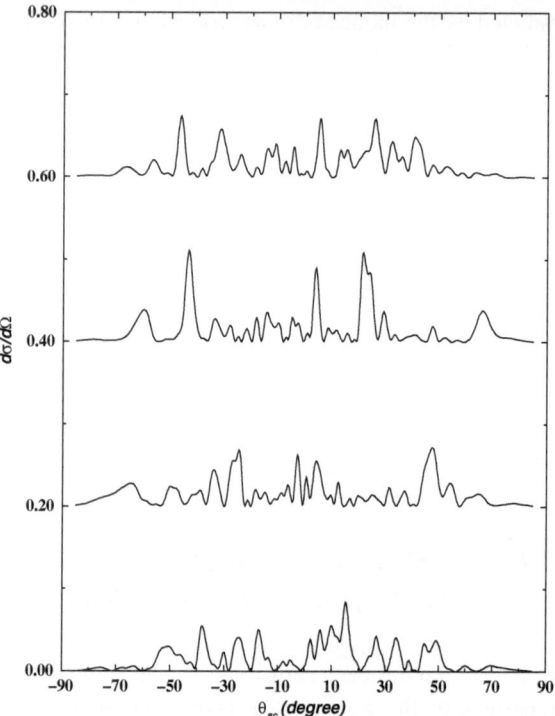

amplitude is important) and most importantly *when the size of the coherence domains is close to that of the illuminated area* so that the speckle is not averaged in the detector. To retrieve the precise angular behaviour of the intensity, one needs an accurate deterministic description of the surface [14]. In Fig. 2.4 the surfaces $s_n$ present totally different intensity patterns even though they have the same statistical properties. However, some similarities can be found in the curves plotted in Fig. 2.4. For example, the typical angular width of the spikes is the same for all surfaces. Indeed, in our numerical experiment it is linked to the width $L$ of the illuminated area (which is here equivalent to the coherence domain). The smallest angular period of the fringes formed by the (farthest-off) coherent point-source pair on the surface determines the minimal angular width $\lambda/L$ of the speckle spikes. This is clearly illustrated in Fig. 2.5, the larger the coherently illuminated area the thinner the angular speckle structures. In optics and radar imaging, sufficiently coherent incident beams (lasers) combined with detectors with fine angular resolution permit to study this phenomenon [14]. In x-ray experiments, the speckle effect can also be visualised in certain configurations. At grazing angles (e.g. $\theta_{sc} = 1$ mrad), the apparent resolution of the detector $\delta q_x = k_0 \theta \delta \theta$ (see Sect. 4.5.2.1) may be better than $10^{-7} k_0$ m$^{-1}$. The size of the illuminated area being 5 mm, the speckle structures are resolved in the detector.

We now suppose that the illuminated area is increased enough so that the typical angular width of the speckle structures will be much smaller than the angular

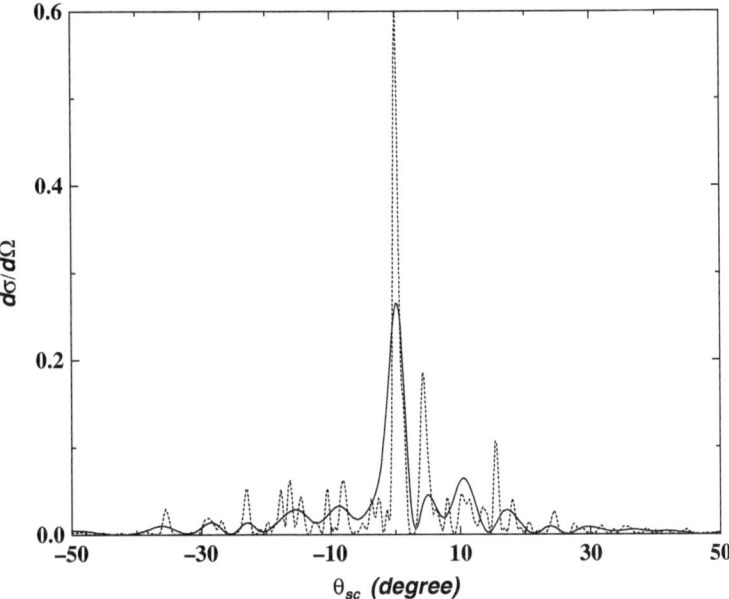

**Fig. 2.5** Illustration of the dependence of the angular width of the speckle structures on the size of the illuminated area. Simulation of the intensity angular distribution for one rough surface illuminated in the first case over 60 μm and in the second case over 30 μm. The incident wavelength is 1 μm, the refractive index is $n = 1.5$, normal incidence

resolution of the detector. The detector integrates the intensity over a certain solid angle and, as a result, the fine structures disappear. One notices then that the smooth intensity patterns obtained for all the different surfaces $s_n$ are quite similar. This is not surprising. Indeed, we have seen in the previous paragraph that the finite angular resolution of the detector is equivalent to the introduction of a coherence domain $S_{\text{coh}}$ (that is smaller than the illuminated area $A$). The measured intensity can be considered the incoherent sum of intensities stemming from the different subsurfaces of size $S_{\text{coh}}$ that constitute the sample. We now suppose that the illuminated area is big enough to cover many "coherent" subsurfaces, $A > 30S_{\text{coh}}$. Moreover, we suppose that the coherence domain is large enough so that each subsurface presents the same statistical properties $L_{\text{coh}} > 30\xi$, where $\xi$ is the correlation length and $L_{\text{coh}}$ the coherence length. If the set of surfaces $\{s_n\}$ can be described by an ergodic stationary process, the ensemble of subsurfaces obtained from one particular realisation $s_j$ will define the same random process with the same ensemble averaging as that created from any other realisation $s_k$. Consequently, the scattered intensity from one "big" surface $s_j$ can be seen as the ensemble average of the "subsurface" $S_{\text{coh}}$ scattered intensity which should be the same for all $s_k$. This assertion is supported by a comparison between two different numerical treatments of the same scattering experiment [13, 15].

In Fig. 2.6 we have plotted the diffuse intensity obtained from a deterministic rough surface $S_j$ illuminated by a perfectly coherent Gaussian beam, with a detector

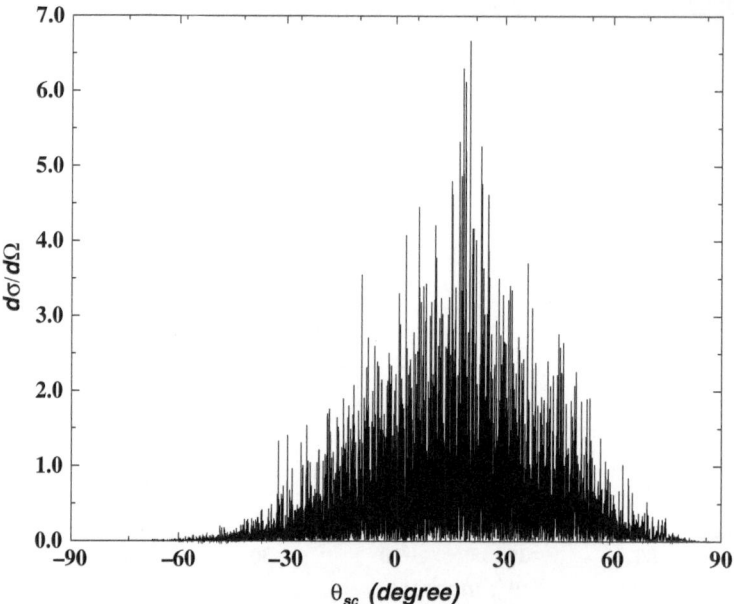

**Fig. 2.6** Simulation of the differential scattering cross-section of a rough deterministic surface which is one realisation of a random process. The illuminated area covers 3 mm (roughly several thousands of optical wavelengths). The statistics of the random process are Gaussian height distribution with $\sigma = 0.2\,\mu m$ and Gaussian correlation function with $\xi = 1\,\mu m$. The incident wavelength is 1 $\mu m$. Courtesy of Prof. M. Saillard [13]

of infinite resolution. The rough surface is one realisation of a random process with Gaussian height distribution function and Gaussian correlation function with correlation length $\xi$. The incident beam is chosen wide enough so that the illuminated part of $S_j$ is representative of the ergodic random process. In other words, $S_j$ can be divided into many subsurfaces (with similar statistical properties) whose set describes accurately the random process. The total length of the illuminated spot is $5000\xi$. It is seen in Fig. 2.6 that the scattered intensity exhibits a very thin speckle pattern. In general these fine structures are not visible. In Fig. 2.7 we have averaged the diffuse intensity over an angular width of 5°, corresponding to the angular resolution of a detector. We compare in Fig. 2.7 the *angular averaged pattern* with the *ensemble average* of the scattered intensity from subsurfaces that are generated with the same random process as $S_j$ but whose coherent illuminated domain is now restricted to $30\xi$ (i.e. to the coherence domain induced by the finite resolution of the detector). We obtain a perfect agreement between the two scattering patterns. In this example, we no longer need the precise value of the characteristic function $z(\mathbf{r}_\parallel)$ but solely the statistical properties of the random process that describe conveniently these particular surfaces. The integration of the intensity over the solid angle $\Delta\Omega$ will then be replaced by the calculation of the ensemble average of the intensity. This ensemble averaging appears also naturally in the case of surfaces varying with time (such as liquid surfaces like ocean) by recording the intensity during a sufficiently long amount of time.

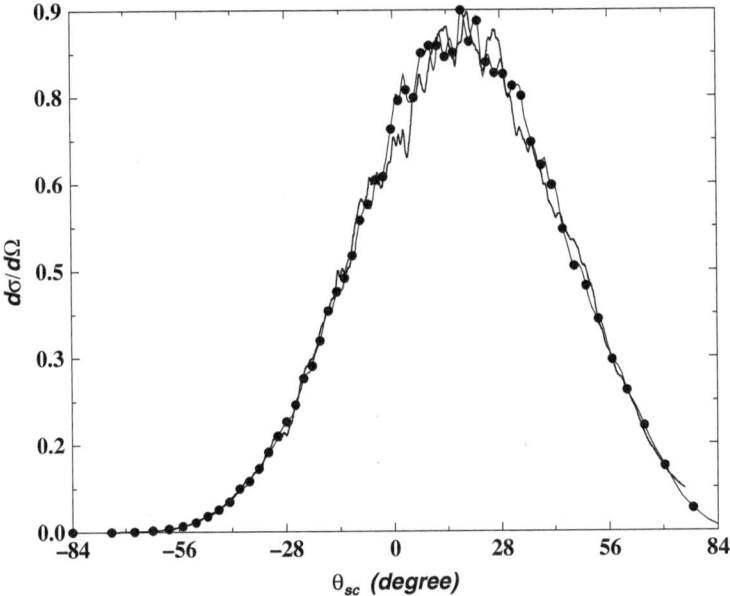

**Fig. 2.7** *Solid line*: Angular average over 5° of the differential scattering cross-section of the "big surface" presented in Fig. 2.6; *dotted line*: ensemble average of the differential scattering cross-section of rough surfaces with the same statistics as the "big surface". Size of each realisation is 30 μm, no angular averaging. Courtesy of Prof. Saillard [13]

Each subsurface (either spread spatially via the coherence domains or temporally) generates an electric field **E**. The latter can be viewed as a function of the random process $z$. The intensity measured by the detector is then related to the mean (in the ensemble averaging sense) square of the field, $\langle |\mathbf{E}|^2 \rangle$. The purpose of most wave scattering theories is to evaluate the various moments of **E**. More precisely, the random field can be divided into a mean and a fluctuating part,

$$\mathbf{E} = \langle \mathbf{E} \rangle + \delta \mathbf{E}. \tag{2.45}$$

We usually study separately the different contributions to the intensity.

### 2.4.2 Notions on Coherent (Specular) and Incoherent (Diffuse) Intensity

In the far field, the scattered electric field $\mathbf{E}_{sc}$ behaves like a plane wave with wave vector $\mathbf{k}_{sc}$ and amplitude $\mathbf{E}(\mathbf{k}_{sc})$, see Eq. (2.32). It can be written as the sum of a mean part and a fluctuating part,

$$\mathbf{E}_{sc} = \langle \mathbf{E}_{sc} \rangle + \delta \mathbf{E}_{sc}. \tag{2.46}$$

The previous discussions have shown that the measured scattered intensity from a rough sample (whose deterministic surface profile is assumed to be one realisation of a given ergodic random process) can be evaluated with the ensemble average of the intensity $\langle |\mathbf{E}_{sc}(\mathbf{k}_{sc})|^2 \rangle$,

$$\langle |\mathbf{E}_{sc}|^2 \rangle = |\langle \mathbf{E}_{sc} \rangle|^2 + \langle |\delta \mathbf{E}_{sc}|^2 \rangle. \tag{2.47}$$

The first term on the right-hand side of Eq. (2.47) is called the coherent intensity while the second term is known as the incoherent intensity. It is sometimes useful to tell the coherent and incoherent processes in the scattered intensity. In the following, we show that the coherent part is a Dirac function that contributes solely to the specular direction [4] if the randomly rough surface is statistically homogeneous in the $(Oxy)$ plane.

In most approximate theories, the random rough surface is of infinite extent and illuminated by a plane wave. Suppose we know the scattered far-field $\mathbf{E}_{sc}$ from a rough surface of defining equation $z = z(\mathbf{r}_{\parallel})$. We now address the issue of how $\mathbf{E}_{sc}$ is modified when the whole surface is shifted horizontally by a vector $\mathbf{d}$. It is clear that such a shift will not modify the physical problem. However, the incident wave amplitude acquires an additional phase factor $\exp(i\mathbf{k}_{in}.\mathbf{d})$ and similarly each scattered plane wave $\mathbf{E}_{sc}$ acquires, when returning to the primary coordinates, the phase factor $\exp(-i\mathbf{k}_{sc}.\mathbf{d})$. Thus we obtain,

$$\mathbf{E}_{sc}^{z(\mathbf{r}_{\parallel}-\mathbf{d})} = e^{-i(\mathbf{k}_{sc}-\mathbf{k}_{in}).\mathbf{d}} \mathbf{E}_{sc}^{z(\mathbf{r}_{\parallel})}. \tag{2.48}$$

We now suppose that the irregularities of the rough surface stem from a random spatially homogeneous process. In this case, the ensemble average is invariant under any translation in the $(xOy)$ plane,

$$\left\langle \mathbf{E}_{sc}^{z(\mathbf{r}_{\parallel}-\mathbf{d})} \right\rangle = \left\langle \mathbf{E}_{sc}^{z(\mathbf{r}_{\parallel})} \right\rangle. \tag{2.49}$$

This equality is only possible if

$$\langle \mathbf{E}_{sc} \rangle = A\delta(\mathbf{k}_{sc\parallel} - \mathbf{k}_{in\parallel}). \tag{2.50}$$

Hence, when the illuminated domain (or coherence domain) is infinite, the coherent intensity is a Dirac distribution in the Fresnel reflection (or transmission) direction. For this reason it is also called specular intensity. Note that unlike the coherent term, the incoherent intensity is a function in the $\mathbf{k}_{sc\parallel}$ plane and its contribution in specular direction tends to zero as the detector acceptance is decreased. In real life, the incident beam is space limited, the coherence domain is finite, thus the specular component becomes a function whose angular width is roughly given by $\lambda/L_{coh}$.

In many x-ray experiments, one is solely interested in the specularly reflected intensity. This configuration allows the determination of the z-dependent electron density profile and is often used for studying stratified interfaces (amphiphilic or polymer-adsorbed film). The modelisation of the coherent intensity requires the evaluation of the single integral Eq. (2.32) that gives the field amplitude while the incoherent intensity requires the evaluation of a double integral Eq. (2.37). It is thus much simpler to calculate only the coherent intensity and many elaborate theories have been devoted to this issue [4]. Chapter 3 of this book gives a thorough description of the main techniques developed for modelling the specular intensity from rough multilayers. However, it is important to bear in mind that the energy measured by the detector about the specular direction comes from both the coherent and incoherent processes inasmuch as the solid angle of collection is non-zero. The incoherent part is not always negligible as compared to the coherent part especially when one moves away from the grazing angles. An estimation of both contributions is then needed to interpret the data.

## 2.5 Statistical Formulation of the Scattered Intensity Under the Born Approximation

In this last section, we illustrate the notions introduced previously with a simple and widely used model that permits to evaluate the scattering crosssection of random rough surfaces within a probabilistic framework. We discuss the relationship between the scattered intensity and the statistics of the surfaces. The main principles of the Born development have been introduced in Chap. 1, Appendix 1.A, and a complementary approach of the Born approximation is given in Chap. 4 with some insights on the electromagnetic properties of the scattered field.

### 2.5.1 The Differential Scattering Cross-Section

We start from Eq. (2.32) that gives the scattered far field as the sum of the fields radiated by the induced dipoles in the sample. The main difficulty of this integral is to evaluate the exact field $\mathbf{E}$ inside the scattering object. In the x-ray domain, the permittivity contrast is very small ($\approx 10^{-6}$) and one can assume that the incident field is not drastically perturbed by surrounding radiating dipoles. Hence, a popular assumption (known as the Born approximation) is to approximate $\mathbf{E}$ by $\mathbf{E}_{\text{in}}$. With this approximation the integrand is readily calculated. For an incident plane wave $\mathbf{E}_{\text{in}}e^{-i\mathbf{k}_{\text{in}}\cdot\mathbf{r}}$, the differential scattering cross-section can be expressed as

$$\frac{d\sigma}{d\Omega} = \frac{1}{16\pi^2}\frac{|\mathbf{E}_{\text{in}\perp}|^2}{|\mathbf{E}_{\text{in}}|^2}\int d\mathbf{r}\int d\mathbf{r}'[k^2(\mathbf{r}) - k_0^2][k^2(\mathbf{r}') - k_0^2]e^{i\mathbf{q}\cdot(\mathbf{r}-\mathbf{r}')}, \qquad (2.51)$$

where $\mathbf{E}_{\text{in}\perp}$ is the projection of the incident electric field on the plane normal to the direction of observation of the differential cross-section. Denoting the unit vectors in direction $\mathbf{E}_{\text{in}}$ and $\mathbf{E}_{\text{sc}}$, $\widehat{\mathbf{e}}_{\text{in}} = \mathbf{E}_{\text{in}}/E_{\text{in}}$ and $(\widehat{\mathbf{e}}_{\text{sc}})^2 = \mathbf{E}_{\text{sc}}/E_{\text{sc}}$, respectively, we have $|\mathbf{E}_{\text{in}\perp}| = E_{\text{in}}(\widehat{\mathbf{e}}_{\text{in}}\cdot\widehat{\mathbf{e}}_{\text{sc}})^2$. In x-ray experiments, the incident field impinges on the surface at grazing angle and one studies the scattered intensity in the vicinity of the specular component. In this configuration, the orthogonal component of the incident field with respect to the scattered direction is close to the total incident amplitude. Yet, we retain the projection term $(\widehat{\mathbf{e}}_{\mathrm{in}}.\widehat{\mathbf{e}}_{\mathrm{sc}})^2$ in the differential scattering cross-section for completeness and coherence with the results of Chap. 1. Bearing in mind the value of the permittivity contrast as a function of the electronic density, Eq. (2.31), Eq. (2.51) simplifies to

$$\frac{d\sigma}{d\Omega} = r_e^2 (\widehat{\mathbf{e}}_{\mathrm{in}}\widehat{\mathbf{e}}_{\mathrm{sc}})^2 \int d\mathbf{r} \int d\mathbf{r}' \rho_{el}(\mathbf{r})\rho_{el}(\mathbf{r}')e^{i\mathbf{q}.(\mathbf{r}-\mathbf{r}')}, \qquad (2.52)$$

with $\rho_{el}$ the electron density and $r_e$ the classical electron radius.[4] In the case of a rough interface separating two semi-infinite homogeneous media one gets,

$$\frac{d\sigma}{d\Omega} = r_e^2 \rho_{el}^2 (\widehat{\mathbf{e}}_{\mathrm{in}}.\widehat{\mathbf{e}}_{\mathrm{sc}})^2 \int_{-\infty}^{z(\mathbf{r}_\parallel)} dz \int_{-\infty}^{z(\mathbf{r}'_\parallel)} dz' \int d\mathbf{r}_\parallel \int d\mathbf{r}'_\parallel e^{i\mathbf{q}.(\mathbf{r}-\mathbf{r}')}. \qquad (2.53)$$

Integrating Eq. (2.53) over $(z, z')$ (with the inclusion of a small absorption term to ensure the convergence at $-\infty$) yields,

$$\frac{d\sigma}{d\Omega} = \frac{\rho_{el}^2 r_e^2}{q_z^2} (\widehat{\mathbf{e}}_{\mathrm{in}}.\widehat{\mathbf{e}}_{\mathrm{sc}})^2 \int d\mathbf{r}_\parallel \int d\mathbf{r}'_\parallel e^{i\mathbf{q}_\parallel.(\mathbf{r}_\parallel - \mathbf{r}'_\parallel)} e^{iq_z[z(\mathbf{r}_\parallel)-z(\mathbf{r}'_\parallel)]}. \qquad (2.54)$$

This equation concerns a priori the scattering from any (deterministic or not) object. In this chapter, we are mostly interested by the scattering from surfaces whose surface profile $z$ is unknown or of no interest. We have seen in the preceding sections that if $z$ is described by a random homogeneous ergodic process, the intensity measured by the detector can be approximated by the ensemble average of the scattering cross-section. It amounts to replacing in Eq. (2.54) the integration over the surface by an ensemble average, $\int f(\mathbf{r}_\parallel)d\mathbf{r}_\parallel = L_x L_y \langle f \rangle$, where $L_x, L_y$ are the dimensions of the surface along $Ox$ and $Oy$. One obtains,

$$\frac{d\sigma}{d\Omega} = \frac{\rho_{el}^2 r_e^2 L_x L_y}{q_z^2} (\widehat{\mathbf{e}}_{\mathrm{in}}.\widehat{\mathbf{e}}_{\mathrm{sc}})^2 \int d\mathbf{r}_\parallel e^{i\mathbf{q}_\parallel.\mathbf{r}_\parallel} \left\langle e^{iq_z[z(\mathbf{r}_\parallel)-z(\mathbf{0}_\parallel)]} \right\rangle. \qquad (2.55)$$

---

[4] One can make a general presentation of elastic scattering under the Born approximation from the scattering by an isolated object as presented in Sect. 1.2.4 and Appendix 1.A. The differential scattering cross-section can be cast in the form

$$\frac{d\sigma}{d\Omega} = \left| \sum_j b e^{i\mathbf{q}.\mathbf{r}_j} \right|^2 = \left| \int d\mathbf{r} \rho b e^{i\mathbf{q}.\mathbf{r}} \right|^2,$$

where $\rho$ is the density of scattering objects and $b$ their scattered length as introduced in Eq. (1.34). The complex exponential is the result of the phase shift between waves scattered in the $\mathbf{k}_{\mathrm{sc}}$ direction by scatterers separated by a vector $\mathbf{r}$ as shown in Fig. 2.8. For neutrons, $b$ is the scattering length which takes into account the strong interaction between the neutrons and the nuclei (we do not consider here magnetic materials); for x-rays, $b = r_e = (e^2/4\pi\epsilon_0 m_e c^2) = 2.8 \times 10^{-15}$ m which is the classical radius of the electron.

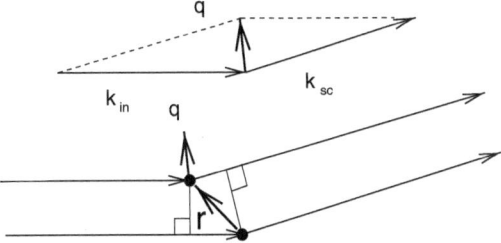

**Fig. 2.8** Phase shift between the waves scattered by two point scatterers separated by a vector **r**. The phase shift is $(\mathbf{k}_{sc} - \mathbf{k}_{in}) \cdot \mathbf{r} = \mathbf{q} \cdot \mathbf{r}$

Note that the expression (2.55) of the differential scattering cross-section accounts for both the coherent and incoherent processes. Hence, this integral does not converge in the function sense, it contains a Dirac distribution if the surface is infinite. This property will be illustrated with various examples in the following. If the probability density of $z$ is Gaussian, we can write the differential cross-section as

$$\frac{d\sigma}{d\Omega} = \frac{\rho_{el}^2 r_e^2 L_x L_y}{q_z^2} (\widehat{\mathbf{e}}_{in} \cdot \widehat{\mathbf{e}}_{sc})^2 \int d\mathbf{r}_{\parallel} e^{i\mathbf{q}_{\parallel} \cdot \mathbf{r}_{\parallel}} e^{-\frac{1}{2} q_z^2 \langle [z(\mathbf{r}_{\parallel}) - z(\mathbf{0}_{\parallel})]^2 \rangle}. \qquad (2.56)$$

We see that, under the Born approximation (where we neglect multiple scattering) the scattered intensity is related to the Fourier transform of the exponential of the pair-correlation function, $g(r_{\parallel}) = \left\langle \left[ z(\mathbf{r}_{\parallel}) - z(\mathbf{0}_{\parallel}) \right]^2 \right\rangle$. In the following we illustrate this result by studying the differential scattering crosssection for various pair-correlation functions. We start by the expression of the scattering differential cross-section in the case of a flat surface.

## 2.5.2 Ideally Flat Surfaces

For ideally flat surfaces $g(r_{\parallel})$ is zero everywhere at the surface and the scattering cross-section yields

$$\frac{d\sigma}{d\Omega} = \frac{r_e^2 \rho_{el}^2 L_x L_y}{q_z^2} (\widehat{\mathbf{e}}_{in} \cdot \widehat{\mathbf{e}}_{sc})^2 \int d\mathbf{r}_{\parallel} e^{i\mathbf{q}_{\parallel} \cdot \mathbf{r}_{\parallel}}. \qquad (2.57)$$

The integral is the Fourier transform of a constant so that,[5]

$$\frac{d\sigma}{d\Omega} = \frac{4\pi^2 r_e^2 \rho_{el}^2 L_x L_y}{q_z^2} (\widehat{\mathbf{e}}_{in} \cdot \widehat{\mathbf{e}}_{sc})^2 \delta(\mathbf{q}_{\parallel}). \qquad (2.58)$$

The scattered intensity is thus a Dirac distribution in the Fresnel reflection direction. As expected, for a perfectly flat surface, the reflectivity comes solely from a

---

[5] Let us recall that $\delta(\mathbf{q}_{\parallel}) = \frac{1}{4\pi^2} \int e^{-i\mathbf{q}_{\parallel} \cdot \mathbf{r}_{\parallel}} d\mathbf{r}_{\parallel}$.

coherent process (Sect. 2.4.2), the incoherent scattering is null $\langle \delta E^2 \rangle = 0$. Note that the reflectivity decreases as a power law with $q_z$. We now turn to the more complicated problem of scattering from rough surfaces that are described statistically by a homogeneous ergodic random process.

## 2.5.3 Self-Affine Rough Surfaces

### 2.5.3.1 Surfaces Without Cut-Off

We first consider self-affine rough surfaces with pair-correlation function $g$ given by Eq. (2.24), $g(r_\parallel) = A_0 r_\parallel^{2h}$. With this pair-correlation function, the roughness cannot be determined since there is no saturation. The scattering cross-section is in this case,

$$\frac{d\sigma}{d\Omega} = \frac{r_e^2 \rho_{el}^2 L_x L_y}{q_z^2} (\hat{\mathbf{e}}_{in} \cdot \hat{\mathbf{e}}_{sc})^2 \int d\mathbf{r}_\parallel e^{-\frac{q_z^2}{2} A R^{2h}} e^{i \mathbf{q}_\parallel \cdot \mathbf{r}_\parallel}, \tag{2.59}$$

and can be expressed in polar coordinates as

$$\frac{d\sigma}{d\Omega} = \frac{r_0^2 \rho_e^2 L_x L_y}{q_z^2} (\hat{\mathbf{e}}_{in} \cdot \hat{\mathbf{e}}_{sc})^2 \int dr_\parallel e^{-\frac{q_z^2}{2} A R^{2h}} J_0(q_\parallel r_\parallel), \tag{2.60}$$

with $q_\parallel$ being the modulus of the in-plane scattering wave vector and $J_0$ the zeroth order Bessel function. The above integral has analytical solutions for $h = 0.5$ and $h = 1$ and has to be calculated numerically in other cases. For $h = 1$, the integration yields,

$$\frac{d\sigma}{d\Omega} = \frac{r_e^2 \rho_{el}^2 L_x L_y}{q_z^2} (\hat{\mathbf{e}}_{in} \cdot \hat{\mathbf{e}}_{sc})^2 e^{-q_\parallel^2 / q_z^4}, \tag{2.61}$$

and for $h = 0.5$,

$$\frac{d\sigma}{d\Omega} = (\hat{\mathbf{e}}_{in} \cdot \hat{\mathbf{e}}_{sc})^2 \frac{r_e^2 \rho_{el}^2 L_x L_y}{q_z^2} \frac{\pi A}{\left( q_\parallel^2 + \left( \frac{A}{2} \right)^2 q_z^4 \right)^{3/2}}. \tag{2.62}$$

The above expressions clearly show that for surfaces of this kind the scattering is purely diffuse (no Dirac distribution, no specular component).

### 2.5.3.2 Surfaces with Cut-Off

Rough surfaces are said to present a cut-off length when the correlation function $C_{zz}(\mathbf{r}_\parallel)$ tends to zero when $r_\parallel$ increases (for example see Eq. (2.26), when $C_{zz}(r_\parallel) = \sigma^2 \exp\left( -\frac{r_\parallel^{2h}}{\xi^{2h}} \right)$, the cut-off is $\xi$). In this general case an analytical calculation is not possible and the scattering cross-section becomes,

$$\frac{d\sigma}{d\Omega} = \frac{r_e^2 \rho_{el}^2 L_x L_y}{q_z^2} e^{-q_z^2 \sigma^2} (\hat{\mathbf{e}}_{\text{in}} \cdot \hat{\mathbf{e}}_{\text{sc}})^2 \int d\mathbf{r}_{\parallel} e^{q_z^2 C_{zz}(\mathbf{r}_{\parallel})} e^{i\mathbf{q}_{\parallel} \cdot \mathbf{r}_{\parallel}}. \tag{2.63}$$

The integrand in Eq. (2.63) does not tend to 0 when $\mathbf{r}_{\parallel}$ is increased. The integration over an infinite surface does not exist in the function sense. Indeed, $d\sigma/d\Omega$ accounts for both the coherent and incoherent contributions to the scattered power. It is possible to extract the specular (coherent) and the diffuse (incoherent) components by writing the integrand in the form,

$$e^{q_z^2 C_{zz}(\mathbf{r}_{\parallel})} = 1 + \left( e^{q_z^2 C_{zz}(\mathbf{r}_{\parallel})} - 1 \right). \tag{2.64}$$

The distributive part (or Dirac function) characterises the coherent or specular reflectivity while the regular part gives the diffuse power. Equation (2.63) is then cast in the form,

$$\frac{d\sigma}{d\Omega} = \left( \frac{d\sigma}{d\Omega} \right)_{\text{coh}} + \left( \frac{d\sigma}{d\Omega} \right)_{\text{incoh}}, \tag{2.65}$$

with

$$\left( \frac{d\sigma}{d\Omega} \right)_{\text{coh}} = \frac{r_e^2 \rho_{el}^2 L_x L_y}{q_z^2} e^{-q_z^2 \sigma^2} (\hat{\mathbf{e}}_{\text{in}} \cdot \hat{\mathbf{e}}_{\text{sc}})^2 \int d\mathbf{r}_{\parallel} e^{i\mathbf{q}_{\parallel} \cdot \mathbf{r}_{\parallel}}$$
$$= \frac{4\pi^2 r_e^2 \rho_{el}^2 L_x L_y}{q_z^2} e^{-q_z^2 \sigma^2} \delta(\mathbf{q}_{\parallel}) (\hat{\mathbf{e}}_{\text{in}} \cdot \hat{\mathbf{e}}_{\text{sc}})^2 \tag{2.66}$$

and

$$\left( \frac{d\sigma}{d\Omega} \right)_{\text{incoh}} = \frac{r_e^2 \rho_{el}^2 L_x L_y}{q_z^2} e^{-q_z^2 \sigma^2} (\hat{\mathbf{e}}_{\text{in}} \cdot \hat{\mathbf{e}}_{\text{sc}})^2 \int d\mathbf{r}_{\parallel} \left( e^{q_z^2 C_{zz}(\mathbf{r}_{\parallel})} - 1 \right) e^{i\mathbf{q}_{\parallel} \cdot \mathbf{r}_{\parallel}}. \tag{2.67}$$

The specular part is similar to that of a flat surface except that it is reduced by the roughness Debye–Waller factor $e^{-q_z^2 \sigma^2}$. The diffuse scattering part may be determined numerically if one knows the functional form of the correlation function. When $q_z^2 C_{zz}(\mathbf{r}_{\parallel})$ is small, the exponential can be developed as $1 + q_z^2 C_{zz}(\mathbf{r}_{\parallel})$. In this case, the differential scattering cross-section appears to be proportional to the power spectrum of the surface $P(\mathbf{q}_{\parallel})$,

$$\left( \frac{d\sigma}{d\Omega} \right)_{\text{incoh}} = r_e^2 \rho_{el}^2 L_x L_y e^{-q_z^2 \sigma^2} 4\pi^2 P(\mathbf{q}_{\parallel}) (\hat{\mathbf{e}}_{\text{in}} \cdot \hat{\mathbf{e}}_{\text{sc}})^2. \tag{2.68}$$

We see with Eqs. (2.66) and (2.68) that the Born assumption permits to evaluate both the coherent and incoherent scattering cross-sections of rough surfaces in a relatively simple way. This technique can be applied without additional difficulties to more complicated structures such as multilayers or inhomogeneous films. Unfortunately, in many configurations, the Born assumption proves to be too restrictive and one can miss major features of the scattering process. More accurate models such as the distorted-wave Born approximation have been developed and are presented in Chap. 4 of this book. Yet, the expressions of the coherent and incoherent scattering cross-sections given here by the first Born approximation provide useful insights on how the measured intensity relates to the shape (statistics) of the sample. The coherent reflectivity, Eq. (2.66), does not give direct information on the surface lateral fluctuations, except for the overall roughness $\sigma$, but it provides the electronic density of the plane substrate. Hence, reflectivity experiments are used in general to probe, along the vertical axis, the electronic density of samples that is roughly homogeneous in the $(xOy)$ plane but varies in a deterministic way along $Oz$ (e.g. typically multilayers). Chapter 3 of this book is devoted to this issue. On the other hand the incoherent scattering Eq. (2.68) is directly linked to the height–height correlation function of the surface. Bearing in mind the physical meaning of the power spectrum, Sect. 2.2.3, we see that measuring the diffuse intensity at increasing $q_\parallel$ permits to probe the surface state at decreasing lateral scales. Hence, scattering experiments can be a powerful tool to characterise the rough sample in the lateral $(Oxy)$ plane. This property will be developed and detailed in Chap. 4.

# References

1. Beckmann, P., Spizzichino, A.: The Scattering of Electromagnetic Waves from Rough Surfaces. Pergamon Press, Oxford, UK (1963).
2. Bass, F.G., Fuks, I.M.: Wave Scattering from Statistically Rough Surfaces. Pergamon, New York (1979).
3. Ogilvy, J.A.: Theory of Wave Scattering from Random Rough Surfaces. Adam Hilger, Bristol, UK (1991).
4. Voronovich, G.: Wave Scattering from Rough Surfaces. Springer-Verlag, Berlin (1994).
5. Mandel, L., Wolf, E.: Optical Coherence and Quantum Optics. Cambridge University Press, Cambridge, USA (1995).
6. Guérin, C.A., Holschneider, M., Saillard, M.: Waves Random Media, **7**, 331–349 (1997).
7. Mandelbrodt, B.B.: The Fractal Geometry of Nature. Freeman, New York (1982).
8. Born, M., Wolf, E.: Principle of Optics. Pergamon Press, New York (1980).
9. Oxtoby, D.W., Novack, F., Rice, S.A.: J. Chem. Phys. **76**, 5278 (1982).
10. Sinha, S.K., Tolan, M., Gibaud, A.: Phys. Rev. B **57**, 2740 (1998)
11. Nieto-Vesperinas, M., Dainty, J.C.: Scattering in Volume and Surfaces. Elsevier Science Publishers, B. V. North-Holland (1990).
12. Petit, R. (ed.): Electromagnetic Theory of Gratings. Topics in Current Physics. Springer Verlag, Berlin (1980).
13. Saillard, M., Maystre, D.: J. Opt. Soc. Am. A, **7**(6), 982–990 (1990).
14. Dainty, J.C. (ed.): Laser speckle and Related Phenomena. Topics in Applied Physics. Springer-Verlag, New York (1975).
15. Saillard, M., Maystre, D.: J. Opt. **19**, 173–176, (1988).

# Chapter 3
# Specular Reflectivity from Smooth and Rough Surfaces

A. Gibaud and G. Vignaud

It is well known that light is reflected and transmitted with a change in the direction of propagation at an interface between two media which have different optical properties. The effects known as reflection and refraction are easy to observe in the visible spectrum but more difficult when x-ray radiation is used (see the introduction for a historical presentation). The major reason for this is the fact that the refractive index of matter for x-ray radiation does not differ very much from unity, so that the direction of the refracted beam does not deviate much from the incident one. The reflection of x-rays is however of great interest in surface science, since it allows the structure of the uppermost layers of a material to be probed. In this chapter, we present the general optical formalism used to calculate the reflectivity of smooth or rough surfaces and interfaces which is also valid for x-rays.[1]

## 3.1 The Reflected Intensity from an Ideally Flat Surface

### 3.1.1 Basic Concepts

The interaction of x-rays with matter can be understood in a quantitative manner if the index of refraction for x-ray radiation of the investigated material is known. A basic determination of this quantity can be obtained in the framework of the phenomenological model of electrons elastically bounded to the nucleus. In this model, only the electrons are moving and the nuclei are considered to be fixed. Fundamental equation of motion applied to an electron in an electromagnetic field E yields

$$m\frac{d^2r}{dt^2} + h\frac{dr}{dt} + kr = -eE, \qquad (3.1)$$

A. Gibaud (✉)
Laboratoire de Physique de l'Etat Condensé, UMR 6087, Université du Maine Faculté des sciences, 72085 Le Mans Cedex 9, France

[1] The basic concepts used to determine the reflection and transmission coefficients of an electromagnetic wave at an interface were first developed by A. Fresnel [1] in his mechano-elastic theory of light.

Gibaud, A., Vignaud, G.: *Specular Reflectivity from Smooth and Rough Surfaces.* Lect. Notes Phys. **770**, 85–131 (2009)
DOI 10.1007/978-3-540-88588-7_3

where $h$ is a phenomenologic friction coefficient and $k$ is a spring constant.

Solution of this equation assuming that the electron follows the oscillations of the field gives the displacement of an electron with respect to its average position

$$r = \frac{-e}{m(\omega_0^2 - \omega^2) + i\omega\frac{h}{m}} E e^{i\omega t}, \tag{3.2}$$

where $\omega_0 = \sqrt{\frac{k}{m}}$ is the eigen pulsation of the electron bounded to the nucleus. This value is much smaller than the pulsation $\omega$ of hard x-ray radiations since $\omega_0 \approx 10^{15}\,\mathrm{rad/s} \ll \omega = 1.2 \times 10^{19}\,\mathrm{rad/s}$.

Therefore one can assume that far from an absorption edge

$$r \approx \frac{eE}{m\omega^2}. \tag{3.3}$$

The dielectric polarisation becomes

$$P = \epsilon_0 \chi E = -\frac{\rho_e^2 E}{m\omega^2}. \tag{3.4}$$

In this expression $\rho_e$ is the number of electrons per unit volume, i.e. the electron density, and $\chi$ is the dielectric susceptibility that can be written as

$$\chi = \epsilon_r - 1 = -\frac{\rho_e e^2}{\epsilon_0 m \omega^2}. \tag{3.5}$$

The index of refraction is thus

$$n = \sqrt{\epsilon_r} = \sqrt{1 + \chi}, \tag{3.6}$$

and since $\chi \ll 1$, one can write

$$n = 1 - \frac{\rho_e e^2}{2m\epsilon_0 \omega^2}. \tag{3.7}$$

Introducing the classical radius $r_e$ of the electron

$$r_e = \frac{e^2}{4\pi\epsilon_0 mc^2} = 2.8 \times 10^{-15}\,\mathrm{m}, \tag{3.8}$$

one finally obtains for no absorption

$$n = 1 - \frac{r_e \rho_e \lambda^2}{2\pi}. \tag{3.9}$$

For well-crystallised materials the volume of the unit cell is known and the electron density $\rho_e$ can be written as

$$\rho_e = \sum_k \frac{Z_k}{V_m}, \tag{3.10}$$

where $V_m$ is the unit cell volume, $Z_k$ is the number of electrons of atom $k$ in the unit cell. In the more elaborate quantum mechanics description (see Chap. 1) the atomic number $Z_k$ is modified by the real $f'$ and imaginary $f''$ part of the anomalous atomic form factor at the wavelength $\lambda$. The sum is carried out over all atoms contained in the unit cell.

Alternatively when the stoichiometric composition and the mass density $\mu$ of a material are known, the electron density is also given by

$$\rho_e = \mathcal{N}\mu \frac{\sum_k x_k \frac{Z_k + f'_k + if''_k}{M_k}}{\sum_k x_k}, \qquad (3.11)$$

with $\mathcal{N}$ the Avogadro number, $x_k$ the number of atoms $k$ and $M_k$ the molar mass of atom $k$.

Generally the index of refraction can be written as (see Sect. 1.4.2 for more details)

$$n = 1 - \delta - i\beta, \qquad (3.12)$$

with

$$\delta = \frac{r_e}{2\pi}\lambda^2 \rho_e = \frac{r_e}{2\pi}\lambda^2 \mu \mathcal{N} \frac{\sum_i \frac{x_i(Z_i + f'_i)}{M_i}}{\sum_i x_i} \qquad (3.13)$$

and

$$\beta = \frac{r_e}{2\pi}\lambda^2 \mu \mathcal{N} \frac{\sum_i x_i \frac{f''_i}{M_i}}{\sum_i x_i}. \qquad (3.14)$$

As an example we consider bulk silicon that crystallises in a cubic structure (see Fig. 3.1) containing eight atoms per unit cell of atomic number $Z = 14$.

The lattice parameter is $a = 5.43\ \text{Å}$. This yields $\rho_e = 0.71e/\text{Å}^3$, $\delta = 7.6 \times 10^{-5}$ and $\beta = 2 \times 10^{-7}$ at the copper $K\alpha$ radiation. When the lattice parameter is

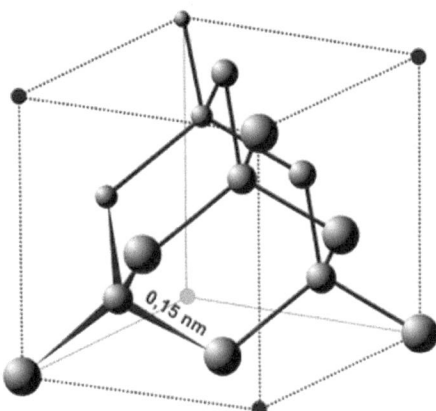

**Fig. 3.1** Crystalline structure of silicon. The unit cell is cubic and contains eight atoms

unknown, as for instance for an amorphous material, one can still determine the electron density provided the stoichiometry and the density of the material are known. Typical values for $\delta$ are usually in the range $10^{-5} - 10^{-6}$ and $\beta$ is about 10 times smaller. A similar equation holds for neutrons where $r_e \rho_e$ has to be replaced by $\rho b$ (see Chap. 5, Eq. (5.24)).

A specific property of x-rays and neutrons is that since the refractive index is slightly less than 1, a beam impinging on a flat surface can be totally reflected. The condition to observe total external reflection is that the angle of incidence $\theta$ (defined here as the angle between the incident ray and the surface) must be less than a critical angle $\theta_c$. This angle can be obtained by applying Snell–Descartes' law with $\cos \theta_{tr} = 1$, yielding in absence of absorption

$$\cos \theta_c = n = 1 - \delta. \tag{3.15}$$

Since $\delta$ is of the order of $10^{-5}$, the critical angle for total external reflection is clearly extremely small. At small angles, $\cos \theta_c$ can be approximated as $1 - \theta_c^2/2$ and (3.15) becomes

$$\theta_c^2 = 2\delta. \tag{3.16}$$

The total external reflection of an x-ray (or neutron) beam is therefore only observed at grazing angles of incidence below about $\theta < 0.5°$. At larger angles, the reflectivity decreases very rapidly as mentioned above.

In this chapter, we will calculate the reflectivity as a function of the incident angle $\theta$ or alternatively as a function of the modulus $q = 4\pi \sin \theta / \lambda$ of the wave-vector transfer $\mathbf{q}$ (see Eq. (2.27) and Fig. 2.3 with $\psi_{sc} = 0$). This means that the following ratios,

$$R(\theta) = \frac{I(\theta)}{I_0}, \tag{3.17}$$

$$R(\mathbf{q}) = \frac{I(\mathbf{q})}{I_0}, \tag{3.18}$$

will be determined, where $I(\theta)$ or $I(\mathbf{q})$ is the reflected intensity (flux of Poynting's vector through the detector area) for an angle of incidence $\theta$ (or wave-vector transfer $\mathbf{q}$), and $I_0$ is the intensity of the incident beam. The theory of x-ray reflectivity is valid under the assumption that it is possible to consider the electron density as continuous (see Chap. 1). Under this approximation, the reflection is treated like in optics, and the reflected amplitude is obtained by writing down the boundary conditions at the interface, i.e. the continuity of the electric and magnetic fields at the interface, leading to the classical Fresnel relations.

### 3.1.2 Fresnel Reflectivity

The reflection and transmission coefficients can be derived by writing the conditions of continuity of the electric and magnetic fields at the interface. The reflected

**Fig. 3.2** Reflection and refraction of an incident wave polarised along $y$ and travelling in the $xOz$ plane of incidence

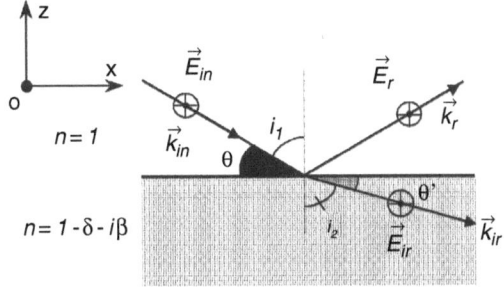

intensity, which is the square of the modulus of the reflection coefficient, is the quantity measured in an experiment. Let us consider an electromagnetic **plane wave** propagating in the $xOz$ plane of incidence, with its electric field polarised normal to this plane along the $Oy$ direction. The interface between air and the reflecting medium which is located at $z = 0$ as shown in Fig. 3.2 will be assumed to be abrupt. In order to better emphasise that the same formalism applies for x-rays and visible optics we use in this section the angles defined from the surface normal as in optics, together with the grazing angles usually used in x-ray or neutron reflectivity.

The expression for the electric field in a homogeneous medium is derived from Maxwell's equations which when combined lead to the propagation equation of the electric field known as Helmholtz's equation (see Chap. 1, Eqs. (1.12) and (1.15) for details)

$$\Delta \mathbf{E} + k_j^2 \mathbf{E} = 0, \tag{3.19}$$

where $k_j$ is the wave vector in medium $j$. The electric field which is solution of Helmholtz's equation is given for the incident (in), reflected (r) and transmitted (tr) plane waves by

$$\mathbf{E}_j = A_j e^{i(\omega t - \mathbf{k}_j \cdot \mathbf{r})} \hat{\mathbf{e}}_y, \tag{3.20}$$

with $j =$ in, r or tr, $k_0 = |\mathbf{k}_{in}| = |\mathbf{k}_r| = 2\pi/\lambda = |\mathbf{k}_{tr}|/n$ and $\hat{\mathbf{e}}_y$ is a unit vector along the $y$ axis (see Fig. 3.2). Note that the convention of signs used in crystallography is adopted here (see Part I, Chap. 1 by F. de Bergevin for details). It is straightforward to show that the components of the (in), (tr) and (r) wave vectors are

$$\begin{cases} \mathbf{k}_{in} = k_0(\sin i_1 \hat{\mathbf{e}}_x - \cos i_1 \hat{\mathbf{e}}_z) \\ \mathbf{k}_r = k_0(\sin i_1 \hat{\mathbf{e}}_x + \cos i_1 \hat{\mathbf{e}}_z) \\ \mathbf{k}_{tr} = k_0 n(\sin i_2 \hat{\mathbf{e}}_x - \cos i_2 \hat{\mathbf{e}}_z). \end{cases} \tag{3.21}$$

The tangential component of the electric field must be continuous at the interface ($z = 0$). In air, the field is the sum of the incident and reflected fields. Assuming that the medium is sufficiently thick for the transmitted beam to be completely absorbed, the following relation must be fulfilled,

$$A_{in} e^{i(\omega t - k_0 \sin i_1 x)} + A_r e^{i(\omega t - k_0 \sin i_1 x)} = A_{tr} e^{i(\omega t - k_0 n \sin i_2 x)}. \tag{3.22}$$

Equation (3.22) must be valid for any value of $x$, so that the following condition must hold,

$$\sin i_1 = n \sin i_2. \tag{3.23}$$

This condition is simply the well-known Snell–Descartes' second law. As a result of this, the conservation of the perpendicular component of the electric field leads to

$$A_{\text{in}} + A_r = A_{tr}. \tag{3.24}$$

It will be assumed that the media are non-magnetic so that the tangential component of the magnetic field must also be continuous. According to the Maxwell–Faraday equation,

$$\nabla \times \mathbf{E} = -\frac{\partial \mathbf{B}}{\partial t} = -i\omega \mathbf{B}. \tag{3.25}$$

The tangential component $B_t$ is the dot product of the magnetic field with the unit vector $\hat{\mathbf{e}}_x$, i.e.

$$B_t = \frac{(\nabla \times \mathbf{E}).\hat{\mathbf{e}}_x}{i\omega}. \tag{3.26}$$

Since the electric field is normal to the incident plane, it is polarised along the $y$ axis and the curl of the field gives

$$\nabla \times \mathbf{E} = \frac{\partial E_y}{\partial x}\hat{\mathbf{e}}_z - \frac{\partial E_y}{\partial z}\hat{\mathbf{e}}_x. \tag{3.27}$$

The tangential component of the magnetic field is then given by

$$B_t = -\frac{1}{i\omega}\frac{\partial E_y}{\partial z}, \tag{3.28}$$

and from Eq. (3.22) it is easy to show that the conservation of this quantity yields

$$(A_{\text{in}} - A_r)\cos i_1 = nA_{tr}\cos i_2. \tag{3.29}$$

Writing the reflected amplitude $r = A_r/A_{\text{in}}$ and the transmitted one $t = A_{tr}/A_{\text{in}}$, the following relations are obtained:

$$1 + r = t$$
$$1 - r = nt\frac{\cos i_2}{\cos i_1}. \tag{3.30}$$

Combining these two equations, the reflected amplitude coefficient in the case of a $(s)$ polarisation is found to be

$$r^{(s)} = \frac{\cos i_1 - n\cos i_2}{\cos i_1 + n\cos i_2}, \tag{3.31}$$

which by the use of the Snell–Descartes' relation leads to

$$r^{(s)} = \frac{\sin(i_2 - i_1)}{\sin(i_1 + i_2)}. \tag{3.32}$$

In the case of an electric field parallel to the plane of incidence, a similar calculation leads to

$$r^{(p)} = \frac{\tan(i_2 - i_1)}{\tan(i_2 + i_1)}. \tag{3.33}$$

Those equations are known as the Fresnel equations [1]. It is easy to show that at small grazing angles of incidence for x-rays $r^{(p)} \approx r^{(s)} \approx r$. Only $(s)$ polarisation (electric field polarised perpendicular to the plane of incidence) will be considered in detail below but some results will also be given for $(p)$ polarisation.

The grazing angle of incidence $\theta$ that the incident beam makes with the reflecting surface is usually the experimental variable in a reflectivity measurement. It is therefore important to express the coefficient of reflection as a function of this angle $\theta$ and also of the refractive index $n$. Starting from

$$r = \frac{\cos i_1 - n \cos i_2}{\cos i_1 + n \cos i_2}, \tag{3.34}$$

and using the fact that the $\theta$ and $i_1$ and the $\theta_{tr}$ and $i_2$ are complementary angles as shown in Fig. 3.2, Eq. (3.34) becomes

$$r = \frac{\sin \theta - n \sin \theta_{tr}}{\sin \theta + n \sin \theta_{tr}}. \tag{3.35}$$

Applying the Snell–Descartes' law,

$$\cos \theta = n \cos \theta_{tr} \tag{3.36}$$

produces the following coefficient of reflection:

$$r(\theta) = \frac{\sin \theta - \sqrt{n^2 - \cos^2 \theta}}{\sin \theta + \sqrt{n^2 - \cos^2 \theta}}. \tag{3.37}$$

In the case of small incident angles (for which $\cos \theta = 1 - \theta^2/2$) and for electromagnetic x-ray waves the refractive index (in the absence of absorption) is given by

$$n^2 = 1 - 2\delta = 1 - \theta_c^2. \tag{3.38}$$

The general equation (3.37) becomes,

$$r(\theta) = \frac{\theta - \sqrt{\theta^2 - \theta_c^2}}{\theta + \sqrt{\theta^2 - \theta_c^2}}. \tag{3.39}$$

The reflectivity, which is the square of the modulus of the reflection coefficient, is given by

$$R(\theta) = rr^* = \left| \frac{\theta - \sqrt{\theta^2 - \theta_c^2}}{\theta + \sqrt{\theta^2 - \theta_c^2}} \right|^2. \tag{3.40}$$

Finally, if the absorption of the x-ray beam by the material is accounted for, the refractive index takes a complex value and the Fresnel reflectivity is then written as

$$R(\theta) = rr^* = \left| \frac{\theta - \sqrt{\theta^2 - \theta_c^2 - 2i\beta}}{\theta + \sqrt{\theta^2 - \theta_c^2 - 2i\beta}} \right|^2. \tag{3.41}$$

The reflectivity can equally well be given in terms of the wave-vector transfer $q$:

$$R(\mathbf{q}) = \left| \frac{q_z - \sqrt{q_z^2 - q_c^2 - \frac{32i\pi^2\beta}{\lambda^2}}}{q_z + \sqrt{q_z^2 - q_c^2 - \frac{32i\pi^2\beta}{\lambda^2}}} \right|^2 . \tag{3.42}$$

When the wave-vector transfer is very large compared to $q_c$, i.e. $q \gtrsim 3q_c$, the following asymptotic behaviour is observed:

$$R = \frac{q_c^4}{16q^4} . \tag{3.43}$$

It can be seen from Fig. 3.3 that the reflectivity curve or reflectivity profile consists of three different regimes:

– A plateau of total external reflection $R = 1$ when $q < q_c$
– A very steep decrease when $q = q_c$
– A $1/q^4$ power law when $q > 3q_c$

It is worth noting that if the value of $q_c$ is measured experimentally, this immediately yields the value of the electron density in the material (see Part I, Chap. 1 by F. de Bergevin) since,

$$q_c = 3.75 \times 10^{-2} \sqrt{\rho_e}, \tag{3.44}$$

where $\rho_e$ is the electron density in the units $e^-/\text{Å}^3$.

Finally, remembering that the reflectivity is observed under specular conditions, reference to the system of axes defined in Fig. 3.2 shows that the Fresnel reflectivity $R(\mathbf{q})$ can be written as

$$R(\mathbf{q}) = \left| \frac{q_z - \sqrt{q_z^2 - q_c^2 - \frac{32i\pi^2\beta}{\lambda^2}}}{q_z + \sqrt{q_z^2 - q_c^2 - \frac{32i\pi^2\beta}{\lambda^2}}} \right|^2 \delta q_x \delta q_y, \tag{3.45}$$

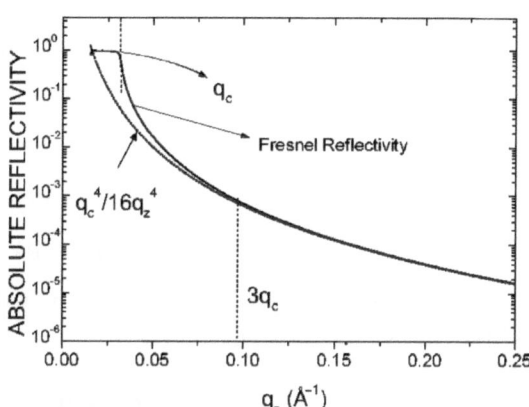

**Fig. 3.3** Calculated reflectivity of a flat silicon wafer and asymptotic law

since $q = q_z$ in Eq. (3.45), and the reflectivity of a flat surface is only measurable in the specular direction. Equation (3.45) completely describes the reflectivity of a homogeneous material, showing in particular that the reflectivity differs from zero only for wave-vector transfers normal to the surface of the sample.[2]

Figure 3.3 illustrates the calculated reflectivity curve for a silicon wafer in the power law regime and also in the case of a more complete dynamical calculation. The deviation from unity due to the absorption of the x-rays in the material can be seen to play a major role in determining the form of the curve in the region close to the critical edge at $q = q_c$. Equation (3.45) shows quite clearly that the calculation of a reflectivity curve requires only the electron density and the absorption of the material (for the wavelength used). Table 3.1 gives some useful data for calculating the reflectivity of various elements and compounds. A much wider database of quantities relevant to reflectivity measurements can be found at the following web site, "http://www-cxro.lbl.gov/optical_constants/".

As a conclusion of this section we wish to stress some points concerning the validity of Eq. (3.45). It is important to realise that in a real experiment we never measure the theoretical reflectivity as given by Eq. (3.45) since the incident beam is not necessarily strictly monochromatic, is generally divergent and the detector has a finite acceptance. For any instrument, the effects of the divergence of the x-ray source, of the slit settings or of the angular acceptance of the monochromator and analyser crystals used to collimate the incident and scattered beams must be taken into account. Those effects can be described using a three-dimensional resolution function which is never a Dirac distribution but a three-dimensional function having a certain width (see Chaps. 4 and 7) which precisely depends on the set-up characteristics detailed above. The value of the measured reflectivity can be estimated through the convolution of Eq. (3.45) with the resolution function of the instrument. For measurements made in the incidence plane and under specular conditions, a first effect is that the convolution smears out the $q_z$ dependence of the reflectivity. This can generally be accounted for by convolving $R(q_z)$ with a Gaussian function. Another most important effect of the finite resolution is that beams outside the specular direction are accepted by the detector (in other words, the specular condition $\delta(q_x)\delta(q_y)$ is replaced by a function having a finite width $\Delta q_x \times \Delta q_y$). Then, if the surface to be analysed is rough, the convolution with the resolution function drastically changes the problem because part of the diffuse intensity which arises from the roughness is contained in the resolution volume. It may even happen for very rough surfaces that the diffuse intensity becomes as intense as the specular reflectivity.

---

[2] For this reason, the reflectivity of a flat surface is described as "*specular*", a term which is more normally used to describe the reflection by an ordinary mirror. It seems that Compton [2] was the first to have foreseen the possibility of totally reflecting x-rays in 1923 and that Forster [3] introduced Eq. (3.41). Prins [4] carried out some experiments to illustrate the predictions of this equation in 1928, using an iron mirror. He also used different anode targets to study the influence of the x-ray wavelength on the absorption. Kiessig also made similar experiments in 1931 [5] using a nickel mirror. An account of the historical development of the subject can be found in the pioneering work of L.G. Parrat [6] in 1954 and of Abélès [7]. The fundamental principles are discussed in the textbook by James [8].

**Table 3.1** A few examples of useful data used in reflectivity analysis. The table contains the electron density $\rho_e$, the critical wave vector $q_c$, the parameter $\delta$, the absorption coefficient $\beta$, the structure of the material and its specific mass ($\delta$ and $\beta$ are given at $\lambda = 1.54$ Å). A useful formula for calculating the critical wave-vector transfer is $q_c(\text{Å}^{-1}) = 0.0375\sqrt{\rho_e(e^-/\text{Å}^3)}$, and conversely $\rho_e = 711q_c^2$

| Material | $\rho_{el}$ | $q_c$ | $\delta$ | $\beta$ | Structure | $\rho$ |
|---|---|---|---|---|---|---|
| | $e^-/\text{Å}^3$ | $\text{Å}^{-1}$ | $10^6$ | $10^7$ | | $kg/m^3$ |
| Si | 0.7083 | 0.0316 | 7.44 | 1.75 | Cubic, diamond a = 5.43 Å, Z = 8 | 2330 |
| SiO$_2$ | 0.618 | 0.0294 | 6.5 | 1.7 | | 2200 |
| Ge | 1.425 | 0.0448 | 15.05 | 5 | Cubic, diamond a = 5.658 Å, Z = 8 | 5320 |
| AsGa | 1.317 | 0.0431 | 13.9 | 4.99 | Cubic, diamond a = 5.66 Å, Z = 8 | 5730 |
| Glass crown | 0.728 | 0.0328 | 8.1 | 1.36 | 67.5% SiO$_2$,12% B$_2$O$_3$, 9% Na$_2$O, 9.5% K$_2$O, 2% BaO | 2520 |
| Float glass | 0.726 | 0.0320 | 7.7 | 1.3 | | – |
| Nb | 2.212 | 0.056 | 24.5 | 15.1 | Cubic, bcc a = 3.03 Å, Z = 2 | 8580 |
| Cu | 2.271 | 0.0566 | 24.1 | 5.8 | Cubic, fcc a = 3.61 Å, Z = 4 | 8930 |
| Au | 4.391 | 0.0787 | 46.5 | 49.2 | Cubic, fcc a = 4.078 Å | 19280 |
| Ag | 2.760 | 0.0624 | 29.25 | 28 | Cubic, fcc a = 4.09 Å | 10500 |
| ZrO$_2$ | 1.08 | 0.0395 | 11.8 | | – | – |
| WO$_3$ | 1.723 | 0.0493 | 18.25 | 12 | – | – |
| H$_2$O | 0.334 | 0.0217 | 3.61 | 0.123 | – | 1000 |
| CH$_3$CH$_2$– | 0.32 | 0.0212 | | | – | – |
| –COOH | 0.53 | 0.0273 | | | – | – |
| CCl$_4$ | 0.46 | 0.0254 | | | – | – |
| CH$_3$OH | 0.268 | 0.0194 | | | – | – |
| PS-PMMA | 0.377 | 0.0233 | | | – | – |

When this is occurring, the only way to use Eq. (3.45) is to subtract the diffuse part from the reflected intensity to obtain the true specular reflectivity (see Sect. 4.4 for details).

### 3.1.3 Measuring the Reflectivity

#### 3.1.3.1 Footprint Effect

The size of the beam is an important parameter because x-ray reflectivity measurements usually start below the critical angle of external reflection. Assuming a rectangular beam with dimensions $t_1t_2$ ($t_2$ the dimension parallel to the surface of the sample) and an incident angle $\alpha$, it is straightforward to show that the footprint of the beam on the surface of the sample is

$$F = \frac{t_1}{\sin \alpha} t_2.$$

At the critical angle of silicon ($\alpha \simeq 0.2°$ for E = 8 keV) and for $t_1 = 100\,\mu m$, we find the footprint along the direction of propagation of the beam is about $\frac{t_1}{\alpha} = 30\,mm$. This shows that the sample size must be at least 30 mm long to totally reflect the incident beam at this angle of incidence. This condition is the minimum condition to be fulfilled to observe total external reflection at the critical angle.

### 3.1.3.2 Divergence of the Beam

The measurement of an x-ray reflectivity curve necessitates the use of a well-collimated parallel incident beam. The divergence, $\Delta \alpha$, of the incident beam needs to be small enough to precisely define the incident angle. A similar condition stands for the divergence $\Delta \beta$ of the reflected beam. The incident and exit angular divergences define the angular width, $\omega = \sqrt{\Delta \alpha^2 + \Delta \beta^2}$, of the direct beam. The lower this value the better the resolution. Needless to say that improving the resolution gives a better estimation of some typical features like the Kiessig fringes minima and the critical angle. However, working in a high-resolution mode has the drawback to narrow the angular width of the reflectivity ridge. The better the resolution the more difficult it is to track the reflectivity curve over a wide range of wave-vector transfers.

The classical reflectivity set-up on a tube laboratory source (Philips X-pert reflectometer) gives $\omega = 0.07°$ (full width half maximum). This yields a width of the ridge equal to 0.035 °. A reflectometer must therefore work with positioning angles better than 0.002° (FWHM).

### 3.1.3.3 In Practice

The first step when measuring a reflectivity curve is to precisely define the direct beam. This is made by scanning the detector into the direct beam after having carefully attenuated the direct beam to avoid detector saturation. Let $I_0$ be the direct beam intensity. The second step consists in cutting the direct beam in half. This is made by translating sample into the direct beam. If the sample surface is parallel to the direct the alignment procedure is over. Nevertheless this is never the case and after cutting the direct beam in half, one usually carries out a rotation of the sample (called a sample scan or sometimes an omega scan) about an axis normal to the incident plane (i.e. the plane containing the normal to the sample surface and the direct beam). During such a scan the detector remains at a fixed zero position and the sample rotates. In a sample scan the intensity in the detector has a typical triangular shape that is peaked at the zero position of the sample. One must iteratively halve the direct beam intensity and look for the zero position of the sample. When the sample is well aligned the intensity goes from 0 to $I_0/2$ and exhibits a symmetric triangular shape. This measurement is a bit crude since part of the direct beam can be totally reflected by the sample surface. The alignment can be further refined by performing a sample scan at different positions along the reflectivity ridge (i.e. at non-zero q wave-vector transfers). The peak position in the scan yields an incident angle $\alpha$ that must be reset to $2\theta/2$. A sample scan about an axis parallel to the incident beam is also recommended at synchrotron facilities where the beam extension can be quite small.

### 3.1.4 The Transmission Coefficient

As shown in Eq. (3.30), the amplitude of the transmission coefficient satisfies the relation $1 + r = t$. It is straightforward to show by combining Eqs. (3.30) and (3.41) that the transmitted intensity must be given by

$$T\left(\theta\right) = tt^* = \left| \frac{2\theta}{\theta + \sqrt{\theta^2 - \theta_c^2 - 2i\beta}} \right|^2, \tag{3.46}$$

$$T\left(q_z\right) = tt^* = \left| \frac{2q_z}{q_z + \sqrt{q_z^2 - q_c^2 - \frac{32i\pi^2\beta}{\lambda^2}}} \right|^2. \tag{3.47}$$

The transmitted intensity has a maximum at $\theta = \theta_c$ as shown in Fig. 3.4 which gives the actual variation of the transmitted intensity as a function of the incident angle $\theta$ (or $q_z$) in the case of silicon, germanium and copper samples irradiated with the copper $K\alpha$ radiation. The transmitted intensity is nearly zero at very small angles in the regime of total reflection. It increases strongly at the critical angle and finally levels off towards a limit equal to unity at large angles of incidence. The maximum in the transmission coefficient, which is also a maximum in the field

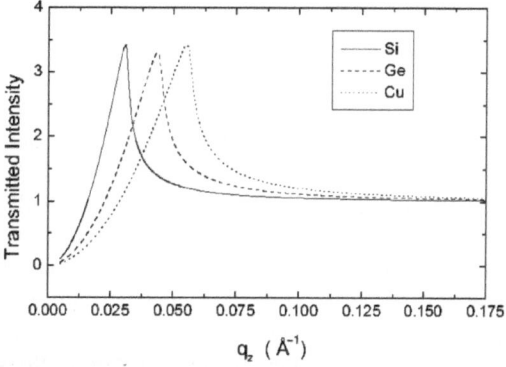

**Fig. 3.4** Transmission coefficient in intensity in different materials, silicon, copper and germanium; the maximum appears at the critical wave-vector transfer of the material

at the interface, is the origin of the so-called Yoneda wings which are observed in transverse off-specular scans (see Sect. 4.3.1).

### 3.1.5 The Penetration Depth

The absorption of a beam in a medium depends on the complex part of the refractive index and limits the penetration of the beam inside the material. The refractive index for x-rays, defined in Eq. (1.6), is $n = 1 - \delta - i\beta$. The amplitude of the electric field polarised along the $y$ direction (($s$) polarisation) and propagating inside the medium of refractive index $n$ is given by

$$E = E_0 e^{i(\omega t - k_0 n \cos \theta_{tr} x + k_0 n \sin \theta_{tr} z)}. \tag{3.48}$$

Since $n \cos \theta_{tr} = \cos \theta$ (the Snell–Descartes' law) and $\sin \theta_{tr} \approx \theta_{tr}$, this equation can be written as

$$E = E_0 e^{+i(\omega t - k_0 \cos \theta x)} e^{i k_0 n \theta_{tr} z}. \tag{3.49}$$

The absorption is governed by the real part of $e^{i k_0 n \theta_{tr} z}$, with

$$n\theta_{tr} = (1 - \delta - i\beta) \sqrt{\theta^2 - 2\delta - 2i\beta} = A + iB. \tag{3.50}$$

The coefficients $A$ and $B$ can be deduced from the above equation and $B$ is given by

$$B(\theta) = -\frac{1}{\sqrt{2}} \sqrt{\sqrt{(\theta^2 - 2\delta)^2 + 4\beta^2} - (\theta^2 - 2\delta)}. \tag{3.51}$$

It follows that the electric field is

$$E = E_0 e^{i(\omega t - k_0 \cos \theta x + k_0 A z)} e^{-k_0 B(\theta) z}. \tag{3.52}$$

Taking the modulus of this electric field shows that the variation of the intensity $I(z)$ with depth into the material is given by

$$I(z) \propto EE^* = I_0 e^{-2k_0 B(\theta) z} \tag{3.53}$$

The absorption coefficient is therefore

$$\mu(\theta) = -2k_0 B(\theta) = \frac{-4\pi B(\theta)}{\lambda}, \tag{3.54}$$

and the penetration depth which is the distance for which the beam is attenuated by $1/e$ is given by

$$z_{1/e}(\theta) = \frac{1}{\mu(\theta)} = \frac{-\lambda}{4\pi B(\theta)} = \frac{1}{2\mathscr{I}m k_{z,1}}. \tag{3.55}$$

**Fig. 3.5** Evolution of the
penetration depth in Si, Ge
and Cu irradiated with the
Kα *line of a copper tube as
a function of the wave-vector
transfer. Note that the figure
is presented as a function of*
$q_z = 4\pi \sin\theta/\lambda$

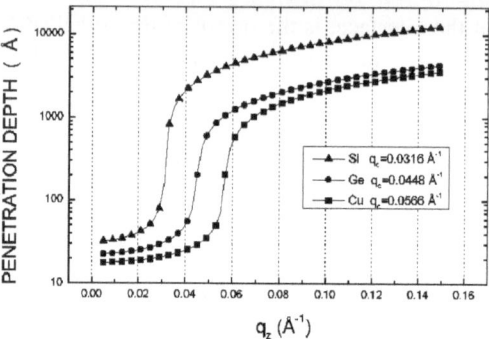

Note that this quantity depends on the incident angle $\theta$ through the value of $B(\theta)$.
In particular, in the limit $\theta \to 0$, neglecting absorption,

$$z_{1/e}(\theta_c) = \frac{\lambda}{4\pi\theta_c}.$$

(3.56)

In addition, the penetration depth is wavelength dependent since $\beta$ depends on
the wavelength. Values of $\beta$ are tabulated in the International Tables for Crystallography, vol. IV [9], or they can also be found at the web site which has already been
referred to, "http://www-cxro.lbl.gov/optical_constants/".

Figure 3.5 shows the variation of the penetration depth as a function of the incident angle in silicon, germanium and copper for the case of CuKα radiation. The
penetration depth remains small, that is, below about 30 Å when $\theta$ is smaller than the
critical angle. This is the property which is exploited in surface diffraction, where
only the first few atomic layers are analysed. The penetration depth increases steeply
at the critical angle and finally slowly grows when $\theta \gg \theta_c$.

## 3.2 X-Ray Reflectivity in Stratified Media

The simple case of a uniform substrate exhibiting a constant electron density was
considered in the previous section. This situation is of course not the most general one. For example, stratified media and multilayers are frequently encountered.
Moreover, interfaces generally cannot be considered as steps, but are rough and
thick. Thick interfaces may be approximated by dividing them into as many slabs of
constant electron density as necessary to describe their (continuous) density profile.
Again, it is not possible in this case to use the Fresnel coefficients directly to calculate the reflectivity. The calculation must be performed by applying the boundary
conditions for the electric and magnetic fields at each of the interfaces between the
slabs of constant electron density. The result is usually presented as the product of
matrices, and multiple reflections are taken into account in the calculation known
as the dynamical theory of reflection. Several excellent descriptions of this kind of
calculation can be found in [10–14].

### 3.2.1 The Matrix Method

Let us consider a plane wave polarised in the direction perpendicular to the plane of incidence ($(s)$ polarisation) and propagating into a stratified medium. The axes are chosen so that the wave is travelling in the $xOz$ plane as shown in Fig. 3.6.

The air is labelled as medium 0 and the strata or layers with different electron densities are identified by $1 \leqslant j \leqslant n$ downwards. In this notation the depth $Z_{j+1}$ marks the interface between the $j$ and $j+1$ layers. The wave travelling through the material will be transmitted and reflected at each interface and the amplitudes of the upwards and downwards travelling waves will be defined as $A^+$ and $A^-$, respectively. The electric field $\mathbf{E}^-$ of the downwards travelling wave in the $j$th stratum, for example, is given by the solution of the Helmholtz's equation,

$$\mathbf{E}^- = A^- e^{+i(\omega t - k_{\text{inx},j}x - k_{\text{inz},j}z)}\hat{\mathbf{e}}_y. \tag{3.57}$$

The following notation will be adopted in the derivation:

$$k_{\text{inx},j} = k_j \cos\theta_j$$
$$k_{\text{inz},j} = -k_j \sin\theta_j = -\sqrt{k_j^2 - k_{\text{inx},j}^2}. \tag{3.58}$$

Note that the value of $k_{\text{inx},j}$ is conserved at each interface since this condition is imposed by the Snell–Descartes' law of refraction. The upwards and downwards travelling waves are obviously superimposed at each interface so that at a depth $z$ from the surface the electric field in medium $j$ is

$$E_j(x,z) = \left(A_j^+ e^{ik_{\text{inz},j}z} + A_j^- e^{-ik_{\text{inz},j}z}\right)e^{+i(\omega t - k_{\text{inx},j}x)}. \tag{3.59}$$

As $k_{\text{inz},j}$ takes a complex value, the magnitude of the upwards and downwards electric fields in layer $j$ will be denoted by

$$U(\pm k_{\text{inz},j},z) = A_j^{\pm} e^{\pm ik_{\text{inz},j}z} \tag{3.60}$$

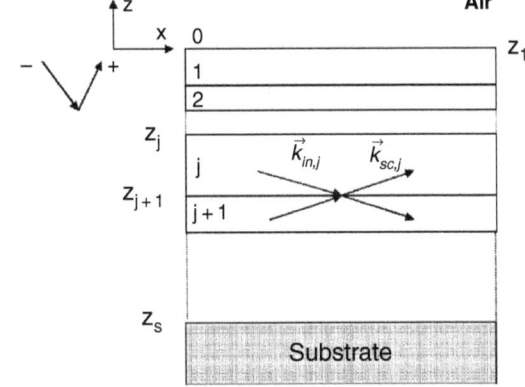

**Fig. 3.6** Illustration of the plane of incidence for a stratified medium. The signs $-$ and $+$ label the direction of propagation of the wave; air is labelled medium 0 and the strata are identified by $1 \leq j \leq n$ layers in which upwards and downwards waves travel

to simplify the notation. In addition, the quantity $k_{\mathrm{inz},j}$ will be replaced by $k_{z,j}$. The condition of continuity of the tangential component of the electric field and the conservation of $k_{x,j}$ at the depth $Z_{j+1}$ of the interface $j, j+1$ lead to

$$U(k_{z,j}, Z_{j+1}) + U(-k_{z,j}, Z_{j+1}) = U(k_{z,j+1}, Z_{j+1}) + U(-k_{z,j+1}, Z_{j+1}). \quad (3.61)$$

It was shown in (3.28) that the tangential component of the magnetic field is continuous when the first derivative of the electric field is conserved. This leads to the equality below, at the $j, j+1$ interface,

$$k_{z,j} \left[ U(k_{z,j}, Z_{j+1}) - U(-k_{z,j}, Z_{j+1}) \right] = k_{z,j+1} \left[ U(k_{z,j+1}, Z_{j+1}) - U(-k_{z,j+1}, Z_{j+1}) \right]. \quad (3.62)$$

The combination of these two equations can be written in a matrix form so that the magnitudes of the electric field in media $j, j+1$ at depth $Z_{j+1}$ must satisfy

$$\begin{bmatrix} U(k_{z,j}, Z_{j+1}) \\ U(-k_{z,j}, Z_{j+1}) \end{bmatrix} = \begin{bmatrix} p_{j,j+1} & m_{j,j+1} \\ m_{j,j+1} & p_{j,j+1} \end{bmatrix} \begin{bmatrix} U(k_{z,j+1}, Z_{j+1}) \\ U(-k_{z,j+1}, Z_{j+1}) \end{bmatrix}, \quad (3.63)$$

with

$$p_{j,j+1} = \frac{k_{z,j} + k_{z,j+1}}{2k_{z,j}},$$

$$m_{j,j+1} = \frac{k_{z,j} - k_{z,j+1}}{2k_{z,j}}. \quad (3.64)$$

The matrix which transforms the magnitudes of the electric field from the medium $j$ to the medium $j+1$ will be called the refraction matrix $\mathcal{R}_{j,j+1}$. It is worth noting that $\mathcal{R}_{j,j+1}$ is not unimodular and has a determinant equal to $k_{z,j+1}/k_{z,j}$. In addition, the amplitude of the electric field within the medium $j$ varies with depth as follows:

$$\begin{bmatrix} U(k_{z,j}, z) \\ U(-k_{z,j}, z) \end{bmatrix} = \begin{bmatrix} e^{-ik_{z,j}h} & 0 \\ 0 & e^{ik_{z,j}h} \end{bmatrix} \begin{bmatrix} U(k_{z,j}, z+h) \\ U(-k_{z,j}, z+h) \end{bmatrix}. \quad (3.65)$$

The matrix which is involved here will be denoted the translation matrix $\mathcal{T}$. The amplitude of the electric field at the surface (depth $Z_1 = 0$) of the layered material in Fig. 3.6 is obtained by multiplying all the refraction and the translation matrices in each layer starting from the substrate (at $z = Z_s$) as follows:

$$\begin{bmatrix} U(k_{z,0}, Z_1) \\ U(-k_{z,0}, Z_1) \end{bmatrix} = \mathcal{R}_{0,1} \mathcal{T}_1 \mathcal{R}_{1,2} \ldots \mathcal{R}_{N,s} \begin{bmatrix} U(k_{z,s}, Z_s) \\ U(-k_{z,s}, Z_s) \end{bmatrix}. \quad (3.66)$$

All the matrices involved in the above product are $2 \times 2$ matrices so that their product which is called the transfer matrix $\mathcal{M}$ is also a $2 \times 2$ matrix. We thus have,

$$\begin{bmatrix} U(k_{z,0}, Z_1) \\ U(-k_{z,0}, Z_1) \end{bmatrix} = \mathcal{M} \begin{bmatrix} U(k_{z,s}, Z_s) \\ U(-k_{z,s}, Z_s) \end{bmatrix} = \begin{bmatrix} M_{11} & M_{12} \\ M_{21} & M_{22} \end{bmatrix} \begin{bmatrix} U(k_{z,s}, Z_s) \\ U(-k_{z,s}, Z_s) \end{bmatrix}. \quad (3.67)$$

The reflection coefficient is defined as the ratio of the reflected electric field to the incident electric field at the surface of the material and is given by

$$r = \frac{U(k_{z,0}, Z_1)}{U(-k_{z,0}, Z_1)} = \frac{M_{11}U(k_{z,s}, Z_s) + M_{12}U(-k_{z,s}, Z_s)}{M_{21}U(k_{z,s}, Z_s) + M_{22}U(-k_{z,s}, Z_s)}. \tag{3.68}$$

It is reasonable to assume that no wave will be reflected back from the substrate if the x-rays penetrate only a few microns, so that

$$U(k_{z,s}, Z_s) = 0, \tag{3.69}$$

and therefore the reflection coefficient is simply defined as

$$r = \frac{M_{12}}{M_{22}}. \tag{3.70}$$

The transmission coefficient is defined as the ratio of the transmitted electric field to the incident electric field

$$t = U(-k_{z,s}, Z_s)/U(-k_{z,0}, Z_1), \tag{3.71}$$

and is given by

$$t = 1/M_{22}. \tag{3.72}$$

This method for the derivation of the reflection and transmission coefficients is known as the matrix technique. It is a general method which is valid for any kind of electromagnetic wave. However, it should be noted that for a plane wave of polarisation $(p)$, the $p_{j,j+1}$ and $m_{j,j+1}$ coefficients must be modified in (3.64) by changing the wave vector $k_{z,j}$ in medium $j$ by $k_{z,j}/n_j^2$. One obtains

$$p_{j,j+1}^{(p)} = \frac{n_{j+1}^2 k_{z,j} + n_j^2 k_{z,j+1}}{2n_{j+1}^2 k_{z,j}}$$

$$m_{j,j+1}^{(p)} = \frac{n_{j+1}^2 k_{z,j} - n_j^2 k_{z,j+1}}{2n_{j+1}^2 k_{z,j}}. \tag{3.73}$$

Let us remark that instead of considering the passage from $U(\pm k_{z,j}, Z_{j+1})$ to $U(\pm k_{z,j+1}, Z_{j+1})$, it is also possible to directly consider the passage from $A_j^{\pm}$ to $A_{j+1}^{\pm}$. The corresponding matrix [12] is

$$\begin{bmatrix} A_j^+ \\ A_j^+ \end{bmatrix} = \begin{bmatrix} p_{j,j+1}e^{i(k_{z,j+1}-k_{z,j})Z_{j+1}} & m_{j,j+1}e^{-i(k_{z,j+1}+k_{z,j})Z_{j+1}} \\ m_{j,j+1}e^{i(k_{z,j+1}+k_{z,j})Z_{j+1}} & p_{j,j+1}e^{-i(k_{z,j+1}-k_{z,j})Z_{j+1}} \end{bmatrix} \begin{bmatrix} A_{j+1}^+ \\ A_{j+1}^+ \end{bmatrix}. \tag{3.74}$$

In this case, it is no longer necessary to introduce the translation matrix. A third alternative consists in defining a matrix which links the electric field and its first derivative at a depth $Z_j$ to the same quantities at a depth $Z_{j+1}$. The matrix is unimodular and is defined for an $(s)$ polarised wave as [11, 13]

$$
\begin{vmatrix} \cos \delta_{j+1} & \dfrac{\sin \delta_{j+1}}{k_{z,j+1}} \\ -k_{z,j+1} \sin \delta_{j+1} & \cos \delta_{j+1} \end{vmatrix}, \tag{3.75}
$$

with $\delta_{j+1} = k_{z,j+1} \left( Z_j - Z_{j+1} \right)$. The application of this general electromagnetic formalism to the case of x-ray reflectivity is discussed in the next section.

### 3.2.2 The Refraction Matrix for X-Ray Radiation

As shown in the previous section (Eqs. (3.63) and (3.64)), the refraction matrix is defined as

$$
\mathscr{R}_{j,j+1} = \begin{bmatrix} p_{j,j+1} & m_{j,j+1} \\ m_{j,j+1} & p_{j,j+1} \end{bmatrix},
$$

with

$$
p_{j,j+1} = \frac{k_{z,j} + k_{z,j+1}}{2k_{z,j}} \quad m_{j,j+1} = \frac{k_{z,j} - k_{z,j+1}}{2k_{z,j}}. \tag{3.76}
$$

Equation (3.58) shows that $k_{z,j}$ is the component of the wave vector normal to the surface and that it is equal to

$$
k_{z,j} = -k_j \sin \theta_j = -\sqrt{k_j^2 - k_{x,j}^2}. \tag{3.77}
$$

$k_{x,j}$ is conserved and is equal to $k \cos \theta$. As a result of this, the $z$ component of $k_0$ in medium $j$ is

$$
k_{z,j} = -\sqrt{k_0^2 n_j^2 - k_0^2 \cos^2 \theta}, \tag{3.78}
$$

where $k_0$ is the wave vector in air. In the limit of small angles and substituting the expression of the refractive index for x-rays, this becomes

$$
k_{z,j} = -k_0 \sqrt{\theta^2 - 2\delta_j - 2i\beta_j}. \tag{3.79}
$$

A similar expression can be obtained for $k_{z,j+1}$ so that the coefficients $p_{j,j+1}$ and $m_{j,j+1}$, and as a consequence the refraction matrix $\mathscr{R}_{j,j+1}$, are entirely determined by the incident angle and by the value of $\delta$ and $\beta$ in each layer.

### 3.2.3 Reflection from a Flat Homogeneous Material

For a homogeneous material, the transfer matrix between air (medium 0) and medium 1 is simply the refraction matrix, which means that $\mathscr{M} = \mathscr{R}_{0,1}$ so that the reflection coefficient $r$ becomes

$$
r = r_{0,1} = \frac{U(k_{z,0},0)}{U(-k_{z,0},0)} = \frac{M_{12}}{M_{22}} = \frac{m_{0,1}}{p_{0,1}} = \frac{k_{z,0} - k_{z,1}}{k_{z,0} + k_{z,1}}, \tag{3.80}
$$

or (neglecting absorption)

$$r = \frac{-k_0\theta + k_0\sqrt{\theta^2 - 2\delta - 2i\beta}}{-k_0\theta - k_0\sqrt{\theta^2 - 2\delta - 2i\beta}} = \frac{\theta - \sqrt{\theta^2 - 2\delta - 2i\beta}}{\theta + \sqrt{\theta^2 - 2\delta - 2i\beta}}. \tag{3.81}$$

Equation (3.81) is of course identical to the one obtained by using the familiar expression for the Fresnel reflectivity (see Eq. (3.41)). Similarly, the transmission coefficient is simply given by

$$t_{0,1} = \frac{U(-k_{z,1},0)}{U(-k_{z,0},0)} = \frac{1}{M_{22}} = \frac{1}{p_{0,1}} = \frac{2k_{z,0}}{k_{z,0} + k_{z,1}}, \tag{3.82}$$

which is the same result as the one obtained earlier in Eq. (3.46). It should be realised that these reflection and transmission coefficients have been derived for an incident wave impinging on the surface of the material with a wave vector $k_{in}$. In some cases (for example in the next chapter when treating the distorted-wave Born approximation), it is important to label these coefficients to indicate which is the incident wave (Fig. 3.7). The detailed notation for the reflection and transmission coefficients will then be $r_{0,1}^{in}$ and $t_{0,1}^{in}$ when $k_{in}$ is concerned and $r_{0,1}^{sc}$ and $t_{0,1}^{sc}$ for the wave vector $k_{sc}$. The explicit expressions for those coefficients are[3]

$$t_{0,1}^{in} = \frac{2k_{inz,0}}{k_{inz,0} + k_{inz,1}}, \tag{3.83}$$

$$t_{0,1}^{sc} = \frac{2k_{scz,0}}{k_{scz,0} + k_{scz,1}}. \tag{3.84}$$

**Fig. 3.7** Definition of the angles for the calculation of $t_{0,1}^{in}$ and $t_{0,1}^{sc}$

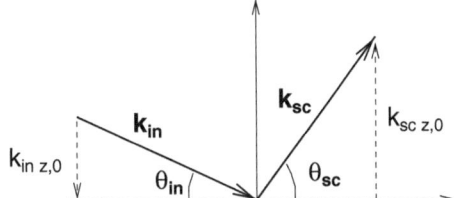

### 3.2.4 A Single Layer on a Substrate

The transfer matrix for the case of a layer of thickness $-h = Z_1 - Z_2$ ($h$ and $k_{z,1}$ are negative) deposited on a substrate is given as

$$\mathcal{R}_{0,1} \mathcal{T}_1 \mathcal{R}_{1,2} = \begin{vmatrix} p_{0,1} & m_{0,1} \\ m_{0,1} & p_{0,1} \end{vmatrix} \begin{vmatrix} e^{-ik_{z,1}h} & 0 \\ 0 & e^{+ik_{z,1}h} \end{vmatrix} \begin{vmatrix} p_{1,2} & m_{1,2} \\ m_{1,2} & p_{1,2} \end{vmatrix}, \tag{3.85}$$

---

[3] Let us also point out that the field in medium 1 associated with a plane wave travelling with a wave vector $\mathbf{k}_{in,1}$ is $E_1(\mathbf{k}_{in,1},\mathbf{r}) = U(\mathbf{k}_{in,1},z)e^{-k_{in,x}x} = E_0 t_{0,1}^{in} e^{-k_{in,x}x}e^{-k_{in,z,1}z}$.

and the reflection coefficient is

$$r = \frac{M_{12}}{M_{22}} = \frac{m_{0,1}p_{1,2}e^{ik_{z,1}h} + m_{1,2}p_{0,1}e^{-ik_{z,1}h}}{m_{0,1}m_{1,2}e^{ik_{z,1}h} + p_{1,2}p_{0,1}e^{-ik_{z,1}h}}. \tag{3.86}$$

Dividing numerator and denominator by $p_{0,1}p_{1,2}$ and introducing the reflection coefficients $r_{i-1,i} = m_{i-1,i}/p_{i-1,i}$ for the two media $i$ and $i-1$, the reflection coefficient of the electric field at the layer is then found to be

$$r = \frac{r_{0,1} + r_{1,2}e^{-2ik_{z,1}h}}{1 + r_{0,1}r_{1,2}e^{-2ik_{z,1}h}}. \tag{3.87}$$

It is worth noting that the denominator of this expression differs from unity by a term which corresponds to multiple reflections in the material, as shown by the product of the two reflection coefficients $r_{01}r_{12}$.
It is also straightforward to determine the transmission coefficient since its value is given by $1/M_{22}$; this yields,

$$t = \frac{t_{0,1}t_{1,2}e^{-ik_{z,1}h}}{1 + r_{0,1}r_{1,2}e^{-2ik_{z,1}h}}. \tag{3.88}$$

In the case when the absorption can be neglected, the reflected intensity is therefore,

$$R = \frac{r_{0,1}^2 + r_{1,2}^2 + 2r_{0,1}r_{1,2}\cos 2k_{z,1}h}{1 + r_{0,1}^2 r_{1,2}^2 + 2r_{0,1}r_{1,2}\cos 2k_{z,1}h}. \tag{3.89}$$

The presence of the cosine terms in Eq. (3.89) indicates clearly that the reflectivity curve will exhibit oscillations in reciprocal space whose period will be defined by the equality

$$2k_{z,1}h \approx q_{z,1}h = 2p\pi, \tag{3.90}$$

or

$$q_{z,1} = \frac{2p\pi}{h}. \tag{3.91}$$

These oscillations are the result of the constructive interference between the waves reflected at interfaces 1 and 2. The difference in path length which separates the two waves is

$$\delta = 2h\sin\theta_1 = p\lambda, \tag{3.92}$$

so that

$$q_{z,1} = \frac{2\pi p}{h}. \tag{3.93}$$

Figure 3.8 which shows the experimental reflectivity of a copolymer deposited onto a silicon substrate provides a good illustration of this type of interference phenomena. The experimental curve is presented in open circles and the calculated one as a solid line. The calculation is made by using the matrix technique in which we use Eq. (3.85) as starting point. The fact that the reflectivity is less than 1 below

**Fig. 3.8** Measured and calculated reflectivities of a thin film of a diblock copolymer PS-PBMA deposited on a silicon wafer

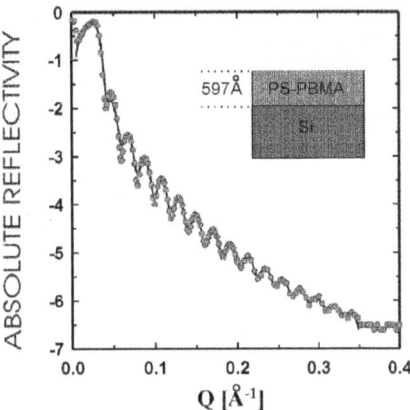

the critical angle is related to a surface effect. At very shallow angles, it frequently happens that the footprint of the beam is larger than the sample surface so that only part of the intensity is reflected. A correction must then be applied to describe this part of the reflectivity curve. The roughness of the interfaces is also included in the calculation as discussed below.

### 3.2.5 Two Layers on a Substrate

The calculation of the reflectivity can also be made by the matrix technique in the case of two layers deposited on a substrate. After multiplying the five matrices of refraction and translation it is possible to express the reflection coefficient as

$$r = \frac{r_{0,1} + r_{1,2}e^{-2ik_{z,1}h_1} + r_{2,s}e^{-2i(k_{z,2}h_2+k_{z,1}h_1)} + r_{0,1}r_{1,2}r_{2,s}e^{-2ik_{z,2}h_2}}{1 + r_{0,1}r_{1,2}e^{-2ik_{z,1}h_1} + r_{1,2}r_{2,s}e^{-2ik_{z,2}h_2} + r_{2,s}r_{0,1}e^{-2i(k_{z,1}h_1+k_{z,2}h_2)}}. \qquad (3.94)$$

The above expression clearly shows that for two layers on a substrate, multiple reflections at each interface appear in the matrix calculation. The phase shifts depend on the path difference calculated in each medium and therefore on the thickness of each layer, and indirectly on the angle of incidence. Examples of this kind are encountered in metallic thin films which tend to oxidise when placed in air. The reflectivity curve of an oxidised niobium thin film deposited on a sapphire substrate [15] is shown in Fig. 3.9. The upper layer obviously corresponds to the niobium oxide. The oxide layer grows as a function of time of exposure to air and reaches a maximum thickness of around 15 Å after a few hours. The reflectivity curve presented in Fig. 3.9 displays a very characteristic shape, which includes the following features:

- Short wavelength oscillations which can be identified with the interferences within the (thick) niobium layer.
- A beating of the oscillations with a longer wavelength in $q$, which comes from the presence of the oxide on top of the niobium layer; this leads to two interfaces at nearly the same altitude from the surface of the sapphire substrate.

**Fig. 3.9** Reflectivity of Nb thin film on sapphire showing the beating of spatial frequency between two comparable thicknesses which are the thickness of the niobium film and the thickness of the entire film (niobium and niobium oxide)

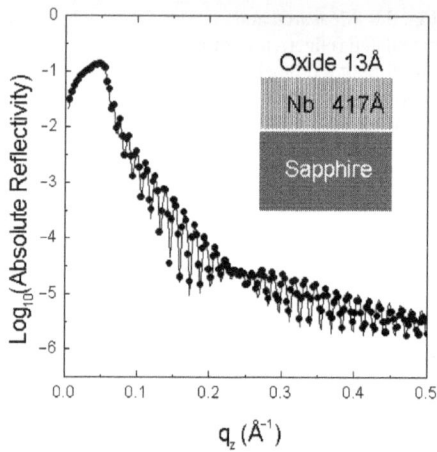

There are similarities between this phenomenon and the characteristic beating of acoustic waves of similar frequency.

## 3.2.6 Organic Multilayers

The reflectivity curve of multilayered systems exhibit additional interesting features due to the repetition of a layered motif. This repetition produces a specific period $\Lambda$ that gives rise to Bragg peaks located every $2\pi/\Lambda$. We show here measurements carried out on a series of different organic film samples prepared on smooth silicon wafer substrates (covered with their native oxide layer), going from a single monolayer of OTS (n-octadecyltrichlorosilane, $CH_3(CH_2)_{17}SiCl_3$) to multilayer OTS/(NTSOH)$x$ films, where NTSOH is OH-terminated NTS, obtained by the oxidation of the terminal ethylenic function of NTS (18-nonadecenyltrichlorosilane, $CH_2 = CH(CH_2)_{17}SiCl_3$), and the number of NTSOH layers, $x$, varies between 1 and 11. Although NTSOH and OTS are similar, both with a silane head group and a long hydrocarbon tail, there are significant differences. NTSOH has a terminal $CH_2OH$ alcohol group whereas OTS (18 carbon atoms) is one carbon atom shorter than NTSOH (19 carbon atoms) and is terminated with a $CH_3$ methyl group. As shown in Fig. 3.10 the reflectivity extends to $0.7\,\text{Å}^{-1}$, over a dynamic range of 9 orders of magnitude. The relatively high overall intensity and the large modulation intensity indicate that both the film and substrate have Angstrom scale roughness. Reflectivity curves of samples with variable $x$ exhibit similarly well-defined Bragg peaks and Kiessig fringes. The Kiessig fringes spacing is inversely proportional to the total film thickness whereas the Bragg peak position is inversely proportional to the layer spacing (more details are available in [16]).

**Fig. 3.10** Calculated (*full line*) and measured (*symbols*) absolute reflectivity curves for the films with $x = 0, 1, 4, 7$ and 11 (for clarity, each curve is offset by $10^2$ with respect to the previous one)

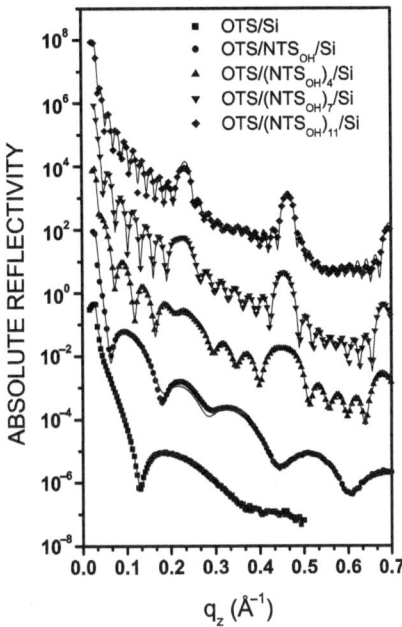

## 3.3 From Dynamical to Kinematical Theory

The full dynamical theory described above is exact but does not clearly show the physics of scattering because numerical calculations are necessary. Sometimes, one can be more interested in an approximated analytical expression. Different approximations can be done [13, 17–19], the simplest one being the Born approximation.[4] We will start from the dynamical expression of the reflected amplitude calculated in the previous section (Eq. (3.89)) for a thin film of thickness $h$ deposited on a substrate

$$r = \frac{r_{0,1} + r_{1,2}\,e^{-2ik_{z,1}h}}{1 + r_{0,1}r_{1,2}\,e^{-2ik_{z,1}h}}, \tag{3.95}$$

and degrade it to obtain approximate expressions. Here the phase shift between the reflected waves on the substrate and the layer denoted by $\varphi = -2k_{z,1}h = q_{z,1}h$ can be written as a function of either $k$ or $q$. The term $r_{0,1}\,r_{1,2}\,e^{i\varphi}$ in this equation represents the effect of multiple reflections in the layer and a first step in the approximation consists in neglecting this term.

This is illustrated in Fig. 3.11 which shows a comparison between the reflectivities calculated for a diblock copolymer film on a silicon wafer with the matrix method taking into account or not the multiple reflections at the interfaces. It can be seen that the two curves are almost identical showing that this approximation

---

[4] This kind of approach was first made by Rayleigh in 1912 in the context of the reflection of electromagnetic waves [17] but has since become known as the Born approximation since Born generalised it to different types of scattering processes.

(a)                                              (b)

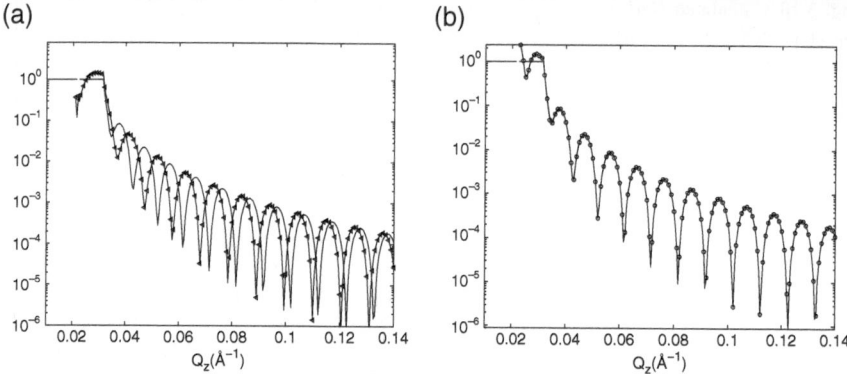

**Fig. 3.11** Comparison between reflectivities calculated with the matrix technique (full line in Figures **a** and **b**) and after neglecting the multiple reflections (*triangles in* (**a**)) and in addition the refraction (*circles in* **b**). Calculations are performed for a diblock copolymer (qc $= 0.022 \,\text{Å}^{-1}$) 600 Å thick on a silicon substrate

is quite good. Under this approximation, the reflection coefficient $r$ for a stratified medium composed of $N$ layers is

$$r = r_{0,1} + r_{1,2}e^{iq_{z,1}d_1} + r_{2,3}e^{i(q_{z,1}d_1 + q_{z,2}d_2)} + \cdots + r_{j,j+1}e^{i\sum_{k=0}^{j} q_{z,k} d_k} + \cdots \quad (3.96)$$

In Eq. (3.96) the ratio $r_{j,j+1}$ of the amplitudes of the reflected to the incident waves at interface $j, j+1$ is

$$r_{j,j+1} = \frac{q_{z,j} - q_{z,j+1}}{q_{z,j} + q_{z,j+1}}, \quad (3.97)$$

with the wave-vector transfer in medium $j$:

$$q_{z,j} = (4\pi/\lambda)\sin\theta_j = \sqrt{q_z^2 - q_{c,j}^2}. \quad (3.98)$$

Finally,

$$R(q_z) = \left| \sum_{j=0}^{n} r_{j,j+1}e^{iq_{z}z_j} \right|^2 \quad \text{with } r_{j,j+1} = \frac{q_{z,j} - q_{z,j+1}}{q_{z,j} + q_{z,j+1}}.$$

A further approximation consists in neglecting the refraction and the absorption in the material in the phase factor in Eq. (3.96):

$$r = \sum_{j=0}^{n} r_{j,j+1}e^{iq_z \sum_{m=0}^{j} d_m}. \quad (3.99)$$

In this case the approximation is more drastic and this can be seen in Fig. 3.11(b) showing that the region of the curve just after the critical angle is most affected, and in particular the positions of the interference fringes.

A final approximation consists in assuming that the wave vector $q_z$ does not change significantly from one medium to the next so that the sum in the denominator of $r_{j,j+1}$ may be simplified:

$$r_{j,j+1} = \frac{q_{z,j}^2 - q_{z,j+1}^2}{(q_{z,j} + q_{z,j+1})^2} = \frac{q_{c,j+1}^2 - q_{c,j}^2}{4q_z^2} = \frac{4\pi r_e(\rho_{j+1} - \rho_j)}{q_z^2}, \tag{3.100}$$

with $q_{c,j} = \sqrt{16\pi r_e \rho_j}$ in which $r_e$ stands for the classical radius of the electron. These approximations lead to the following expression for the reflection coefficient,

$$r = 4\pi r_e \sum_{j=1}^{n} \frac{(\rho_{j+1} - \rho_j)}{q_z^2} e^{iq_z \sum_{m=0}^{j} d_m}. \tag{3.101}$$

If the origin of the $z$ axis is chosen to be at the upper surface (medium 0 at a depth of $Z_1 = 0$), then the sum over $d_m$ in the phase factor can be replaced by the depth $Z_{j+1}$ of the interface $j, j+1$ and the equation becomes

$$r = 4\pi r_e \sum_{j=1}^{n} \frac{(\rho_{j+1} - \rho_j)}{q_z^2} e^{iq_z Z_{j+1}}. \tag{3.102}$$

Finally, if we consider that the material is made of an infinite number of thin layers, the sum may then be transformed into an integral over $z$, and the reflection coefficient $r$ has the form

$$r = \frac{4\pi r_e}{q_z^2} \int_{-\infty}^{+\infty} \frac{d\rho(z)}{dz} e^{iq_z z} dz. \tag{3.103}$$

A very useful, less-drastic approximation is obtained by replacing $(4\pi r_e \rho_s)^2 / q_z^4$ by $R_F(q_z)$ in Eq. (3.103). Under this approximation the reflectivity can be written as [19]

$$R(q_z) = r.r^* = R_F(q_z) \left| \frac{1}{\rho_s} \int_{-\infty}^{+\infty} \frac{d\rho(z)}{dz} e^{iq_z z} dz \right|^2. \tag{3.104}$$

The above expression for $R(q_z)$ is not rigorous but it has the advantage of being easily handled in analytical calculations. In addition, if the Wiener–Khintchine theorem is applied to this result, we find

$$\frac{R(q_z)}{R_F(q_z)} = \frac{1}{\rho_s^2} TF \left[ \rho'(z) \otimes \rho'(z) \right], \tag{3.105}$$

so that the data inversion gives the autocorrelation function of the first derivative of the electron density [20] or the Patterson function [21, 22].

Figure 3.12 illustrates the main features of this data inversion. It is based on a calculation with a model structure [19] for a sample consisting of two layers, a

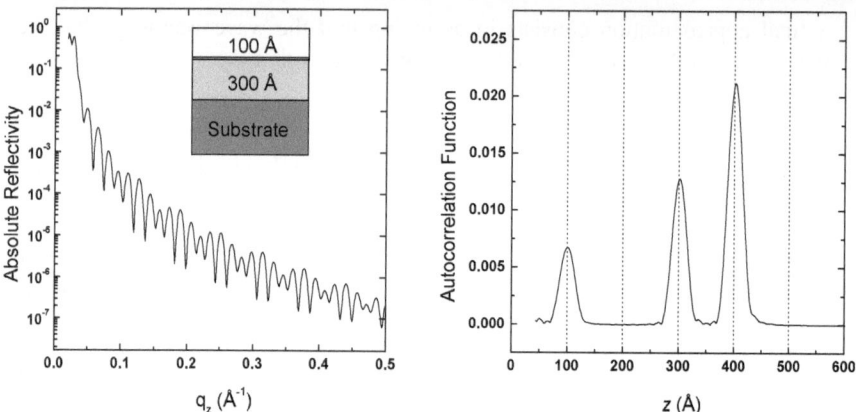

**Fig. 3.12** Calculated reflectivity of a two-layer system and its Fourier transform after division of the data by the Fresnel reflectivity of the substrate. In the calculation the two layers of different electron densities are 300 and 100 Å thick. The Fourier transform immediately gives the thickness of each layer without relying on any model. One can also note the expected peak at $z = 400$ Å in the autocorrelation function

lower one of 300 Å and an upper one of 100 Å on a substrate. The left-hand side diagram gives the calculated reflectivity curve which shows a feature similar to the "beating" effect seen in Fig. 3.9, arising here because of the similar thicknesses of the two layers. The right-hand side diagram gives the autocorrelation function, which has intense peaks at the interfaces where the derivative of the electron density is maximum. In an ideally flat sample these peaks would be delta functions, but for a real case their width depends on factors such as the roughness, the degree of interdiffusion at the interfaces and the instrumental resolution.

Equation (3.104) is a good starting point to introduce a last formulation for the reflected intensity. Starting from Eq. (3.104)

$$R(q_z) = rr^* = R_F(q_z) \left| \frac{1}{\rho_s} \int_{-\infty}^{+\infty} \frac{d\rho(z)}{dz} e^{iq_z z} dz \right|^2 = \frac{(4\pi r_e \rho_s)^2}{q_z^4} \left| \frac{1}{\rho_s} \int_{-\infty}^{+\infty} \frac{d\rho(z)}{dz} e^{iq_z z} dz \right|^2$$

$$(3.106)$$

and using the general relation between the Fourier transform of a function and the Fourier transform of its first derivative, we have

$$R(q_z) = \frac{(4\pi r_e)^2}{q_z^2} \left| \int_{-\infty}^{+\infty} \rho(z) e^{iq_z z} dz \right|^2$$

$$= \frac{(4\pi r_e)^2}{q_z^2} \iint \rho(z)\rho(z') e^{iq_z(z-z')} dz dz'.$$

$$(3.107)$$

To summarise, it has been shown in this section that the kinematic theory is derived from the dynamical theory by three approximations:

(1) No multiple reflections at the interfaces
(2) The effects of refraction can be neglected
(3) The reflection coefficient at each interface is proportional to the difference of electron density

All the expressions discussed above have been derived under the assumption of ideally flat interfaces in the samples. In such a case, the lateral position of reflecting points at the interfaces is unimportant, since all of the points are at the same depth from the surface. It is thus implicit that the intensity is localised along the specular direction. This means that the expression above can be considered as valid over the entire reciprocal space after multiplication by the delta functions $\delta q_x$ and $\delta q_y$ which characterise the specular character of the reflected intensity. Therefore, the last equation of (3.107) for example may as well be written as

$$R(\mathbf{q}) = \frac{(4\pi r_e)^2}{q_z^2} \iint \rho(z)\rho(z')e^{iq_z(z-z')}dzdz'\delta q_x\delta q_y. \qquad (3.108)$$

Note that

$$R(\mathbf{q}) = \frac{(4\pi r_e)^2}{q_z^2} \iint \rho(z)\rho(z')e^{iq_z(z-z')}dzdz'\delta q_x\delta q_y \qquad (3.109)$$

is the well-known Born approximation (or kinematical) expression for x-ray scattering. It can be recovered from the integration of the scattering cross-section

$$\frac{d\sigma}{d\Omega} = \frac{4r_e^2}{L_xL_yq_z^2} \iint drdr'\rho(\mathbf{r})\rho(\mathbf{r}')e^{i\mathbf{q}\cdot(\mathbf{r}-\mathbf{r}')}, \qquad (3.110)$$

as shown in Sect. 4.2.2.[5]

## 3.4 Influence of the Roughness on the Matrix Coefficients

It was shown in Chap. 2 that scattering from a rough surface/interface can be separated into two contributions, coherent and incoherent scattering. In this chapter, we are only interested in the specular intensity, i.e. the coherent intensity given by the average value of the field. We give here a simple method to take roughness into account in the reflection by a rough multilayer using the matrix method. We rely on a more complete and rigorous treatment of the case of a single interface given in Appendix 1.A to this chapter. In this appendix, it is shown that for roughnesses with in-plane characteristic lengths smaller than the extinction length $\approx 1\,\mu$m for x-rays, introduced in Chap. 1,

$$r_{0,1}^{\text{rough}} = r_{0,1}^{\text{flat}}e^{-2k_{z,0}k_{z,1}\sigma_1^2}. \qquad (3.111)$$

---

[5] We may notice that if applied to a flat surface this expression would lead to $R(\mathbf{q}) = q_c^4\delta(q_x)\delta(q_y)/16q_z^4$.

The exponential in Eq. (3.111) is known as the Croce–Névot factor [23]. We now apply the method of Sect. 3.A.3 to the matrix method. Starting from Eq. (3.74)

$$
\begin{bmatrix} A_j^+ \\ A_j^- \end{bmatrix} = \begin{bmatrix} p_{j,j+1}e^{i\left(k_{z,j+1}-k_{z,j}\right)Z_{j+1}} & m_{j,j+1}e^{-i\left(k_{z,j+1}+k_{z,j}\right)Z_{j+1}} \\ m_{j,j+1}e^{i\left(k_{z,j+1}+k_{z,j}\right)Z_{j+1}} & p_{j,j+1}e^{-i\left(k_{z,j+1}-k_{z,j}\right)Z_{j+1}} \end{bmatrix} \begin{bmatrix} A_{j+1}^+ \\ A_{j+1}^- \end{bmatrix}, \quad (3.112)
$$

which links the amplitudes of the electric field in two adjacent layers, we assume that the position of the interface $Z_{j+1}$ between the $j$ and $j+1$ layers fluctuates vertically as a function of the lateral position because of the interface roughness. Following a method proposed by Tolan [24], we replace the quantity $Z_{j+1}$ by $Z_{j+1} + z_{j+1}(x,y)$ in the above matrix and we take the average value of the matrix over the whole area coherently illuminated by the incident x-ray beam (in the spirit of Sect. 3.A.3, this amounts to averaging the phase relationship between the fields above and below the interface). This leads to (as shown in Appendix 1.A, such expressions are only valid at first order in $\langle z_j^2 \rangle$)

$$
\left\langle \begin{bmatrix} A_j^+ \\ A_j^- \end{bmatrix} \right\rangle = \left\langle \begin{bmatrix} p_{j,j+1}e^{i(k_{z,j+1}-k_{z,j})Z_{j+1}}e^{i(k_{z,j+1}-k_{z,j})z_{j+1}(x,y)} \\ m_{j,j+1}e^{i(k_{z,j+1}+k_{z,j})Z_{j+1}}e^{i(k_{z,j+1}+k_{z,j})z_{j+1}(x,y)} \end{bmatrix} \cdots \right.
$$

$$
\left. \begin{matrix} m_{j,j+1}e^{-i(k_{z,j+1}+k_{z,j})Z_{j+1}}e^{-i(k_{z,j+1}+k_{z,j})z_{j+1}(x,y)} \\ p_{j,j+1}e^{-i(k_{z,j+1}-k_{z,j})Z_{j+1}}e^{-i(k_{z,j+1}-k_{z,j})z_j(x,y)} \end{matrix} \begin{bmatrix} A_{j+1}^+ \\ A_{j+1}^- \end{bmatrix} \right\rangle. \quad (3.113)
$$

For Gaussian statistics, or at lowest order in $\sigma_j^2$, we have, assuming the independence of the different interface roughnesses:

$$
\left\langle \begin{bmatrix} A_j^+ \\ A_j^- \end{bmatrix} \right\rangle = \begin{bmatrix} p_{j,j+1}e^{i(k_{z,j+1}-k_{z,j})Z_{j+1}}e^{-(k_{z,j+1}-k_{z,j})^2\sigma_{j+1}^2/2} \\ m_{j,j+1}e^{i(k_{z,j+1}+k_{z,j})Z_{j+1}}e^{-(k_{z,j+1}+k_{z,j})^2\sigma_{j+1}^2/2} \end{bmatrix}
$$

$$
\begin{matrix} m_{j,j+1}e^{-i(k_{z,j+1}+k_{z,j})Z_{j+1}}e^{-(k_{z,j+1}+k_{z,j})^2\sigma_{j+1}^2/2} \\ p_{j,j+1}e^{-i(k_{z,j+1}-k_{z,j+1})Z_{j+1}}e^{-(k_{z,j+1}-k_{z,j})^2\sigma_{j+1}^2/2} \end{matrix} \left\langle \begin{bmatrix} A_{j+1}^+ \\ A_{j+1}^- \end{bmatrix} \right\rangle. \quad (3.114)
$$

The influence of the interface roughness is apparent from this result. The coefficients $m_{j,j+1}$ and $p_{j,j+1}$ are, respectively, reduced by the factors $e^{-\left(k_{z,j+1}+k_{z,j}\right)^2\sigma_{j+1}^2/2}$ and $e^{-\left(k_{z,j+1}-k_{z,j}\right)^2\sigma_{j+1}^2/2}$. It was shown in the previous section that the ratio $m_{j,j+1}/p_{j,j+1}$ is the relevant quantity in the expression of the reflected intensity. This ratio which is the Fresnel coefficient of reflection at the altitude $Z_{j+1}$ is therefore reduced by the amount

$$
\frac{r_{j,j+1}^{\text{rough}}}{r_{j,j+1}^{\text{flat}}} = e^{-2k_{z,j+1}k_{z,j}\sigma_{j+1}^2} = e^{-q_{z,j+1}q_{z,j}\sigma_{j+1}^2/2} \quad (3.115)
$$

in the presence of interface roughness. In the particular case where the Born approximation holds ($k_{z,j} = k_{z,j+1} = (1/2)q_z$), the Fresnel coefficient is reduced by the amount

**Fig. 3.13** Influence of roughness on the specular reflectivity of a 600 Å thin layer deposited on a substrate

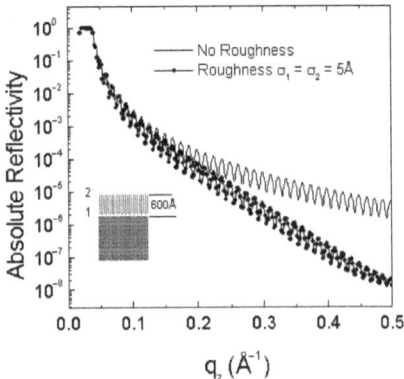

$$\frac{r_{j,j+1}^{\text{rough}}}{r_{j,j+1}^{\text{flat}}} = e^{-2k_{z,j+1}k_{z,j}\sigma_{j+1}^2} = e^{-q_z^2\sigma_{j+1}^2/2}, \tag{3.116}$$

which is the Debye–Waller factor.

We present in Fig. 3.13 how the introduction of the roughness at the interfaces modifies the reflectivity curve. In particular, this figure shows that the reflectivity curve falls faster for rough interfaces and that the amplitude of the fringes is significantly reduced at high wave-vector transfers.

As a conclusion we have shown in this chapter that the calculation of the reflectivity can be properly handled by the matrix technique. This technique is the most widely used in the calculation of the specular reflectivity for the reason that it is simple and exact. However, the main drawback of this technique is that it is only valid in specular conditions, which is an important restriction. Incoherent scattering is discussed in the next chapter, using in particular the matrix formalism described above.

## 3.A Appendix: The Treatment of Roughness in Specular Reflectivity

François de Bergevin, Jean Daillant, Alain Gibaud and Anne Sentenac

The aim of this appendix is to give an overview of the different methods which can be used to take roughness into account in specular reflectivity. We first present the second-order Rayleigh calculation for a sinusoidal grating in order to introduce the main ideas. Then, we discuss the distorted-wave Born approximation (DWBA) results (see Chap. 4 for a presentation of this approximation). Finally, we shortly discuss a simple method that allows one to retrieve the Debye–Waller and Croce–Névot factors which are the limiting laws for, respectively, large and small in-plane correlation lengths. We consider scalar waves in all this appendix.

## 3.A.1 Second-Order Rayleigh Calculation for a Sinusoidal Grating

Let us consider the problem of the reflection by a rough interface (here simplified as a one-dimensional sinusoidal grating of period $\Lambda$) separating two media and illuminated by a plane wave $e^{i(\omega t - k_{inx}x - k_{inz}z)}$. The Rayleigh method [25, 26] consists in expanding the fields in both media as sets of plane waves and in writing the boundary conditions for the field and its first derivative. In order to write these boundary conditions, one has to calculate the values of the field and of its first derivative on the surface, as a series of terms like

$$a_\eta \exp -i\left(k_{\eta,x}x + k_{\eta,z}z(x)\right),$$

where $\eta$ refers to both the medium (above or below the interface) and to the plane wave in the expansion (in particular, the component of its wave vector parallel to the surface describing the scattering order). One then expands

$$a_\eta \exp -i\left(k_{\eta,x}x + k_{\eta,z}z(x)\right) \approx a_\eta \exp -i\left(k_{\eta,x}x\right)\left[1 - ik_{\eta,z}z(x) - \frac{1}{2}k_{\eta,z}^2 z^2(x) + \cdots\right].$$

$$(3.A1)$$

Since $z(x)$ can be expressed as a sum of two exponentials $(z_0/2)\exp(\pm 2i\pi x/\Lambda)$ (in the general case this would be a particular term in the Fourier expansion of the roughness), the expressions in the boundary conditions consist of sums of exponentials in $x$. For the boundary conditions to be satisfied for all $x$, it is necessary and sufficient that they are satisfied for each of these exponentials separately. We now have a series of equations, each corresponding to a scattering order:

$$k_{inx}, k_{inx} \pm 2\pi/\Lambda, \ldots.$$

We define in medium 0 or 1

$$k_{(0,1)z}^{\pm 1} = \sqrt{k_{(0,1)}^2 - \left(k_{inx} \pm \frac{2\pi}{\Lambda}\right)^2},$$

and similarly

$$k_0 \cos\theta_{\pm 1} = k_0 \cos\theta_{in} \pm 2\pi/\Lambda.$$

A series in $z_0$ appears in each equation, and the system will be solved perturbatively at each order. At zeroth order we get the Fresnel coefficients. At first order in $z_0$ (in amplitude), we get for the intensities in the $\pm 1$ scattering orders [27]

$$I_{\pm 1}^{(1)} = I_0 z_0^2 k_{0,z} k_{0,z}^{\pm 1} \sqrt{R_F(\theta_{in})R_F(\theta_{\pm 1})}, \qquad (3.A2)$$

where $R_F(\theta_{in})$ and $R_F(\theta_{\pm 1})$ are the Fresnel reflection coefficients in intensity for the angles $\theta_{in}$ and $\theta_{\pm 1}$, respectively.

At second order (in amplitude) in $z_0$, we get in the specular

$$I_0^{(2)} = -I_0 z_0^2 k_{0,z} R_F(\theta_{in}) \mathcal{R}e \left( 2k_{1,z} + k_{0,z}^{+1} - k_{1,z}^{+1} + k_{0,z}^{-1} - k_{1,z}^{-1} \right). \qquad (3.A3)$$

We now try to find the change in reflectivity coefficient in the limiting cases of large and small $\Lambda$ values.

- *Large $\Lambda$ values* For large $\Lambda$ values, the diffracted orders in both media get close to the specular and transmitted beams:

$$k_{0,z}^{\pm 1} \approx k_{0,z}, \ k_{1,z}^{\pm 1} \approx k_{1,z}.$$

Then

$$R(\theta_{in}) \approx R_F(\theta_{in}) + I_0^{(2)}/I_0 \approx R_F(\theta_{in})(1 - 2z_0^2 k_{0,z}^2).$$

Since one has $\langle z^2 \rangle = z_0^2/2$,

$$R(\theta_{in}) \approx R_F(\theta_{in})(1 - 4k_{0,z}^2 \langle z^2 \rangle)$$

which is the first-order expansion of the Debye–Waller factor in $\langle z^2 \rangle$.
- *Small $\Lambda$ values* One has

$$k_{0,z}^{\pm 1} - k_{1,z}^{\pm 1} = \frac{k_c^2}{k_{0,z}^{\pm 1} + k_{1,z}^{\pm 1}}, \qquad (3.A4)$$

where $k_c = k_0 \sqrt{1 - n^2}$ is the critical wave vector. For small $\Lambda$ values, $k_{0,z}^{\pm 1}, k_{1,z}^{\pm 1} \gg k_c$ and therefore, using Eq. (3.A4), $k_{0,z}^{\pm 1} - k_{1,z}^{\pm 1} \ll k_c$. Since $k_{1,z}$ is never much smaller than $k_c$, it is the only term that survives in the sum in Eq. (3.A3). Therefore,

$$R(\theta_{in}) \approx R_F(\theta_{in})(1 - 4\mathcal{R}e(k_{0,z}k_{1,z}) \langle z^2 \rangle),$$

which is the first-order expansion of the Croce–Névot factor [23] in $\langle z^2 \rangle$.

### 3.A.2 The Treatment of Roughness in Specular Reflectivity Within the DWBA

The issue of the modification of the specular intensity due to surface scattering has been considered within the distorted-wave Born approximation, in particular in [28, 29]. The results of [28, 29] agree with the Rayleigh treatment given in the previous section. It is nevertheless interesting to note that contrary to what is sometimes assumed, the specular intensity can be affected in the first-order DWBA. This is because the basis for this approximation includes both the reflected and transmitted fields. It is therefore possible that *single* scattering events transfer energy from one field to the other (in fact, energy would be conserved at this level of approximation for the sum of the reflected and transmitted fields, see Appendix 1.A). In particular, the first-order result of the DWBA Eq. (4.47) or [28] yields the Croce–Névot factor at first order in $\langle z^2 \rangle$.

*Exercise:* Show this.

The second-order DWBA [29] shows, as did the Rayleigh calculation discussed above, that the Debye–Waller factor is obtained for large $\Lambda$ values whereas the Croce–Névot factor is obtained for small $\Lambda$ values.

### 3.A.3 Simple Derivation of the Debye–Waller and Croce–Névot Factors

The accuracy of approximated expressions for the reflectivity coefficient mainly relies on the quality of the approximations made on the local value of the electric field at the interface. Let us consider the two limiting cases of roughnesses with very small and very large in-plane characteristic length scales. If the characteristic length scale of the roughness is much larger than the extinction length (we have a slowly varying interface height), the field can be written locally for the well-defined interface at a scale smaller than the roughness characteristic scale (this is the so-called tangent plane approximation):

$$E_j(x,z) = \left( A_j^+ e^{ik_{j,z}z} + A_j^- e^{-ik_{j,z}z} \right) e^{i\omega t - \mathbf{k}_i \text{in} \|.\mathbf{r}_\|}. \qquad (3.A5)$$

The field will be reflected at different heights depending on $x$, and the reflection coefficient is

$$r^{\text{rough}} = \frac{\langle A_0^+ \rangle_x}{A_0^-},$$

where the average value is taken over the surface. Writing the boundary conditions, one obtains with the notations of Chap. 3 for a surface located at $z$

$$\begin{cases} A_0^+ e^{ik_{z,0}z} + A_0^- e^{-ik_{z,0}z} & = A_1^- e^{-k_{z,1}z} \\ k_{z,0} A_0^+ e^{ik_{z,0}z} - k_{z,0} A_0^- e^{-ik_{z,0}z} & = -k_{z,1} A_1^- e^{-k_{z,1}z}. \end{cases} \qquad (3.A6)$$

One obtains

$$r^{\text{rough}} = \frac{\langle A_0^+ \rangle_x}{A_0^-} = r_{0,1} \langle e^{2ik_{0,z}z} \rangle = r_{0,1} e^{-2k_{0,z}\langle z^2 \rangle},$$

which is the Debye–Waller factor as expected. We obtain this factor because the roughness characteristic length is large enough for the incident and reflected fields to have a precise phase relationship.

We now assume the characteristic length to be much smaller than the extinction length, which is on the order of $1\,\mu\text{m}$. Then, the electric field is not perturbed at the roughness scale (in other words, there are no short-scale correlations between the field and the roughness). There is only an overall perturbation of the electric field which can be written to a good approximation as the combination of upwards and downwards propagating plane waves, whose amplitude will however depend on the roughness:

$$E_j(x,z) = \left( A_{j,\text{eff}}^+ e^{ik_{j,z}z} + A_{j,\text{eff}}^- e^{-ik_{j,z}z} \right) e^{i\omega t - \mathbf{k}_i \text{in} \| . \mathbf{r}_\|}, \qquad (3.A7)$$

where the $A_{j,\text{eff}}$ ($j = 0,1$) are unknown effective amplitudes for the rough interface. The reflection coefficient is defined as

$$r^{\text{rough}} = \frac{A_{0,\text{eff}}^+}{A_{0,\text{eff}}^-}.$$

Now, we assume that the phase relationships between the field above and below the interface are only valid **on average** because the field "does not see" the local roughness (this is of course not a rigorous argument, the justification for Eq. (3.A8) is the calculations given in the two previous sections):

$$\begin{cases} 2k_{0,z}A_{0,\text{eff}}^+ = (k_{0,z} - k_{1,z})A_{1,\text{eff}}^- \langle e^{-i(k_{0,z}+k_{z,1})z} \rangle \\ 2k_{0,z}A_{0,\text{eff}}^- = (k_{0,z} + k_{1,z})A_{1,\text{eff}}^- \langle e^{i(k_{0,z}-k_{z,1})z} \rangle. \end{cases} \qquad (3.A8)$$

Equation (3.A8) can be obtained from Eq. (3.A6) or directly using the matrix method Eq. (3.74). Then

$$r^{\text{rough}} = \frac{A_{0,\text{eff}}^+}{A_{0,\text{eff}}^-} = r^{\text{flat}} e^{-2k_{z,0}k_{z,1}\langle z^2 \rangle}, \qquad (3.A9)$$

one obtains the Croce–Névot factor. Note that in this case, the transition layer method would give an equally good result. Note also that the method could be applied to the averaging of transfer matrices, as it is done in Chaps. 3 and 6.

## 3.B Appendix: Inversion of Reflectivity Data

François Rieutord

### 3.B.1 Introduction

Inversion of scattering data is a – if not *the* – general problem of crystallography. It amounts generally to the problem of phase determination. In a scattering experiment (x-ray or neutron), we have access to the intensity of the scattered radiation only, not to its phase. Techniques exist that provide also the phase (e.g. x-ray holography) but they are not (yet) standard techniques. Having lost the phase data, one cannot directly reconstruct the structure of the scattering object from the scattering measurements. Looking for phase determination essentially amounts to a better understanding of the origin of the scattered signals within simple or complex systems. Paradoxically, one problem of x-ray or neutron reflectivity is the simplicity of an exact calculation of the reflectivity (i.e. taking into account multiple scattering or absorption). One just has to split the electron density profile into a series of boxes and perform the product of matrices describing light propagation through these boxes [30–33]. The formalisms, derived from multilayer optics, are very powerful, easy to set up on a computer and include in standard parameter refinements routines [34]. As a consequence, the temptation is strong to use such formalisms in automatic fitting routines of electron density profiles. However, these calculations often hinder the origin of scattering and do not allow one to estimate the confidence degree one can have in a given solution. Moreover, the splitting of a profile into a stack of boxes (layers) is often not the best way to describe the main features of a reflection curve as the different features of a curve are not equally sensitive to the different profile parameters. Our approach is here to provide alternative descriptions of profiles and analyse in detail the structure scattering data relation, be it with approximations. In the following we shall be mainly concerned with the so-called kinematical approximation (single scattering and low absorption, i.e. use of Fourier transforms).

Mathematically, the scattered amplitude reads, in the single scattering approximation, as a Fourier transform of the structure:

$$I(\mathbf{q}) = |a(\mathbf{q})|^2 = \left| \int d\mathbf{r} e^{i\mathbf{q}\mathbf{r}} \rho(\mathbf{r}) \right|^2. \qquad (3.B1)$$

The reflection geometry is only a special scattering geometry where the problem is rather simpler being one dimensional:

$$\frac{R(q_z)}{R_F(q_z)} = \left| \int dz e^{iq_z z} \frac{1}{\rho_{-\infty}} \frac{\partial \rho(z)}{\partial z} \right|^2, \qquad (3.B2)$$

where $R_F(q_z)$ is the Fresnel reflectivity of the substrate with electron density $\rho_{-\infty}$.

In the following, we shall review different means that can be worked out to recover the structure, keeping in mind that we shall always need additional information to the strict scattering intensity to help solve part or all of the missing information due to the lack of phase information. We shall stay within the kinematical approximation where the reflectivity is proportional to the FT of the derivative of the index profile. (Note: In principle we could think of using the part of the reflectivity close to total external reflection to separate between two profiles having the same Fourier transform (and hence the same reflectivity at large angles). Even though accurate measurements of $\theta_c$ are not always straightforward to perform, this part was sometimes used in the past to raise degenerescence between two solutions.)

## 3.B.2 A Few Examples

In this section we shall give a few examples proving that strict inversion of reflection data is impossible, showing that two dissimilar profiles can yield the same reflection curve.

The first case (Fig. 3.14) would correspond to a thin film deposited on a substrate and bound by rough interfaces. The figure shows that the reflection curves are identical, although the profiles are very different. If one plots the derivative of the index profiles, the reason of the similarity immediately shows up: the two profiles are symmetrical to each other and the two scattered amplitudes have only different phase differences. Indeed, we write the profile derivative as

$$\rho'(z) = \Delta\rho_1 \exp\left(-\frac{z^2}{2\sigma_1^2}\right) + \Delta\rho_2 \exp\left(-\frac{(z-d)^2}{2\sigma_2^2}\right), \tag{3.B3}$$

with a reflected amplitude equal to

$$a(\mathbf{q}) \propto \Delta\rho_1 \exp(-q^2\sigma_1^2) + \Delta\rho_2 \exp(-q^2\sigma_2^2)\exp(iqd). \tag{3.B4}$$

The complex amplitude reads in one case

$$a_1(q) = A_1(q) + A_2(q)\exp(iqd), \tag{3.B5}$$

and in the other

$$a_2(q) = A_2(q) + A_1(q)\exp(iqd), \tag{3.B6}$$

with $A_1(q)$ and $A_2(q)$ real. One can write

$$a_1(q) = F(q)\exp(i\Phi(q)) \tag{3.B7}$$

$$a_2(q) = \overline{a_1(q)}\exp(iqd) = F(q)\exp(-i\Phi(q))\exp(iqd). \tag{3.B8}$$

The two complex amplitudes differ only by a phase term.

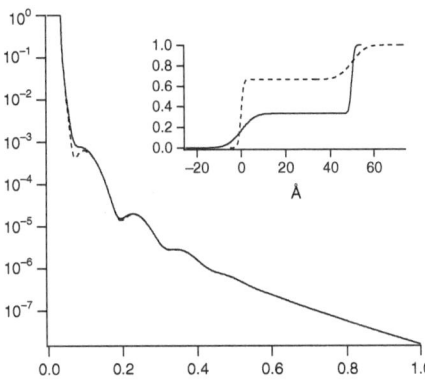

**Fig. 3.14** Two profiles giving the same reflectivity curve

**Fig. 3.15** Two profiles giving
the same reflection with
interfaces at different
locations

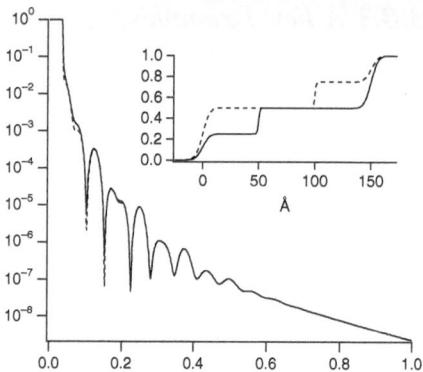

In the same way, we can imagine more complex situations and profiles, where the same type of symmetries are present ending with different interface positions (see Fig. 3.15 for example).

In a still more general way, we see how to generate, from a given profile, other profiles with the same reflection curves. If $a(q)$ is the complex amplitude scatted by profile $\rho(z)$,

$$a(q) \propto \int_{-\infty}^{\infty} \rho'(z) \exp(iqz) dz. \qquad (3.B9)$$

[*Note: as $\rho(z)$ is a physical quantity, we have $\overline{a(q)} = a(-q)$*]. We get by inverse Fourier transform,

$$\rho'(z) \propto \int_{-\infty}^{\infty} a(q) \exp(iqz) dq. \qquad (3.B10)$$

We see that any profile obtained replacing $a(q)$ by another complex function with the same module will give the same reflectivity. We just need to add any phase $\phi(q)$ to $a(q)$. The only constraint on $\phi(q)$ comes from the condition $\overline{a(q)} = a(-q)$ that requests that $\phi(q)$ be odd ($\phi(-q) = -\phi(q)$). On Fig. 3.16, we see two profiles generated this way, adding a phase of type $\phi(q) = q^3 \sigma^3$ to the FT of the first profile and back transforming again. As the choice of $\phi(q)$ is open, we can generate this way a wide variety of different profiles yielding identical reflections. For the Fig. 3.14 example, one writes $a(q) = |a(q)| \exp(i\phi_a(q))$, and the phase difference between the two complex scattered amplitudes reads

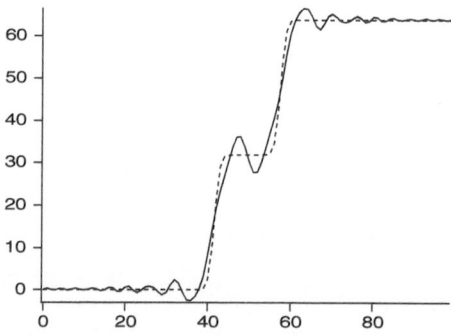

**Fig. 3.16** Two profiles giving
the same reflectivity curve,
obtained by phase changes of
the complex amplitude and
inverse Fourier transform

$$\phi(q) = -2\phi_a(q) + qd. \tag{3.B11}$$

We will show in the following a few special cases where we use a special physical knowledge about the profile to possibly choose only one solution among the set of mathematically possible solutions.

## 3.B.3 Multilayers

We shall first study the multilayer case. Here the additional knowledge brought into the problem is the periodic nature of the unknown index profile. This case is interesting as it is close to the case of traditional crystallography where crystals have periodic structures and where the question is to find the distribution of material within the elementary cell.

As the system is periodic, the intensity is concentrated into periodic Bragg peaks and the internal structure within the cell is reconstructed by Fourier series rather than continuous integrals. The phase problem is now discrete: we need to get the phase of each individual Bragg peak. Then the structure may be worked out by simple inverse Fourier series of the scattered amplitude.

Crystallographers have developed numerous methods to deal with the phase loss or to get back this phase. In a general way, methods that are used for this purpose are based on the interference between a reference wave and the unknown signal whose phase is wanted. Most popular methods are

- The heavy atom method: within the structure, it acts as a reference. The phase of the reference signal is obtained by changing the wavelength close to an absorption edge.
- The multiple diffraction method: one looks on the effect on a Bragg peak when a second Bragg peak is brought in reflection position. The change in intensity of the different peaks allows an estimate of the phase difference between the peaks. A series of phase difference between peaks (triplets) can then be obtained which may allow data inversion.

In the method below, we follow the same ideas but the phase reference is a surface reflection: we shall be interested in the case of a periodic structure deposited on a substrate while we try to determine the multilayer structure (i.e. the structure factor of a period of the multilayer). Our method is based on the statement that in a reflectivity signal from such a system, we do not only see the FT of the multilayer structure (a series of Bragg peaks with secondary maxima) but we also see a contribution from the external interfaces bounding the film. This appears as a variation in the intensity of secondary maxima on the side of the main Bragg peaks. Actually, we find in the literature several names for these secondary maxima that may be confusing:

- They are no Kiessig fringes that are equal inclination fringes due to interference between outer surfaces and that we observe even if the film has no inner structure.

– They are not pure secondary maxima due to the finite-size structure (and that we would see even if the film had no outer surfaces, e.g. if it was embedded in a matrix with the same average density).

We are dealing with an interference between these two quantities. This interference may be constructive on one side and destructive on the other side due to possible sign change of structure factor at Bragg peak position, explaining the large asymmetry that is often observed around them. We explain in the following how to take advantage of this interference to get the phase of the structure factor. The idea was first worked out on Langmuir–Blodgett layers [35, 36] but it is of course applicable to a broader range of multilayer systems.

Here we shall present the principles on a simple case, using the kinematical expression of the reflected intensity:

$$R_N(q) = \frac{R(q)}{R_F(q)} = \left| T.F.(\frac{\rho(z)}{\rho_{-\infty}}) \right|^2. \tag{3.B12}$$

In the case of a multilayer deposited on a substrate

$$R_N(q) = |k(q) + p(q)f(q)|^2, \tag{3.B13}$$

where $k(q)$ is the amplitude reflected by our reference, $p(q)$ and $f(q)$ the form and structure factor, respectively. In the present case, the reference is the reflected amplitude from the substrate, which is a constant (=1 since we normalised the curves to the Fresnel reflection). To improve the description one can use an amplitude of type $a = a\exp(-q^2\sigma^2/2)$ to account for attenuation due to a rough interface.

The form factor $p(q)$ can be simply expressed for a finite-size multilayer:

$$p(q) = \exp[iqd/2] \sum_{n=0}^{N} (\exp(iqd))^n, \tag{3.B14}$$

i.e.

$$p(q) = \frac{\sin(Nqd/2)}{\sin(qd/2)} \exp(iqNd/2). \tag{3.B15}$$

In a similar way as for the reference, we can introduce an additional damping to the amplitude, adding an imaginary part to the form factor phase:

$$\varphi(q) = qd + i\frac{q^2\sigma^2}{2}. \tag{3.B16}$$

We get for $p(q)$, including absorption

$$p(q) = \frac{i}{2} \frac{\exp(\frac{q^2\sigma^2}{4})}{\sin(qd/2 + iq^2\sigma^2/4)} (1 - \exp(iNqd)\exp(-Nq^2\sigma^2/2)). \tag{3.B17}$$

If absorption is strong, we get

$$p(q) = \frac{i}{2} \frac{\exp(\frac{q^2\sigma^2}{4})}{\sin(qd/2 + iq^2\sigma^2/4)}.$$  (3.B18)

The expressions give a maximum amplitude for $q = n2\pi/d$ and determine the shape of Bragg peaks.

The structure factor $f(q)$ is the data we are looking for. It reads

$$f(q) = \int_{-d/2}^{d/2} \frac{\rho'(z)}{\rho_{-\infty}} \exp[iqz]dz.$$  (3.B19)

Taken at Bragg peak positions ($q = n2\pi/d$), this expression provides the Fourier coefficients of the structure factor $f_n = |f_n|\exp(i\varphi_n)$. The problem set is the determination of not only $|f_n|$ but also $\varphi_n$. If we can measure these two data, for a large number of Bragg peaks, we can then get back to the structure ($\rho'(z)$) through its Fourier series expansion,

$$\frac{\rho'(z)}{\rho_{-\infty}} = \frac{2}{d} \sum_n |f_n| \cos\left(n\frac{2\pi}{d}z + \varphi_n\right),$$  (3.B20)

and, integrating

$$\frac{\rho(z)}{\rho_{-\infty}} = \frac{\rho(0)}{\rho_{-\infty}} + \sum_n \frac{|f_n|}{\pi n} \sin\left(n\frac{2\pi}{d}z + \varphi_n\right).$$  (3.B21)

We have represented on Fig. 3.17 the interference figures of a Bragg peak with a reflection signal coming from the substrate, for different values of the structure factor phase. We notice strong shape differences close to peak feet as a function of phase showing we have a way to measure the phase data, e.g. fitting expression (1) to experimental data.

We put into practice the method on several types of multilayers as Langmuir–Blodgett layers [35, 36]. As an example, we represent on Fig. 3.18 a reflectivity curve obtained on a protein multilayer (bacteriorhodopsin) deposited on a silicon

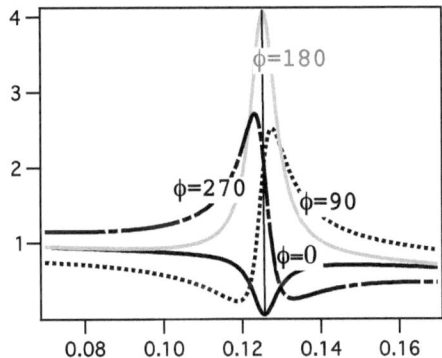

**Fig. 3.17** The four interference structures for a Bragg peak interfering with a surface reflection for phases in the four quadrants

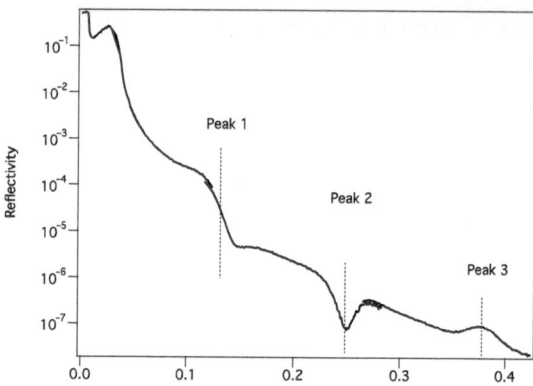

**Fig. 3.18** Reflectivity curve of a bacteriorhodopsin multilayer including $N \approx 5$ layers

substrate. The reflection curve obtained is of poor quality showing only three Bragg peaks, as the multilayer included only a few layers ($N \leq 5$) with raw deposition conditions. However, we can recognise easily on the curve the different shapes of Fig. 3.17 from which we can extract three (amplitude, phase) doublets for the structure factor at the Bragg peak locations. The Fourier reconstruction of the period has some similarity with the structure along c of the protein, so that the measurements are already accurate enough to indicate under which orientation the protein in the purple membrane deposits on silicon. It is clear that in these cases the interest of such measurement is only anecdotic and that a much larger number of peaks would be necessary to get interesting data. This may be achievable as for Langmuir–Blodgett layers for instance, several tens of peaks are observable. For protein layers and bacteriorhodopsin in particular, the same is true. This type of method using a surface as a reference has a special interest for thin layer only. For bulk crystals, they seem hardly applicable unless one is able to achieve extremely flat outer surfaces (the outer surface works also as a reference).

Note that other references may be considered not using a specular reflection but a Bragg line of the substrate (if it is a single crystal, e.g. silicon). Rotating the sample around the normal to the Bragg plane, we can bring adsorbate line in reflection position whose phase may be determined with respect to the substrate line. This kind of method may find an application for polycrystalline samples on substrates. The method is then similar to the triplet method mentioned above, with the advantage of a known substrate.

### 3.B.4 Simple Profiles

In the case of a simple interface, for instance the external surface of a material, the first parameter that can be extracted from the reflection curve is the index change at the vacuum/material interface (this scales the reflectivity curve). Looking in more details, we can also extract the roughness, i.e. the width parameter for the interface. Looking further we can track the presence of modulations, etc. which may put into evidence a surface layer, a composition profile, etc. We propose here to formalise the relation between the reflectivity and the structure to check how far we can go in such description. To do so, we have expanded the derivative of the electron density profile in moments that naturally account for these different terms (index difference, roughness, etc.)

By definition, one gets

$$M_i = \frac{1}{\rho_{-\infty}} \int_{-\infty}^{\infty} \left( \frac{\partial \rho}{\partial z} \right) z^i dz. \tag{3.B22}$$

This definition may be adjusted to use normalised quantities. The first three moments give the scale of the problem:

$M_0$ is the electron density jump (the index jump).
$M_1$ is the profile centre position, which can always be set to zero choosing the origin of the x-scale.
$M_2$ is the mean quadratic width of the profile.
Taking $z_0 = M_1$, we can define the $M_i$ so that $M_0 = 1$, $M_1 = 0$ and $M_2 = \sigma^2$.

We can define dimensionless quantities for higher order moments:
$r = M_3/M_2^{3/2}$, $K = M_4/M_2^2 - 3$.
These data quantify the degree of asymmetry and extent at large distances for the profile. The data of all the different moment allow one in principle to reconstruct the profile. A convenient way of doing so is to decompose the profile (or its derivative) (which is square summable) in a series of Hermite functions $n$th derivative of a Gaussian. This type of expansion (used in quantum mechanics or probability theory) has the advantage of being almost invariant upon Fourier transform. One reads

$$\rho'(z) = \sum_{n=0}^{\infty} c_n \phi^{(n)}(z) = \phi(z) \sum_{n=0}^{\infty} c_n(-1/\sqrt{2})^n H_n(z/\sqrt{2}), \tag{3.B23}$$

where

$$\phi(z) = \frac{1}{2\pi} \exp(-z^2/2)$$

$$H_n(z) = \exp(z^2)(-1)^n \frac{d^n}{dz^n} \exp(-z^2). \tag{3.B24}$$

We have

$$c_0 = 1$$
$$c_1 = c_2 = 0$$
$$c_3 = -r/6$$
$$c_4 = K/4!, \tag{3.B25}$$
$$c_5 = (-M_5 + 10M_3)/5!, \tag{3.B26}$$
$$c_6 = (M_6 - 15M_4 + 30)/6!. \tag{3.B27}$$

Upon Fourier transform, the Gaussian is invariant and we get the following expansion for the reflectivity:

$$\frac{R(q)}{R_F(q)} = |T.F.(\frac{\rho'(z)}{\rho_{-\infty}})|^2$$

$$= \exp(-q^2\sigma^2)\left|\sum_{n=0}^{\infty} c_n(iq\sigma)^n\right|^2$$

$$= \exp(-q^2\sigma^2)\left(1 + 2c_4(q\sigma)^4 + (c_3^2 - 2c_6)(q\sigma)^6 + \cdots\right). \quad (3.B28)$$

We have applied this formalism to the free surface of liquid helium. At low temperatures, this surface is subject to quantum capillary wave excitation ("ripplons") that tend to smooth the liquid/vapour interface. Functional density calculations of this liquid/vapour interface have predicted the following shape for the profile [37]:

$$\rho(z) = \frac{\rho(-\infty)}{[1 + \exp(z/a)]^\nu}, \quad (3.B29)$$

with $a = 0.196$ nm and $\nu = 5/2$ for $^4$He. Numerically for this profile (represented in Fig. 3.19) one gets s = 0.286 nm; r = −0.695 and K = 1.474.

The series of $c_n$ coefficients is 1, 0, 0, 0.061, 0.031, 0.024 for $c_0, c_1, \ldots, c_5$. In the present case the Fourier transform can be calculated exactly. One finds

$$\frac{R(q)}{R_F(q)} = F(qa), \quad (3.B30)$$

with

$$F(x) = \pi x(1 + 4/9x^2)(1 + 4x^2)/(\sinh(\pi x)\cosh(\pi x)). \quad (3.B31)$$

Hence, we can evaluate precisely the quality of the approximations performed. We can see in this case that exact calculation and expansion yield similar results up to $q = 4$ nm$^{-1}$ where the reflectivity is typically $10^{-9}$, which is the limit value for the technique using e.g. synchrotron sources. The discrepancy between the pure Gaussian description is visible at lower q, but the first corrective term is due to parameter $c_4$ and not due to asymmetry $c_3$ (Fig. 3.20).

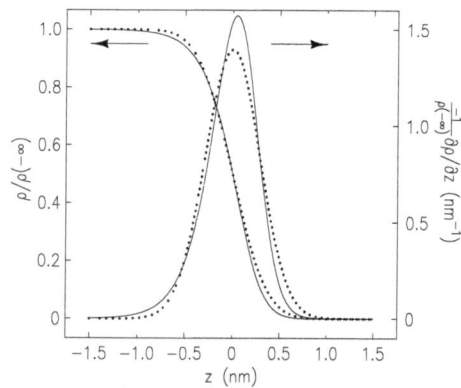

**Fig. 3.19** Plot of the asymmetric profile expected for the free surface of liquid helium (profile and derivative). The *dashed curves* represent the standard Gaussian approximation (2nd moment)

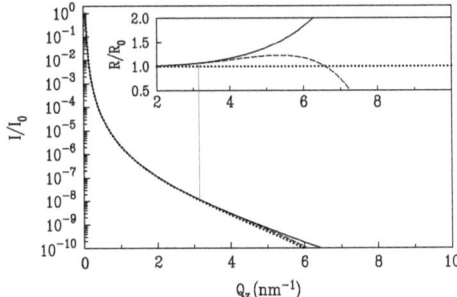

**Fig. 3.20** Reflectivity curve calculations for the different approximations. *Solid line*: exact calculation, *dotted curve*: Gaussian approximation, *dashed curve*: expansion using higher order moments. Insert: Reflectivities normalised to the Fresnel reflectivity with a Gaussian attenuation, showing up to which wave vector value one needs to measure to detect asymmetry

In principle this technique can be applied to more complex density profiles such as a thin film bound by two interfaces. Describing such a profile by a series of moments is not efficient yet the previous development can be used for each individual interface profile. The FT will display terms involving not only individual interface parameters but also crossed interference terms. We can see then that in some cases (e.g. if one interface is perfectly known), the asymmetry of the other interface appears more rapidly in the reflection curve due to the existence of these crossed terms [38].

## 3.B.5 *Methods Based on Several Measurements*

The phase problem can be viewed as due to the fact that at each point on the reflection curve, we have only one data on the complex amplitude whereas two data are necessary to get the complex amplitude.

If one introduces in the scattering length density profile an element whose scattering length can be varied in a controlled manner, we shall then have, for each point at a given $q$, several values for the modulus which could be used to extract the complex amplitude.

These methods, similar to that of the heavy atom, whose scattering length is varied by changing the energy close to an absorption edge, can be also operated with neutrons [39] varying beam polarisation on a magnetic substrate, or the substrate or its isotope composition.

If we assume we have the same profile $\rho(z)$ deposited on two substrates $a$ and $b$ whose composition and top interface are known, we will have for the reflectivities of the film/substrate system, in either case

$$r_a(q) = |r_F(q) + r_{sa}(q)|^2, \tag{3.B32}$$

$$r_b(q) = |r_F(q) + r_{sb}(q)|^2. \tag{3.B33}$$

The unknown is here $\mathbf{r_F(q)}$, the complex amplitude reflected by the film, data are $r_a(q)$ and $r_b(q)$ while $r_{sa}(q)$ and $r_{sb}(q)$ are known.

$$r_a(q) = |r_F(q)|^2 + |r_{sa}(q)|^2 + 2r_{sa}(q)r_F(q)\cos(\phi_F(q)), \qquad (3.B34)$$

$$r_b(q) = |r_F(q)|^2 + |r_{sb}(q)|^2 + 2r_{sb}(q)r_F(q)\cos(\phi_F(q)). \qquad (3.B35)$$

We have two equations in two unknowns, $|r_F(q)|$ and $\phi_F(q)$, that we can solve for any $q$ if $r_{sa}(q) \neq r_{sb}(q)$. A third measurement may improve the phase determination and raise possible uncertainties on $\phi_F(q)$ phase sign [40].

## 3.B.6 Direct Methods

They are based on the fact that the amplitude and phase of the scattered wave derive from the real and imaginary parts of a complex function, Fourier transform of the scattering length density profile. This profile has special properties that can be used to limit the number of solutions to the phase problem. For example, the profile function is continuous, limited and generally considered as zero outside a finite range (in the top and bottom media, e.g. air and substrate $\rho(z) = cst \rightarrow \rho'(z) = 0$). Choosing the origin, one may set $\rho'(z) = 0$ when $z < 0$. The complex amplitude may be considered as the Fourier transform of a response function and, taking into account the properties of this function, we can use complex variable function theory (e.g. Cauchy relations) to obtain integral relations between real and imaginary parts of the Fourier transform of this function ("dispersion" or Kramers–Kronig relations).

In practice the problem is slightly more complicated. We do not want to determine the imaginary part of the scattered amplitude from the knowledge of the real part, but the phase knowing the modulus. The idea is to consider the logarithm of the complex amplitude:

$$\ln(a(q)) = \ln(|r(q)|) + i\phi(q), \qquad (3.B36)$$

with $r(q) = \sqrt{(R(q))}$. Making the assumption that properties of the logarithm of the FT of the response function are the same as those of the FT itself, we can write Kramers–Kronig relations between real and imaginary parts of the log of the complex amplitude, which are the log of the modulus and the phase [41]:

$$\phi(q) = \frac{-q}{\pi}P\int_{-\infty}^{\infty}\frac{\ln[r[\chi]/r[q]]}{\chi(\chi-q)}d\chi. \qquad (3.B37)$$

Actually this equation, obtained by an application of the Cauchy formula to the function $\ln(r(\chi))/\chi$ along a contour that excludes $\chi = 0$ and $\chi = q$ points, presumes a number of assumptions that are not necessarily obvious for functions corresponding to physical profiles (Fig. 3.21).

It requires for instance that the integral along the half-circle whose radius tends towards infinity in the upper half complex plane be zero. As $r(q) \rightarrow 0$ when

Fig. 3.21 Integration contour
in the complex plane

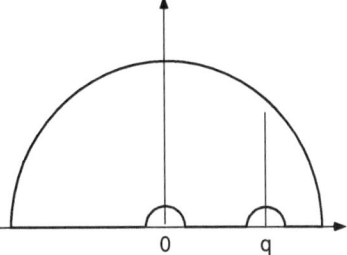

$q \to \infty$, $\ln(r(q)) \to -\infty$ and this is not a priori obvious. This is why one considers $\ln(r(\chi))/\chi$ rather than $\ln(r(\chi))$.

Moreover, Cauchy formula gives a zero integral for a complex variable function only when it has no zero inside the contour (the upper half complex plane in the present case). In practice, with simple density profiles, this is not always true and making this assumption amounts to choosing one solution among various possible phase solutions, while such a choice may not be very clear from a physical point of view. Clinton has discussed these methods in a few simple cases [42]. When the profile is dominated by an interface located at $z = 0$, i.e.

$$a(\chi) = a_0 + \sum_i a_i \exp i\chi d_i, \qquad (3.B38)$$

with $d_i > 0$ and $a_0 > \sum_i |a_i|$, then $|a(\chi)| > 0, \forall \chi$ verifying Im $\chi \geq 0$. This case is close to those studied in Sects. 3.B.3 and 3.B.4. A few examples where this technique has been applied to experimental profiles are given by A. van der Lee [43]. The main interest of this technique is to provide a starting point independent of any model. One is free in a second step to start from this model and change the phases (i.e. introduce zeroes in the complex plane) to make the profile correspond to physical data obtained independently.

## 3.B.7 Case of Symmetrical Profiles

A special case that is worth being considered is the case where the profile is symmetrical. This is the case for example when considering free standing films (e.g. liquid crystal or surfactant films) or interfacial profiles in the case of bonding between similar materials. In these cases, the special knowledge we have about the profile function $\rho(z)$ is that it is even. As a consequence, the Fourier transform of this function is a real function and the phase problem reduces now to a sign problem. The scattered amplitude being a continuous function, the sign changes can occur only at zeroes of the function that are generally easily located on the intensity curve.

This method has been used practically on reflection data taken on bonding interfaces (Fig. 3.22). When bonding identical substrates together, we can expect the density profile to be an even function. Then the inversion procedure can be worked out in a straightforward way and completely automated.

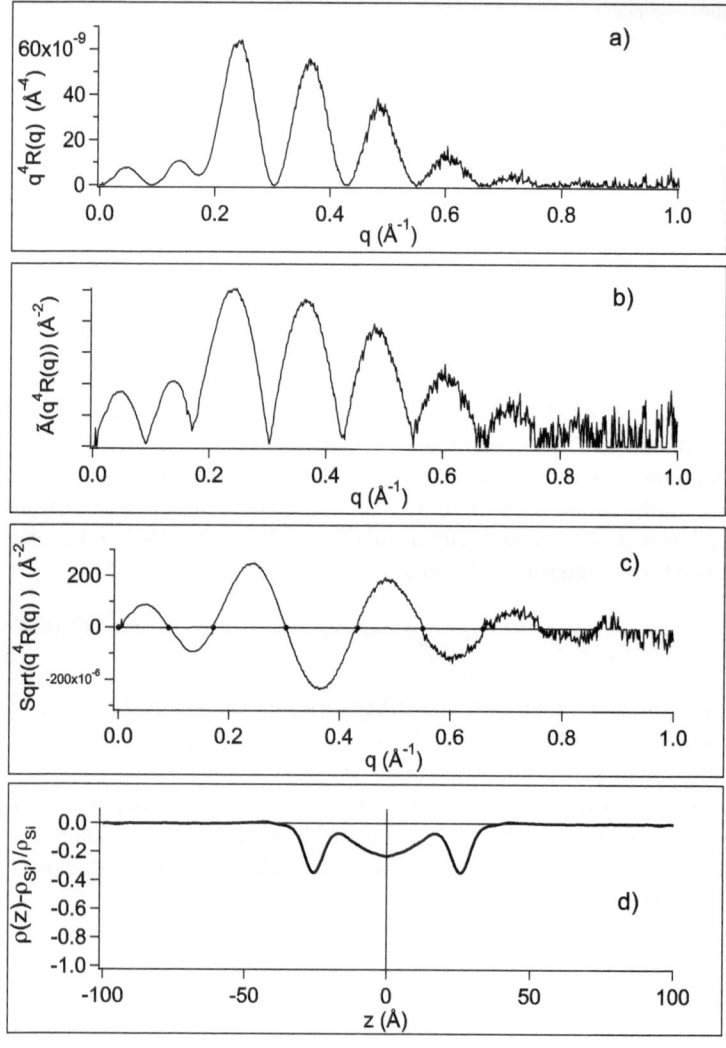

**Fig. 3.22** Sequence for a direct inversion of reflection data in the case of a symmetrical profile (here the interface of a silicon direct bonding). (**a**) $q^4 R(q)$ reflectivity data. (**b**) Square root of the previous curve. (**c**) Sign allocation assuming minima corresponding to a phase sign change. (**d**) Inverse FT of curve c, giving the electron density profile

## 3.B.8 Conclusion

It is difficult to provide a general method for interpreting reflectivity data. We proposed a few examples allowing the recovery of the "lost" phase when additional information on the scattering length profile is available. It is important to keep in mind the uncertainties related to this phase loss when interpreting experiments.

# References

1. Fresnel, A.: Mémoires de l'Académie **11**, 393 (1823).
2. Compton, A.H.: Phil. Mag. **45**, 1121 (1923).
3. Forster, R.: Helv. Phys. Acta. **1**, 18 (1927).
4. Prins, J.A.: Z. Phys. **47**, 479 (1928).
5. Kiessig, H.: Ann. Der Physik **10**, 715 (1931).
6. Parrat, L.G.: Phys. Rev. **95**, 359 (1954).
7. Abélès, F.: Ann. de Physique **5**, 596 (1950).
8. James, R.W.: The optical principles of the diffraction of x-rays, G. Bell and Sons, London, (1967).
9. International tables for x-ray crystallography, vol. IV. The Kynoch Press, Birmingham, (1968).
10. Petit, R.: Ondes Electromagnétiques en radioélectricité et en optique, Ed. Masson (1989).
11. Born, M., Wolf, E.: Principles of Optics, 6th edn. Pergamon, London (1980).
12. Vidal, B., Vincent, P.: Appl. Opt., **23**, 1794 (1984).
13. Lekner, J.: Theory of Reflection of Electromagnetic and Particle Waves, Martinus Nijhoff Publishers (1987).
14. Russel, T.P.: Mater. Sci. Rep. **5**, 171 (1990).
15. Gibaud, A., McMorrow, D., Swadling, P.P.: J. Phys. Condens. Matter **7**, 2645 (1995).
16. Baptiste, A., Gibaud, A., Bardeau, J.F., Wen, K., Maoz, R., Sagiv, J., Ocko, B.M.: Langmuir **18**(10), 3916–3922 (2003).
17. Rayleigh, J.W.S.: Proc. Roy. Soc. **86**, 207 (1912).
18. Hamley, I.W., Pedersen, J.S.: Appl. Cryst. **27**, 29 (1994).
19. Vignaud, G.: Thèse de l'Université du Maine (1997).
20. Als-Nielsen, J.: Z. Phys. B **61**, 411 (1985).
21. Tidswell, I.M., Ocko, B.M., Pershan, P.S., Wassermann, S.R., Whitesides, G.M., Axe, J.D.: Phys. Rev. B **41**, 1111 (1990).
22. Vignaud, G., Gibaud, A., Grübel, G., Joly, S., Ausserré, D., Legrand, J.F., Gallot, Y.: Physica B **248**, 250 (1998).
23. Névot, L., Croce, P.: Revue de Physique appliquée, **15**, 761 (1980).
24. Tolan, M.: Rontgenstreuung an strukturierten Oberflächen Experiment &Theorie Ph.D. Thesis, Christian-Albrechts Universität, Kiel, (1993).
25. Rayleigh, L.: Proc. R. Soc. London A **79**, 339 (1907).
26. Rice, S.O.: Com. Pure Appl. Math. **4**, 351 (1951).
27. Roschchupkin, D.V., Brunel, M., de Bergevin, F., Erko, A.I.: Nucl. Inst. Meth. B **72**, 471 (1992).
28. Sinha, S.K., Sirota, E.B., Garroff, S., Stanley, H.B.: Phys. Rev. B, **38**, 2297 (1988).
29. de Boer, D.K.G.: Phys. Rev. B **49**, 5817 (1994).
30. Abéles, F.: Annales de physique **3**, 504 (1948).
31. Abéles, F.: Annales de physique **3**, 504 (1950).
32. Born M., Wolf E.: Principles of Optics, 6th edn. Pergamon, London (1980).
33. Gibaud A., Vignaud G.: *X-ray and Neutron Reflectivity*. Lect. Notes Phys. 770. Springer, Heidelberg (2009).
34. Parratt, L.G.: Phys. Rev. **95**, 359 (1954).
35. Rieutord, F., Benattar, J.J., Bosio, L., Robin, P., Blot, C., de Kouchkovsky, R.: J. Physique **48**, 679 (1987).
36. Rieutord, F., Benattar, J.J., Rivoira, R., Lepetre, Y., Blot, C., Luzet, D.: J. Acta Cryst. **7**, 445 (1989).
37. Stringari, S., Treiner, J.: Phys. Rev. B **36**, 8369 (1987).
38. Rieutord, F., Braslau, A., Simon, R., Pasyuk, V., Lauter, H.: Physica B **221**, 538 (1996).
39. Majkrzak C.F., Berk N.F., et al.: Physica B **221**, 520 (1996).
40. de Haan, V.O., van Vell, A.A., Sacks, P.E., Adenwalla, S., Felcher, G.P.: Physica B **221**, 524 (1996).
41. Burge, R.E., Fiddy, M.A., Greenaway, A.H., Ross, G.: Proc. R. Soc. Lond. A **350**, 191 (1976).
42. Clinton, W.J.: Phys. Rev. B **48**, 1 (1993).
43. van der Lee, A.: Eur. Phys. J. B **13**, 755 (2000).

# Chapter 4
# Diffuse Scattering

J. Daillant, S. Mora and A. Sentenac

Specular reflectivity, as described in Chap. 3, is sensitive to the average density profile along the normal $(Oz)$ to a sample surface. Very often, one would also like to
determine the statistical properties of surfaces or interfaces (i.e. the "lateral" structures in the $(xOy)$ plane). We have seen in Chap. 2 that the scattered intensity depends on the roughness statistics of the sample (when the coherence domains are
much smaller than the illuminated area). More precisely, under several simplifying
assumptions, the differential scattering cross-section is related to the power spectrum of the surface. More generally, we will see that x-ray scattering experiments
allow the determination of the lateral lengths of surface morphologies and of the
correlations between buried interfaces over more than 5 orders of magnitude from
Ångströms to tens of microns in plane (see Sect. 4.5.1). Specific methods known as
GISAXS (grazing-incidence small angle x-ray scattering) have been developed in
the small angle regime, in particular when one is interested in the size, shape and
distribution of particles. They are described in Chap. 7.

In this chapter we present the theory of scattering by random media from an electromagnetic point of view. Starting from Maxwell equations we establish the volume integral equation giving the scattered field as the field radiated by the dipoles
induced in the material (part of these results have been used without demonstration
in Chap. 2). We then describe several perturbation techniques (Born approximation
and distorted-wave Born approximation, DWBA) which allow one to obtain simple expressions for the differential scattering cross-section. Then, special attention
is paid to the resolution function and to the determination of absolute (measured)
intensities. This is necessary if one wants to draw quantitative information from
an experiment. These methods are finally applied to scattering problems of increasing complexity: scattering by a single rough surface, surface scattering in a thin film,
scattering by rough inhomogeneous multilayers and scattering by surface crystalline
structure.

Most of the chapter is devoted to the discussion of the so-called distorted-wave
Born approximation (DWBA) which presently provides the most accurate analysis
of x-ray and neutron data.

J. Daillant (✉)
SCM/LIONS, bât. 125, CEA Saclay, 91191 Gif sur Yvette Cedex, France

Daillant, J. et al.: *Diffuse Scattering.* Lect. Notes Phys. **770**, 133–182 (2009)
DOI 10.1007/978-3-540-88588-7_4

## 4.1 Differential Scattering Cross-Section for X-Rays

In this section, we first establish the propagation equation for the electric field and show that its solution can be put in the form of an integral equation using Green functions. This integral equation is the basis for the Born development (see Sect. 4.2 for the Born approximation and Sect. 4.3 for the (first-order) DWBA). The definition of the distorted-wave Born approximation then amounts to the choice of an unperturbated (ideal) state for which the field in the sample and the Green functions have to be evaluated exactly. The evaluation of the Green functions is the main difficulty of the technique and we give here a simple method, based on the reciprocity theorem, to calculate them for various reference states: an infinite homogeneous medium (for the Born approximation) and a planar multilayer (for the DWBA).

The developments made in this section are valid for complicated systems like multilayers with rough interfaces and possibly density inhomogeneities. However, for simplicity the reader can refer to the case of a single rough interface separating two material media (0) and (1) depicted in Fig. 4.1.

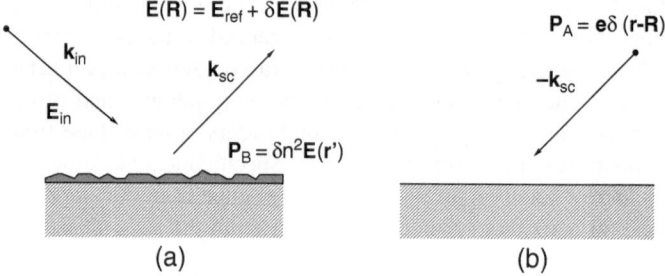

**Fig. 4.1** Illustration of the reciprocity theorem in the case of a single rough surface. The rough surface (real state in (**a**)) is viewed as a perturbation (*in grey*) of the reference state (in (**b**)). The location of the planar interface in the reference case is arbitrary (it is here situated below the deepest incursion of the roughness, see Footnote 9 on this subject). The total electric field $\mathbf{E}(\mathbf{R})$ existing in the real (rough) state is the sum of the specular field $\mathbf{E}_{ref}(\mathbf{R})$ coming from the reference medium and of the scattered field $\delta\mathbf{E}(\mathbf{R})$ radiated by the dipole density $\delta\mathbf{P}(\mathbf{r}') = \epsilon_0 \delta n^2(\mathbf{r}')\mathbf{E}(\mathbf{r}')$, which is nonzero only within the *grey region*. In order to use the reciprocity theorem to calculate $\delta\mathbf{E}(\mathbf{R})$ we consider two distributions of sources: (1) A unit dipole with moment $\hat{\mathbf{e}}$ placed at $\mathbf{R}$ (detector location) creating a field distribution $\mathbf{E}_{A}(\mathbf{r}') = \mathbf{E}_{det}^{\hat{\mathbf{e}}}(\mathbf{R}, \mathbf{r}')$ in the reference state in (**b**); (2) A dipole density representing the perturbation brought by the roughness (*grey region*) $\delta\mathbf{P}(\mathbf{r}') = \epsilon_0 \delta n^2(\mathbf{r}')\mathbf{E}(\mathbf{r}')$ creating a field distribution $\mathbf{E}_{B}(\mathbf{R}) = \delta\mathbf{E}(\mathbf{R})$ in (**a**). The reciprocity theorem yields $\delta\mathbf{E}(\mathbf{R}).\hat{\mathbf{e}} = \int d\mathbf{r}' \delta\mathbf{P}(\mathbf{r}')\mathbf{E}_{det}^{\hat{\mathbf{e}}}(\mathbf{R}, \mathbf{r}')$

### 4.1.1 Propagation Equation

Using Maxwell's equations

$$\nabla \times \mathbf{E} = -\frac{\partial \mathbf{B}}{\partial \mathbf{t}},$$

(4.1)

$$\nabla \times \mathbf{H} = \mathbf{j} + \frac{\partial \mathbf{D}}{\partial \mathbf{t}}, \qquad (4.2)$$

one obtains the propagation equation for the electric field in the homogeneous media (0) and (1) containing no charges or currents:

$$\nabla \times \nabla \times \mathbf{E} - n^2(\mathbf{r})\frac{\omega^2}{c^2}\mathbf{E} = -\nabla^2\mathbf{E} - n^2(\mathbf{r})\frac{\omega^2}{c^2}\mathbf{E}$$
$$= \nabla^2\mathbf{E} - \frac{\omega^2}{c^2}\left(\mathbf{E} + \frac{\mathbf{P}}{\epsilon_0}\right)$$
$$= 0, \qquad (4.3)$$

for waves having a $e^{i\omega t}$ time dependence. $n$ is the refractive index; the dielectric constant and the refractive index are related by $\epsilon = n^2$; $k = n\omega/c$ is the wavevector; in vacuum $k_0 = \omega/c = 2\pi/\lambda$ ($\lambda$ is the wavelength). Note that all the possible complexity (roughness, inhomogeneities) of a sample is contained in $n^2(\mathbf{r})$.

In the case of neutrons, one has to solve Schrödinger equation which has a similar structure:

$$\left(-\frac{\hbar^2}{2m}\nabla^2 + \frac{2\pi\hbar^2}{m}\sum_i b_i\rho_i\right)\psi(r) = \mathscr{E}\psi(r), \qquad (4.4)$$

where $b_i$ is the scattering length of nuclei $i$ whose (number) density in the sample is $\rho_i$.

In the following we will work out a perturbative solution for the problem of surface scattering. We will start from a reference case, close to the real case of interest (for example the corresponding flat surface in the case of a rough surface) for which the electromagnetic field can be exactly calculated using the methods of Chap. 3. Then from the linearity of Eq. (4.3) the electric field in the real case can be calculated as the electric field in the reference case plus the field radiated by dipoles induced in the perturbation (for example a very thin rough layer in the case of a rough surface, see Fig. 4.1). Due to the weak interaction of x-rays with matter (the index is only slightly different from 1), the perturbation is weak and a good approximation is obtained. More precisely, we decompose the index as

$$n^2(\mathbf{r}) = n_{\text{ref}}^2(\mathbf{r}) + \delta n^2(\mathbf{r}), \qquad (4.5)$$

where $n_{\text{ref}}$ is the refractive index in the reference case. The reference state is simple enough for the electromagnetic field to be calculated exactly (vacuum, plane interface, planar multilayers). It represents the basis (zeroth order) of the perturbation development and should be as close as possible to the real medium in order to minimise the influence of the perturbation. $n_{\text{ref}}^2$ yields a specular reflection, and $\delta n^2$ will give incoherent scattering. Equations (4.3) and (4.5) are rewritten as

$$\nabla \times \nabla \times \mathbf{E}(\mathbf{r}) - n_{\text{ref}}^2(\mathbf{r})k_0^2\mathbf{E}(\mathbf{r}) = \delta n^2(\mathbf{r})k_0^2\mathbf{E}(\mathbf{r}) = \frac{\omega^2}{\epsilon_0 c^2}\delta\mathbf{P}. \qquad (4.6)$$

The right-hand side of Eq. (4.6) can be considered as a **dipole source** $\delta P(r') = \epsilon_0 \delta n^2(r') E(r')$ *in the reference medium.*[1] Maxwell's equations and Eq. (4.3) being linear, the electric field can be written as $E = E_{ref} + \delta E$ where $E_{ref}$ is the field in the reference case and $\delta E$ the perturbation in the field radiated by the dipole source $\delta P$.

### 4.1.2 Perturbation Theory

The field radiated at the detector by the dipole sources $\delta P$ can be calculated using Green functions, where the Green tensor $\mathscr{G}(R, r')$ for the propagation equation in the reference (ideal) case is defined as the solution of[2]

$$\nabla \times \nabla \times \mathscr{G}(R, r') - n_{ref}^2(R) k_0^2 \mathscr{G}(R, r') = \frac{k_0^2}{\epsilon_0} \delta(R - r'), \qquad (4.7)$$

which satisfies *outgoing* wave boundary conditions (the $\nabla \times \nabla$ operator acts on $R$).[3]

It follows from the linearity of Maxwell's and propagation equations that (insert Eq. (4.8) in Eq. (4.6))

$$\begin{aligned}
E(R) &= E_{ref}(R) + \delta E(R) \\
&= E_{ref}(R) + \int dr \mathscr{G}(R, r) . \delta P(r) \\
&= E_{ref}(R) + \epsilon_0 \int dr \delta n^2(r) \mathscr{G}(R, r) . E(r) \qquad (4.8)
\end{aligned}$$

is the solution of Eq. (4.6). Equation (4.8) is formally equivalent to Eq. (1.A4) and will be the basis for the Born (or DWBA) development. However, we first need an expression for the Green function. This can in particular be done using an elegant method due to P. Croce [9–13, 31] based on the reciprocity theorem [28].

### 4.1.3 Derivation of the Green Functions

#### 4.1.3.1 The Reciprocity Theorem

In this paragraph we determine the Green function $\mathscr{G}(R, r')$ introduced in Eq. (4.8) for a vacuum, a simple interface and a planar multilayer. For this, we use the

---

[1] Note that $E(r')$ is the real *unknown* field at $r'$.

[2] We have the identity $\nabla \times \nabla = \mathbf{grad} \operatorname{div} - \Delta$. In a vacuum, $\operatorname{div} E = 0$ and the propagation equation reduces to a set of three Helmholtz equations $-\Delta E - k_0^2 E = 0$. With this sign convention, the outgoing Green tensor $\mathscr{G}(r)$ reduces to $-G_-(r)$ defined in Chap. 1 for the scalar field obeying the Helmholtz equation $\Delta E + k_0^2 E = 0$.

[3] The solution (in the sense of distributions) of Eq. (4.7) satisfies by construction the boundary conditions in the system (e.g. the saltus conditions at each interface if the reference state is a planar multilayer).

reciprocity theorem demonstrated for example in Appendix 4.A and [27, 28, 41]. The reciprocity theorem states that in a given reference medium, two different distributions of dipole sources $\mathbf{P_A}$ and $\mathbf{P_B}$ creating the fields $\mathbf{E_A}$ and $\mathbf{E_B}$ are linked by the relation,

$$\int d\mathbf{r} \mathbf{E_A}(\mathbf{r}).\mathbf{P_B}(\mathbf{r}) = \int d\mathbf{r} \mathbf{E_B}(\mathbf{r}).\mathbf{P_A}(\mathbf{r}). \tag{4.9}$$

In order to calculate the perturbation in the field at the detector $\delta\mathbf{E}(\mathbf{R})$, we consider the following sources and field distributions (see Fig. 4.1):

- The source with polarisation vector $\delta\mathbf{P}(\mathbf{r}') = \epsilon_0 \delta n^2(\mathbf{r}')\mathbf{E}(\mathbf{r}')$ creating an unknown field $\delta\mathbf{E}(\mathbf{R})$ at the detector in the real case of the rough interface
- The unit dipole $\hat{\mathbf{e}}\delta(\mathbf{R} - \mathbf{r}')$ located at the detector, creating a known field $\mathbf{E}_{\mathrm{det}}^{\hat{\mathbf{e}}}(\mathbf{R}, \mathbf{r}')$ at point $\mathbf{r}'$ in the roughness region (the field can be calculated exactly since the unit dipole radiates in the simple reference geometry)

$$\mathbf{E_A} = \mathbf{E}_{\mathrm{det}}^{\hat{\mathbf{e}}}(\mathbf{R}, \mathbf{r}') \quad \mathbf{P_A} = \delta(\mathbf{R} - \mathbf{r}')\hat{\mathbf{e}}$$
$$\mathbf{E_B} = \delta\mathbf{E}(\mathbf{R}) \qquad \mathbf{P_B} = \epsilon_0 \delta n^2 \mathbf{E}(\mathbf{r}'),$$

and the reciprocity theorem Eq. (4.9) yields

$$\delta\mathbf{E}(\mathbf{R}).\hat{\mathbf{e}} = \int d\mathbf{r}' \epsilon_0 \delta n^2(\mathbf{r}')\mathbf{E}(\mathbf{r}').\mathbf{E}_{\mathrm{det}}^e(\mathbf{R}, \mathbf{r}'). \tag{4.10}$$

Equation (4.10) is equivalent to

$$\mathbf{E}(\mathbf{R}).\hat{\mathbf{e}} = \mathbf{E}_{\mathrm{ref}}(\mathbf{R}).\hat{\mathbf{e}} + \int d\mathbf{r}' \epsilon_0 \delta n^2(\mathbf{r}')\mathbf{E}(\mathbf{r}').\mathbf{E}_{\mathrm{det}}^{\hat{\mathbf{e}}}(\mathbf{R}, \mathbf{r}'). \tag{4.11}$$

The unit vector $\hat{\mathbf{e}}$ being arbitrary, Eq. (4.11) is in fact a vector (and not scalar) equation (choose $\hat{\mathbf{e}}$ equal, respectively, to $\hat{x}$, $\hat{y}$ and $\hat{z}$ to calculate the different field components). We retrieve formally Eq. (4.8),[4]

$$\mathbf{E}(\mathbf{R}) = \mathbf{E}_{\mathrm{ref}}(\mathbf{R}) + \epsilon_0 \int d\mathbf{r}' \delta n^2 \mathscr{G}(\mathbf{R}, \mathbf{r}').\mathbf{E}(\mathbf{r}').$$

Comparing Eqs. (4.8) and (4.11), we see that *the Green function required to calculate the scattered field can therefore be simply calculated as the field in r' due to a unit dipole in R (detector) in the reference case*. In practice, we will directly use this property in Eq. (4.10) to calculate the scattered field. This is particularly convenient in the far-field approximation within which the dipole field is easy to calculate. Note that Eq. (4.11) is an exact relation [15] from which approximations can be made. That Eq. (4.11) is exact is verified in Appendix 4.B in the particular case of the reflection on a film.

---

[4] Note that the reciprocity theorem gives Eq. (4.11) but does not tell us that the Green tensor is reciprocal in the sense that $\mathscr{G}(\mathbf{R}, \mathbf{r}') = \mathscr{G}(\mathbf{r}', \mathbf{R})$. In fact the symmetry relations on the Green tensor involve transpositions, see e.g. [41].

In the far-field approximation, the polarisation vector $\widehat{\mathbf{e}}_{sc}$ is necessarily perpendicular to the (meaningful in far-field) sample-to-detector direction $\widehat{\mathbf{u}} = \mathbf{R}/R$. It appears convenient to introduce two main polarisation states. In polarisation $(s)$, the field direction is normal to the scattering plane (defined by the normal to the sample and the sample-to-detector direction $(Oz, \widehat{\mathbf{u}})$), in polarisation $(p)$ the field direction lies in the scattering plane. Polarisation effects however generally remain negligible at grazing incidence. In any case, the scattered field amplitude is

$$\delta E(\mathbf{R}) = \int d\mathbf{r}' \epsilon_0 \delta n^2 \mathbf{E}(\mathbf{r}') . \mathbf{E}_{det}^{\widehat{\mathbf{e}}_{sc}}(\mathbf{R}, \mathbf{r}'). \tag{4.12}$$

We now explicitly calculate the Green function in a vacuum (which is the reference state in the Born approximation), a single interface (which also gives a simple approximation for thin films, Sect. 4.5.2) and for a planar multilayer (which is a natural reference state for the DWBA).

### 4.1.3.2 Green Function in a Vacuum

The electric field at point $\mathbf{r}'$ created by a dipole moment $\widehat{\mathbf{e}}$ located at $\mathbf{R}$ in a homogeneous infinite medium (vacuum) can be written as (neglecting the $1/|\mathbf{R} - \mathbf{r}'|^2$ and $1/|\mathbf{R} - \mathbf{r}'|^3$ terms in the limit of large $|\mathbf{R} - \mathbf{r}'|$) [25]

$$\mathbf{E}_{det}^{\widehat{\mathbf{e}}}(\mathbf{R}, \mathbf{r}') = k_0^2 (\widehat{\mathbf{u}} \times \widehat{\mathbf{e}}) \times \widehat{\mathbf{u}} \frac{e^{-ik_0|\mathbf{R} - \mathbf{r}'|}}{4\pi\epsilon_0 |\mathbf{R} - \mathbf{r}'|}, \tag{4.13}$$

where $\widehat{\mathbf{u}} = |\mathbf{R} - \mathbf{r}'|/|\mathbf{R} - \mathbf{r}'|$ is the unit vector in the direction of observation. Hereafter we assume that $R \gg r'$ so that $\widehat{\mathbf{u}} \approx \mathbf{R}/R$. Note that the dipole is here located at the detector position and that the field is observed at the sample. In the far-field approximation, one can develop

$$|\mathbf{R} - \mathbf{r}'| = |R\widehat{\mathbf{u}} - \mathbf{r}'| \approx R - \widehat{\mathbf{u}}.\mathbf{r}' \approx R - \mathbf{k}_{sc}.\mathbf{r}'/k_0$$

(see Fig. 4.1).[5] The spherical wave can therefore be developed on the tangent plane wave,

---

[5] The far-field conditions (or Fraunhofer diffraction) are more restricting than only $R \gg r'$. Indeed, to neglect the quadratic term in the expansion of $e^{-ik_0|\mathbf{R} - \mathbf{r}'|}$ one needs $r'^2/\lambda$ to be small compared to $R$. Applying this approximation in Eq. (4.8) yields a condition on the whole size of the scattering object (since $\mathbf{r}'$ covers all the perturbated region). The discussion in Chap. 2 has shown that the support of the integral appearing in Eq. (4.8) can actually be restricted to the domain of coherence (induced by the incident beam and detector acceptance) of the scattering processes. In this case the far-field conditions can be written as $l_{coh}^2/\lambda \ll R$. In a typical x-ray experiment, the sample-to-detector distance is $R = 1$ m, the wavelength is $\lambda = 1$ Å. The total illuminated area is a few mm but the coherence length is $l_{coh} \approx 1$ μm, hence the far-field approximation is valid. When the coherence length is too important (very small detector acceptance) for the far-field conditions to be satisfied, we are in the frame of the Fresnel diffraction and one needs to retain the quadratic terms in the expansion of $|\mathbf{R} - \mathbf{r}'|$ [3, 40].

$$\mathbf{E}_{\mathrm{det}}^{\hat{\mathbf{e}}}(\mathbf{R},\mathbf{r}') = k_0^2 (\hat{\mathbf{u}} \times \hat{\mathbf{e}}) \times \hat{\mathbf{u}} \frac{e^{-ik_0R}}{4\pi\epsilon_0 R} e^{i\mathbf{k}_{\mathrm{sc}}\cdot\mathbf{r}'}. \tag{4.14}$$

If we choose $\hat{\mathbf{e}}$ normal to the direction of scattering, for example along the $(s)$ and $(p)$ polarisation directions $\hat{\mathbf{e}}^{(s)}$ or $\hat{\mathbf{e}}^{(p)}$, one simply has

$$\mathbf{E}_{\mathrm{det}}^{(s),(p)}(\mathbf{R},\mathbf{r}') = k_0^2 \frac{e^{-ik_0R}}{4\pi\epsilon_0 R} e^{i\mathbf{k}_{\mathrm{sc}}\cdot\mathbf{r}'} \hat{\mathbf{e}}^{(s),(p)}. \tag{4.15}$$

### 4.1.3.3 Green Function for a Single Interface

A second example is that of a single interface between medium 0 and 1. The same calculation can be repeated for a dipole moment $\hat{\mathbf{e}}$ located at the detector position $\mathbf{R}$ and $\mathbf{r}'$ in medium 1, the only difference being the coefficient $t_{0,1}^{\mathrm{sc}}$ giving account of the transmission at the interface. One obtains

$$\mathbf{E}_{\mathrm{det}}^{(s),(p)}(\mathbf{R},\mathbf{r}') = k_0^2 \frac{e^{-ik_0R}}{4\pi\epsilon_0 R} t_{0,1}^{\mathrm{sc}} e^{i\mathbf{k}_{\mathrm{sc}}\cdot\mathbf{r}'} \hat{\mathbf{e}}^{(s),(p)}. \tag{4.16}$$

### 4.1.3.4 Green Function for a Stratified Medium

We finally consider a planar multilayer as reference state and consider the expression of the electric field created at point $\mathbf{r}'$ by a unit dipole placed in $\mathbf{R}$. The point $\mathbf{r}'$ can be taken anywhere in the stratified medium, see Fig. 4.6. Using the same plane wave limit in the general case of a stratified medium, Eq. (4.15) can be generalised for $(s)$ or $(p)$ polarisation for $\mathbf{r}'$ lying in layer $j$:[6]

---

[6] The electric field is the solution of the inhomogeneous differential equation $\nabla \times \nabla \times \mathbf{E}_{\mathrm{det}}^{\hat{\mathbf{e}}}(\mathbf{r}') - n_{\mathrm{ref}}^2(z')k_0^2\mathbf{E}_{\mathrm{det}}^{\hat{\mathbf{e}}}(\mathbf{r}') = \hat{\mathbf{e}}\delta(\mathbf{R}-\mathbf{r}')$ that satisfies *outgoing* wave boundary conditions. The unit dipole at the detector position lies in medium 0 as depicted in Fig. 4.1. In the homogeneous region 0, the electric field can be written as the sum of a particular solution and a homogeneous solution. The particular solution is given in Eq. (4.13) while the general homogeneous solutions are simply up-going plane waves with wavevector modulus $k_0$. In media $j$ with $j \neq 0 \neq s$, the electric field is solution of the homogeneous vectorial Helmholtz equation and it can be written as a sum of up-going and down-going plane waves with wavevector modulus $k_j$. In the substrate the general solutions are down-going plane waves with wavevector modulus $k_s$. To obtain the amplitudes of these plane waves we write the boundary conditions at each interface. The far-field approximation permits to simplify greatly the problem. In this case, the expression of the particular solution at $z = Z_1$ is given by Eq. (4.15). The dipole field close to the first interface can be approximated by an "incident" plane wave with wavevector $\mathbf{k}_{\mathrm{sc}}$. Hence, the amplitudes of the other plane waves (that are the general solutions of the homogeneous Helmholtz equations) are calculated easily with the transfer matrix technique presented in Chap. 3. The problem has been reduced to the calculation of the electric field in a stratified medium illuminated by a plane wave. The meaning of superscripts $(s)$ or $(p)$ is always unambiguous: it indicates the direction of the radiating unit dipole in a vacuum for a given position $\mathbf{R}$ of the detector. In other words, it indicates the polarisation state of the scattered plane wave with wavevector $\mathbf{k}_{\mathrm{sc}}$. Note that the directions of $\hat{\mathbf{e}}_{\mathrm{sc}}^{(p)}$ and $\mathbf{k}_{\mathrm{sc}}$ will vary from layer to layer due to refraction whereas the directions given by $\mathbf{k}_{\mathrm{sc}\parallel}$ and $\hat{\mathbf{e}}_{\mathrm{sc}}^{(s)}$ do not change.

$$\mathbf{E}_{\text{det}}{}^{(s),(p)}(\mathbf{R},\mathbf{r}') = k_0^2 \frac{e^{-ik_0 R}}{4\pi\epsilon_0 R} E_j^{PW\,(s),(p)}(-k_{\text{scz,j}},z')\widehat{\mathbf{e}}_{\text{sc}}^{(s),(p)} e^{i\mathbf{k}_{\text{sc}\parallel}\cdot\mathbf{r}'_{\parallel}} \qquad (4.17)$$

is the field in medium $j$ for an incident plane wave with polarisation $(s)$ or $(p)$ which can be computed by using standard iterative procedures [3, 24]. Using the notations of Chap. 3, Eqs. (3.59) and (3.60), one has,

$$E_j^{PW\,(s),(p)}(-k_{z,j},z) = U_j^{(s),(p)}(-k_{z,j},Z_j)e^{-ik_{z,j}z} + U_j^{(s),(p)}(k_{z,j},Z_j)e^{ik_{z,j}z}, \qquad (4.18)$$

where $\mathbf{r} = (\mathbf{r}_{\parallel},z)$ with $z$ is the $z$ coordinate with the origin taken at $z = Z_j$ and where the superscript "PW" has been used to emphasise that $E^{PW}$ is calculated for an *incident plane wave*.

## 4.1.4 The Differential Scattering Cross-Section

Summarising the previous expressions, one can always write for the different polarisations of the electric field

$$\mathbf{E}(\mathbf{R}) = \mathbf{E}_{\text{ref}}(\mathbf{R}) + \epsilon_0 k_0^2 \frac{e^{-ik_0 R}}{4\pi\epsilon_0 R} (\widehat{\mathbf{e}}_{\text{in}} \cdot \widehat{\mathbf{e}}_{\text{sc}}) \dots$$

$$\times \widehat{\mathbf{e}}_{\text{sc}} \int d\mathbf{r}'\, \delta n^2(\mathbf{r}') E^{PW}(k_{\text{inz}},z')\, E^{PW}(-k_{\text{scz}},z') e^{i\mathbf{q}_{\parallel}\cdot\mathbf{r}'_{\parallel}}, \qquad (4.19)$$

where $E_j^{PW\,(s),(p)}(-k_{\text{scz}},z')$ is the field at $\mathbf{r}'$ in the medium for an incident plane wave at detector position. Let us note here that the $-k_{\text{sc}}$ orientation of $E_j^{PW\,(s),(p)}(-k_{z,j},z)$ which results from the reciprocity theorem is necessary to give account of the phase advance of the field radiated by a dipole located at $\mathbf{r}'$ compared to one at origin.

To calculate the differential scattering cross-section we proceed by deriving the Poynting vector expression. In the far-field approximation,

$$\mathbf{B}_{\text{sc}} = \frac{1}{c}\widehat{\mathbf{u}} \times \mathbf{E},$$

and the Poynting's vector is

$$\mathbf{S} = \frac{|\mathbf{E}|^2}{2\mu_0 c}\widehat{\mathbf{u}}.$$

The differential scattering cross-section is obtained by calculating the flux $cR^2\mathbf{S}.\mathbf{u}$ of Poynting's vector (power radiated) per unit solid angle in direction $\mathbf{k}_{\text{sc}}$ across a sphere of radius $R$ for a unit incident flux. One gets,

$$\frac{d\sigma}{d\Omega} = \frac{k_0^4}{16\pi^2 |\mathbf{E}_{\text{in}}|^2} (\widehat{\mathbf{e}}_{\text{in}} \cdot \widehat{\mathbf{e}}_{\text{sc}})^2 \left| \int d\mathbf{r}'\, \delta n^2(\mathbf{r}') E^{PW}(k_{\text{inz}},z')\, E^{PW}(-k_{\text{scz}},z') e^{i\mathbf{q}_{\parallel}\cdot\mathbf{r}'_{\parallel}} \right|^2.$$

$$(4.20)$$

If one considers the issue of scattering from random media, we have seen in Chap. 2 that scattering can be separated into a coherent process and an incoherent process. The latter is the usual quantity of interest in a scattering experiment and it is given by

$$\left(\frac{d\sigma}{d\Omega}\right)_{\text{incoh}} = \frac{k_0^4}{16\pi^2|\mathbf{E}_{\text{in}}|^2}(\widehat{\mathbf{e}}_{\text{in}}.\widehat{\mathbf{e}}_{\text{sc}})^2$$

$$\times \left\{ \left| \int d\mathbf{r}' \delta n^2(\mathbf{r}') E^{PW}(k_{\text{inz}},z') E^{PW}(-k_{\text{scz}},z') e^{i\mathbf{q}_{\parallel}.\mathbf{r}'_{\parallel}} \right|^2 \right.$$

$$\left. - \left| \left\langle \int d\mathbf{r}' \delta n^2(\mathbf{r}') E^{PW}(k_{\text{inz}},z') E^{PW}(-k_{\text{scz}},z') e^{i\mathbf{q}_{\parallel}.\mathbf{r}'_{\parallel}} \right\rangle \right|^2 \right\}. \quad (4.21)$$

If we had chosen a vacuum as the reference state, we would have obtained

$$\frac{d\sigma}{d\Omega} = \frac{k_0^4}{16\pi^2|\mathbf{E}_{\text{in}}|^2} \left| \int d\mathbf{r}' \delta n^2(\mathbf{r}') \mathbf{E}_{\perp}(\mathbf{r}') e^{i\mathbf{k}_{\text{sc}}.\mathbf{r}'} \right|^2. \quad (4.22)$$

This exact expression has been used in Chap. 2 to discuss the effect of an extended detector on the measured scattered intensity in relation with the statistical properties of a surface. In general, it will however be less easy to develop an accurate approximation from this expression.

## 4.2 First Born Approximation

The first Born approximation which neglects multiple reflections can only be used far from the critical angle for total external reflection or Bragg peaks. Close to these points, the scattering cross-sections are large and the contribution to the measured intensity of multiple reflections cannot be neglected. The main advantage of presenting this approximation here is that it makes the structure of the scattered intensity very transparent. It has already been presented in Chap. 2 in a different context with the aim of illustrating how statistical information about surfaces or interfaces can be obtained in a scattering experiment.

### 4.2.1 Expression of the Differential Scattering Cross-Section

In the Born approximation, both the Green function and the electric field are evaluated in a vacuum, Eq. (4.11).

$$\mathbf{E}_{\text{det}}^{\widehat{\mathbf{e}}}(\mathbf{R},\mathbf{r}') = k_0^2(\widehat{\mathbf{u}} \times \widehat{\mathbf{e}}) \times \widehat{\mathbf{u}} \frac{e^{-ik_0R}}{4\pi\epsilon_0R} e^{i\mathbf{k}_{\text{sc}}.\mathbf{r}'} \quad (4.23)$$

$$\mathbf{E}(\mathbf{r}') \approx \mathbf{E}_{in} e^{-i\mathbf{k}_{in} \cdot \mathbf{r}'}. \qquad (4.24)$$

$-\mathbf{k}_{sc}$ is the wavevector oriented from the detector to the surface which gives the dipole field of Eq. (4.10). Then, substituting into (4.11),

$$\mathbf{E}^{(s)} = \mathbf{E}_{in} e^{-i\mathbf{k}_{in} \cdot \mathbf{r}'} + \frac{k_0^2 e^{-ik_0 R}}{4\pi R} (\mathbf{E}_{in} \cdot \widehat{\mathbf{e}}_{sc}) \widehat{\mathbf{e}}_{sc} \int d\mathbf{r} \delta n^2 e^{i\mathbf{q} \cdot \mathbf{r}}, \qquad (4.25)$$

with the wavevector transfer

$$\mathbf{q} = \mathbf{k}_{sc} - \mathbf{k}_{in}. \qquad (4.26)$$

For such a field dependence, we obtain for the differential scattering cross-section (power scattered per unit solid angle per unit incident flux) [25]

$$\frac{d\sigma}{d\Omega} = \frac{k_0^4}{16\pi^2} (\widehat{\mathbf{e}}_{in} \cdot \widehat{\mathbf{e}}_{sc})^2 \left| \int d\mathbf{r}\, \delta n^2 e^{i\mathbf{q} \cdot \mathbf{r}} \right|^2 = r_e^2 (\widehat{\mathbf{e}}_{in} \cdot \widehat{\mathbf{e}}_{sc})^2 \left| \int d\mathbf{r}\, \delta \rho e^{i\mathbf{q} \cdot \mathbf{r}} \right|^2. \qquad (4.27)$$

### 4.2.2 Single Flat Surface

As a first example we now calculate the differential scattering cross-section in the case of a single flat surface.

Integrating Eq. (4.27) for a perfect dioptre (index of refraction $n_0$ and $n_1$, see Fig 4.2), one obtains

$$\frac{d\sigma}{d\Omega} = \frac{k_0^4}{16\pi^2} (n_1^2 - n_0^2)^2 (\widehat{\mathbf{e}}_{in} \cdot \widehat{\mathbf{e}}_{sc})^2 \left| \int d\mathbf{r}_{\parallel} e^{i\mathbf{q}_{\parallel} \cdot \mathbf{r}_{\parallel}} \int_{-\infty}^{0} dz e^{iq_z \cdot z} \right|^2. \qquad (4.28)$$

The upper medium is medium 0, and the substrate (medium 1) is made slightly absorbing in order to make the integral over $z$ converge ($\mathscr{I}m(q_z) < 0$):

$$\int_{-\infty}^{0} dz e^{iq_z \cdot z} = \frac{1}{iq_z} \left[ e^{iq_z \times 0} - \underbrace{e^{iq_z \times -\infty}}_{\to 0} \right].$$

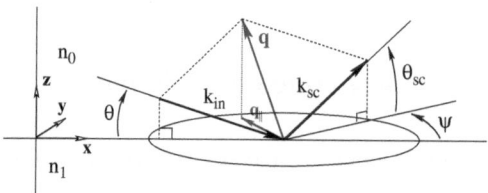

**Fig. 4.2** Scattering geometry. The components of the wavevector transfer are $q_x = k_0(\cos\theta_{sc}\cos\psi - \cos\theta_{in})$, $q_y = k_0\cos\theta_{sc}\sin\psi$ and $q_z = k_0(\sin\theta_{sc} + \sin\theta_{in})$. The surface between medium 0 ($z > 0$) and medium 1 ($z < 0$) is located at $z = 0$

The differential scattering cross-section can then be written as

$$\frac{d\sigma}{d\Omega} = \frac{k_0^4}{16\pi^2 q_z^2} (n_1^2 - n_0^2)^2 (\hat{\mathbf{e}}_{in}.\hat{\mathbf{e}}_{sc})^2 \left| \int d\mathbf{r}_\| e^{i\mathbf{q}_\|.\mathbf{r}_\|} \right|^2$$

$$= \frac{k_0^4}{16\pi^2 q_z^2} (n_1^2 - n_0^2)^2 (\hat{\mathbf{e}}_{in}.\hat{\mathbf{e}}_{sc})^2 \int d\mathbf{r}_\| \int d\mathbf{r}'_\| e^{i\mathbf{q}_\|.(\mathbf{r}_\| - \mathbf{r}'_\|)}. \qquad (4.29)$$

Making the change of variables $\mathbf{R}_\| = \mathbf{r}_\| - \mathbf{r}'_\|$ and integrating over $\mathbf{R}_\|$,

$$\frac{d\sigma}{d\Omega} = \frac{k_0^4}{16\pi^2 q_z^2} (n_1^2 - n_0^2)^2 (\hat{\mathbf{e}}_{in}.\hat{\mathbf{e}}_{sc})^2 \overbrace{\int d\mathbf{r}_\|}^{=A} \overbrace{\int d\mathbf{R}_\| e^{i\mathbf{q}_\|.\mathbf{R}_\|}}^{=4\pi^2 \delta(\mathbf{q}_\|)},$$

we obtain

$$\frac{d\sigma}{d\Omega} = \frac{k_0^4 A}{4q_z^2} (n_1^2 - n_0^2)^2 (\hat{\mathbf{e}}_{in}.\hat{\mathbf{e}}_{sc})^2 \delta(\mathbf{q}_\|), \qquad (4.30)$$

where $A$ is the illuminated area and the identity

$$\frac{1}{4\pi^2} \int d\mathbf{R}_\| e^{i\mathbf{q}_\|.\mathbf{R}_\|} = \delta(\mathbf{q}_\|) \qquad (4.31)$$

has been used. The condition $\delta(\mathbf{q}_\|)$ yields specular reflection. We can now calculate the reflectivity $R$ which has been defined in Chap. 3 as

$$R(\mathbf{q}) = \frac{I(\mathbf{q})}{I_0},$$

where $I(\mathbf{q})$ is the scattered intensity (flux of Poynting's vector through the detector area) and $I_0$ is the incident beam intensity. By definition, the differential scattering cross-section is the power scattered per unit solid angle per unit incident flux.

$$I(\mathbf{q}) = I_0 \int \frac{d\sigma}{d\Omega} (\mathbf{q}) d\Omega.$$

The angular acceptance of the detector is $\Delta\Omega = \Delta\theta \times \Delta\psi$, $\Delta\theta$ and $\Delta\psi$ being the vertical and horizontal angular spread of the detector, respectively. The detector being in the specular direction, $dq_x = k_0 \sin\theta_{sc} d\theta_{sc}$ and $dq_y = k_0 d\psi$. Since $q_z = 2k_0 \sin\theta$, we have

$$d\theta d\phi = \frac{2d\mathbf{q}_\|}{k_0 q_z}.$$

The incident flux is equal to $I_0/(A\sin\theta)$, and one has

$$R = \frac{1}{I_0} \times \frac{I_0}{A} \underbrace{\frac{1}{\sin\theta}}_{=q_z/(2k_0)} \iint \frac{k_0^4 A}{4q_z^2} (n_1^2 - n_0^2)^2 (\hat{\mathbf{e}}_{in}.\hat{\mathbf{e}}_{sc})^2 \delta(\mathbf{q}_\|) \times \frac{2d\mathbf{q}_\|}{k_0 q_z}. \qquad (4.32)$$

One finally obtains in (s) polarisation

$$R = \frac{k_0^4}{q_z^4}(n_1^2 - n_0^2)^2 = \frac{16\pi^2}{q_z^4}r_e^2(\rho_1 - \rho_0)^2 = \frac{q_c^4}{16q_z^4}, \qquad (4.33)$$

which shows the well-known $q_z^{-4}$ decay in reflectivity. Equation (4.32) also shows that within the Born approximation, the Brewster angle, for which the reflection coefficient is 0 in $(p)$ polarisation, is $45°$.

## 4.2.3 Single Rough Surface

We now consider a rough surface. This example is mainly detailed here to show how height–height correlation functions arise as average surface quantities in the scattering cross-section. The scheme of the calculations is similar to the previous one (case of a flat surface). It will always be the same within the Born or distorted-wave Born approximations, whatever the kind of surface or interface roughness considered. We start from

$$\frac{d\sigma}{d\Omega} = \frac{k_0^4}{16\pi^2}(n_1^2 - n_0^2)^2 (\hat{e}_{in}.\hat{e}_{sc})^2 \left| \int d\mathbf{r}_\| \int_{-\infty}^{z(\mathbf{r}_\|)} dz \, e^{i\mathbf{q}.\mathbf{r}} \right|^2. \qquad (4.34)$$

Following the same method as for a flat surface (Sect. 4.2.2), the integration over $z$ yields

$$\frac{d\sigma}{d\Omega} = \frac{k_0^4}{16\pi^2 q_z^2}(n_1^2 - n_0^2)^2 (\hat{e}_{in}.\hat{e}_{sc})^2 \left| \int d\mathbf{r}_\| \, e^{iq_z z(\mathbf{r}_\|)} e^{i\mathbf{q}_\| . \mathbf{r}_\|} \right|^2. \qquad (4.35)$$

Equation (4.35) can be written as

$$\frac{d\sigma}{d\Omega} = \frac{k_0^4(n_1^2 - n_0^2)^2}{16\pi^2 q_z^2}(\hat{e}_{in}.\hat{e}_{sc})^2 \int d\mathbf{r}_\| \int d\mathbf{r}'_\| e^{iq_z(z(\mathbf{r}_\|)-z(\mathbf{r}'_\|))} e^{i\mathbf{q}_\| . \mathbf{r}_\|} e^{-i\mathbf{q}_\| . \mathbf{r}'_\|}. \qquad (4.36)$$

Making the change of variables $\mathbf{R}_\| = \mathbf{r}_\| - \mathbf{r}'_\|$ and integrating over $\mathbf{R}_\|$:

$$\frac{d\sigma}{d\Omega} = \frac{k_0^4 A}{16\pi^2 q_z^2}(n_2^2 - n_1^2)^2 (\hat{e}_{in}.\hat{e}_{sc})^2 \int d\mathbf{R}_\| \, \langle e^{iq_z(z(\mathbf{R}_\|)-z(\mathbf{0}))} \rangle e^{i\mathbf{q}_\| . \mathbf{R}_\|}, \qquad (4.37)$$

where $A$ is the illuminated area and we have simply used the definition of the average over a surface.[7] Assuming Gaussian statistics of the height fluctuations $z(\mathbf{r}_\|)$ (see Chap. 2), or in any case expanding the exponential to the lowest (second) order, we have

$$\left\langle e^{iq_z(z(\mathbf{R}_\|)-z(\mathbf{0}))} \right\rangle = e^{-\frac{1}{2}q_z^2 \langle z(\mathbf{R}_\|)-z(\mathbf{0})\rangle^2}. \qquad (4.38)$$

---

[7] In general, this average over the surface will not be known and we will use an ensemble average as discussed in Chap. 2.

We then obtain

$$\frac{d\sigma}{d\Omega} = \frac{k_0^4 A}{16\pi^2 q_z^2} \left(n_1^2 - n_0^2\right)^2 (\hat{\mathbf{e}}_{\text{in}}.\hat{\mathbf{e}}_{\text{sc}})^2 e^{-q_z^2 \langle z \rangle^2} \int d\mathbf{R}_\parallel e^{q_z^2 \langle z(\mathbf{R}_\parallel) z(0) \rangle} e^{i\mathbf{q}_\parallel .\mathbf{R}_\parallel}. \quad (4.39)$$

This equation also includes specular (coherent) components because it has been constructed from the general solution of an electromagnetic field in a vacuum. The diffuse intensity can be obtained by removing the specular component (calculated by applying exactly the same scheme of calculation as for the case of the flat surface):

$$\left(\frac{d\sigma}{d\Omega}\right)_{\text{coh}} = \frac{k_0^4 A}{4 q_z^2} \left(n_1^2 - n_0^2\right)^2 e^{-q_z^2 \langle z \rangle^2} (\hat{\mathbf{e}}_{\text{in}}.\hat{\mathbf{e}}_{\text{sc}})^2 \delta(\mathbf{q}_\parallel). \quad (4.40)$$

The diffuse (incoherent) intensity is then

$$\left(\frac{d\sigma}{d\Omega}\right)_{\text{incoh}} = \frac{k_0^4 A}{16\pi^2 q_z^2} (n_1^2 - n_0)^2 (\hat{\mathbf{e}}_{\text{in}}.\hat{\mathbf{e}}_{\text{sc}})^2$$
$$\times e^{-q_z^2 \langle z^2 \rangle} \int d\mathbf{R}_\parallel \left( e^{q_z^2 \langle z(\mathbf{R}_\parallel) z(0) \rangle} - 1 \right) e^{i\mathbf{q}_\parallel .\mathbf{R}_\parallel}. \quad (4.41)$$

## 4.3 Distorted Wave Born Approximation

We will now present an approximation with a further order of complexity. We choose as reference state the same system as the real one, but with perfectly flat interfaces (step index profiles). The Green function and the field in Eq. (4.11) are therefore those for flat steep interfaces, and the iterative methods discussed in Chap. 3 can be used to calculate the field and the Green function. This approximation yields better results than the first Born approximation near the critical angle for total external reflection. It is currently the most popular approximation for the treatment of x-ray surface scattering data.

A first change due to the new choice of reference state is that, because refraction is taken into account, the normal component of the wavevector now depends on the local index. Using Snell–Descartes law,

$$k_{z,i} = k_0 \sqrt{\sin^2 \theta - \sin^2 \theta_{ci}}, \quad (4.42)$$

where $\theta_{ci}^2 = 2(1 - n_i)$ is the critical angle for total external reflection between vacuum and medium $i$ with $n_i = 1 - \delta_i - i\beta_i$. More precisely, the real and imaginary parts of the wavevector in medium $i$ are

$$\mathscr{R}e(k_{z,i}) = \frac{1}{\sqrt{2}} k_0 \sqrt{[(\theta^2 - 2\delta_i)^2 + 4\beta_i^2]^{1/2} + (\theta^2 - 2\delta_i)}, \quad (4.43)$$

$$\mathscr{I}m(k_{z,i}) = \frac{1}{\sqrt{2}} k_0 \sqrt{[(\theta^2 - 2\delta_i)^2 + 4\beta_i^2]^{1/2} - (\theta^2 - 2\delta_i)}. \quad (4.44)$$

As mentioned above, refraction also implies that the direction of the polarisation vector in $(p)$ polarisation changes from layer to layer. To avoid the complications related to this point, unless otherwise specified, we will always limit ourselves to the case of scattering of a $(s)$ polarised wave into $(s)$ polarisation in the rest of this section. Then, one has $(\hat{\mathbf{e}}_{in}.\hat{\mathbf{e}}_{sc}) = \cos\psi$ in every layer.

### 4.3.1 A Single Interface

#### 4.3.1.1 Scattering Cross-Section for a Single Rough Interface Between Two Homogeneous Media

Considering only one rough interface between media (0) and (1) and placing the reference plane above the real rough interface (Fig. 4.1),[8] we have, for $(s)$ or $(p)$ polarisation

$$
\mathbf{E}_{det}^{\hat{\mathbf{e}}_{sc}}(\mathbf{R},\mathbf{r}') = \frac{k_0^2 e^{-ik_0 R}}{4\pi\epsilon_0 R} E_1^{PW}(-k_{scz,1},z')e^{i\mathbf{k}_\parallel \cdot \mathbf{r}'}\hat{\mathbf{e}}_{sc}
$$

$$
= \frac{k_0^2 e^{-ik_0 R}}{4\pi\epsilon_0 R} t_{0,1}^{sc} e^{i\mathbf{k}_{sc,1} \cdot \mathbf{r}'}\hat{\mathbf{e}}_{sc}, \tag{4.45}
$$

$$
\mathbf{E}(\mathbf{r}') = \mathbf{E}_{in} E_1^{PW}(k_{inz,1},z')e^{-i\mathbf{k}_\parallel \cdot \mathbf{r}'}\hat{\mathbf{e}}_{in}
$$

$$
= \mathbf{E}_{in} t_{0,1}^{in} e^{-i\mathbf{k}_{in,1} \cdot \mathbf{r}'}\hat{\mathbf{e}}_{in}, \tag{4.46}
$$

where $t_{0,1}^{in}$ and $t_{0,1}^{sc}$ are the Fresnel transmission coefficients for polarisation $(s)$ for the angle of incidence $\theta_{in}$ and the scattering angle in the scattering plane $\theta_{sc}$, respectively. Explicit expressions for those coefficients are given by Eqs. (3.83) and (3.84). Putting Eqs. (4.45) and (4.46) in Eq. (4.11) and following the same treatment of the integrals as in Sect. 4.2.3, we obtain a generalisation of Eq. (4.41):

$$
\left(\frac{d\sigma}{d\Omega}\right)_{incoh} = A\frac{k_0^4}{16\pi^2}(n_1^2 - n_0^2)^2 \left|t_{0,1}^{in}\right|^2 \left|t_{0,1}^{sc}\right|^2 \frac{e^{-\frac{1}{2}(q_{z,1}^2 + q_{z,1}^{*2})\langle z^2 \rangle}}{|q_{z,1}|^2}
$$

$$
\times (\hat{\mathbf{e}}_{in}.\hat{\mathbf{e}}_{sc})^2 \int d\mathbf{R}_\parallel \left[e^{|q_{z,1}|^2 \langle z(\mathbf{R}_\parallel)z(0)\rangle} - 1\right] e^{i\mathbf{q}_\parallel \cdot \mathbf{R}_\parallel}, \tag{4.47}
$$

or, using $n = 1 - \lambda^2/2\pi r_e \rho$,

$$
\left(\frac{d\sigma}{d\Omega}\right)_{incoh} = A r_e^2 (\rho_1 - \rho_0)^2 \left|t_{0,1}^{in}\right|^2 \left|t_{0,1}^{sc}\right|^2 \frac{e^{-\frac{1}{2}(q_{z,1}^2 + q_{z,1}^{*2})\langle z^2 \rangle}}{|q_{z,1}|^2}
$$

$$
\times (\hat{\mathbf{e}}_{in}.\hat{\mathbf{e}}_{sc})^2 \int d\mathbf{R}_\parallel \left[e^{|q_{z,1}|^2 \langle z(\mathbf{R}_\parallel)z(0)\rangle} - 1\right] e^{i\mathbf{q}_\parallel \cdot \mathbf{R}_\parallel}. \tag{4.48}
$$

---

[8] See Sect. 4.3.1.

This expression is explicitly symmetrical in the source and detector positions as required by the reciprocity theorem. It differs from Eq. (4.41) by the additional transmission coefficients.

### 4.3.1.2  Density Inhomogeneities in the Substrate

Only surface scattering has been considered up to this point. However, the dielectric index inhomogeneities leading to scattering can also be density fluctuations. This should always be borne in mind when interpreting experiments. The scattering due to density inhomogeneities can be treated using a formalism similar to that used for surface scattering. The relevant correlation functions will be of the form $\langle \delta \rho(\mathbf{0}, z') \delta \rho(\mathbf{r}_\parallel, z) \rangle$. This problem was considered in the early paper of Bindell and Wainfan [2]. Again we limit the discussion to the scattering of a $(s)$ polarised incident wave into a $(s)$ polarised wave.

The surface is assumed to be perfectly smooth in this analysis. The upper medium 0 is homogeneous whereas density inhomogeneities occur in medium 1. Within the DWBA the differential scattering cross-section is

$$\frac{d\sigma}{d\Omega} = A r_e^2 \left| t_{0,1}^{\text{in}} \right|^2 \left| t_{0,1}^{\text{sc}} \right|^2 (\hat{\mathbf{e}}_{\text{in}} . \hat{\mathbf{e}}_{\text{sc}})^2 \tag{4.49}$$

$$\times \left\langle \int d\mathbf{r}_\parallel \int d\mathbf{r}'_\parallel e^{i\mathbf{q}_\parallel \cdot (\mathbf{r}_\parallel - \mathbf{r}'_\parallel)} \int dz \int dz \delta \rho_1(\mathbf{r}_\parallel, z) \delta \rho_1(\mathbf{r}'_\parallel, z') e^{i(q_z z - q_z^* z')} \right\rangle,$$

where we have used $n = 1 - (\lambda^2/2\pi) r_e \rho$. Using the identity

$$i\left(q_z z - q_z^* z'\right) = iq_z \left(z - z'\right) - 2\mathscr{I}m\left(q_z\right) z', \tag{4.50}$$

and making the change of variables $z \mapsto z + z'$ one obtains

$$\frac{d\sigma}{d\Omega} = A r_e^2 \left| t_{0,1}^{\text{in}} \right|^2 \left| t_{0,1}^{\text{sc}} \right|^2 (\hat{\mathbf{e}}_{\text{in}} . \hat{\mathbf{e}}_{\text{sc}})^2 \int_{-\infty}^0 dz' e^{-2\mathscr{I}m(q_z) z'}$$

$$\times \int d\mathbf{R}_\parallel \int_{-\infty}^{-z'} dz\, e^{i\mathbf{q}_\parallel \cdot \mathbf{r}} e^{iq_z z} \left\langle \delta \rho_1(\mathbf{R}_\parallel, z) \delta \rho_1(\mathbf{0}, 0) \right\rangle. \tag{4.51}$$

Equation (4.51) usually leads to a shorter equation when $\langle \delta \rho(\mathbf{0}, z') \delta \rho(\mathbf{r}_\parallel, z) \rangle$ is known. The example of liquids is treated in Sect. 4.5.1 [see Eq. (4.78)].

### 4.3.1.3  DWBA Versus First Born Approximation

The amplitude of the real and imaginary parts of the unperturbated electric field (reference case) is plotted in Fig. 4.3 for a grazing angle of incidence $\theta_{\text{in}}$ fixed below the critical angle of reflection. For $z > 0$ (in the upper medium), the electric field is sinusoidal (standing waves). For $z < 0$ (below the surface) the field decays exponentially (evanescent waves). Within the Born approximation, the field amplitude is

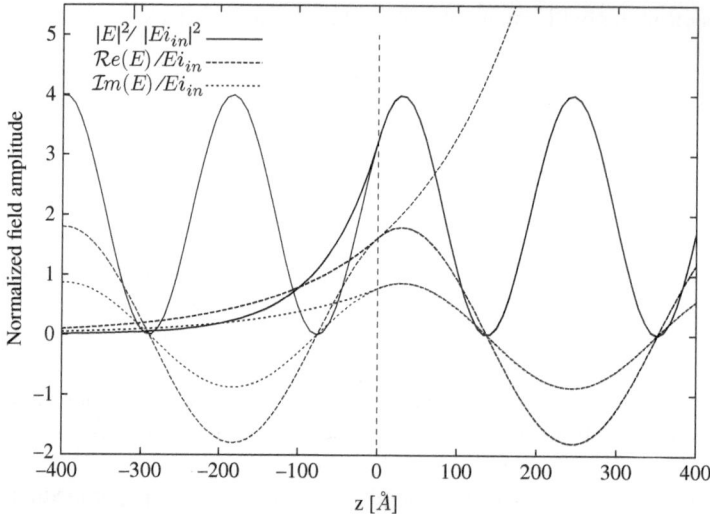

**Fig. 4.3** Amplitude of the real and imaginary parts and modulus of the electric field at the silicon–air interface. The surface is at $z = 0$, silicon at $z < 0$. The angle of incidence is $\theta_{in} = 0.5\theta_c$ below the critical angle for total reflection. The prolongation of either the transmitted or reflected field in, respectively, air or silicon is shown using *thinner lines*

a constant, and at any time, the field is a sine wave (propagating at speed c). The DWBA is a better approximation than the Born approximation as its reference case is a better approximation to the real system. It obviously leads to very different results for inhomogeneities deeply buried in the sample as there is no total reflection within the Born approximation. Even for surface roughness below the critical angle for total external reflection the result will be quite different as the amplitude of the field at the surface is larger than $E_{in}$.[9]

Another important point which is clear on Fig. 4.3 is that not only the field but also its first derivative is continuous at the surface. For this reason, for moderate roughnesses (less than a few nm), one can use either the expression of the field below or above the surface to calculate the scattered intensity. As there is only one wave propagating in the substrate, it is easier to use the transmitted field as has been done in Eq. (4.47). The full calculation using the reflected field is given for example in [15]. The exact position of the reference surface is also unimportant for the same reason.

At the critical angle for total external reflection $\theta_{in} = \theta_{ci}$, the transmission coefficients have a peak value of 2 (Fig. 4.4) as the incident and reflected fields are in phase at $z = 0$. As the dipole source equivalent to roughness $\epsilon_0 \delta n^2 \mathbf{E}$ is proportional to $\mathbf{E}$, there is a maximum in the scattered intensity. By using the reciprocity theorem, one can see that the Green function is also peaked near $\theta_{sc} = \theta_c$.[10] Those

---

[9] This is not a problem from the conservation of energy point of view as there is no energy flux inside the substrate.

[10] Equivalently, the peak in the Green function can be seen to arise from the angular dependence of the field emitted by a dipole placed below the interface.

**Fig. 4.4** Square of the modulus of the Fresnel transmission coefficient for polarisation $(s)$ $(|t^{in}|^2)$ as a function of $\theta_{in}/\theta_c$. $\theta$ is the angle of incidence and $\theta_c$ the critical angle for total external reflection

**Fig. 4.5** Scattering by the bare water surface. $\theta_{sc}$ detector scan at fixed $\theta_{in}$ and $\psi$. Note the presence of the Yoneda peak

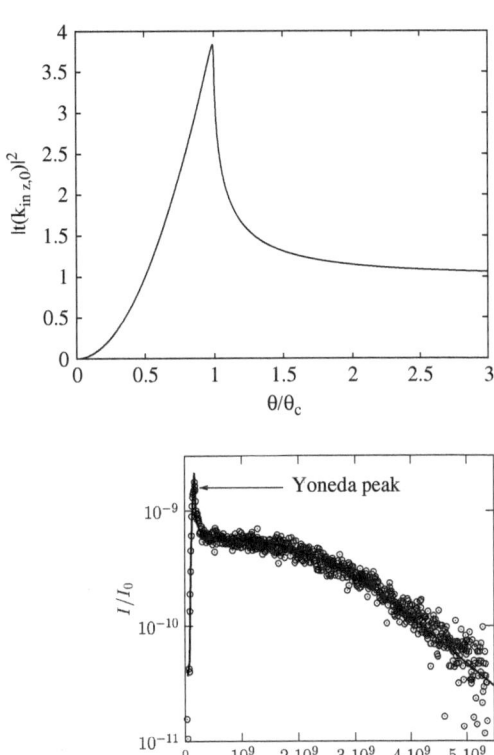

peaks are the so-called Yoneda peaks [46]. The Yoneda peak at $\theta_{sc} = \theta_c$ is shown in Fig. 4.5 for a scan at fixed $\theta_{in}$. The two peaks at $\theta_{sc} = \theta_c$ and $\theta_{in} = \theta_c$ are shown for a different type of scan (rocking curve) on Fig. 4.10.

## 4.3.2 General Case of a Stratified Medium

### 4.3.2.1 Interface Scattering

In the general case of a stratified medium depicted in Fig. 4.6, one has in layer $j$ for $(s)$ or $(p)$ polarisation

$$\mathbf{E}(\mathbf{r}') \approx E_{in} E_j^{PW}(k_{in\,z,j}, z') e^{-i\mathbf{k}_{in\|} \cdot \mathbf{r}'\|} \widehat{\mathbf{e}}_{in\,j}$$

$$\mathbf{E}_{det}^{\widehat{\mathbf{e}}_{sc}}(\mathbf{R}, \mathbf{r}') = \frac{k_0^2 e^{-ik_0 R}}{4\pi\epsilon_0 R} E_j^{PW}(-k_{sc\,z,j}, z') e^{i\mathbf{k}_{sc\|} \cdot \mathbf{r}'\|} \widehat{\mathbf{e}}_{sc\,j}. \qquad (4.52)$$

The DWBA method consists then in developing the $E^{PW}$ functions defined in Eq. (4.18) in each medium as, for example, in $(s)$ polarisation, in layer $j$:

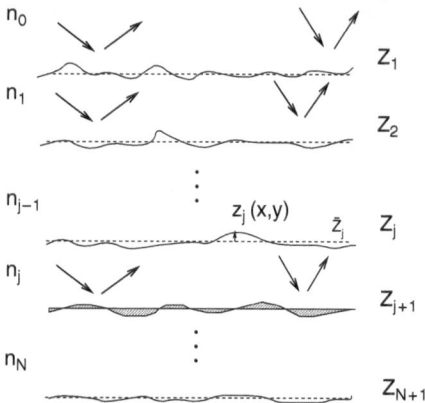

**Fig. 4.6** X-ray surface scattering in a stratified rough medium. Because of multiple reflections, there are waves propagating upwards (with an amplitude $U(k_{z,j},z)$) and downwards (with an amplitude $U(-k_{z,j},z)$) in layer $j$ where the total of the field amplitude is $E_{\text{in},j}^{PW}$ (there is an equivalent dependence of the Green function). Multiple reflections are considered within the DWBA but not within the first Born approximation. The perturbation method consists in evaluating the field scattered by the dipolar density equivalent to the index difference $(n_{j-1} - n_j)$ between the real system where the rough interface profile is $z_j(\mathbf{r}_\parallel)$ and the unperturbated system where the interface is located at $Z_j$ and is placed here at the average interface plane. For interface $j$ the unperturbated and real index distributions differ in the hatched region

$$E_j^{PW\,(s)}(k_{z,j},z) = U^{(s)}(k_{z,j},Z_j)e^{ik_{z,j}z} + U^{(s)}(-k_{z,j},Z_j)e^{-ik_{z,j}z}$$
$$= \sum_{\pm} U^{(s)}(\pm k_{z,j},Z_j)e^{\pm ik_{z,j}z}, \qquad (4.53)$$

where the $U$ coefficients are the magnitudes of the upwards and downwards propagating waves which are explicitly obtained in Chap. 3 of this book, Eq. (3.60), using the "Vidal and Vincent" representation of transfer matrices [44]. The field is then written (put Eqs. (4.52) in Eq. (4.11) and sum over all interfaces) as

$$E^{(s)} = E_{\text{ref}} + E_{\text{in}} \frac{k_0^2 e^{-ik_0 R}}{4\pi\epsilon_0 R} \widehat{\mathbf{e}}_{\text{in}}.\widehat{\mathbf{e}}_{\text{sc}} \sum_{j=0}^{N} \int d\mathbf{r}_\parallel\, e^{i\mathbf{q}_\parallel.\mathbf{r}_\parallel}$$
$$\int_0^{z_{j+1}(\mathbf{r}_\parallel)} dz \epsilon_0 (n_{j+1}^2 - n_j^2) E_{j+1}^{PW}(k_{\text{inz},j+1},z) E_{j+1}^{PW}(-k_{\text{scz},j+1},z), \quad (4.54)$$

where it has been assumed that the reference plane is located above the interface, hence the $E_{j+1}^{PW}$ fields. Then, the generalisation of Eq. (4.47) is

$$\left(\frac{d\sigma}{d\Omega}\right)_{\text{incoh}} = \frac{k_0^4}{16\pi^2} (\widehat{\mathbf{e}}_{\text{in}}.\widehat{\mathbf{e}}_{\text{sc}})^2 \sum_{j=1}^{N} \sum_{k=1}^{N} \sum_{\pm} \sum_{\pm} \sum_{\pm} \sum_{\pm} (n_j^2 - n_{j-1}^2)(n_k^2 - n_{k-1}^2)^*$$
$$U^{(s)}(\pm k_{\text{inz},j},Z_j)U^{(s)}(\pm k_{\text{scz},j},Z_j)U^{(s)*}(\pm k_{\text{inz},k},Z_k)U^{(s)*}(\pm k_{\text{scz},k},Z_k)$$
$$\tilde{Q}_{j,k}(\pm k_{\text{inz},j} \pm k_{\text{scz},j}, \pm k_{\text{inz},k} \pm k_{\text{scz},k}), \qquad (4.55)$$

with

$$
\tilde{Q}_{j,k}(q_z,q_z') = \int d\mathbf{r}_\parallel \int d\mathbf{r}'_\parallel e^{i\mathbf{q}_\parallel \cdot (\mathbf{r}_\parallel - \mathbf{r}'_\parallel)} \left[ \left\langle \int_0^{z_j(\mathbf{r}_\parallel)} dz \int_0^{z_k(\mathbf{r}'_\parallel)} dz' e^{i(q_z z - q_z^* z')} \right\rangle \right.
$$
$$
\left. - \left\langle \int_0^{z_j(\mathbf{r}_\parallel)} dz e^{iq_z z} \right\rangle \left\langle \int_0^{z_k(\mathbf{r}'_\parallel)} dz' e^{-iq_z^* z'} \right\rangle \right],  \tag{4.56}
$$

where the specular (coherent) contribution, obtained as an average over the field as shown in Chap. 2, has been removed. Performing the integrations over $z$ and $z'$ and making the change of variables $\mathbf{r}_\parallel - \mathbf{r}'_\parallel \to \mathbf{R}_\parallel$ as previously,

$$
\tilde{Q}_{j,k}(q_z,q_z') = A \frac{e^{-\frac{1}{2}[q_{j,z}^2 \langle z_j^2 \rangle + (q_{k,z}^*)^2 \langle z_k^2 \rangle]}}{q_{j,z} q_{k,z}^*} \int d\mathbf{R}_\parallel \, e^{i\mathbf{q}_\parallel \cdot \mathbf{R}_\parallel} \left( e^{q_{j,z} q_{k,z}^* \langle z_j(0) z_k(\mathbf{R}_\parallel) \rangle} - 1 \right).  \tag{4.57}
$$

Because reflection at all interfaces is taken into account, all the possible combinations of the incident and scattered wavevectors appear in the formulae.

### 4.3.3 Density Inhomogeneities in a Multilayer

The interfaces are now assumed to be perfectly smooth. Within the DWBA, and assuming effective $U$ functions within the layers,[11] the differential scattering cross-section will be (cf. Eq. (4.55))

$$
\frac{d\sigma}{d\Omega} = r_e^2 (\hat{\mathbf{e}}_{in}.\hat{\mathbf{e}}_{sc})^2 \sum_{j=1}^{N} \sum_{k=1}^{N} \sum_\pm \sum_\pm \sum_\pm \sum_\pm \tilde{B}_{j,k}(\pm k_{inz,j} \pm k_{scz,j}, \pm k_{inz,k} \pm k_{scz,k})
$$
$$
U^{(s)}(\pm k_{inz,j}, Z_j) U^{(s)}(\pm k_{scz,j}, Z_j) U^{(s)*}(\pm k_{inz,k}, Z_k) U^{(s)*}(\pm k_{scz,k}, Z_k),  \tag{4.58}
$$

where now

$$
\tilde{B}_{j,k}(q_z,q_z') = \int d\mathbf{r}_\parallel \int d\mathbf{r}'_\parallel e^{i\mathbf{q}_\parallel \cdot (\mathbf{r}_\parallel - \mathbf{r}'_\parallel)} \int_0^{Z_{j+1}-Z_j} dz \int_0^{Z_{k+1}-Z_k} dz'
$$
$$
\left\langle \delta\rho_j^2(\mathbf{r}_\parallel,z) \delta\rho_k^{2*}(\mathbf{r}'_\parallel,z') \right\rangle e^{i(q_z z - q_z^* z')}.  \tag{4.59}
$$

Making the change of variables $\mathbf{r}_\parallel - \mathbf{r}'_\parallel \to \mathbf{R}_\parallel$,

---

[11] Assuming effective $U$ functions within the layers is only possible if the characteristic size of the inhomogeneities is much smaller than the extinction length, see Appendix 3.A to Chap. 3. This might not be the case for multilayer gratings (see Sect. 6.7) or large copolymer domains [8].

$$\tilde{B}_{j,k}(q_z, q_z') = A \int d\mathbf{R}_\| e^{i\mathbf{q}_\| \cdot \mathbf{R}_\|} \int_0^{Z_{j+1}-Z_j} dz \int_0^{Z_{k+1}-Z_k} dz'$$
$$\left\langle \delta\rho_j^2(\mathbf{0}, z)\delta\rho_k^{2*}(\mathbf{R}_\|, z') \right\rangle e^{i(q_z z - q_z^* z')}. \tag{4.60}$$

In the case of a semi-infinite medium, only $U_1(-k_{\text{in},z,1}, Z_1) = t^{\text{in}}$ and $U_1(-k_{\text{sc}}, z, 1, Z_1) = t^{\text{sc}}$ are different from 0. Writing $q_{z,1} = \mathscr{R}e(q_{z,1}) + i\,\mathscr{I}m(q_{z,1})$, one obtains

$$\tilde{B}_{1,1}(q_{z,1}, q_{z,1}) = A \int d\mathbf{R}_\| e^{i\mathbf{q}_\| \cdot \mathbf{R}_\|}$$
$$\int_{-\infty}^0 \int_{-\infty}^0 dz dz'\, e^{i\mathscr{R}e(q_{z,1})(z-z')} e^{\mathscr{I}m(q_{z,1})(z+z')} \left\langle \delta\rho_1^2(\mathbf{0}, z)\delta\rho_1^{2*}(\mathbf{R}_\|, z') \right\rangle, \tag{4.61}$$

i.e. the bulk fluctuations are integrated over the penetration length of the beam.

Comparing Eq. (4.61) to Eq. (4.57), we note that contrary to bulk scattering, surface scattering is inversely proportional to the square of wavevector transfers. Therefore, surface scattering will generally be dominant at grazing angles whereas bulk scattering will ultimately dominate at large scattering angles.

### 4.3.4 Further Approximations

The distorted-wave Born approximation as presented here does not always allow an accurate enough representation of the scattered intensity close to the critical angle for total external reflection [45]. Understanding scattering at grazing angles is highly desirable because bulk scattering is minimised under such conditions. This is critical because the signal scattered by surfaces or interfaces is generally very low. Different approaches have been attempted to improve the DWBA.

A first approximation consists in taking into account the average interface profile in Eq. (4.11) [20]. The reference medium is now defined by the relative permittivity $\epsilon_{\text{ref}}(z) = \langle\epsilon\rangle(z)$, with $\langle\epsilon\rangle(z) = 1/A \int \epsilon(\mathbf{r}_\|, z) d\mathbf{r}_\|$ in the case of a rough surface defined by an ergodic random process. (In [1] the shape of the average permittivity is approximated by a hyperbolic tangent profile to simplify the calculation of the reference Green tensor.) The main interest of this new reference medium is that the reference field $\mathbf{E}_{\text{ref}}$ is that of the transition layer and thus contains directly the Névot–Croce factor in the reflection coefficient. Moreover the perturbation $\delta n^2(\mathbf{r}_\|)$ is of null average $\langle\delta n^2\rangle = 0$ and we may expect to have minimised its value (and thus extended the validity domain of the perturbative development). This improvement has been shown to yield much better results than the classical DWBA in the optical domain where the permittivity contrasts are important [38]. In the x-ray domain its interest is more questionable since it does not lead to simple expressions for the scattering cross-section. Indeed $\delta n^2(\mathbf{r}_\|)$ is no longer a step function and the integration along the z-axis cannot be done analytically. A possibility is to use the matrix method to describe the transition layer [20].

Another possibility would be to directly take into account multiple surface (roughness) scattering without using the effective medium approximation. It is then necessary to iterate the fundamental Eq. (4.11) [5–7, 20]. This has been done up to the second order in [18] for specular reflectivity, and the corrections might be important close to the critical angle for total external reflection.

Finally there exist many approximate methods that have been developed in totally different contexts (optics, radar). In most methods, the field scattered by the rough surface is evaluated with a *surface integral equation* (given by the Huygens–Fresnel principle (or Kirchhoff integral) [32]). The integrand of the latter contains the field values and its normal derivatives at each point of the surface. The Kirchhoff approximation consists in replacing the field on the surface by the field that would exist if the surface is locally assimilated to its tangent plane. This technique, when applied to the coherent field yields the famous Debye–Waller factor on the reflection coefficient. It is a single scattering approximation (also called physical optics approximation). The perturbative theory (the small parameter is the rms height of the surface) has also been widely used. A possible starting point is writing the boundary conditions on the field and its derivative at the interface under the Rayleigh hypothesis. A brief survey of this method is given in Sect. 3.A.1. Note that the iteration of these methods permit to account for some multiple scattering effects, but the increasing complexity of the calculation limits their interest. It is now also possible to consider the resolution of the surface integral equation satisfied by the field without any approximation (and thus to account for all the multiple scattering). Preliminary results have been already presented in the radar and optical domain. However, in the x-ray domain those techniques have a major drawback: They only consider surface scattering (with a surface integral equation) and the generalisation to both surface and volume scatterings is not straightforward. The differential method [33] which consists in solving the inhomogeneous differential equation satisfied by the Fourier component of the field (in the $\mathbf{k}_\parallel$ space) with a Runge–Kutta algorithm along the z-axis would be more adequate. It has already been used to calculate the diffraction by multilayer gratings and accounts for all multiple scattering (no approximation), but it remains difficult to use it for non-periodic (rough) surfaces because of the computing time and memory required.

## 4.4 From the Scattering Cross-Section to the Scattered Intensity

### 4.4.1 The Different Types of Scans

Different types of scans can be used depending on the information one is looking for. These are

- Constant $q_z$ scans in the plane of incidence (defined by the incident wave-vector and the normal to the surface). They are also known as "Rocking curves", as rocking the sample for a fixed incident beam/detector geometry approximately provides a constant $q_z$. The advantage of such scans is to provide a simple $q_x$ dependence. The disadvantages are a limited $q_x$ range (see Fig. 4.7) and possibly a large background from the substrate as the grazing angle of incidence will generally be larger than the grazing angle for total external reflection.

(a)                                            (b)

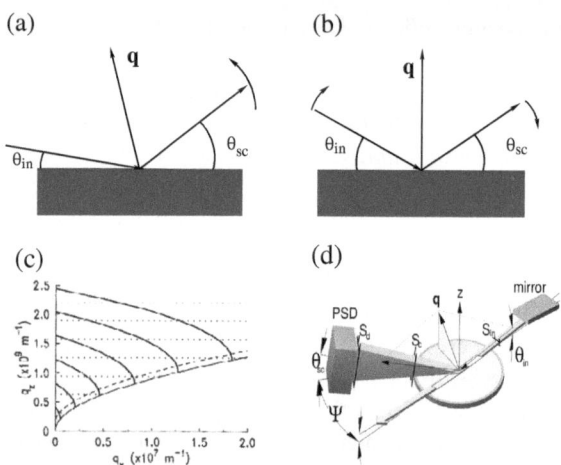

(c)                            (d)

**Fig. 4.7** Scan types. (**a**) Detector scan in the plane of incidence. (**b**) Constant $q_z$ scan in the plane of incidence. (**c**) Constant $q_z$ (*horizontal dotted lines*) and detector scans (*thick continuous lines*) in the ($q_x,q_z$) plane of incidence. The *right-bottom* area is below the surface and cannot be reached in the plane of incidence. (**d**) $\psi$ scan using a linear detector around the surface normal

- Detector scans in the plane of incidence for a fixed angle of incidence, generally below the critical angle for total external reflection, known as "detector scans". The advantage of such scans is to always keep the background from the substrate low if the grazing angle of incidence is lower than the critical angle for total external reflection. The scattered intensity however has a more complicated structure than in rocking curves as both $q_x$ and $q_z$ are varied during a scan. Both rocking curves and detector scans give access to low $q_x$ values as $q_x = k_0(\cos \theta_{sc} - \cos \theta_{in})$ (see Sect. 4.5.1).
- Scans in the plane of incidence with a constant $\theta_{sc}$ offset known as longitudinal scans, or with a constant $q_x$ offset. Such scans can be very useful for example for determining the conformality of interfaces in thin films. They can also be used to determine and subtract the background in a reflectivity experiment.
- $\psi$ scans around the normal to the surface, generally performed using a linear position-sensitive detector. Such scans should be performed when one wants access to large $q_{\parallel}$ values. The combination of detector and $\psi$ scans allows the full determination of surface spectra for in-plane wave-vectors in the range $10^5 \, \text{m}^{-1}$ to $10^{10} \, \text{m}^{-1}$.

These different types of scans are illustrated in Fig. 4.7.

## 4.4.2 Expression of the Scattered Intensity

In contrast to specular reflectivity where the specular condition $\delta(\mathbf{q}_{\parallel})$ implies that resolution effects amount to a simple convolution, the scattered intensity in a diffuse scattering experiment is proportional to the resolution volume. It is therefore necessary to have a detailed knowledge of the resolution function in order to get quantitative information from an experiment.

Starting from the definition of the scattering cross-section, we have for the scattered intensity

$$I = \frac{I_0}{h_i w_i} \int \frac{d\sigma}{d\Omega} d\Omega = \frac{I_0}{h_i w_i} \int \left(\frac{d\sigma}{d\Omega}\right) \mathscr{R}es\left(\Omega\right) d\Omega, \qquad (4.62)$$

with $d\Omega = d\theta_{sc} d\psi$. Since each experimental setup is different, it is impossible to give a general expression for the scattered intensity. We will detail here the calculation for detector scans in the plane of incidence which is more complicated to handle. It is in general easier to perform the integrations in the wavevector space. Using

$$q_x = k_0 \left(cos\theta_{in} - cos\theta_{sc}\right) \Rightarrow dq_x = k_0 sin\theta_{sc} d\theta_{sc} \qquad (4.63)$$

$$q_y = k_0 sin\psi \simeq k_0 \psi \Rightarrow dq_y = k_0 d\psi \qquad (4.64)$$

for $\theta_{in} \approx 0$, we get for scattering by a rough surface in the plane of incidence

$$I = \frac{I_0}{h_i w_i} \frac{Ar_e^2 \rho_1^2 |t_{0,1}^{in}|^2 |t_{0,1}^{sc}|^2 (\hat{\mathbf{e}}_{in}.\hat{\mathbf{e}}_{sc})^2}{k_0^2 sin\theta_{sc}} \int dx dy \int dq_x dq_y e^{iq_x x + iq_y y} (e^{q_z^2 \langle z(0)z(\mathbf{r}_\|)\rangle} - 1), \qquad (4.65)$$

where the integration in $q_x$ and $q_y$ is over the q-space resolution. Generally, we will have widely opened slits in the y direction and $\int dq_y e^{iq_y y} = 2\pi\delta(y)$. Concerning the $q_x$ integration, the exact shape of the resolution function is unimportant as long as its area is conserved. Considering a Gaussian resolution function

$$\mathscr{R}es\left(\delta q_x\right) = \frac{1}{\sqrt{2\pi}} e^{-\frac{(\delta q_x - q_x)^2}{2\Delta q_x^2}} \qquad (4.66)$$

around $q_x$, we get

$$I = \frac{I_0}{h_i w_i} \frac{2\pi Ar_e^2 (\rho_1 - \rho_0)^2 |t_{0,1}^{in}|^2 |t_{0,1}^{sc}|^2 (\mathbf{e}_{in} \cdot \mathbf{e}_{sc})^2 \Delta\theta_{sc}}{k_0}$$

$$\int dx e^{iq_x x} e^{-\frac{\Delta q_x^2 x^2}{2}} (e^{q_z^2 \langle z(0)z(x)\rangle} - 1) \qquad (4.67)$$

*Exercise:* Show that for $\langle z_s(\mathbf{r}_\|) z_s(0)\rangle = \sigma_s^2 e^{-\left(\mathbf{r}_\|^2/xi^2\right)}$,

$$I = \frac{I_0}{h_i w_i} Ar_e^2 |t_{0,1}^{in}|^2 |t_{0,1}^{sc}|^2 (\mathbf{e}_{in} \cdot \mathbf{e}_{sc})^2 e^{-q_z^2 \sigma_s^2} (\rho_1 - \rho_0)^2 \frac{2\pi^2 \sigma_s^2}{k_0^2 sin\theta_{sc}}$$

$$\times \left(erf\left(\frac{\xi}{2}\left(q_x + \frac{\Delta q_x}{2}\right)\right) - erf\left(\frac{\xi}{2}\left(q_x - \frac{\Delta q_x}{2}\right)\right)\right), \qquad (4.68)$$

with erf(qu) $= 2q/\sqrt{\pi} \int_0^u \exp(-q^2 x^2) dx$.

It should always be kept in mind that wavevector resolution functions are delicate to handle (see [19]).[12] From a computing point of view, a numerical integration of the scattering cross-section which reduces to a multiplication with the detector solid angle when $\Delta q \ll q$ can be preferable.

### 4.4.3 Illuminated Area

In addition to the solid angle, the scattered intensity depends on the effective illuminated area which also needs to be precisely determined. A standard way to define the scattering direction is through the use of a pair of slits. We first consider detector scans in the plane of incidence. We have a slit $S_d$ ($h_d \times w_d$) in front of the detector ($L_d$ away from the sample) and another one $S_c$ ($h_c \times w_c$) at a distance $L_d - L_c$ from the sample (see Fig. 4.8). The $\theta_{\mathrm{sc}}$ acceptance of the detector is $\Delta\theta_{\mathrm{sc}} = h_d/L_d$, and the $\psi$ acceptance is simply $\Delta\psi = \max(w_d/L_d, \Delta\psi_{\mathrm{in}})$. The illuminated area is $h_i/\theta_{\mathrm{in}}$ and the intensity is proportional to $h_i h_d/L_d\theta_{\mathrm{in}}$. At grazing angles of detection, all points in the footprint "see" the detector, but for $\theta_{\mathrm{sc}} \geq (h_c L_d \theta_{\mathrm{in}})/(h_i L c)$, the slits $S_c$ and $S_d$ play the role of a collimator (which helps in reducing the background), and only a fraction of the points in the footprint can see the detector. The intensity is now proportional to $(w_i h_c h_d)/(\theta_{\mathrm{sc}} L c)$.

For scans in the sample plane (diffraction or diffuse scattering experiments), the footprint of the beam can be discretised as shown in Fig. 4.8b. The $\theta_{\mathrm{sc}}$ acceptance of the detector is $\Delta\theta_{\mathrm{sc}} = h_d/L_d$. If a vertically mounted position-sensitive detector

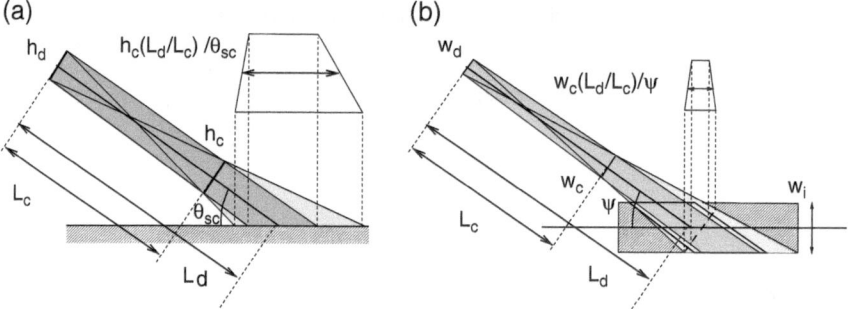

**Fig. 4.8** Resolution effects. The vertical × horizontal dimensions of the detector slit $S_d$ and of the slit placed just after the sample $S_c$ are ($h_d \times w_d$) and ($h_c \times w_c$), respectively. The sample-to-$S_d$ distance is $L_d$, and the $S_c$-to-$S_d$ distance is $L_c$. (**a**) Detector scans in the plane of incidence. A point lying in the *dark-grey area* "sees" the detector under an angle $h_d/L_d$. This angle decreases to 0 in the *light-grey* area, as represented by the trapezoid whose FWHM is $h_c(L_d/L_c)/\theta_{\mathrm{sc}}$. (**b**) Detector scans around a vertical axis. Points along the *thick dotted line* normal to the sample-to-detector axis "see" the detector proportionally to the trapezoid height. The effective illuminated length is $w_c(L_d/L_c)$ and the effective illuminated area is $w_c(L_d/L_c)/\psi \times w_i$

---

[12] The major problem is that the transformation of the angular resolution function into a wavevector resolution function leads to a function which is generally not separable in $q_x$ and $q_z$.

of height $H_d$ and resolution $h_d$ is used, it can be seen as the superposition of $H_d/h_d$ small detectors. The $\psi$ acceptance is $\Delta\psi = w_d/L_d$. Again, for very small $\psi$ values, the entire footprint contributes to the measured intensity, and for increasing $\psi$ values the slit $S_c$ eventually limits the illuminated area. The points along each subsurface contribute to the measured intensity proportionally to the trapezoid on Fig. 4.8b. The effective length on such a subsurface is $w_c L_c/L_d$, and the effective area of all the subsurfaces (the parallelogram of Fig. 4.8b) is $(w_i w_c L_d)/(L_c \psi)$. A cut-off therefore occurs for $\psi = (w_c l_d \theta_{in})/(h_i L_c)$.

The dependence of the scattered intensity on the slit sizes is easily visualised in the case of scattering by a liquid surface since the scattering cross section is well known and has a smooth dependence on $q_{\parallel}$. For example, on Fig. 4.9b for water, one

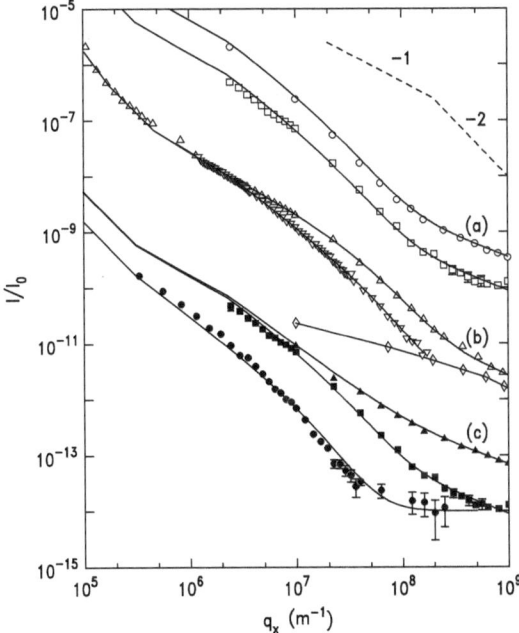

**Fig. 4.9** Resolution effects for detector scans in the plane of incidence in the case of scattering by the water-free surface. $\lambda = 0.096$ nm, $\theta_{in} = 1.41$ mrad, the incident beam size was $0.4 \times 0.2$ nm$^2$ ($H \times V$), the sample-to-detector distance was 650 nm and the collimator distance $S_c - S_d$ was 470 mm. The *symbols* are experimental data, and the *lines* are calculated without adjustable parameters. **(a)** Scaling of the scattered intensity with the height of the detector slit $h_d$ and the collimator slit $h_c$. If both $h_c$ and $h_d$ are multiplied by a factor of 2, the intensity is multiplied by a factor of 4 (2 because the effective area seen by the detector is twice as large and 2 for the angular acceptance). (*Circles:* $w_c \times h_c = 10 \times 1$ mm, $w_d \times h_d = 1 \times 0.5$ mm; *Squares:* $w_c \times h_c = 10 \times 2$ mm, $w_d \times h_d = 1 \times 0.25$ mm.) **(b)** The cut-off between the footprint limited and the collimator limited regimes is determined by the collimator slit height $h_c$ (*Triangles:* $w_c \times h_c = 10 \times 10$ mm, $w_d \times h_d = 1 \times 0.25$ mm, the *diamonds* indicate the background for this scan; *Inverted triangles:* $w_c \times h_c = 10 \times 1$ mm, $w_d \times h_d = 1 \times 0.25$ mm, these curves have been divided by 100.) **(c)** Depending on $\Delta\psi$, there is a full integration $q_y$ ($\propto q_x^{-1}$ law) or a simple convolution ($\propto q_x^{-2}$ law). (*Filled triangles:* $w_c \times h_c = 10 \times 10$ mm, $w_d \times h_d = 1 \times 0.25$ mm; *Filled squares:* $w_c \times h_c = 10 \times 1$ mm, $w_d \times h_d = 1 \times 0.25$ mm; *Filled circles:* $w_c \times h_c = 10 \times 0.1$ mm, $w_d \times h_d = 1 \times 0.25$ mm; these curves have been divided by a factor of $10^4$.) The *broken lines* indicate the $q_x^{-1}$ and $q_x^{-2}$ laws

can identify the wings of the direct beam on the left, a regime where $I \approx q_x^{-1}$ with the resolution being mainly determined by the illuminated area ($d\sigma/d\Omega \propto A/q_x^2 \propto 1/\theta_{sc}^3$; $\Delta\theta_{sc} \propto \theta_{sc}$, and therefore $I \propto 1/\theta_{sc}^2 \propto 1/q_x$) and finally a regime where the resolution is limited by the $S_c - S_d$ collimator.

## 4.4.4 Reflectivity Revisited

Equation (4.48) above shows that the diffuse intensity decreases as $q_z^{-2}$ for small $q_z$ values, whereas it was shown in Chap. 3 that the specular (coherent) intensity decreases as $q_z^{-4}$. One therefore expects that diffuse scattering will eventually dominate over the specular reflectivity. Of course the wavevector at which diffuse scattering becomes dominant will depend on the experimental resolution since the diffuse intensity is proportional to the resolution volume. In fact, for reasonable experimental conditions, the corresponding wavevectors are rather small, on the order of a few nm$^{-1}$, and this leads to major difficulties in the treatment of reflectivity data. A "reflectivity curve" $I(q_z)$ is indeed never a pure specular reflectivity curve. Moreover, the diffuse intensity is often (but not always) peaked in the specular direction (Fig. 4.10), making the separation of the specular and diffuse components very difficult experimentally.

This is a very difficult problem since the $q_z$ dependence of the diffuse intensity depends on the exact interface correlation function. A simple model can therefore

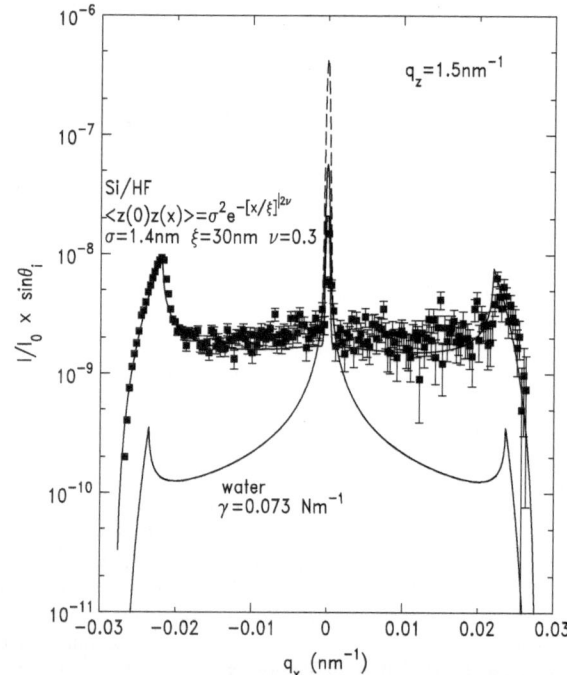

**Fig. 4.10** Diffuse scattering from the water surface which is peaked in the specular direction because capillary waves of longer wavelength cost less energy (only a calculated intensity is presented here because the large background due to bulk scattering prevents from a precise measurement, see below) and a solid surface with a flat power spectrum

**Fig. 4.11** Reflectivity of an octadecyltrichlorosilane film on water. The *broken line* corresponds to specular intensity. It is dominated by diffuse intensity (*grey line*) for wavevectors larger than $2\,\mathrm{nm}^{-1}$. The *black line* is the total (specular + diffuse) intensity. Inset: corresponding electron densities for the complete model Eq. (4.71) (*thick line*) and the simple box model with error function transition layers (*thin line*)

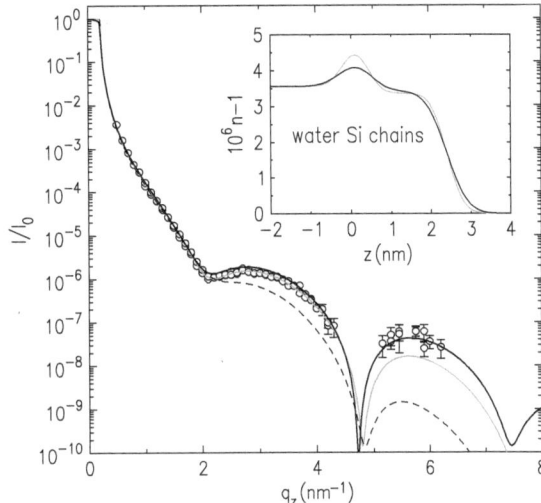

no longer be used for the analysis of "reflectivity" curves. This is the situation found for the system of Fig. 4.11, an octadecyltrichlorosilane Langmuir film on water [4]. In this case, the surface spectrum can be calculated, and the specular and diffuse contributions to the reflectivity can be compared. The roughness spectrum (here thermally excited capillary waves) is obtained from thermodynamic considerations by Fourier decomposition of the free energy, see Sect. 4.5.1:

$$\left\langle z(\mathbf{q}_{\|})z(-\mathbf{q}_{\|})\right\rangle = \frac{1}{A}\frac{k_B T}{\Delta\rho g + \gamma\mathbf{q}_{\|}^2 + \kappa\mathbf{q}_{\|}^4}. \tag{4.69}$$

$A$ is the interfacial area, $\gamma$ is the surface tension and $\kappa$ is the bending rigidity modulus. The correlation function can be obtained by Fourier transformation:

$$\langle z(\mathbf{0})z(\mathbf{r}_{\|})\rangle = k_B T/2\pi\gamma \times \left[K_0\left(r_{\|}\sqrt{\Delta\rho g/\gamma}\right) - K_0\left(r_{\|}\sqrt{\gamma/\kappa}\right)\right], \tag{4.70}$$

where $K_0$ is the modified Bessel function of second kind of order 0. Then, for a wavevector resolution $\Delta q_x$, the intensity measured in the $\theta_{\mathrm{sc}} = \theta_{\mathrm{in}}$ direction is smaller than the reflectivity of a perfectly flat interface by a factor

$$\pi^{-1/2}\Gamma\left[\frac{1}{2} - \frac{k_B T q_z^2}{4\pi\gamma}, \frac{1}{2}\Delta q_x^2\frac{K}{\gamma}\right] \times \exp-\left[\frac{q_z^2 k_B T}{2\pi\gamma}\ln\left(\frac{e^{\gamma_E}}{\sqrt{2}}\frac{\sqrt{\gamma/K}}{\Delta q_x}\right)\right], \tag{4.71}$$

where $\Gamma$ is the incomplete $\Gamma$ function and $\gamma_E$ is the Euler's constant. Note that this factor is larger than $e^{-q_z^2 \langle z^2\rangle}$ because diffuse scattering has been taken into account in addition to specular reflectivity.

It can be seen on Fig. 4.11 that even for relatively small wavevectors the diffuse intensity dominates. It would not have been possible to obtain physically reasonable parameters from the experiment without taking its contribution into account (see Fig. 4.11).

## 4.5 Examples

### 4.5.1 Liquid-Free Surfaces

#### 4.5.1.1 The Scattering Cross-Section

The aim of this paragraph is to show how the DWBA can be used in order to obtain the differential scattering cross-section in a real case. The roughness of liquid interfaces is due to thermally excited capillary waves whose spectrum can be written as [26]

$$\langle z(\mathbf{q}_\parallel) z(-\mathbf{q}_\parallel) \rangle = \frac{1}{A} \frac{k_B T}{\Delta\rho g + \gamma q_\parallel^2}. \tag{4.72}$$

$\Delta\rho$ is the density difference between the lower and upper liquids, $\gamma$ is the surface tension and $A$ the area. Equation (4.72) describes thermally excited capillary waves limited by gravity at large scales and by surface tension at distances smaller than the so-called capillary length ($l_c = \sqrt{\gamma/\Delta\rho g} \approx 2.7$ mm for water). More precisely, the $\gamma q_\parallel^2$ term stems from the increase in interfacial area due to the deformation. As this area increase is less important for long wavelengths, we obtain the $q^{-2}$ characteristic divergence. Fourier transforming, we obtain the height–height correlation function:

$$\langle z(\mathbf{0}) z(\mathbf{r}_\parallel) \rangle = k_B T/(2\pi\gamma) K_0 \left( r_\parallel \sqrt{\Delta\rho g/\gamma} \right). \tag{4.73}$$

$K_0$ is the modified second kind Bessel function of order 0. $K_0(x)_{x\to 0} \approx Log2 - \gamma_E \, Logx$, with $\gamma_E$ Euler's constant, and $limK_0(x)_{x\to\infty} = 0$. Then, the scattering cross-section is [from Eq. (4.47)] is given by

$$\left( \frac{d\sigma}{d\Omega} \right)_{\text{surf}} = A \frac{k_0^4}{16\pi^2} (n_1^2 - n_0^2)^2 \left| t_{0,1}^{\text{in}} \right|^2 \left| t_{0,1}^{\text{sc}} \right|^2 \frac{e^{-\frac{1}{2}(q_{z,1}^2 + q_{z,1}^{*2})\langle z^2 \rangle}}{|q_{z,1}|^2} (\widehat{\mathbf{e}}_{\text{in}}.\widehat{\mathbf{e}}_{\text{sc}})^2$$

$$\times \int d\mathbf{R}_\parallel \left[ e^{|q_{z,1}|^2 k_B T/(2\pi\gamma) K_0(q_{min} R_\parallel)} - 1 \right] e^{i\mathbf{q}_\parallel . \mathbf{R}_\parallel}, \tag{4.74}$$

where $q_{\min} = \sqrt{\Delta\rho g/\gamma}$ is the minimum wavevector in the capillary wave spectrum and $q_{\max}$ is the largest one, on the order $2\pi$/molecular size. The approximation

$$\int d\mathbf{R}_\parallel \frac{\left[ e^{|q_{z,1}|^2 k_B T/(2\pi\gamma) K_0(q_{min} R_\parallel)} - 1 \right] e^{i\mathbf{q}_\parallel . \mathbf{R}_\parallel}}{|q_{z,1}|^2} \simeq \frac{k_B T}{\gamma q_\parallel^2} \left( \frac{q_\parallel}{q_{max}} \right)^\eta, \tag{4.75}$$

where $\eta = (k_B T/2\pi\gamma) q_z^2$, is very accurate for $\eta \ll 1$ and leads to [21,37,43]

$$\left( \frac{d\sigma}{d\Omega} \right)_{\text{surf}} \simeq A \frac{k_0^4}{16\pi^2} (n_1^2 - n_0^2)^2 \left| t_{0,1}^{\text{in}} \right|^2 \left| t_{0,1}^{\text{sc}} \right|^2 (\widehat{\mathbf{e}}_{\text{in}}.\widehat{\mathbf{e}}_{\text{sc}})^2 \frac{k_B T}{\gamma q_\parallel^2} \left( \frac{q_\parallel}{q_{max}} \right)^\eta. \tag{4.76}$$

Another contribution is that of density fluctuations in the bulk (thermally excited acoustic waves). Inserting the density–density correlations for bulk liquid fluctuations,

$$\langle \rho(\mathbf{r})\rho(\mathbf{r}')\rangle = \rho^2 k_B T \, \kappa_T \delta(\mathbf{r}-\mathbf{r}') \tag{4.77}$$

in Eq. (4.51), where $\kappa_T = -1/V(\partial V/\partial P)_T$ is the isothermal compressibility yielding

$$\left(\frac{d\sigma}{d\Omega}\right)_{\text{bulk}} = \frac{k_0^4}{16\pi^2}\frac{A(n_1^2-n_0^2)^2\,|t^{\text{in}}|^2\,|t^{\text{sc}}|^2\,k_B T\,\kappa_T}{2\mathcal{I}m(q_{z,1})}(\hat{\mathbf{e}}_{\text{in}}.\hat{\mathbf{e}}_{\text{sc}})^2, \tag{4.78}$$

where $A/2\mathcal{I}m(q_{z,1})$ is the effective scattering volume. The total diffuse scattering cross-section is therefore

$$\frac{d\sigma}{d\Omega} = \left(\frac{d\sigma}{d\Omega}\right)_{\text{surf}} + \left(\frac{d\sigma}{d\Omega}\right)_{\text{bulk}}$$

$$= \frac{k_0^4}{16\pi^2}A(n_1^2-n_0^2)^2\left|t^{\text{in}}\right|^2|t^{\text{sc}}|^2(\hat{\mathbf{e}}_{\text{in}}.\hat{\mathbf{e}}_{\text{sc}})^2\left[\frac{k_B T}{\gamma q_{\parallel}^2}\left(\frac{q_{\parallel}}{q_{\text{max}}}\right)^{\eta} + \frac{k_B T\,\kappa_T}{2\mathcal{I}m(q_{z,1})}\right]. \tag{4.79}$$

### 4.5.1.2  Experimental Procedures

One could expect that the same scans which are used for the measurement of diffuse scattering from solid surfaces could be also applied to liquid surfaces, and in particular "rocking curves" (equivalent to $q_x$ scans at fixed $q_z$ for small angles). Since $q_z$ is constant, the signal obtained by doing such scans is very simply related to the surface spectrum, and a very good resolution is obtained because of the grazing angle geometry [16, 19]: Using $q_x = k_0(\cos(\theta_{\text{sc}}) - \cos(\theta_{\text{in}}))$,

$$\Delta q_x = k_0\left(\sin\theta_{\text{in}}\Delta\theta_{\text{in}} + \sin\theta_{\text{sc}}\Delta\theta_{\text{sc}}\right), \tag{4.80}$$

and the grazing angle geometry results in an increase in resolution by a factor of the order of $\sin\theta_{\text{in}}$ or $\sin\theta_{\text{sc}}$. In fact such measurements do not yield satisfactory results because bulk scattering dramatically increases when both the grazing angle of incidence and the scattering angle become larger than the critical angle for total external reflection, which is unavoidable in such scans. Indeed, the penetration length which is inversely proportional to $\mathcal{I}m(q_{z,1})$, Eq. (4.78), is then on the order of a μm, whereas it is on the order of a nm if either $\theta_{\text{in}}$ or $\theta_{\text{sc}}$ is smaller than $\theta_c$. In practice, it is therefore necessary to fix the grazing angle of incidence below the grazing angle for total external reflection $\theta_c$. Under this constraint, it is possible to measure the scattered intensity by scanning either in the plane of incidence (detector scan) or in the horizontal sample plane.

In the first case, $q_x$ and $q_z$ are varied together and the scans are therefore sensitive to both the surface roughness (because of the $q_x$ variation) and to the normal structure (because of the $q_z$ variation). For diffuse scattering measurements, the main advantage of this configuration is the geometrical increase in Fourier space resolution for scans performed in the plane of incidence discussed above. Using Eq. (4.80) and neglecting the incident beam divergence, $\Delta q_x = k_0 \sin \theta_{sc} \Delta \theta_{sc}$. Also, using $q_x = k_0 (\cos(\theta_{sc}) - \cos(\theta_{in})) \approx (1/2) k_0 (\theta_{in}^2 - \theta_{sc}^2)$ with $\lambda = 0.14$ nm, $\theta_{in} = 2$ mrad and a minimum accessible $\theta_{sc} = 2.5$ mrad at the tails of the reflected beam, one can see that the minimum accessible $q_x \approx 10^5 \, \mathrm{m}^{-1}$ is in good agreement with the experiments. This kind of measurement should also be considered whenever one is interested in the determination of the normal structure of thin films. It is indeed a valuable alternative to reflectivity experiments when using synchrotron radiation. Its two main advantages over reflectivity are a reduced background because the grazing angle of incidence is fixed below the critical angle for total external reflection and the much lighter experimental setup (only a mirror is required instead of a complicated beam deflector).

When one is only interested in the roughness spectrum at large wavevectors, the second kind of scans (in the horizontal sample plane) which directly yields a signal proportional to the roughness spectrum should be preferred. A PSD mounted perpendicular to the sample surface can be used to determine the normal structure. Alternatively, one can integrate over the detector length to improve the statistics.

The Fourier space resolution for such scans is $\Delta q_y = k_0 \Delta \psi$, and there is no geometrical increase in resolution in this geometry. Denoting $w_d$ the width of the slit placed in front of the detector, and $L_d$ the sample-to-detector distance, the smallest accessible wavevector is $k_0 w_d / L_d \approx 2 \times 10^7 \, \mathrm{m}^{-1}$ limited by the slit size. The counterpart of the coarser resolution is that the measured intensity is larger, and the statistics are also improved by the integration of the intensity along $z$. As a consequence, the surface scattered signal can be recorded up to wavevectors on the order of $10^{10} \, \mathrm{m}^{-1}$. The two kinds of scans (in the plane of incidence and in the plane of the sample) are therefore complementary, and using both of them allows a determination of surface spectra from $10^5 \, \mathrm{m}^{-1}$ to $10^{10} \, \mathrm{m}^{-1}$.

Fixing the grazing angle of incidence below the critical angle for total external reflection, the dominant source of scattering is from the upper phase. Even when working at the liquid–gas interface where the scattering from the subphase-vapour-saturated helium gas is very weak (it will be at least 3 orders of magnitude larger when the upper phase is a liquid), it can be necessary to subtract this background. A very reliable procedure for background subtraction is that of [22]. It consists in lowering the trough and scanning around the direct beam (which is transposed by twice the incident angle). Since the experiment is done under *total* external reflection, even if the reflected beam and the transmitted beam (when the sample is lowered) follow a different path, both paths have exactly the same length, and the beam intensity is the same along them. Only two differences can be noted: x-rays scattered downwards before being reflected will not be totally reflected, but this does not lead to any difference in the region of interest which is above the specular beam. X-rays scattered upwards before being reflected will have a contribution $2 \times \theta_{in}$ below the point where they should contribute. Since the background is perfectly flat in the region of

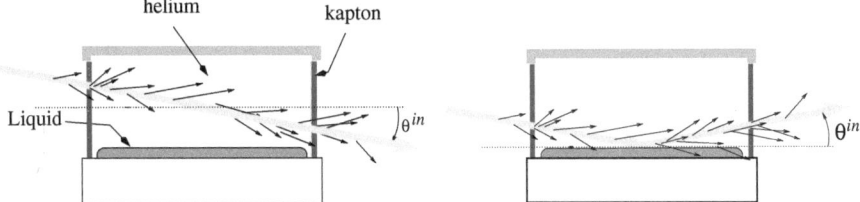

**Fig. 4.12** Procedure for background subtraction

interest, this does again not lead to any noticeable consequence (see Fig. 4.12). This procedure is therefore reliable, even if the background is not negligible.

A last important point not specific to liquid surfaces is that contrary to the specular or Bragg-reflected intensity (presence of $\delta$ distributions in the scattering cross-section), the diffuse intensity (including Bragg singularities in two dimensions) increases with the resolution volume (Fig. 4.9). It is therefore necessary to precisely determine the resolution function as a function of slit openings and the footprint of the beam on the surface to precisely determine the magnitude of this intensity.

Figure 4.7d gives a schematic view of a diffuse scattering experiment at the air–liquid interface. A Teflon trough (inner diameter $\sim 300\,\text{mm}$) is mounted on an active antivibration system. A monochromatic incident beam is first extracted from the polychromatic beam of an undulator source using a two-crystal diamond (111) monochromator. Higher harmonics are eliminated using two platinum-coated glass mirror, also used to set the grazing angle of incidence $\theta$ on the liquid. The size of the incident beam is fixed by a slit $S_{in}$ (for instance $250 \times 250\,\mu\text{m}$). The horizontal resolution of the experiment is fixed by the slits $S_c$ and $S_d$ (for instance $300\,\mu\text{m}$ wide and $500\,\mu\text{m}$ wide, respectively). $S_c$ is placed at 250 mm from the sample, and $S_d$ at 1000 mm. These two last slits also fix the illuminated area $A$ seen by the detector (see Fig. 4.8). The precise knowledge for the illuminated area is needed for the calculation of the scattering cross-section [see for instance Eq. (4.47)]. Intensity is collected by a vertically mounted position-sensitive detector (PSD). When all the scattering sources of the sample are known, it is possible to calculate in absolute units the scattered intensity (see for instance Fig. 4.13 where the wavelength is fixed at $\lambda = 0.154\,\text{nm}$).

The measurements are performed under a vapour-saturated helium atmosphere. Despite the low level of background, it is necessary to subtract the residual background. A rectangular Langmuir trough (with a movable barrier and a surface pressure sensor) is used for experiments on surfactant films at air–water interfaces.

The intensity scattered by the bare water surface is plotted in Fig. 4.13 for a $\psi$ and compared to the intensity calculated from Eq. (4.79). The excess of scattering at $q_\| \geq 10^9\,\text{m}^{-1}$ indicates that Eq. (4.79) does not capture all sources of scattering. It has been shown that this excess comes from a lower effective surface tension for large $q_\|$ [30].

**Fig. 4.13** Scattering by the bare water surface. (**a**) Detector scan. (**b**) $\psi$ scan at fixed $\theta_{in}$. An integration along the linear detector ($\theta_{sc}$) has been performed. Surface scattering dominates for $\theta_{in} =$ 2.01 mrad with the characteristic $q^{-2}$ divergence whereas scattering by density fluctuations dominates $\theta_{in} =$ 4.61 mrad above the critical angle for total external reflection. After *Nature* **403** 871 (2000). Note the structure peak of water around $2 \times 10^{10}\,\mathrm{m}^{-1}$. The measurements extend over more than 5 orders of magnitude in $q_{\parallel}$

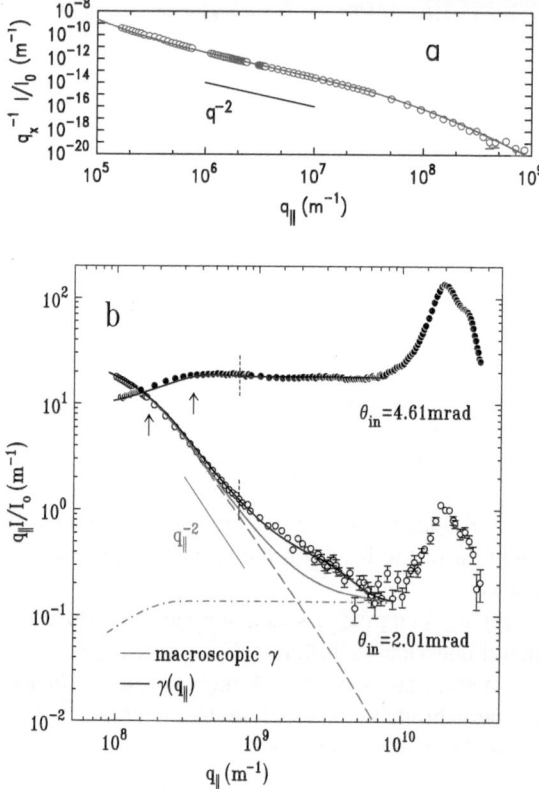

### 4.5.2 Thin Films

In this section we derive expressions for thin films when the refractive index of the film is not too different from that of the substrate. A good approximation consists in taking the substrate as the reference medium. One therefore obtains

$$\frac{d\sigma}{d\Omega} = \left(\frac{d\sigma}{d\Omega}\right)_{ref} + r_e^2 \left|t_{0,1}^{in}\right|^2 \left|t_{0,1}^{sc}\right|^2 (\widehat{\mathbf{e}}_{in}.\widehat{\mathbf{e}}_{sc})^2 \left\langle \left| \int d\mathbf{r}\delta\rho(\mathbf{r})e^{i\mathbf{q}.\mathbf{r}} \right|^2 \right\rangle, \quad (4.81)$$

defining $\rho_{flat}(\mathbf{r}_{\parallel}, z)$ as the electron density for an equivalent flat film, and making the change of variables $z \to z - \zeta(\mathbf{r}_{\parallel})$, we get

$$\frac{d\sigma}{d\Omega} = \left(\frac{d\sigma}{d\Omega}\right)_{ref} + r_e^2 \left|t_{0,1}^{in}\right|^2 \left|t_{0,1}^{sc}\right|^2 (\widehat{\mathbf{e}}_{in}.\widehat{\mathbf{e}}_{sc})^2 \left\langle \left| \int_{z=\zeta(\mathbf{r})}^{z=l+\zeta(\mathbf{r})} d\mathbf{r}(\rho_{flat}(\mathbf{r}_{\parallel}, z) - \rho_1)e^{i\mathbf{q}.\mathbf{r}} \right.\right.$$

$$\left.\left. + \int_{z=l}^{z=l+\zeta(\mathbf{r})} d\mathbf{r}(\rho_1 - \rho_0)e^{i\mathbf{q}.\mathbf{r}} \right|^2 \right\rangle.$$

$$\frac{d\sigma}{d\Omega} = \left(\frac{d\sigma}{d\Omega}\right)_{\text{ref}} + r_e^2 \left|t_{0,1}^{\text{in}}\right|^2 \left|t_{0,1}^{\text{sc}}\right|^2 (\hat{\mathbf{e}}_{\text{in}}.\hat{\mathbf{e}}_{\text{sc}})^2 \left\langle \left| \int_{z=0}^{z=l} d\mathbf{r} (\rho_{\text{flat}}(\mathbf{r}_{\parallel},z) - \rho_1) e^{iq_z \zeta(\mathbf{r}_{\parallel})} e^{i\mathbf{q}.\mathbf{r}} \right. \right.$$

$$\left. \left. + \int d\mathbf{r}_{\parallel} \frac{\rho_1 - \rho_0}{iq_z} e^{iq_z l} \left(e^{iq_z \zeta(\mathbf{r}_{\parallel})} - 1\right) e^{i\mathbf{q}_{\parallel}.\mathbf{r}_{\parallel}} \right|^2 \right\rangle. \tag{4.82}$$

*Exercise:* Check that

$$E = r_{0,1} \frac{ik_0 R}{4\pi R} \frac{iq_z}{(\rho_1 - \rho_0)} \int_{-\infty}^{0} (\rho_1 - \rho_0) e^{i\mathbf{q}.\mathbf{r}} d\mathbf{r}, \tag{4.83}$$

where $r_{0,1}$ is the reflectivity coefficient between media 0 and 1 and is the exact solution for the electric field scattered by a flat substrate. Show that at the same level of approximation as above, one has for a film

$$E = r_{0,1} \frac{ik_0 R}{4\pi R} \int_{-\infty}^{0} \left(iq_z + \frac{\delta\rho(z)}{\rho_1 - \rho_0}\right) e^{i\mathbf{q}.\mathbf{r}} d\mathbf{r}. \tag{4.84}$$

Deduce that at the same level of approximation,

$$R(q_z) = R_F(q_z) \left|1 + iq_z \int \frac{\delta\rho(z)}{\rho_1 - \rho_0} e^{iq_z z} dz\right|^2$$

$$= R_F(q_z) \left|\frac{1}{\rho_1 - \rho_0} \int \left(\frac{\partial\rho}{\partial z}\right) e^{iq_z z} dz\right|^2, \tag{4.85}$$

a very popular formula for x-ray reflectivity.

We now use the general expression Eq. (4.82) to show how all correlations can be determined in a thin film. We will then consider the case of domains and grazing incidence diffraction in a flat film.

### 4.5.2.1 Determination of Correlations in a Film

Let us consider a film of electron density $\rho_{\text{film}}$, thickness $d$ and profile $\zeta_f(\mathbf{r}_{\parallel})$ on a rough substrate of electron density $\rho_{\text{sub}}$ and profile $\zeta_s(\mathbf{r}_{\parallel})$.

$$\frac{d\sigma}{d\Omega} = \frac{A}{q_z^2} r_e^2 \left|t_{0,1}^{\text{in}}\right|^2 \left|t_{0,1}^{\text{sc}}\right|^2 (\hat{\mathbf{e}}_{\text{in}}.\hat{\mathbf{e}}_{\text{sc}})^2 \left[\rho_{\text{film}}^2 e^{-q_z^2 \langle \zeta_f^2 \rangle} \int d\mathbf{r}_{\parallel} \left(e^{q_z^2 \langle \zeta_f(0)\zeta_f(\mathbf{r}_{\parallel}) \rangle} - 1\right) e^{i\mathbf{q}_{\parallel}.\mathbf{r}_{\parallel}} \right.$$

$$+ 2\rho_{\text{film}}^2 (\rho_{\text{sub}}^2 - \rho_{\text{film}}^2) e^{-\frac{1}{2}q_z^2(\langle \zeta_f^2 \rangle + \langle \zeta_s^2 \rangle)} \cos(q_z d) \int d\mathbf{r}_{\parallel} \left(e^{q_z^2 \langle \zeta_f(0)\zeta_s(\mathbf{r}_{\parallel}) \rangle} - 1\right) e^{i\mathbf{q}_{\parallel}.\mathbf{r}_{\parallel}}$$

$$\left. + \rho_{\text{sub}}^2 e^{-q_z^2 \langle \zeta_s^2 \rangle} \int d\mathbf{r}_{\parallel} \left(e^{q_z^2 \langle \zeta_s(0)\zeta_s(\mathbf{r}_{\parallel}) \rangle} - 1\right) e^{i\mathbf{q}_{\parallel}.\mathbf{r}_{\parallel}}\right]. \tag{4.86}$$

From this expression, it is possible to determine the substrate–film $\langle \zeta_s(0)\zeta_f(\mathbf{r}_{\parallel}) \rangle$ and film–film $\langle \zeta_f(0)\zeta_f(\mathbf{r}_{\parallel}) \rangle$ correlation functions, provided the substrate–substrate correlation function $\langle \zeta_s(0)\zeta_s(\mathbf{r}_{\parallel}) \rangle$ has been measured first as the oscillating $\cos(q_z d)$

provides a way to separate them. This has been used by Tidswell et al. [42] in order to study how the surface of a liquid film follows a rough substrate due to van der Waals forces.

### 4.5.2.2 Domains in a Flat Film

For domains in a flat film, we have $\rho_{\text{flat}}(\mathbf{r}_{\parallel}, z) - \rho_1 = \delta\rho_{av} + (\rho_D - \rho_{av})\sum_i \delta(\mathbf{r}_{\parallel} - \mathbf{r}_{i\parallel}) \otimes s(\mathbf{r}_{\parallel})$, with $\rho_{av}$ the average electron density in the film and $\delta\rho_{av} = (1-c)\rho_F + c\rho_D - \rho_1$ for D domains (with coverage c) located in $\mathbf{r}_{i\parallel}$ and of shape $s(\mathbf{r}_{\parallel})$ ($s(\mathbf{r}_{\parallel}) = 1$ inside a domain and 0 outside) in a $F$ film. Integrating over $z$,

$$\left\langle \left| \int d\mathbf{r}(\rho_{\text{flat}} - \rho_1)e^{i\mathbf{q}\cdot\mathbf{r}} \right|^2 \right\rangle = (\rho_D - \rho_{av})^2 \frac{\left| e^{iq_z l} - 1 \right|^2}{q_z^2} |\tilde{s}(\mathbf{q}_{\parallel})|^2$$

$$\times \left\langle \left| \int d\mathbf{r}_{\parallel} \sum_i \delta(\mathbf{r}_{\parallel} - \mathbf{r}_{i\parallel})e^{i\mathbf{q}_{\parallel}\cdot\mathbf{r}_{\parallel}} \right|^2 \right\rangle. \tag{4.87}$$

We can now write

$$\left\langle \sum_i \sum_j \delta(\mathbf{r}_{\parallel} - \mathbf{r}_{i\parallel})\delta(\mathbf{r}_{\parallel} - \mathbf{r}_{j\parallel}) \right\rangle = \left\langle \sum_i \sum_j \delta(\mathbf{r}_{\parallel} - \mathbf{r}_{i\parallel} + \mathbf{r}_{j\parallel}) \right\rangle$$

$$= N\delta(\mathbf{r}_{\parallel}) + \left\langle \sum_{i \neq j} \delta(\mathbf{r}_{\parallel} - \mathbf{r}_{i\parallel})\delta(\mathbf{r}_{\parallel} - \mathbf{r}_{j\parallel}) \right\rangle$$

$$= N\delta(\mathbf{r}_{\parallel}) + Ncg(\mathbf{r}_{\parallel}), \tag{4.88}$$

where $g(\mathbf{r}_{\parallel})$ is the pair correlation function which gives the probability of finding a domain at $\mathbf{r}_{\parallel}$ knowing there is one at origin. One finally obtains for the incoherent part of the scattering cross-section,

$$\left( \frac{d\sigma}{d\Omega} \right)_{\text{incoh}} = Ar_e^2 \left| t_{0,1}^{\text{in}} \right|^2 \left| t_{0,1}^{\text{sc}} \right|^2 (\hat{\mathbf{e}}_{\text{in}}\cdot\hat{\mathbf{e}}_{\text{sc}})^2 (\rho_D - \rho_{av})^2 \frac{\left| e^{iq_z l} - 1 \right|^2}{q_z^2} |\tilde{s}(\mathbf{q}_{\parallel})|^2$$

$$\times \left[ c + c^2 \int d\mathbf{r}_{\parallel}(g(\mathbf{r}_{\parallel}) - 1)e^{i\mathbf{q}_{\parallel}\cdot\mathbf{r}_{\parallel}} \right]. \tag{4.89}$$

For circular domains of radius R

$$|\tilde{s}(\mathbf{q}_{\parallel})|^2 = \left( \frac{2J_1(q_{\parallel}R)}{q_{\parallel}R} \right)^2.$$

### 4.5.2.3 Grazing Incidence Diffraction in a Flat Film

We now consider the case of a film with crystalline order, for example a Langmuir film. $\rho_{\text{uc}}(\mathbf{r})$ is the electron density in the two-dimensional unit cell defined by the vectors $\mathbf{a_1}$ and $\mathbf{a_2}$, and the crystal consists of $N_1$ and $N_2$ unit cells along $x$ and $y$. We therefore have

$$\delta\rho(\mathbf{r}) = \sum_{i_1=1}^{N_1} \sum_{i_2=1}^{N_2} \rho_{uc}(\mathbf{r} - i_1\mathbf{a_1} - i_2\mathbf{a_2}), \qquad (4.90)$$

and for our flat film

$$\left\langle \left| \int d\mathbf{r}(\rho_{flat}(\mathbf{r}_\|,z) - \rho_1)e^{i\mathbf{q}\cdot\mathbf{r}} \right|^2 \right\rangle$$
$$= \left\langle \left| \int d\mathbf{r} \sum_{i_1=1}^{N_1} \sum_{i_2=1}^{N_2} (\rho_{uc}(\mathbf{r}) - \rho_1)e^{i(\mathbf{q}_\|\cdot\mathbf{a_1})i_1 + i(\mathbf{q}_\|\cdot\mathbf{a_2})i_2} e^{i\mathbf{q}\cdot\mathbf{r}} \right|^2 \right\rangle . \qquad (4.91)$$

Summing first, then integrating, we get

$$\frac{d\sigma}{d\Omega} = r_e^2 \left|t_{0,1}^{in}\right|^2 \left|t_{0,1}^{sc}\right|^2 (\hat{\mathbf{e}}_{in}\cdot\hat{\mathbf{e}}_{sc})^2 |\tilde{\rho}_{uc}(\mathbf{q})|^2 \frac{\sin^2\left(\frac{N_1\mathbf{q}\cdot\mathbf{a_1}}{2}\right) \sin^2\left(\frac{N_2\mathbf{q}\cdot\mathbf{a_2}}{2}\right)}{\sin^2\left(\frac{\mathbf{q}\cdot\mathbf{a_1}}{2}\right) \sin^2\left(\frac{\mathbf{q}\cdot\mathbf{a_2}}{2}\right)}, \qquad (4.92)$$

where $\tilde{\rho}_{uc}(\mathbf{q})$ is the Fourier transform of $\rho_{uc}(\mathbf{r}) - \rho_{sub}$. If $N_1$ and $N_2$ are large, an example of calculation using Eq. (4.92) is given in Fig. 4.14,

$$\frac{d\sigma}{d\Omega} \approx 4\pi^2 N_1 N_2 r_e^2 \left|t_{0,1}^{in}\right|^2 \left|t_{0,1}^{sc}\right|^2 (\hat{\mathbf{e}}_{in}\cdot\hat{\mathbf{e}}_{sc})^2 |\tilde{\rho}_{uc}(\mathbf{q})|^2 \sum_n \sum_p \delta(\mathbf{q}\cdot\mathbf{a_1} - 2n\pi)\delta(\mathbf{q}\cdot\mathbf{a_2} - 2p\pi)$$
$$\approx \sum_{\mathbf{G}} 4\pi^2 A r_e^2 \left|t_{0,1}^{in}\right|^2 \left|t_{0,1}^{sc}\right|^2 (\hat{\mathbf{e}}_{in}\cdot\hat{\mathbf{e}}_{sc})^2 \frac{|\tilde{\rho}_{uc}(\mathbf{q})|^2}{|\mathbf{a_1}\times\mathbf{a_2}|^2} \delta(\mathbf{q}_\| - \mathbf{G}), \qquad (4.93)$$

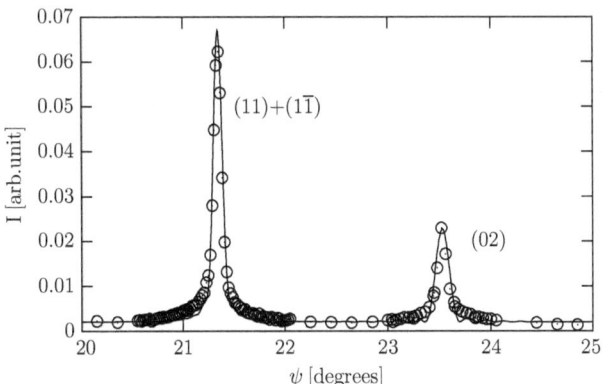

**Fig. 4.14** Grazing incidence diffraction from a behenic acid film in the S phase at $T = 16.1°C$ and $\Pi = 32.1$ mN/m (*circles*). $a = 0.5027$ nm and $b = 0.7265$ nm. Under these conditions F(1,1)= F(0,2). The *line* is a calculation using Eq. (4.92) with $N_1 = 140$ and $N_2 = 80$ plus proper powder averaging [34]

with $\mathbf{G} = (2n\pi/a_1, 2p\pi/a_2)$ a vector of the reciprocal space. Finally,

$$\frac{d\sigma}{d\Omega} = 4\pi^2 r_e |t_{0,1}^{in}|^2 |t_{0,1}^{sc}|^2 (\widehat{\mathbf{e}}_{in}.\widehat{\mathbf{e}}_{sc})^2 \times \frac{A}{|\mathbf{a_1} \times \mathbf{a_2}|^2} \sum_{h,k} |F(h,k,q_z)|^2 \delta(\mathbf{q}_{\parallel} - \mathbf{G}).$$

(4.94)

### 4.5.3 Liquid–Liquid Interfaces and Membranes

The investigation of liquid–liquid or solid–liquid interfaces poses several additional problems.

- Transmission through a bulk phase implies to use high-energy penetrating x-rays. The transmission through a 7 cm wide cell full of hexadecane ($C_{16}H_{34}$) as a function of energy is given for example in Fig. 4.15. One can see that one needs at least an energy of 20 keV to get a transmission of 15%. Energies in the range 20–30 keV offer a good compromise. Much higher energies and very small beams have been successfully used for reflectivity measurements [35]. Due to the specular condition, the background is indeed reduced when using very small beams. For diffuse scattering, signal and "background" share the same dependence on slit sizes and there is no gain in reducing the beam size.
- The long path through the upper liquid requires a careful background subtraction using the procedure of Sect. 4.5.1. An example is given for a double bilayer close to a solid wall in Fig. 4.16.

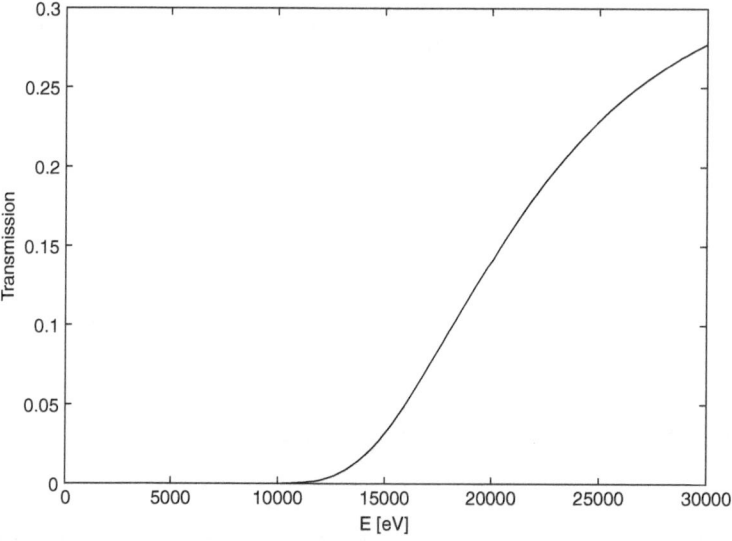

**Fig. 4.15** Transmission by 7 cm of hexadecane ($C_{16}H_{34}$) as a function of energy

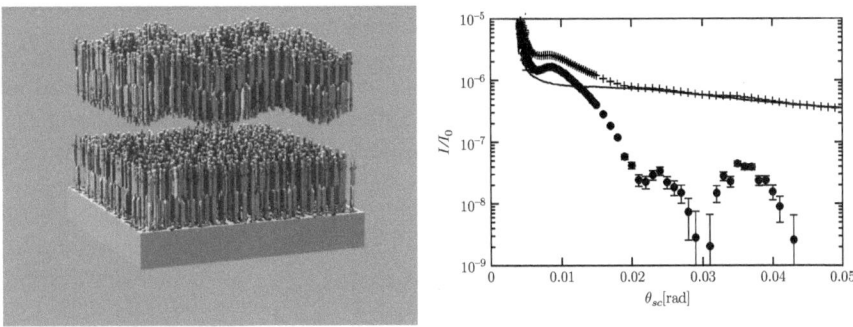

**Fig. 4.16** *Left*: double lipid bilayer close to a solid wall. The first bilayer is absorbed on the silicon wafer, whereas the second one is freely floating a few nms from the first one. *Right*: background subtraction for diffuse scattering from the double bilayer system [14]

- The high energy implies a very low value for the critical angle for external reflection, which in turn implies to work with wide enough cells, in particular when there is a meniscus, if one wants to work on the flat part of the surface.

We give here two examples of such measurements.

### 4.5.3.1 Monolayers at the Oil–Water Interface

The first one is a monolayer of the lipid DSPC (di-stearoyl-phosphatidylcholine) at the interface between oil (hexadecane) and water (Fig. 4.17). The measurements were carried out in the plane of incidence using 22 keV x-rays at the ID10B beamline of the ESRF. The data were analysed using Eqs. (4.72), (4.73), (4.74), (4.75) and (4.76) with $\gamma = 2.5$ mN/m and $\kappa = 40 k_B T$.

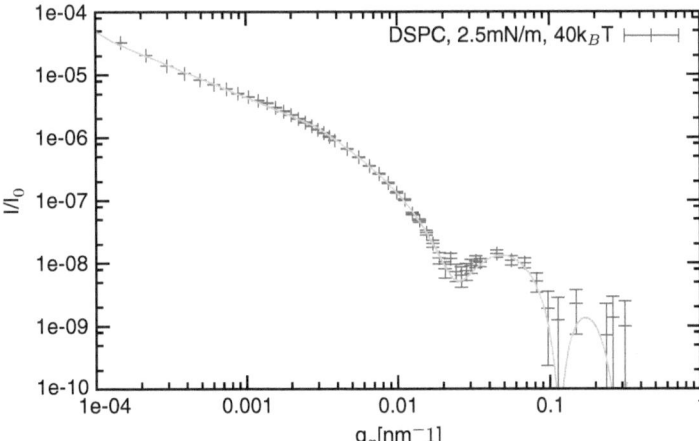

**Fig. 4.17** Diffuse scattering from a DSPC (di-stearoyl-phosphatidylcholine) monolayer at the interface between oil (hexadecane) and water. Measurements in the plane of incidence using 22 keV x-rays at the ID10B beamline of the ESRF

#### 4.5.3.2 Lipid Membranes Close to a Solid Wall

A second example is that of lipid membranes close to a solid wall as depicted in
Fig. 4.16 (left). A double bilayer of, for example, DSPC is deposited using the
Langmuir–Blodgett and the Langmuir–Schaeffer techniques [14]. Whereas the first
bilayer is adsorbed on the substrate (a very flat silicon wafer), the second one is
freely floating in the potential of the wall and the first one. The potential consists of
an attractive van der Waals contribution and repulsive hydration and entropic contributions and can be accurately determined as well as the membrane tension and
bending rigidity. The Hamiltonian writes

$$\mathscr{H}[z(\mathbf{r}_\parallel)] = \int_{\mathscr{R}^2} d^2\mathbf{r}_\parallel \left[ \frac{1}{2}\kappa\left(\nabla^2 z\right)^2 + \gamma(\nabla z)^2 + U(z) \right], \qquad (4.95)$$

and after expansion of $U$ to second order leads to a spectrum equivalent to Eq. (4.72)
with $U''$ instead of $\Delta\rho g$. Let us note here that $U''$ is determined by diffuse scattering
whereas the position of the minimum is determined in complementary reflectivity
experiments. Examples of scattering curves obtained in the plane of incidence using
20 keV x-rays at the ESRF BM32 beamline are given in Fig. 4.18.

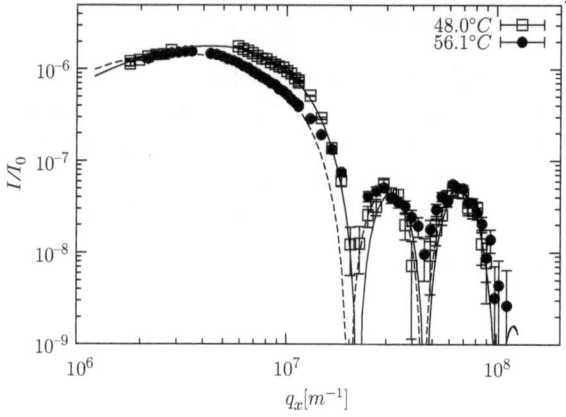

**Fig. 4.18** Non-specular scattering data (*symbols*) and best fits (*lines*) for DSPC double layers at
48°C (*squares* and *solid line*) and 56°C (*filled circles* and *dashed line*)

## 4.A Appendix: The Reciprocity Theorem

We consider a medium described by the relative permittivity $\epsilon_{\text{ref}}(\mathbf{r})$ which is assumed to be different from 1 in a localised region of space. Let two different current
distribution sources $\mathbf{J}_A$, $\mathbf{J}_B$ (with same frequency $\omega$) be placed in this medium. We
denote by the indices $A$, $B$ the fields created by these sources, separately, in the
medium. They satisfy Maxwell's equations,

$$\nabla \times \mathbf{E}_{A(B)} = -i\omega \mathbf{B}_{A(B)}$$
$$\nabla \times \mathbf{H}_{A(B)} = \mathbf{J}_{A(B)} + i\omega \mathbf{D}_{A(B)}, \tag{4.A1}$$

where $\mathbf{D}_{A,(B)}(\mathbf{r}) = \epsilon_0 \epsilon_{\text{ref}}(\mathbf{r}) \mathbf{E}_{A(B)}(\mathbf{r})$ and $\mathbf{B}_{A(B)} = \mathbf{H}_{A(B)}/\mu_0$. Substituting the Maxwell's equations in the vectorial identity,

$$\nabla.(\mathbf{E_A} \times \mathbf{H}_B - \mathbf{E}_B \times \mathbf{H}_A) = \mathbf{H}_B.\nabla \times \mathbf{E}_A - \mathbf{E}_A.\nabla \times \mathbf{H}_B$$
$$-\mathbf{H}_A.\nabla \times \mathbf{E}_B + \mathbf{E}_B.\nabla \times \mathbf{H}_A, \tag{4.A2}$$

leads to

$$\nabla.(\mathbf{E}_A \times \mathbf{H}_B - \mathbf{E}_B \times \mathbf{H}_A) = \mathbf{E}_B.\mathbf{J}_A - \mathbf{E}_A.\mathbf{J}_B + i\omega(\mathbf{E}_B.\mathbf{D}_A - \mathbf{E_A}.\mathbf{D_B})$$
$$+ i\omega(\mathbf{H}_A.\mathbf{B}_B - \mathbf{H}_B.\mathbf{B}_A). \tag{4.A3}$$

The last two terms on the right-hand side are zero so that we get,

$$\nabla.(\mathbf{E}_A \times \mathbf{H}_B - \mathbf{E}_B \times \mathbf{H}_A) = (\mathbf{E}_B.\mathbf{J}_A - \mathbf{E}_A.\mathbf{J}_B). \tag{4.A4}$$

Integrating Eq. (4.A4) over all space gives

$$\int d^3 r \nabla.(\mathbf{E}_A \times \mathbf{H}_B - \mathbf{E}_B \times \mathbf{H}_A) = \int d^3 r (\mathbf{E}_B.\mathbf{J}_A - \mathbf{E}_A.\mathbf{J}_B),$$

and using the divergence theorem

$$\int d^2 r (\mathbf{E}_A \times \mathbf{H}_B - \mathbf{E}_B \times \mathbf{H}_A) = \int d^3 r (\mathbf{E}_B.\mathbf{J}_A - \mathbf{E}_A.\mathbf{J}_B). \tag{4.A5}$$

If now the current sources are limited to a finite volume, the surface of integration in Eq. (4.A5) is infinitely remote from them, and the electromagnetic field can be approximated by a plane wave with $\mathbf{E}$ and $\mathbf{H}$ orthogonal and transverse.

$$\mathbf{H} = \sqrt{\frac{\epsilon_0}{\mu_0}} \hat{\mathbf{n}} \times \mathbf{E}.$$

It follows that,

$$\mathbf{E}_A \times \mathbf{H}_B - \mathbf{E}_B \times \mathbf{H}_A = \mathbf{0},$$

which yields

$$\int d^3 r \mathbf{E}_B.\mathbf{J}_A = \int d^3 r \mathbf{E}_A.\mathbf{J}_B, \tag{4.A6}$$

which is the reciprocity theorem [27, 28, 41]. Equation (4.A6) can also be written for dipole density sources through $\mathbf{P} = (1/i\omega)\mathbf{J}$, one gets

$$\int d^3 r \mathbf{E}_B.\mathbf{P}_A = \int d^3 r \mathbf{E}_A.\mathbf{P}_B. \tag{4.A7}$$

## 4.B Appendix: Verification of the Integral Equation in the Case of the Reflection by a Thin Film on a Substrate

It has been indicated in the main text that the integral equation (4.11) obtained by applying the reciprocity theorem is an exact equation. In this appendix, we verify that this is indeed the case for a single film on a substrate. The reference situation is a homogeneous medium of optical index $n_0$ and we want to calculate the electric field in the case where there is a film (1) of thickness $d$ on a substrate (2). The reflection and transmission coefficients of the smooth film are, respectively, Eqs. (3.87) and (3.88):

$$r = \frac{r_{0,1} + r_{1,2}e^{-2ik_{z,1}d}}{1 + r_{0,1}r_{1,2}e^{-2ik_{z,1}d}}, t = \frac{t_{0,1}t_{1,2}e^{-ik_{z,1}d}}{1 + r_{0,1}r_{1,2}e^{-2ik_{z,1}d}}.$$

The real case differs from the ideal one within the substrate where the refractive index difference between the real and ideal case is $(n_2^2 - n_0^2)$ and in the film where the difference is $(n_1^2 - n_0^2)$. In Eq. (4.11), we need the field in the real case, which is the transmitted field in the substrate, and is

$$E_1^{PW}(z) = \frac{t_{0,1}}{1 + r_{0,1}r_{1,2}e^{2ik_{inz,1}d}} \left[ e^{-ik_{inz,1}z} + r_{1,2}e^{-ik_{inz,1}(2d-z)} \right]$$

in the film. We also need the Green function in vacuum (medium (0)),

$$\frac{k_0^2 e^{-ik_0R}}{4\pi\epsilon_0 R} e^{i\mathbf{k}_{sc,0}\cdot\mathbf{r}}.$$

Using Eq. (4.11), the electric field can be written as

$$E = E_0 + \frac{e^{-ik_0R}}{4\pi R} \left\{ \int d\mathbf{r}_\| e^{i\mathbf{q}_\|\cdot\mathbf{r}_\|} \left[ \dots \right. \right.$$

$$(k_{1z}^2 - k_{0z}^2) \int_0^d e^{-ik_{scz,0}z} \frac{t_{01}(e^{-ik_{inz,1}z} + r_{12}e^{ik_{inz,1}(2d-z)})}{1 + r_{01}r_{12}e^{-2ik_{inz,1}d}} dz$$

$$\left. \left. + (k_{2z}^2 - k_{0z}^2) \int_{-\infty}^d e^{-k_{scz,0}z} t e^{-ik_{inz,2}(z-d)} \right] \right\}. \tag{4.B1}$$

In Eq. (4.B1), we have used that $k_{z,j}^2 = k_j^2 - k_x^2 = n_j^2 k_0^2 - k_x^2$ implies $k_0^2(n_j^2 - n_0^2) = k_{z,j}^2 - k_{z,0}^2$. Medium (2) is considered to be slightly absorbing in order to ensure the convergence of the integration. One obtains

$$E = E_0 - \frac{e^{-ikR}}{4\pi R} \int d\mathbf{r}_\| e^{i\mathbf{q}_\|\cdot\mathbf{r}_\|} (2ik_0\sin\theta_0 r). \tag{4.B2}$$

The differential scattering cross-section is thus

$$\frac{d\sigma}{d\Omega} = \frac{k_0^2}{4\pi^2}|r|^2 \sin^2 \theta_0 4\pi^2 \mathscr{A} \, \delta_{\mathbf{k}_{sc\parallel}, \mathbf{k}_{in\parallel}} = k_0^2 \sin^2 \theta_0 |r|^2 \mathscr{A} \, \delta(\mathbf{q}_\parallel). \qquad (4.B3)$$

And we find for the reflection coefficient

$$R = \frac{1}{\mathscr{A} \sin \theta_0} \int \frac{d\sigma}{d\Omega} d\Omega = |r|^2 \qquad (4.B4)$$

as expected.

## 4.C Appendix: Interface Roughness in a Multilayer Within the Born Approximation

In this appendix we treat the case of a rough multilayer within the Born approximation in order to show some simple properties of the scattered intensity. In the case of the rough multilayer depicted in Fig. 4.6, Eq. (4.27) gives

$$\frac{d\sigma}{d\Omega} = \frac{k_0^4}{16\pi^2} \cos^2 \psi \left| \sum_{i=1}^{N} \int d\mathbf{r}_\parallel \int_{z_i}^{z_{i+1}} dz \, (n_i^2 - 1) e^{i\mathbf{q}\cdot\mathbf{r}} \right|^2. \qquad (4.C1)$$

The upper medium (air or vacuum) is medium 0, and the substrate (medium $s$) is slightly absorbing in order to make the integrals converge.

$$\frac{d\sigma}{d\Omega} = \frac{k_0^4}{16\pi^2} \cos^2 \psi \left| \sum_{i=1}^{s} \int d\mathbf{r}_\parallel \frac{1}{iq_z} \left[ e^{iq_z z_{i+1}} - e^{iq_z z_i} \right] n_i^2 e^{i\mathbf{q}_\parallel \cdot \mathbf{r}_\parallel} \right|^2, \qquad (4.C2)$$

which can be written as

$$\frac{d\sigma}{d\Omega} = \frac{k_0^4}{16\pi^2 q_z^2} \cos^2 \psi \left| \sum_{i=0}^{N} \int d\mathbf{r}_\parallel \left[ n_{i+1}^2 - n_i^2 \right] e^{iq_z z_i} e^{i\mathbf{q}_\parallel \cdot \mathbf{r}_\parallel} \right|^2. \qquad (4.C3)$$

Let us then define

$$z_i = Z_i + z_i(\mathbf{r}_\parallel), \qquad (4.C4)$$

where $Z_i$ is the height of the flat interface in the reference case. Equation (4.C3) can be written as

$$\frac{d\sigma}{d\Omega} = \frac{k_0^4}{16\pi^2 q_z^2} \cos^2 \psi \times \sum_{i=0}^{N} \sum_{j=0}^{N} \int d\mathbf{r}_\parallel \int d\mathbf{r}'_\parallel \left[ n_{i+1}^2 - n_i^2 \right] \left[ n_{j+1}^2 - n_j^2 \right]$$

$$e^{iq_z(Z_i - Z_j)} e^{iq_z(z_i(\mathbf{r}_\parallel) - z_j(\mathbf{r}'_\parallel))} e^{i\mathbf{q}_\parallel \cdot (\mathbf{r}_\parallel - \mathbf{r}'_\parallel)}. \qquad (4.C5)$$

Making the change of variables $\mathbf{r}_\parallel - \mathbf{r}'_\parallel \to \mathbf{r}_\parallel$ and integrating over $\mathbf{r}'_\parallel$,

$$\frac{d\sigma}{d\Omega} = \frac{k_0^4 A}{16\pi^2 q_z^2} \cos^2 \psi \times \sum_{i=0}^{N} \sum_{j=0}^{N} \int d\mathbf{r}_{\parallel} \left[ n_{i+1}^2 - n_i^2 \right] \left[ n_{j+1}^2 - n_j^2 \right]$$

$$e^{iq_z(Z_i - Z_j)} \left\langle e^{iq_z(z_i(\mathbf{r}_{\parallel}) - z_j(0))} \right\rangle e^{i\mathbf{q}_{\parallel} \cdot \mathbf{r}_{\parallel}}, \tag{4.C6}$$

where $A$ is the illuminated area. Assuming Gaussian statistics of the height fluctuations $z_i(\mathbf{r}_{\parallel})$, or in any case expanding the exponential to the lowest (second) order, we have

$$\left\langle e^{iq_z(z_i(\mathbf{r}_{\parallel}) - z_j(0))} \right\rangle = e^{-\frac{1}{2} q_z^2 \langle z_i(\mathbf{r}_{\parallel}) - z_j(0) \rangle^2}. \tag{4.C7}$$

We then obtain

$$\frac{d\sigma}{d\Omega} = \frac{k_0^4 A}{16\pi^2 q_z^2} \cos^2 \psi$$

$$\sum_{i=0}^{N} \sum_{j=0}^{N} \left[ n_{i+1}^2 - n_i^2 \right] \left[ n_{j+1}^2 - n_j^2 \right] e^{iq_z(Z_i - Z_j)} e^{-\frac{1}{2} q_z^2 \langle z_i \rangle^2 - \frac{1}{2} q_z^2 \langle z_j \rangle^2}$$

$$\int d\mathbf{r}_{\parallel} e^{q_z^2 \langle z_i(\mathbf{r}_{\parallel}) z_j(0) \rangle^2} e^{i\mathbf{q}_{\parallel} \cdot \mathbf{r}_{\parallel}}. \tag{4.C8}$$

This equation also includes specular components because it has been constructed from the general solution of an electromagnetic field in a vacuum. The diffuse intensity can be obtained by removing the specular component:

$$\frac{d\sigma}{d\Omega} = \frac{k_0^4 A}{4q_z^2} \sum_{i=0}^{N} \sum_{j=0}^{N} \left[ n_{i+1}^2 - n_i^2 \right] \left[ n_{j+1}^2 - n_j^2 \right]$$

$$e^{iq_z(Z_i - Z_j)} e^{-\frac{1}{2} q_z^2 \langle z_i \rangle^2 - \frac{1}{2} q_z^2 \langle z_j \rangle^2} \delta(\mathbf{q}_{\parallel}). \tag{4.C9}$$

The diffuse intensity is then

$$\left( \frac{d\sigma}{d\Omega} \right)_{\text{incoh}} = \frac{k_0^4 A}{16\pi^2 q_z^2}$$

$$\sum_{i=0}^{N} \sum_{j=0}^{N} \left[ n_{i+1}^2 - n_i^2 \right] \left[ n_{j+1}^2 - n_j^2 \right] e^{iq_z(Z_i - Z_j)} e^{-\frac{1}{2} q_z^2 \langle z_i \rangle^2 - \frac{1}{2} q_z^2 \langle z_j \rangle^2}$$

$$(\widehat{\mathbf{e}}_{\text{in}} \cdot \widehat{\mathbf{e}}_{\text{sc}})^2 \int d\mathbf{r}_{\parallel} \left( e^{q_z^2 \langle z_i(\mathbf{r}_{\parallel}) z_j(0) \rangle^2} - 1 \right) e^{i\mathbf{q}_{\parallel} \cdot \mathbf{r}_{\parallel}}. \tag{4.C10}$$

For a single surface, we get Eq. (4.41). It is remarkable that Eq. (4.C10) has exactly the same structure as the reflectivity coefficient,

$$\left( \frac{d\sigma}{d\Omega} \right)_{\text{coh}} = \frac{k_0^4 A}{4q_z^2} \sum_{i=0}^{N-1} \sum_{j=0}^{N-1} \left[ n_{i+1}^2 - n_i^2 \right] \left[ n_{j+1}^2 - n_j^2 \right]$$

$$e^{iq_z(Z_i - Z_j)} e^{-\frac{1}{2} q_z^2 \langle z_i \rangle^2 - \frac{1}{2} q_z^2 \langle z_j \rangle^2} \delta(\mathbf{q}_{\parallel}), \tag{4.C11}$$

each term simply being multiplied by a "transverse" coefficient:

$$\frac{1}{4\pi^2} \int d\mathbf{r}_\| \left( e^{q_z^2 \langle z_i(\mathbf{r}_\|) z_j(\mathbf{0}) \rangle^2} - 1 \right) e^{i\mathbf{q}_\| \cdot \mathbf{r}_\|} .$$

## 4.D  Appendix: Quantum Mechanical Approach of Born and Distorted-Wave Born Approximations

T. Baumbach and P. Mikulík

In this appendix we treat the formal quantum mechanical approach to scattering by multilayers with random fluctuations. That can be interface roughness, but also porosity or density fluctuations. In particular we develop the differential scattering cross-section in the kinematical approximation (first Born approximation) and in the distorted-wave Born approximation in terms of the structure amplitudes of the individual layers and of their disturbances. This approach is written in a general way. In Chap. 6 it will be applied to the reflection and to diffraction under conditions of specular reflection under grazing incidence by rough multilayers and multilayered gratings. We would like to notice that we adopted here the *phase-sign* notation of this book, with plane waves $e^{-ikr}$ and Fourier transforms $e^{+iqr}$, which is contrary to that used in most publications using this formalism.

### 4.D.1  Formal Theory

Here we develop formally the incoherent approach for the scattering by multilayers with defects independently of the specific scattering method. We make use of the (scalar) quantum mechanical scattering theory and its approximations, in particular the first-order Born approximation (kinematical theory) and the distorted-wave Born approximation (semi-dynamical theory).

Scattering of the incident wave $|K_0\rangle$ by the potential $V$ produces the total wave field $|E\rangle$, described by the integral equation [17]

$$|E\rangle = |K_0\rangle + \hat{G}_0 \hat{V} |E\rangle , \qquad (4.D1)$$

where $\hat{G}_0$ is the Green function operator of the free particle. We define the *transition operator* by $\hat{T}|K_0\rangle \equiv \hat{V}|E\rangle$ and the *transition matrix* by the matrix elements $T_{0S} = \langle K_S | \hat{T} | K_0 \rangle$, characterising the scattering from $|K_0\rangle$ into $|K_S\rangle$. The *differential scattering cross-section* $\sigma$ into an elementary solid angle $\delta\Omega$ can be expressed by the matrix elements of the transition matrix

$$d\sigma = \frac{1}{16\pi^2} |T_{0S}|^2 d\Omega . \qquad (4.D2)$$

Scattering by a Randomly Disturbed Potential

Including a random spatial fluctuation of the scattering potential, the differential
cross-section averages over the statistical ensemble of all microscopic configurations

$$d\sigma = \frac{1}{16\pi^2} \left\langle |T_{0S}|^2 \right\rangle d\Omega. \tag{4.D3}$$

We divide $d\sigma$ into coherent and incoherent contributions

$$d\sigma = \left\{ \frac{1}{16\pi^2} |\langle T_{0S} \rangle|^2 + \frac{1}{16\pi^2} |\mathrm{Cov}(T_{0S}, T_{0S})|^2 \right\} d\Omega \equiv d\sigma_{\mathrm{coh}} + d\sigma_{\mathrm{incoh}} \tag{4.D4}$$

by denoting the covariance

$$\mathrm{Cov}(a,b) = \langle ab^* \rangle - \langle a \rangle \langle b \rangle^* . \tag{4.D5}$$

Defining the non-random part of the scattering potential by $V^A$ (unperturbed potential) and the random (perturbed) potential by $V^B$, the *coherent* part of the differential cross-section writes

$$d\sigma_{\mathrm{coh}} = \frac{1}{16\pi^2} \left| T^A + \langle T^B \rangle \right|^2 d\Omega \tag{4.D6}$$

and the *incoherent* differential cross-section

$$d\sigma_{\mathrm{incoh}} = \frac{1}{16\pi^2} \mathrm{Cov}(T^B, T^B) d\Omega . \tag{4.D7}$$

If the random part $V^B$ causes only a small disturbance to the scattering by $V^A$,
we can calculate $T^B$ within the distorted-wave Born approximation (DWBA). It is
worth noting that in contrast to the widely spread opinion it is not a small potential
$V^B \ll V^A$ which defines the validity of the DWBA, but rather the *scattering by $V^B$*
which has to be weak.

Scattering by a Randomly Disturbed Multilayer

In a *multilayer* we represent each layer by the product of its volume polarisability
$\chi_{\infty j}(r)$ and the layer size function $\Omega_j(r)$

$$\chi(r) = \sum_{j=1}^{N} \chi_j(r) = \sum_{j=1}^{N} \chi_{\infty j}(r)\Omega_j(r) . \tag{4.D8}$$

The optical (or scattering) potential for x-rays can be expressed by the polarisability: $V(r) = -k_0^2 \chi(r)$. The contribution of the different layers to the scattering cross-section is distinguished by considering each layer as an independent
scatterer

$$\hat{V}(r) = \sum_j v_j(r) . \tag{4.D9}$$

Then Eq. (4.D4) writes

$$d\sigma = \frac{1}{16\pi^2} \left\{ \left| \sum_{j=1}^N \langle \tau_j \rangle \right|^2 + \sum_{j=1}^N \sum_{k=1}^N \mathrm{Cov}\,(\tau_j, \tau_k) \right\} d\Omega, \tag{4.D10}$$

with $\tau_j = \langle K_S | v_j | E \rangle$.

Separating the non-random and the random part of each layer, $v_j = v_j^A + v_j^B$, we obtain

$$d\sigma = \frac{1}{16\pi^2} \left\{ \left| \sum_{j=1}^N \tau_j^A + \sum_{j=1}^N \langle \tau_j^B \rangle \right|^2 + \sum_{j=1}^N \sum_{k=1}^N \mathrm{Cov}\,(\tau_j^B, \tau_k^B) \right\} d\Omega, \tag{4.D11}$$

where the $\tau_j^A$ are the contributions of the non-perturbed layers to scattering, and $\tau_j^B$ are those of the layer disturbances. The first term is the coherent part $d\sigma_{\mathrm{coh}}$, which consists of the contribution of the ideal multilayer and of the averaged transition elements of the layer disturbances. The second, incoherent part $d\sigma_{\mathrm{incoh}}$ contains the covariance functions of all single-layer transition elements.

Formally the division of $\hat{V}$ into a sum of scatterers $\sum_j v_j$ is arbitrary. The sticking point is to find a set of eigenstates, which is convenient to serve as basis for calculation of the transition elements. Finally, we remind the reader that until now no approximation has been made.

### 4.D.2  Formal Kinematical Treatment by First-Order Born Approximation

Within the kinematical treatment (first-order Born approximation) we approximate the transition operator by the operator of the scattering potential $\hat{T}|K\rangle \approx \hat{V}|K\rangle$. The set of vacuum wavevectors $|K\rangle = e^{-ikr}$ provides an orthogonal basis for the calculation of the differential scattering cross-section. The transition elements of the individual layers are

$$\tau_j = \langle K_S | v_j | K_0 \rangle = -k_0^2 \int dr\, \chi_j(r)\, e^{iqr} , \tag{4.D12}$$

where $q = k_S - k_0$. Defining the *structure factor of the layer*

$$S_j(q) = \int_S dr_{\parallel}\, F_j(q_z, r_{\parallel})\, e^{iq_{\parallel}r_{\parallel}}, \tag{4.D13}$$

with the random one-dimensional *layer form factor*

$$F_j(q_z, r_{\parallel}) = \int dz\, \chi_j(r)\, e^{iq_z(z - Z_j)}, \tag{4.D14}$$

the transition element becomes

$$\tau_j = -k_0^2 e^{iq_z Z_j} S_j(q) .$$ (4.D15)

The coherent scattering cross-section (4.D9) uses the statistical averages $\langle \tau_j \rangle_{av}$, and so we search for the *mean layer form factor* $\langle F_j(q_z, r_\parallel) \rangle_{av}$. The incoherent differential scattering cross-section contains the *covariance functions*

$$\tilde{q}_{jk} = \mathrm{Cov}(S_j, S_k)$$ (4.D16)
$$= \int dr_\parallel \int dr_\parallel' e^{iq_\parallel(r_\parallel - r_\parallel')} \mathrm{Cov}\left(F_j(q_z, r_\parallel), F_k(q_z, r_\parallel')\right) .$$

Substituting (4.D13) and (4.D16) into (4.D10), we obtain the kinematical differential scattering cross-section of an arbitrary multilayer

$$d\sigma = \frac{k_0^4}{16\pi^2} \left\{ \left| \sum_{j=1}^{N} \langle S_j \rangle_{av} e^{iq_z Z_j} \right|^2 + \sum_{j=1}^{N} \sum_{k=1}^{N} \tilde{q}_{jk} e^{iq_z(Z_j - Z_k)} \right\} d\Omega .$$ (4.D17)

## 4.D.3 Formal Treatment by a Distorted-Wave Born Approximation

The distorted-wave Born approximation takes all those effects of multiple scattering into account which are caused by the unperturbed potential $V^A$. It is less the method itself, but rather the right choice of $V^A$, which decides about the success in order to be enough transparent and sufficiently precise. We search for such a $V^A$ which enables to explain the essential multiple scattering effects. However, it should provide the simplest possible solutions $E_K^A$ used as orthonormal basis for the representation of scattering by the disturbance (perturbed potential) $V^B$.

Scattering by *planar multilayers with sharp interfaces* produces such simple solutions. It has been shown that rough multilayers as well as intentionally laterally patterned multilayers and gratings can be treated advantageously by starting with an ideal potential of a planar (laterally averaged) multilayer, splitting the polarisability in

$$\chi = \chi^A + \chi^B \qquad \text{with} \qquad \chi^A = \sum_{j=1}^{N} \chi_j^{A \, \text{planar}} .$$ (4.D18)

Coherent scattering by the non-perturbed multilayer generates a wave field $E_K^A$, which can be decomposed into a small number of plane waves within each plane homogeneous layer, both with constant complex amplitudes and wavevectors,

$$E_{K,j}^A(r) = \left[ \sum_{n=1}^{I} E_{k_n,j} e^{-ik_{n\parallel,j} r_\parallel} e^{-ik_{nz,j}(z - Z_j)} \right] \Omega_j^A(z) .$$ (4.D19)

In case of specular reflection it is $I = 2$ (one transmitted and one reflected wave), for grazing incidence diffraction and strongly asymmetric x-ray diffraction $I = 8$. The $E_K^A(r)$ are used as non-perturbed states for the estimation of $T^B$ (4.D6).

Within the first-order DWBA one obtains

$$T^{B,\text{DWBA}} = \langle E_S^{A*} | \hat{V}^B | E_0^A \rangle = -k_0^2 \int dr \, E_S^A(r) \chi^B(r) E_0^A(r) \,. \tag{4.D20}$$

Again, it is recommendable to describe the contribution of the disturbance within each plane layer separately by

$$\tau_j^B = -k_0^2 \int dr \, E_S^A(r) \chi_j^B E_0^A(r), \tag{4.D21}$$

with

$$\chi_j^{B\,\text{planar}} = \chi \Omega_j^A - \chi_j^{A\,\text{planar}} \,. \tag{4.D22}$$

We define $F_j^{mn}$ and $S_j^{mn}$ formally similar to the expressions (4.D13) and (4.D14); however, now with respect to the *disturbance* $\chi_j^B$ and corresponding to the actual scattering vector

$$q_j^{mn} = k_{Sj}^m - k_{0j}^n, \tag{4.D23}$$

*inside* the layer

$$F_j^{mn}(q_{z,j}^{mn}, r_{\parallel}) = \int dz \, \chi_j^B(r) \, e^{i q_{z,j}^{mn}(z - Z_j)} \,. \tag{4.D24}$$

Each $\tau_j^B$ consists of $I \times I$ terms

$$\tau_j^B = -k_0^2 \sum_{m=1}^{I} \sum_{n=1}^{I} E_{Sj}^m(z) S_j^{mn} E_{0j}^n(z) \,, \tag{4.D25}$$

or using the matrix formalism

$$\tau_j^B = -k_0^2 \mathbf{E}_{Sj}^m \hat{S}_j \mathbf{E}_{0j}^n \,, \tag{4.D26}$$

where the column vector $\mathbf{E}_{K,j}^m$ contains the amplitudes of the $I$ plane waves of one non-perturbed state in the $j$th layer and $\hat{S}_j$ is the structure factor matrix of the layer disturbance, respectively. Each term in (4.D25) represents the contribution of the disturbance to the scattering from one plane wave of the initial state $E_{K_0}^A$ to another plane wave of the final state $E_{K_S}^A$. Each scattering process is characterised by the product of the according wave amplitudes $E_{Sj}^m E_{0j}^n$ and by the disturbance structure factors $S_j^{mn}$.

In order to determine the *coherent scattering cross-section* we average $F$ and $S$ over the statistical ensemble and substitute these terms in (4.D10). The *incoherent cross-section* contains the covariance functions for each layer pair

$$\text{Cov}\left(\tau_j^B, \tau_k^B\right) = k_0^4 \sum_{m,n,o,p} E_{Sj}^m E_{Sj}^n \tilde{Q}_{jk}^{mnop} E_{0k}^o E_{0k}^p, \tag{4.D27}$$

with

$$\tilde{Q}_{jk}^{mnop} = \text{Cov}\left(S_j^{mn}, S_k^{op}\right)$$

$$= \int dr_{\parallel} \int dr_{\parallel}' e^{iq_{\parallel}(r_{\parallel} - r_{\parallel}')} \text{Cov}\left(F_j(q_{z,j}^{mn}, r_{\parallel}), F_k(q_{z,k}^{op}, r_{\parallel}')\right). \qquad (4.D28)$$

Each term represents the covariance of one scattering process in layer $j$ and a second scattering process in layer $k$. Adding up the contributions of all scattering processes and all layers we obtain finally

$$d\sigma = \frac{k_0^4}{16\pi^2}\left\{ \left| \sum_{j=1}^{N} \tau_j^A + \sum_{j=1}^{N} \sum_{m,n=1}^{I} E_{Sj}^m \langle S_j^{mn} \rangle E_{0j}^n \right|^2 \right. \qquad (4.D29)$$

$$\left. + \sum_{j,k=1}^{N} \sum_{m,n,o,p=1}^{I} E_{Sj}^m (E_{Sj}^n)^* \tilde{Q}_{jk}^{mnop} E_{0k}^o (E_{0k}^p)^* \right\} d\Omega .$$

In *x-ray reflectivity*, each eigenstate of the unperturbed potential consists of a transmitted and reflected wave, thus $I = 2$. The four wavevector transfers $q^{11}, \dots, q^{22}$, corresponding to $(k_{\text{sc}\parallel} - k_{\text{in}\parallel}, \pm k_{\text{sc},z} \pm k_{\text{in},z})$ in (4.56), (4.57) or (6.48), are represented in the reciprocal space in Fig. 6.40. Further, the above expressions are written explicitly for diffuse scattering in Eqs. (6.46), (6.47), (6.48) and (6.49) and for coherent reflectivity for deterministic (i.e. non-random) grating potential $V^B$ in (6.72). The covariance for grazing incidence diffraction is presented by (6.62).

## Simpler DWBA for Multilayers

The expressions simplify enormously if we can approximate the non-perturbed polarisability by its mean value in the multilayer, averaging vertically over the whole multilayer stack. We obtain a homogeneous "non-perturbed layer". The splitting of the potential in this way gives

$$\chi^A(r) = \langle \chi^{\text{ML}}(r) \rangle_{\text{av}}$$

$$\chi^B(r) = \sum_{j=1}^{N} \chi_j^{B\text{layer}}(r) \quad \text{with} \quad \chi_j^{B\text{layer}}(r) = \left(\chi(r) - \langle \chi^{\text{ML}}(r) \rangle_{\text{av}}\right) \Omega_j^{\text{id}}(r) . \qquad (4.D30)$$

Now the non-perturbed wave field below the sample surface consists of the transmitted wave only. In consequence exclusively the *primary scattering processes*

$$\text{Cov}(\tau_j^B, \tau_k^B) = K^4 t_S t_S^* \tilde{Q}_{jk}^{11} r_0 r_0^* \qquad (4.D31)$$

and the transmission function of the sample surface are considered. Also the effect of refraction is included.

# References

1. Atyukov, I.A., Karabekov, A.Yu., Kozhevnikov, I.V., Alaudinov, B.M., Asadchikov, V.E.: Physica B **198**, 9 (1994)
2. Bindell, J.B., Waifan, N.: J. Appl. Phys. **3**, 503 (1970)
3. Born, M., Wolf, E.: Principles of Optics, 6th edn, p. 51. Pergamon, London (1980)
4. Bourdieu, L., Daillant, J., Chatenay, D., Braslau, A., Colson, D.: Phys. Rev. Lett. **72**, 1502 (1994)
5. Brown, G., Celli, V., Haller, M., Maradudin, A.A., Marvin, A.: Phys. Rev. B **31**, 4993 (1985)
6. Brown, G.C., Celli, V., Coopersmith, M. Haller, M.: Surf. Sci. **129**, 507 (1983)
7. Brown, G.C., Celli, V., Haller, M., Marvin, A.: Surf. Sci. **136**, 381 (1984)
8. Cai, Z.-h., Huang, K., Montano, P.A., Russel, T.P., Bai, J.M., Zajac, G.W.: J. Chem. Phys. **98**, 2376 (1993)
9. Croce, P.: J. Opt. (Paris) **8**, 127 (1977)
10. Croce, P.: J. Opt. (Paris) **14**, 213 (1983)
11. Croce, P., Névot, L.: Revue Phys. Appl. **11**, 113 (1976)
12. Croce, P., Névot, L., Pardo, B.: C.R. Acad. Sci. Paris **274 B**, 803 (1972)
13. Croce, P., Névot, L., Pardo, B.: C.R. Acad. Sci. Paris **274 B**, 855 (1972)
14. Daillant, J., Bellet-Amalric, E., Braslau, A., Charitat, T., Fragneto, G., Graner, F., Mora, S., Rieutord, F., Stidder, B.: PNAS **102**, 11639 (2005)
15. Daillant, J., Bélorgey, O.: J. Chem. Phys. **97**, 5824 (1992)
16. Daillant, J., Bélorgey, O.: J. Chem. Phys. **97**, 5837 (1992)
17. Davydov, A.S.: Quantum Mechanics. Pergamon Press, London (1969)
18. de Boer, D.K.G.: Phys. Rev. B **49**, 5817 (1994)
19. de Jeu, W.H., Schindler, J.D., Mol, E.A.L.: J. Appl. Cryst. **29**, 511 (1996)
20. Dietrich, S., Haase, A.: Phys. Rep. **260**, 1 (1995)
21. Fukuto, M., Heilmann, R.K., Pershan, P.S., Griffiths, J.A., Yu, S.M., Tirrell, D.A.: Phys. Rev. Lett. **81**, 3455 (1998)
22. Gourier, C., Daillant, J., Braslau, A., Alba, M., Quinn, K., Luzet, D., Blot, C., Chatenay, D., Grübel, G., Legrand, J.F., Vignaud, G.: Phys. Rev. Lett. **78**, 3157 (1997)
23. Helfrich, W.: Z. Naturforschung **28 c**, 693 (1973)
24. Herpin, A.: C.R. Acad. Sci. Paris **225**, 182 (1947)
25. Jackson, J.D.: Classical Electrodynamics, 2nd edn. Wiley, New York (1975)
26. Kayser, R.F.: Phys. Rev. A **33**, 1948 (1986)
27. Landau, L.D., Lifshitz, E.M.: Electrodynamics of Continuous Media, Course of Theoretical Physics, vol. 8, §69. Pergamon Press, Oxford (1960)
28. Lorrain, P., Corson, D.R.: Electromagnetic Fields and Waves, p. 629. W.H. Freeeman and Company, San Francisco (1970)
29. Mora, S., Daillant, J., Luzet, D., Struth, B.: Europhys. Lett. **66**, 694 (2004)
30. Mora, S., Daillant, J., Mecke, K., Luzet, D., Alba, M., Braslau, A., Struth, B.: Phys. Rev. Lett. **90**, 216101 (2003)
31. Névot, L., Croce, P.: Revue Phys. Appl. **15**, 761 (1980)
32. Nieto-Vesperinas, M., Dainty, J.C.: Scattering in Volume and Surfaces. Elsevier Science Publishers, B.V. North-Holland (1990)
33. Petit, R. (ed): Electromagnetic Theory of Gratings, Topics in Current Physics. Springer Verlag, Berlin (1980)
34. Pignat, J., Daillant, J., Leiserowitz, L., Perrot, F.: J. Phys. Chem. B **110**, 22178 (2006)
35. Reichert, H., Honkimäki, V., Snigirev, A., Engemann, S., Dosch, H.: Physica B **336**, 46 (2003)
36. Saint Martin, E., Konovalov, O., Daillant, J.: Thin Solid Films **515**, 5687 (2007)
37. Sanyal, M.K., Sinha, S.K., Huang, K.G., Ocko, B.M.: Phys. Rev. Lett. **66**, 628 (1991)
38. Sentenac, A., Greffet, J.J.: J. Opt. Soc. Am. A **15**, 528 (1998)
39. Sinha, S.K., Sirota, E.B., Garoff, S., Stanley, H.B.: Phys. Rev. B **38**, 2297 (1988)
40. Sinha, S.K., Tolan, M., Gibaud, A.: Phys. Rev. B **57**, 2740 (1998)
41. Tai, C.-T.: Dyadic Green Functions in Electromagnetic Theory. IEEE Press, New York (1994)

42. Tidswell, I.M., Rabedeau, T.A., Pershan, P.S., Kosowsky, S.D.: Phys. Rev. Lett. **66**, 2108 (1991)
43. Tostmann, H., DiMasi, E., Pershan, P.S., Ocko, B.M., Shpyrko, O.G., Deutsch, M.: Phys. Rev. B **59**, 783 (1999)
44. Vidal, B., Vincent, P.: Appl. Opt. **23**, 1794 (1984)
45. Weber, W., Lengeler, B.: Phys. Rev. B **46**, 7953 (1992)
46. Yoneda, Y.: Phys. Rev. **131**, 2010 (1963)

# Chapter 5
# Neutron Reflectometry

C. Fermon, F. Ott and A. Menelle

## 5.1 Introduction

Neutron reflectometry is a relatively new technique [1, 2]. In the last years, it has been extensively used for solving soft matter problems like polymer mixing [3–5] or the structure of liquids at the surface [6, 7], for example. The key property of neutrons for polymer studies is their large contrast between $^1$H and $^2$H which allows selective "labeling" by deuteration.

In the late 1980s, a new field of application of neutron reflectometry emerged. Following the discovery of giant magneto-resistance in anti-ferromagnetically coupled multilayered films [8] and new magnetic phenomena in ultra-thin films, there has been an interest in the precise measurement of the magnetic moment direction in each layer of a multilayer and at the interface between layers. Owing to the large magnetic coupling between the neutron and the magnetic moment, neutron reflectometry has proved to be a powerful tool for obtaining information about these magnetic configurations and for measuring magnetic depth profiles.

At grazing incidence, it is possible to distinguish three scattering geometries (Fig. 5.1): specular reflection, scattering in the incidence plane (off-specular scattering) and scattering perpendicular to the incidence plane (grazing incidence SANS). These different scattering geometries probe different length scales $x$ and directions in the sample surface. Specular reflectivity probes the structure along the depth in the film ($3\,\text{nm} < \xi < 100\,\text{nm}$). Off-specular scattering probes surface features at a micrometric scale ($600\,\text{nm} < \xi < 60\,\mu\text{m}$). Finally, grazing incidence SANS probes surface features in the range $3\,\text{nm} < \xi < 100\,\text{nm}$. These different scattering geometries allow the study of a very wide range of length scales $\xi$, ranging from a few nm up to several μm.

In this chapter, we give an overview of the experimental and theoretical methods used in neutron reflectometry, focusing mainly on specular reflectivity. The corresponding theory is partly derived from previous work on x-rays, and we emphasize on the aspects specific to neutrons.

C. Fermon (✉)
Service de Physique de l'Etat Condensé, Orme des Merisiers, CEA Saclay,
91191 Gif sur Yvette Cedex, France

Fermon, C. et al.: *Neutron Reflectometry.* Lect. Notes Phys. **770**, 183–234 (2009)
DOI 10.1007/978-3-540-88588-7_5

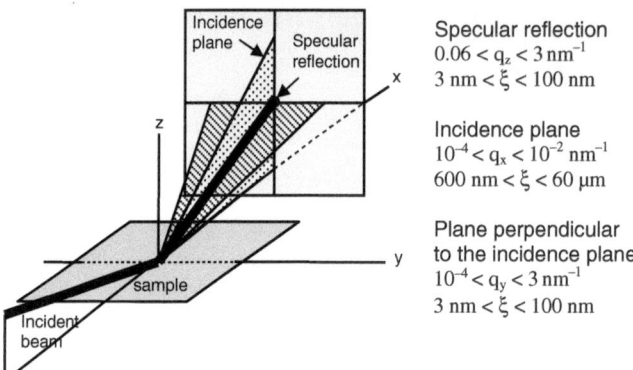

**Fig. 5.1** The different surface scattering geometries. (*Black line*) specular reflectivity geometry; (*dotted plane*) off-specular scattering plane, corresponding to the incidence plane; (*hashed plane*) GISANS scattering plane, perpendicular to the incidence plane. These different scattering geometries probe a very wide range of length scales and directions in the sample surface

In a first part, we will review the neutron–matter interactions. We then describe the non-magnetic scattering. In this case, it is possible to introduce an optical index and give a treatment which is similar to x-ray reflectometry (Chap. 3).

In a second part, the neutron spin is introduced. In this case, optical indices cannot be used any longer and it is necessary to completely solve the Schrödinger equation. A detailed matrix formalism is presented.

We then discuss the different aspects of data processing and the problems related to the surface roughness. Two types of neutron reflectometers are described: fixed-wavelength two-axis reflectometers and time-of-flight spectrometers.

The use of neutron reflectivity in the field of polymer films and magnetic layers is then illustrated by several examples.

Finally, we present the use of off-specular scattering and grazing incidence SANS applied to the study of magnetic surfaces. However, the theoretical aspects of magnetic off-specular scattering are not discussed since it would require a dedicated chapter.

## Notations used in this chapter

$b, b_j$     bound scattering length of a nucleus, mean scattering length of a layer $j$

$b_c$     bound coherent scattering length

$b_i$     incoherent scattering length

$b_N$     spin-dependent scattering length

$b'$     real part of the scattering length

$b''$     imaginary part of the scattering length

$\mathscr{E}_0, \mathscr{E}_j$     energy of the neutron in the vacuum and in layer $j$

$e$     charge of the electron

| $d, d_j$ | thickness of a layer |
|---|---|
| $g$ | Landé factor $(g = 2)$ |
| $\hbar$ | Planck constant |
| $\mathbf{I}$ | nuclear spin operator |
| $\mathbf{k}$ | wave vector |
| $\mathbf{M}, \mathbf{M_j}$ | magnetic moment of an electron and of a layer |
| $m$ | neutron mass |
| $m_e$ | electron mass |
| $n_j$ | refractive index of layer $j$ |
| $p$ | $p = 2.696\,\text{fm}$, conversion factor of magnetization to an effective scattering length |
| $\mathbf{q}$ | scattering vector |
| $\mathbf{s}$ | spin operator of the electron |
| $\sigma$ | Pauli operator associated with the neutron spin |
| $V_j$ | volume of layer $j$ |
| $V(r)$ | interaction Hamiltonian |
| $g_n$ | $g_n = -1.9132$, nuclear Landé factor of the neutron |
| $\lambda, \lambda_0$ | neutron wavelength |
| $\mu_B$ | Bohr magneton $(\mu_B = e\hbar/(2m_e) = 9.27 \times 10^{-24}\,\text{JT}^{-1})$ |
| $\mu_n$ | nuclear magneton $(\mu_n = e\hbar/(2m_p) = 5.05 \times 10^{-27}\,\text{JT}^{-1})$ |
| $\rho_j$ | atomic density of layer $j$ (atoms per $\text{cm}^3$) |
| $\sigma_j$ | absorption |
| $\theta_j, \varphi_j$ | spherical angles of the magnetization of layer $j$ |
| $\theta_{\text{in}}, \theta_{\text{r}}$ | incident and reflected angles of the neutron beam |

We call "up" (resp. "down") the neutron polarization parallel (resp. anti-parallel) to the external applied magnetic field.

"Down–up" designates a polarized "down" incident beam and polarized "up" detected beam.

"Down–up" and "up–down" are called spin-flip processes.

## 5.2 Schrödinger Equation and Neutron–Matter Interactions

### 5.2.1 Schrödinger Equation

The neutron can be described by a wave of wavelength $\lambda$, of wave vector:

$$k_0 = \frac{2\pi}{\lambda}, \tag{5.1}$$

and of energy

$$\mathscr{E}_0 = \frac{\hbar^2 k_0^2}{2m}. \tag{5.2}$$

Its wave function verifies the Schrödinger equation (1.17):

$$\frac{\hbar^2}{2m}\frac{d^2\psi}{dr^2} + [\mathscr{E} - V(r)]\,\psi = 0,\qquad(5.3)$$

where $m$ is the neutron mass, $\mathscr{E}$ its energy and $V$ the interaction potential. The neutron is a spin 1/2 particle so that $\psi(r)$ can be expressed on the base of the two spin states:

$$\psi_+(r)|+\rangle + \psi_-(r)|-\rangle.\qquad(5.4)$$

When there is an external or internal magnetic field, an "up" (resp. "down") neutron designates a neutron in the eigenstate $|+\rangle$ (resp. $|-\rangle$). In the following, the space dependence ($\mathbf{r}$) of the index will often be dropped.

## 5.2.2 Neutron–Matter Interaction

The two main interactions are the strong interaction with the nuclei and the magnetic interaction with the existing magnetic moments (nuclear and electronic). There are a large number of second-order interactions which are described in [9].

### 5.2.2.1 Neutron–Nucleus Interaction: Fermi Pseudo-Potential

The scattering of a neutron by a nucleus comes mainly from the strong interaction. The interaction potential is large but its extension is much smaller than the wavelength of the neutron. Hence this interaction can be considered as punctual and isotropic. Within the Born approximation, it can be described by the Fermi pseudo-potential [10]:

$$V_F(r) = b\left(\frac{2\pi\hbar^2}{m}\right)\delta(\mathbf{r}),\qquad(5.5)$$

where $b$ is the scattering length and $\mathbf{r}$ is the position of the neutron. The value of the scattering length $b$ depends on the nucleus and on the nuclear spin of the nucleus. Formally it can be written as

$$b = b_c + \frac{1}{2}b_N\mathbf{I}.\boldsymbol{\sigma},\qquad(5.6)$$

*N.B.*: the scattering length is generally a complex number: $b = b' + ib''$. The first term $b_c$ is called the coherent scattering length. The second term corresponds to the strong interaction of the spin of the neutron (described by the operator $1/2\boldsymbol{\sigma}$) with that of the nucleus (operator $\mathbf{I}$). The total spin $J = 1/2\boldsymbol{\sigma} + I$ is a good quantum number for the neutron spin – nucleus spin interaction $1/2\boldsymbol{\sigma}.\mathbf{I}$. In the manifold $\{I \pm 1/2\}$, the eigenvalues of the spin-dependent operator $\mathbf{I}.\boldsymbol{\sigma}$ are $I$ (for $J = I + 1/2$) and $-(I+1)$ (for $J = I - 1/2$). We name $b^+$ and $b^-$ the two scattering lengths associated with these two eigenvalues, corresponding to the two states $|+\rangle$ and $|-\rangle$ of the neutron spin. The nucleus spin-dependent scattering lengths can then be written as follows [11]:

$$\begin{cases} b^+ = b_0 + \frac{1}{2}b_n I \\ b^- = b_0 - \frac{1}{2}b_n(I+1) \end{cases}, \tag{5.7}$$

where $I$ is the nuclear spin quantum number.

We remind that the total scattering cross section is given by (see Eq. (1.35))

$$\sigma_{tot} = 4\pi\langle|b|^2\rangle, \tag{5.8}$$

in which the brackets designate the statistical average over the neutron and nuclear spins.

### 5.2.2.2 Neutron Absorption

The absorption of neutrons is described by the imaginary part of the scattering length $b''$. The absorption cross section is given by

$$\sigma_{abs} = (4\pi/k_0)\,b''. \tag{5.9}$$

The absorption is negligible for thin films except for some elements: Gd, Sm, B and Cd. These elements present $(n,\gamma)$ nuclear resonances at thermal neutron energies which strongly increase the absorption.

### 5.2.2.3 Incoherent Scattering

Incoherent scattering comes from the random distribution of isotopes or nuclear spin states in a material. In this case, the total scattering cross section (see Eq. (5.8)) can be written as

$$\sigma_{tot} = 4\pi\langle|b|^2\rangle = 4\pi\left(\langle|b|\rangle^2 + \left(\langle|b|^2\rangle - \langle|b|\rangle^2\right)\right) = \sigma_{coh} + \sigma_{incoh}, \tag{5.10}$$

where $\sigma_{coh}$ and $\sigma_{incoh}$ are called the coherent and incoherent scattering lengths. In the presence of isotope or spin disorder, the second term in Eq. (5.10) is not zero. If, for example, the nucleus carries a spin (see 5.8) we have a spatial distribution $b^+$ and $b^-$ of scattering lengths. In the case of an isotope distribution $b_\alpha$ in the material, the incoherent cross section is given by

$$\sigma_{inc,isotope} = 4\pi \sum_{\alpha<\beta} c_\alpha c_\beta |b_\alpha - b_\beta|^2, \tag{5.11}$$

where $c_\alpha$ designates the fraction of isotope $\alpha$ in the material. Incoherent scattering appears as a $q$-independent background in the experiments and can be treated as an absorption plus a flat background. The incoherent scattering is particularly important for hydrogenated layers but it is small for deuterated layers. A more detailed discussion of incoherent scattering can be found in [12–14]. *Tables of the different scattering lengths (coherent, incoherent, absorption) of the different elements can be found in [13, 14].*

### 5.2.2.4 Magnetic Interaction

The main magnetic interaction is the dipolar interaction of the neutron spin with the magnetic field created by the unpaired electrons of the magnetic atoms. This field contains two terms, the spin part and the orbital part:

$$\mathbf{B} = \frac{\mu_0}{4\pi} \left( \nabla \times \left\{ \frac{\mu_e \times \mathbf{R}}{|\mathbf{R}|^3} \right\} - \frac{e \mathbf{v_e} \times \mathbf{R}}{|\mathbf{R}|^3} \right), \tag{5.12}$$

where $\mu_e = -2\mu_B \sigma$ is the magnetic moment of the electron, $\mu_B$ is the Bohr magneton, $\mathbf{v}_e$ is the speed of the electron.

The neutron magnetic moment is equal to

$$\mu = g_n \mu_n \sigma. \tag{5.13}$$

The magnetic interaction is expressed as

$$V_M(\mathbf{r}) = -\mu.\mathbf{B} = -g_n \mu_n \sigma.\mathbf{B}. \tag{5.14}$$

Neutron reflectivity does not allow the separation of the orbital and spin contributions, it is only sensitive to the internal magnetic field.

### 5.2.2.5 The Zeeman Interaction

It is the interaction of the neutron spin with an external magnetic field $\mathbf{B}_0$:

$$V_Z(\mathbf{r}) = -g_n \mu_n \sigma.\mathbf{B}_0. \tag{5.15}$$

## 5.3 Reflectivity on Non-Magnetic Systems

For non-magnetic systems we can introduce the notion of optical indices. It is an approach similar to the x-ray formalism (Chaps. 1 and 2). It can be applied to neutron reflectometry on soft matter [15] and non-magnetic systems.

We consider a neutron beam, reflected by a perfect surface with an incident angle $\theta$. As in Chap. 3, the surface is defined by the interface between the air ($n = 1$) and a material with an optical refractive index $n$. In vacuum, the energy of the neutron is given by

$$\mathscr{E} = \frac{\hbar^2 k_0^2}{2m} = \frac{h^2}{2m\lambda^2}. \tag{5.16}$$

Let $\mathbf{q} = \mathbf{k}_r - \mathbf{k}_{in}$ be the scattering wave vector. The projection of the scattering wave vector on the $z$-axis (perpendicular to the surface) is given by

$$q_z = \frac{4\pi}{\lambda} \sin \theta_{in}. \tag{5.17}$$

## 5.3.1 Neutron Optical Indices

The neutron indices are very different from the x-ray indices and we will determine their expression from the Schrödinger equation. We suppose that the interaction potential $V(\mathbf{r})$ in the medium is independent of the in-plane coordinates $x$ and $y$. The mean potential $V$ in the medium is given by the integration of the Fermi pseudo-potential:

$$V = \frac{1}{v} \int_v V(\mathbf{r}) d^3 \mathbf{r} = \frac{2\pi\hbar^2}{m} \rho b, \tag{5.18}$$

where $\rho$ is the number of atoms per unit volume. The atomic details at the interface are smoothed out (Fig. 5.2b) and the interaction potential $V$ is only a function of the depth in the film $z$.

In the absence of any magnetic field, the Schrödinger equation can be written as

$$\frac{\hbar^2}{2m} \frac{d^2\psi}{dr^2} + [\mathscr{E} - V]\psi = 0. \tag{5.19}$$

Equation (5.19) can be written in the form of a Helmholtz propagation equation similar to the electromagnetic case:

$$\frac{d^2\psi}{dr^2} + k^2\psi = 0, \tag{5.20}$$

with

$$k^2 = \frac{2m}{\hbar^2} [\mathscr{E} - V]. \tag{5.21}$$

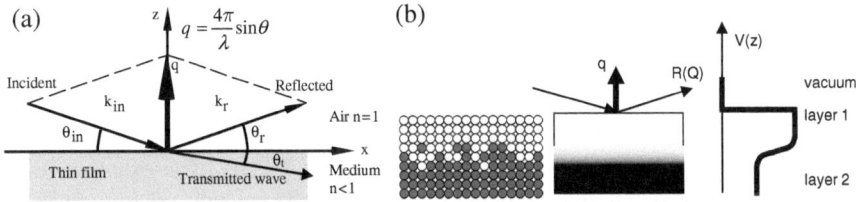

**Fig. 5.2** (a) Specular reflectivity geometry. (b) Interface between two surfaces. In the optical approximation, the interface is approximated as a continuous medium

We define the optical index as follows:

$$n^2 = \frac{k^2}{k_0^2} \tag{5.22}$$

The optical index $n$ can be written as[1]

$$n^2 = 1 - \frac{V}{\mathscr{E}} = 1 - \frac{\lambda^2}{\pi}\rho b. \tag{5.23}$$

It is in most cases smaller than 1 except for materials with a negative scattering length (e.g., Ti and Mn). The quantity $1 - n$ is of the order of $10^{-5}$ and thus $n$ can be written as

$$n \approx 1 - \frac{\lambda^2}{2\pi}\rho b. \tag{5.24}$$

### 5.3.2 Critical Angle for Total External Reflection

At the interface between two media, the Snell's law applies:

$$\cos\theta_{\text{in}} = n\cos\theta_{\text{tr}}. \tag{5.25}$$

Since we have shown that the index is smaller than 1, for angles $\theta \leq \theta_c$, there is a total reflection of the incident wave like in the case of x-ray reflection. The critical angle $\theta_c$ is given by the condition $\theta_{\text{tr}} = 0$, i.e.,

$$\cos\theta_c = n. \tag{5.26}$$

Since $\theta_c$ is very small it is possible to use a Taylor expansion. Using (5.24) and (5.26) the expression of $\theta_c$ is given by

$$\theta_c = \sqrt{\frac{\rho b}{\pi}}\lambda. \tag{5.27}$$

The corresponding critical wave vector is

$$q_c = \frac{4\pi\sin\theta_c}{\lambda} = 4\sqrt{\pi\rho b}. \tag{5.28}$$

### 5.3.3 Determination of Scattering Lengths and Optical Indices

In the case of pure materials, the knowledge of $b$ and $\rho$ fully characterizes the material. In the case of crystalline solids of the type $A_x B_y$, for example, unit cells must be considered and the density $\rho$ of unit cells per unit volume must be

---

[1] This expression is similar to that for x-ray (see Sect. 3.1.1) where the classical electron radius $r_e$ is the scattering length density for x-rays.

**Table 5.1** Scattering length, atomic density, optical index $\delta = 1 - n$ (at $\lambda = 0.4\,\text{nm}$) and critical wave vector of some common materials. More exhaustive data can be found at *"www.neutron.anl.gov"*

| Material | $b_n$ (fm) | $\rho$ ($10^{28}\text{m}^{-3}$) | $\rho b$ ($10^{13}\text{m}^{-2}$) | $\delta$ ($10^{-6}$) | $q_c$ (nm$^{-1}$) |
|---|---|---|---|---|---|
| H (hydrogen) | −3.73 | | | | |
| D (deuterium) | 6.67 | | | | |
| C (graphite) | 6.64 | 11.3 | 75 | 19.1 | 0.19 |
| C (diamond) | 6.64 | 17.6 | 117 | 29.8 | 0.24 |
| O | 5.80 | | | | |
| Si | 4.15 | 5.00 | 20.8 | 5.28 | 0.10 |
| Ti | −3.44 | 5.66 | −19.5 | −5.0 | − |
| Al | 3.45 | 6.02 | 20.8 | 6.11 | 0.10 |
| Fe | 9.45 | 8.50 | 80.3 | 20.45 | 0.20 |
| Co | 2.49 | 8.97 | 22.34 | 5.69 | 0.11 |
| Ni | 10.3 | 9.14 | 94.1 | 24.0 | 0.22 |
| Cu | 7.72 | 8.45 | 65.2 | 16.6 | 0.18 |
| Ag | 5.92 | 5.85 | 34.6 | 8.82 | 0.13 |
| Au | 7.63 | 5.90 | 45 | 11.5 | 0.15 |
| $H_2O$ | −1.68 | 3.35 | −5.63 | −1.43 | − |
| $D_2O$ | 19.1 | 3.34 | 63.8 | 16.2 | 0.18 |
| $SiO_2$ | 15.8 | 2.51 | 39.7 | 10.1 | 0.14 |
| GaAs | 13.9 | 2.21 | 30.7 | 7.82 | 0.12 |
| $Al_2O_3$ (sapphire) | 24.3 | 2.34 | 56.9 | 14.5 | 0.17 |
| Pyrex | | | 42 | 10.7 | 0.14 |
| Polystyrene | 23.2 | 0.61 | 14.2 | 3.6 | 0.084 |
| Polystyrene (deuterated) | 106.5 | 0.61 | 65 | 16.5 | 0.18 |

calculated. The average scattering length $b_{av}$ in the unit cell is simply given by $b_{av} = (x b_A + y b_B)/(x + y)$. The value $\rho b_{av}$ can then be used to calculate the index of the material. The case of liquids and polymers is more complex since it is usually more difficult to define a "unit cell". Thus, the best method is to calibrate the optical index of each polymer or liquid that one wants to study by measuring a thick film, for example. Table 5.1 gives the scattering length, optical index $\delta = 1 - n$ and critical wave vector $q_c$ (at $\lambda = 0.4$ nm) for various elements and compounds.

## 5.3.4 Reflection on a Homogeneous Medium

As shown by Eq. (5.19) the problem of the reflection of a neutron beam on a non-magnetic medium can be treated exactly in the same way as the reflection of x-rays. Since the potential $V$ is only $z$ dependent, the Schrödinger Eq. (5.19) reduces to the one-dimensional equation:

$$\frac{\hbar^2}{2m}\frac{d^2\psi_z}{dz^2} + [\mathcal{E}_z - V_z]\,\psi_z = 0, \tag{5.29}$$

with a wave function of the form $\psi = e^{i(k_{\text{in}x}x + k_{\text{in}y}y)}\,\psi_z$.

In the medium, the general solution is given by

$$\psi_z = A e^{ik_{trz}z} + B e^{-ik_{trz}z}. \tag{5.30}$$

The transmitted wave vector can be related to the incident wave vector using (5.21):

$$k^2 = \frac{2m}{\hbar^2}[\mathscr{E} - V] = k_{in}^2 - 4\pi\rho b. \tag{5.31}$$

At the interface we have to apply the continuity condition on $\psi$ and $\nabla\psi$. In a way similar to the x-ray case, it is then possible to show that the parallel components of the incident and reflected waves are continuous [10]. The continuity of the parallel components allows us to write

$$k_{trz}^2 = k_{inz}^2 - 4\pi b\rho. \tag{5.32}$$

Considering Eqs. (5.20) and (5.32), the problems of neutron and x-ray reflectivity are formally the same. It is possible to use the same formalism as the one developed in Chap. 3 for x-ray reflectivity.

In particular, it is possible to use the classical Fresnel formulae. The reflected and transmitted amplitudes are given by

$$r = \frac{\sin\theta_{in} - n\sin\theta_{tr}}{\sin\theta_{in} + n\sin\theta_{tr}}, \tag{5.33}$$

$$t = \frac{2\sin\theta_i}{\sin\theta_i + n\sin\theta_{tr}}. \tag{5.34}$$

In terms of scattering wave vector, the reflected intensity is given by

$$R = \left| \frac{k_{0z} - k_{trz}}{k_{0z} + k_{trz}} \right|^2. \tag{5.35}$$

Figure 5.3 shows a typical curve calculated for a perfect surface. Below the critical wave vector $q_c$ the reflectivity is exactly 1. Beyond this region, the signal decays as $1/q^4$.

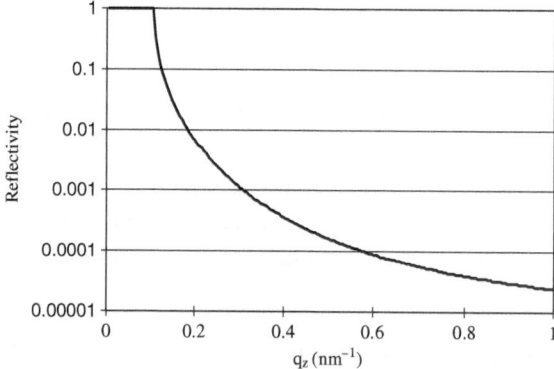

**Fig. 5.3** Reflected intensity as a function of $q_z$ for a silicon substrate (at $\lambda = 0.4$ nm)

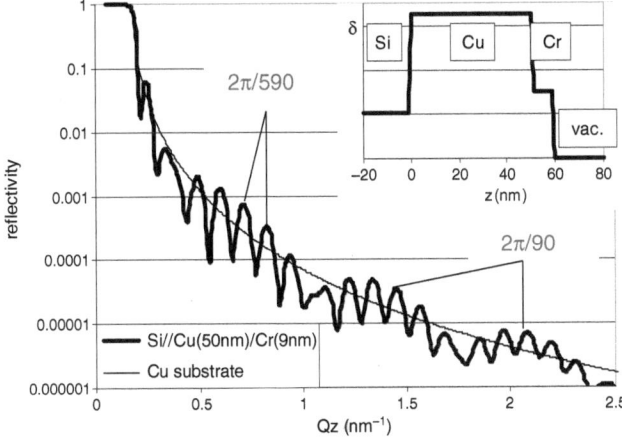

**Fig. 5.4** Reflectivity on a multilayer system Si//Cu (50 nm)/Cr (9 nm). The short period oscillations are characteristic of the total thickness of the layer (59 nm); the long range modulation is characteristic of the thin Cr layer (9 nm). (*Insert*) Optical index profile as a function of the depth in the film

Figure 5.4 presents the situation of the reflection of a neutron beam on a multilayer $Si \parallel Cu \mid Cr$. Modulations of the reflected intensities are observed. They correspond to constructive and destructive interferences of the neutron waves scattered by the different interfaces of the multilayer system. These oscillations are called Kiessig fringes. Their pattern is characteristic of the multilayer system.

## 5.4 Neutron Reflectivity on Magnetic Systems

If the system is magnetic or if there is an external magnetic field on the sample, we need to take into account the spin of the neutron. In the simplest case where the magnetization of the different layers is collinear to the applied magnetic field defining the neutron polarization direction, it is possible to introduce a magnetic optical index. However, in the general case, when the magnetizations of the layers are not parallel to the applied field, it is not possible to use optical indices and it is always necessary to completely solve the Schrödinger equation [16–19]. In the case of homogeneous, infinite magnetic layers, the problem can be solved using a formalism very similar to the non-magnetic case developed in the previous part.

### 5.4.1 Interaction of the Neutron with an Infinite Homogeneous Layer

We consider a magnetic layer of thickness $d$, where the neutron interacts with the different unpaired electrons. We perform a direct integration on the layer in order to obtain the potential $V$ for the Schrödinger equation.

### 5.4.1.1 The Magnetic Interaction

A first approach is to assume that the neutron is sensitive to the internal magnetic field in the magnetic layer. The interaction potential is then written as

$$-g_n \mu_n \sigma . [\mu_0 (1 - D) \mathbf{M} + \mathbf{B_0}], \tag{5.36}$$

where $\mathbf{M}$ is the magnetization of the layer, $D$ is the demagnetizing factor and $\mathbf{B_0}$ is the external magnetic field. In the case of an infinite magnetic thin film, $(1 - D)\mathbf{M}$ is equal to the in-plane component of the magnetization $\mathbf{M}_{\parallel}$. It is possible to demonstrate this result but the calculations are somewhat lengthy. This is developed below for the interested reader but it can be skipped at the first reading.

The magnetic interaction can be written as

$$-g_n \mu_n \sigma . \mathbf{B} = -g_n \mu_n \sigma . \left( \nabla \times \left\{ \frac{\mu_e \times \mathbf{R}}{|\mathbf{R}|^3} \right\} - \frac{e}{c} \frac{\mathbf{v_e} \times \mathbf{R}}{|\mathbf{R}|^3} \right) \tag{5.37}$$

or

$$-g_n \mu_n \left\{ \sigma . \nabla \times \left( \frac{\mu_e \times \mathbf{R}}{|\mathbf{R}|^3} \right) - \frac{e}{2m_e c} \left( \mathbf{p}_e . \frac{\sigma \times \mathbf{R}}{|\mathbf{R}|^3} + \frac{\sigma \times \mathbf{R}}{|\mathbf{R}|^3} . \mathbf{p}_e \right) \right\}, \tag{5.38}$$

with

$$\mathbf{p}_e = -i\hbar \nabla_e. \tag{5.39}$$

If we first consider only the spin-dependent part of the interaction, we can write

$$\nabla \times \left( \frac{\mu_e \times \mathbf{r}}{r^3} \right) = -\nabla \times \left( \mu_e \times \nabla \left( \frac{1}{\mathbf{r}} \right) \right) = \frac{1}{2\pi^2} \int \frac{1}{q^2} (\mathbf{q} \times (\mu_e \times \mathbf{q})) \exp(i\mathbf{q}.\mathbf{r}) d\mathbf{q}. \tag{5.40}$$

### 5.4.1.2 Integration on a Homogeneous Layer

We suppose a constant atomic density $\rho$. We replace $\mathbf{r}$ by $\mathbf{r} + \mathbf{r_0}$, where $\mathbf{r_0}$ is the distance between the neutron and the center of the layer. $\mathbf{r}$ is the distance between the center and the volume $d\mathbf{r}$ in the layer. The spin-dependent part of the interaction is

$$2g_n \mu_n \mu_B \sigma \frac{1}{2\pi^2} \int \frac{1}{q^2} \int_V \rho(r)(\mathbf{q} \times \bar{s}(\mathbf{r}) \times \mathbf{q}) \exp(i\mathbf{q}.\mathbf{r_0}) \exp(i\mathbf{q}.\mathbf{r}) d\mathbf{r} d\mathbf{q}, \tag{5.41}$$

where $\rho(\mathbf{r})$ is the density and $\bar{s}(\mathbf{r}) = \bar{s}$ is the mean value of the spin magnetization in the volume $d\mathbf{r}$. The two first integrations over $x$ and $y$ give Dirac distributions:

$$4g_n \mu_n \mu_B \rho \sigma \int \frac{1}{q_z^2} \int_{-L/2}^{L/2} dr_z (\mathbf{q}_z \times \hat{s} \times \mathbf{q}_z) \exp(iq_z.r_{0z}) \exp(iq_z r_z) dq_z. \tag{5.42}$$

The third integration gives

$$8\pi g_n \mu_n \mu_B \rho \sigma.\bar{s}_\parallel \left[\theta\left(r_{0z}+L/2\right)-\theta\left(r_{0z}-L/2\right)\right].\tag{5.43}$$

where $\theta(r)$ is the Heaviside function. We can do the same calculation on the orbital part and we obtain

$$\frac{2\pi\hbar^2}{m}p\sigma.\mathbf{M}_\parallel \rho \left[\theta\left(r_{0z}+L/2\right)-\theta\left(r_{0z}-L/2\right)\right],\tag{5.44}$$

with $p = 2.696$ fm. $\mathbf{M}_\parallel$ is given in $\mu_B$ per atom and represents the in-plane component of the magnetization and not necessarily the magnetization perpendicular to the wave vector.

### 5.4.1.3 Conclusion

From Eq. (5.44), we can deduce two very important points: it is only possible to measure the in-plane magnetization and the magnetic interaction is zero out of the layer. These two properties are essential, the first is the main limitation to the use of neutrons for the study of magnetic thin films, the second is the justification of solving the Schrödinger equation in each layer, independently of the others. Thus the formalism developed for non-magnetic systems can be adapted to the magnetic case, however, with some complications.

## 5.4.2 Solution of the Schrödinger Equation

The interaction potential for a layer $j$ is given by

$$V_j = \frac{2\pi\hbar^2}{m}\rho_j b_j - \frac{2\pi\hbar^2}{m}\rho_j p \sigma.\mathbf{M}_{j_\parallel} - g_n \mu_n \sigma.\mathbf{B}_0.\tag{5.45}$$

We introduce an effective field $\mathbf{B}_{\text{eff}}$ defined by (Fig. 5.5)

$$\mathbf{B}_{\text{eff}} = \mathbf{B}_0 + \mu_0 \mathbf{M}_\parallel.\tag{5.46}$$

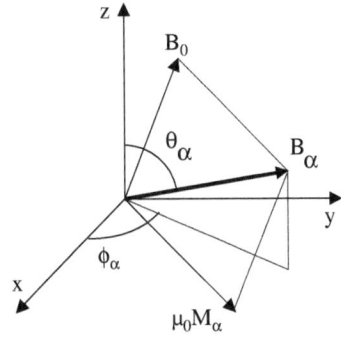

**Fig. 5.5** Effective field $\mathbf{B}_{\text{eff},j}$ in the layer, sum of the external field $\mathbf{B}_0$ and of the magnetization of the layer $\mathbf{M}_j$. Definition of the spherical coordinates $\phi_j$ and $\theta_j$

If we introduce the spherical angles $\theta$ and $\phi$ to describe the effective field:

$$\begin{cases} B_{\text{eff},jx} = B_{0x} + \mu_0 M_{jx} = B_{\text{eff},j} \sin(\theta_\alpha) \cos(\varphi_\alpha) \\ B_{\text{eff},jy} = B_{0y} + \mu_0 M_{jy} = B_{\text{eff},j} \sin(\theta_\alpha) \sin(\varphi_\alpha) \\ B_{\text{eff},jz} = B_{0z} = B_{\text{eff},j} \cos(\theta_j) \end{cases} \tag{5.47}$$

the interaction potential $V$ can then be written in the compact form as

$$V_j = \frac{2\pi\hbar^2}{m} \rho_j b_j - g_n \mu_n \sigma \cdot \mathbf{B}_{\text{eff},\alpha}. \tag{5.48}$$

The Schrödinger equation

$$-\frac{\hbar^2}{2m} \Delta \psi + V(r)\psi = \mathscr{E}\psi \tag{5.49}$$

is a vectorial equation in the basis of the two spin states $|+\rangle$ and $|-\rangle$. We have to solve the Schrödinger equation (5.49) with a wave function expressed with its two spinor components $\psi_+$ and $\psi_-$. It can be written explicitly as

$$-\frac{\hbar^2}{2m} \Delta \begin{pmatrix} \psi_+ \\ \psi_- \end{pmatrix} + \left( \frac{2\pi\hbar^2}{m} \rho_j b_j \right) \begin{pmatrix} \psi_+ \\ \psi_- \end{pmatrix}$$

$$- g_n \mu_n \left( B_{\text{eff},x} \sigma_x + B_{\text{eff},y} \sigma_y + B_{\text{eff},z} \sigma_z \right) \begin{pmatrix} \psi_+ \\ \psi_- \end{pmatrix} = \mathscr{E} \begin{pmatrix} \psi_+ \\ \psi_- \end{pmatrix},$$

where the Pauli spin operators $\sigma$ are given by

$$\sigma_x = \begin{pmatrix} 0 & 1 \\ 1 & 0 \end{pmatrix} \quad \sigma_y = \begin{pmatrix} 0 & -i \\ i & 0 \end{pmatrix} \quad \sigma_z = \begin{pmatrix} 1 & 0 \\ 0 & -1 \end{pmatrix}.$$

We obtain the two coupled equations involving the two spinor components $\psi_+$ and $\psi_-$:

$$\left( -\frac{\hbar^2}{2m} \nabla^2 + \frac{2\pi\hbar^2}{m} b_j \rho_j - g_n \mu_n B_{\text{eff},z} \right) \psi_+(r)$$

$$+ \left( -g_n \mu_n B_{\text{eff},x} + i g_n \mu_n B_{\text{eff},y} \right) \psi_-(r) = \mathscr{E}\psi_+(r)$$

$$\left( -\frac{\hbar^2}{2m} \nabla^2 + \frac{2\pi\hbar^2}{m} b_j \rho_j + g_n \mu_n B_{\text{eff},z} \right) \psi_-(r)$$

$$+ \left( -g_n \mu_n B_{\text{eff},x} - i g_n \mu_n B_{\text{eff},y} \right) \psi_+(r) = \mathscr{E}\psi_-(r).$$

### 5.4.3 General Solution

We search solutions of the form $\psi_+(\mathbf{r}) = a_+ \exp(i\mathbf{k}.\mathbf{r})$ and $\psi_-(\mathbf{r}) = a_- \exp(i\mathbf{k}.\mathbf{r})$. The possible values for $\mathbf{k}$ are given by the possibility of finding non-zero solutions of the previous system (i.e., zero determinant condition). These conditions give four possible values for $\mathbf{k}$:

$$k_j^{\pm 2} = \frac{2m}{\hbar^2}\mathcal{E} - 4\pi\rho_j b_j \pm \frac{2mg_n\mu_n}{\hbar^2}|B_j|. \tag{5.50}$$

A general solution of the form

$$\exp(ik_{\|,j}r_\|)\left(a\exp(ik_{z,j}^+z)+b\exp(-ik_{z,j}^+z)\ +c\exp(ik_{z,j}^-z)+d\exp(-ik_{z,j}^-z)\right) \tag{5.51}$$

is not valid when there is an external magnetic field because $k_{\|,0}^+ \neq k_{\|,0}^-$. At this point there are two ways of solving the problem. The first way is to solve the problem for each eigenstate $|+\rangle$ and $|-\rangle$. The second way consists in taking the general solution which is expressed as follows:

$$\begin{aligned}
&\exp(ik_{\|,j}^{++}r_\|)\left(a\exp(ik_{z,j}^{++}z)+b\exp(-ik_{z,j}^{++}z)\right)\\
&+\exp(ik_{\|,j}^{--}r_\|)(c\exp(ik_{z,j}^{--}z)+d\exp(-ik_{z,j}^{--}z))\\
&+\exp(ik_{\|,j}^{+-}r_\|)\left(e\exp(ik_{z,j}^{+-}z)+f\exp(-ik_{z,j}^{+-}z)\right)\\
&+\exp(ik_{\|,j}^{-+}r_\|).(g\exp(ik_{z,j}^{-+}z)+h\exp(-ik_{z,j}^{-+}z)),
\end{aligned} \tag{5.52}$$

with

$$\begin{cases}
k_j^{++2}=k_j^{+2}, & k_{\|,j}^{++2}=k_{\|,0}^{+2}\\
k_j^{+-2}=k_j^{+2}, & k_{\|,j}^{+-2}=k_{\|,0}^{-2}\\
k_j^{-+2}=k_j^{-2}, & k_{\|,j}^{-+2}=k_{\|,0}^{+2}\\
k_j^{--2}=k_j^{-2}, & k_{\|,j}^{--2}=k_{\|,0}^{-2}
\end{cases} \tag{5.53}$$

The solution of Eq. (5.4.2) is the solution (5.52) rotated by the angles of the quantization axis

$$\begin{aligned}
\psi_j^+(r)\\
&= \exp(ik_{\|,j}^{++}r_\|)(a_j^{++}\exp(ik_{z,j}^{++}z)+b_\alpha^{++}\exp(-ik_{z,\alpha z}^{++}z))\cos(\theta_\alpha/2)\\
&+\exp(ik_{\|,j}^{+-}r_\|)(a_j^{+-}\exp(ik_{z,j}^{+-}z)+b_j^{+-}\exp(-ik_{z,jz}^{+-}z))\cos(\theta_j/2)\\
&-\exp(ik_{\|,j}^{-+}r_\|)(a_j^{-+}\exp(ik_{z,j}^{-+}z)+b_j^{-+}\exp(-ik_{z,jz}^{-+}z))e^{-i\varphi_j}\sin(\theta_\alpha/2)\\
&-\exp(ik_{\|,j}^{--}r_\|)(a_j^{--}\exp(ik_{z,j}^{--}z)+b_j^{--}\exp(-ik_{z,jz}^{--}z))e^{-i\varphi_j}\sin(\theta_\alpha/2)
\end{aligned}$$

and

$$\begin{aligned}
\psi_j^-(r)\\
&= \exp(ik_{\|,j}^{++}r_\|)(a_j^{++}\exp(ik_{z,j}^{++}z)+b_j^{++}\exp(-ik_{z,jz}^{++}z))e^{i\varphi_j}\sin(\theta_j/2)\\
&+\exp(ik_{\|,j}^{+-}r_\|)(a_j^{+-}\exp(ik_{z,j}^{+-}z)+b_j^{+-}\exp(-ik_{z,jz}^{+-}z))e^{i\varphi_j}\sin(\theta_j/2)\\
&+\exp(ik_{\|,j}^{-+}r_\|)(a_j^{-+}\exp(ik_{z,j}^{-+}z)+b_j^{-+}\exp(-ik_{z,jz}^{-+}z))\cos(\theta_j/2)\\
&+\exp(ik_{\|,j}^{--}r_\|)(a_j^{--}\exp(ik_{z,j}^{--}z)+b_j^{-+}\exp(-ik_{z,jz}^{--}z))\cos(\theta_j/2).
\end{aligned}$$

## 5.4.4 Continuity Conditions and Matrices

The eight constants $a_j^\pm$ and $b_j^\pm$ are fixed by the continuity of $\psi$ and $\nabla\psi$ at the interface. This gives exactly eight equations. The reflection matrix $\mathcal{M}$ is then a $8 \times 8$ matrix but with two non-zero $4 \times 4$ blocks (there are no cross terms between the group with "$k_{\parallel,0}^+$" components and the group with "$k_{\parallel,0}^-$" components in their wave vector. It is possible to split the problem into two calculations using $4 \times 4$ matrices. The continuity relations can be written as

$$
\mathcal{D}_j(\mathbf{r}_j)
\begin{pmatrix}
a_j^{++} \\
b_j^{++} \\
a_j^{-+} \\
b_j^{-+} \\
a_j^{--} \\
b_j^{--} \\
a_j^{+-} \\
b_j^{+-}
\end{pmatrix}
= \mathcal{D}_{j+1}(\mathbf{r}_j)
\begin{pmatrix}
a_{j+1}^{++} \\
b_{j+1}^{++} \\
a_{j+1}^{-+} \\
b_{j+1}^{-+} \\
a_{j+1}^{--} \\
b_{j+1}^{--} \\
a_{j+1}^{+-} \\
b_{j+1}^{+-}
\end{pmatrix},
\tag{5.54}
$$

where the $8 \times 8$ matrix $\mathcal{D}_j(\mathbf{r}_j)$ is written as

$$
\mathcal{D}_j(\mathbf{r}_j) = \begin{pmatrix} \mathcal{D\!A}_j & 0 \\ 0 & \mathcal{D\!B}_j \end{pmatrix}.
\tag{5.55}
$$

We give here the explicit expression of the two matrices $\mathcal{D\!A}$ and $\mathcal{D\!B}$ (we omit the index $j$ and we write ($\theta' = \theta_j/2$)):

$$
\mathcal{D\!A}_j =
\begin{pmatrix}
e^{ik_\parallel^{++}r}\cos(\theta')e^{ik_z^{++}z} & e^{ik_\parallel^{++}r}\cos(\theta')e^{-ik_z^{++}z} \\
k_z^{++}e^{ik_\parallel^{++}r}\cos(\theta')e^{ik_z^{++}z} & -k_z^{++}e^{ik_\parallel^{++}r}\cos(\theta')e^{-ik_z^{++}z} \\
e^{ik_\parallel^{++}r}e^{i\varphi}\sin(\theta')e^{ik_z^{++}z} & e^{ik_\parallel^{++}r}e^{i\varphi}\sin(\theta')e^{-ik_z^{++}z} \\
k_z^{++}e^{ik_\parallel^{++}r}e^{i\varphi}\sin(\theta')e^{ik_z^{++}z} & -k_z^{++}e^{ik_\parallel^{++}r}e^{i\varphi}\sin(\theta')e^{-ik_z^{++}z}
\end{pmatrix}
\tag{5.56}
$$

$$
\begin{matrix}
-e^{ik_\parallel^{-+}r}e^{-i\varphi}\sin(\theta')e^{ik_z^{-+}z} & -e^{ik_\parallel^{-+}r}e^{-i\varphi}\sin(\theta)e^{-ik_z^{-+}z} \\
-k_z^{-+}e^{ik_\parallel^{-+}r}e^{-i\varphi}\sin(\theta)e^{ik_z^{-+}z} & k_z^{-+}e^{ik_\parallel^{-+}r}e^{-i\varphi}\sin(\theta)e^{-ik_z^{-+}z} \\
e^{ik_\parallel^{-+}r}\cos(\theta')e^{ik_z^{-+}z} & e^{ik_\parallel^{-+}r}\cos(\theta')e^{-ik_z^{-+}z} \\
-k_z^{-+}e^{ik_\parallel^{-+}r}\cos(\theta')e^{ik_z^{-+}z} & k_z^{-+}e^{ik_\parallel^{-+}r}\cos(\theta')e^{-ik_z^{-+}z}
\end{matrix},
$$

$$
\mathcal{D\!B}_j =
\begin{pmatrix}
-e^{ik_\parallel^{--}r}e^{-i\varphi}\sin(\theta')e^{ik_z^{--}z} & -e^{ik_\parallel^{--}r}e^{-i\varphi}\sin(\theta')e^{-ik_z^{--}z} \\
-k_z^{--}e^{ik_\parallel^{--}r}e^{-i\varphi}\sin(\theta')e^{ik_z^{--}z} & k_z^{--}e^{ik_\parallel^{--}r}e^{-i\varphi}\sin(\theta')e^{-ik_z^{--}z} \\
e^{ik_\parallel^{--}r}\cos(\theta')e^{ik_z^{--}z} & e^{ik_\parallel^{--}r}\cos(\theta')e^{-ik_z^{--}z} \\
k_z^{--}e^{ik_\parallel^{--}r}\cos(\theta')e^{ik_z^{--}z} & -k_z^{--}e^{ik_\parallel^{--}r}\cos(\theta')e^{-ik_z^{--}z}
\end{pmatrix}
\tag{5.57}
$$

$$
\left(
\begin{array}{cc}
e^{ik_\parallel^{+-}r}\cos(\theta')e^{ik_z^{+-}z} & e^{ik_\parallel^{+-}r}\cos(\theta')e^{-ik_z^{+-}z} \\
k_z^{+-}e^{ik_\parallel^{+-}r}\cos(\theta')e^{ik_z^{+-}z} & -k_z^{+-}e^{ik_\parallel^{+-}r}\cos(\theta')e^{-ik_z^{+-}z} \\
e^{ik_\parallel^{+-}r}e^{i\varphi}\sin(\theta')e^{ik_z^{+-}z} & e^{ik_\parallel^{+-}r}e^{i\varphi}\sin(\theta')e^{-ik_z^{+-}z} \\
k_z^{+-}e^{ik_\parallel^{+-}r}e^{i\varphi}\sin(\theta')e^{ik_z^{+-}z} & -k_z^{+-}e^{ik_\parallel^{+-}r}e^{i\varphi}\sin(\theta')e^{-ik_z^{+-}z}
\end{array}
\right).
$$

The reflection matrix $\mathscr{M}$ is defined by

$$
\mathscr{M} = \prod_{j=0}^{j=N}\mathscr{D}_j^{-1}(\mathbf{r}_j)\mathscr{D}_{j+1}(\mathbf{r}_j) = \begin{pmatrix} \mathscr{M}\mathscr{A} & 0 \\ 0 & \mathscr{M}\mathscr{B} \end{pmatrix}, \tag{5.58}
$$

where

$$
\mathscr{M}\mathscr{A} = \prod_{j=0}^{N}\mathscr{D}\mathscr{A}_j^{-1}\mathscr{D}\mathscr{A}_{j+1} \quad\text{and}\quad \mathscr{M}\mathscr{B} = \prod_{j=0}^{N}\mathscr{D}\mathscr{B}_j^{-1}\mathscr{D}\mathscr{B}_{j+1}. \tag{5.59}
$$

We have the relation:

$$
\begin{pmatrix} a_0^{++} \\ b_0^{++} \\ a_0^{-+} \\ b_0^{-+} \\ a_0^{--} \\ b_0^{--} \\ a_0^{+-} \\ b_0^{+-} \end{pmatrix} = \mathscr{M} \begin{pmatrix} a_s^{++} \\ b_s^{++} \\ a_s^{-+} \\ b_s^{-+} \\ a_s^{--} \\ b_s^{--} \\ a_s^{+-} \\ b_s^{+-} \end{pmatrix}. \tag{5.60}
$$

In the case of incident "up" neutrons, Eq. (5.60) gives

$$
\begin{pmatrix} 1 \\ r_0^{++} \\ 0 \\ r_0^{-+} \\ 0 \\ 0 \\ 0 \\ 0 \end{pmatrix} = \mathscr{M} \begin{pmatrix} t_s^{++} \\ 0 \\ t_s^{-+} \\ 0 \\ 0 \\ 0 \\ 0 \\ 0 \end{pmatrix}. \tag{5.61}
$$

For "down" neutrons we have

$$
\begin{pmatrix} 0 \\ 0 \\ 0 \\ 0 \\ 1 \\ r_0^{--} \\ 0 \\ r_0^{+-} \end{pmatrix} = \mathscr{M} \begin{pmatrix} 0 \\ 0 \\ 0 \\ 0 \\ t_s^{--} \\ 0 \\ t_s^{+-} \\ 0 \end{pmatrix}. \tag{5.62}
$$

Let $r_0^{++}$, $r_0^{-+}$ be the reflectivity amplitudes for a neutron "up" (resp. "down"), reflected "up" (resp. "down"). The corresponding transmission coefficients are given by $t_s^{++}$, $t_s^{-+}$. We deduce

$$\begin{cases} r_0^{++} = \dfrac{MA_{21}MA_{33} - MA_{23}MA_{31}}{MA_{11}MA_{33} - MA_{13}MA_{31}} \\ r_0^{-+} = \dfrac{MA_{41}MA_{33} - MA_{43}MA_{31}}{MA_{11}MA_{33} - MA_{13}MA_{31}} \end{cases} \tag{5.63}$$

and

$$\begin{cases} t_s^{++} = \dfrac{MA_{33}}{MA_{11}MA_{33} - MA_{13}MA_{31}} \\ t_s^{-+} = \dfrac{-MA_{31}}{MA_{11}MA_{33} - MA_{13}MA_{31}} \end{cases} . \tag{5.64}$$

We find similar relations for the four other coefficients. The reflected intensities are given by

$$R^{++} \propto \left| r^{++} \right|^2, \tag{5.65}$$

and

$$R^{-+} \propto \left| r^{-+} \right|^2. \tag{5.66}$$

In the case of small external magnetic field, we have $R^{+-} \approx R^{-+}$.

### 5.4.5 Reflection on a Magnetic Dioptre

Let $q_z$ be the (Oz) component of the scattering vector $\mathbf{q} = \mathbf{k}_r - \mathbf{k}_{in}$. We will consider the case of a reflection on a magnetic substrate.

To simplify the problem, we assume that the applied magnetic field $\mathbf{B}_0$ is small so that $q_{0z}^+ \approx q_{0z}^- = q_{0z}$. The component of the $\mathbf{q}$ vector in the magnetic medium is given by (see 5.32)[2]

$$q_{sz}^{\pm} \approx \sqrt{q_{0z}^2 - 16\pi\rho(b_n \pm b_m)}. \tag{5.67}$$

We will assume that the external field $\mathbf{B}_0$ and the magnetization $\mathbf{M}$ lie in the layer plane. This corresponds to $\theta = 90°$ (see Fig. 5.6). Let $\phi$ be the angle between $\mathbf{B}_0$ and $\mathbf{M}$. The expressions of the reflection coefficients deduced from the expressions of the $\mathcal{M}$ matrices are given by

$$r^{++} = \frac{\cos^2\frac{\phi}{2} \left(q_{0z} - q_{sz}^+\right)\left(q_{0z} + q_{sz}^-\right) + \sin^2\frac{\phi}{2} \left(q_{0z} - q_{sz}^-\right)\left(q_{0z} + q_{sz}^+\right)}{\cos^2\frac{\phi}{2} \left(q_{0z} + q_{sz}^+\right)\left(q_{0z} + q_{sz}^-\right) + \sin^2\frac{\phi}{2} \left(q_{0z} + q_{sz}^-\right)\left(q_{0z} + q_{sz}^+\right)}, \tag{5.68}$$

---

[2] We remind that "$q_z = 2k_z$", the scattering wave vector in the substrate is equal to twice the projection of the incident wave vector on the (Oz) axis.

**Fig. 5.6** Neutron beam incident on a magnetic substrate of magnetization **M** in an applied field **B₀**

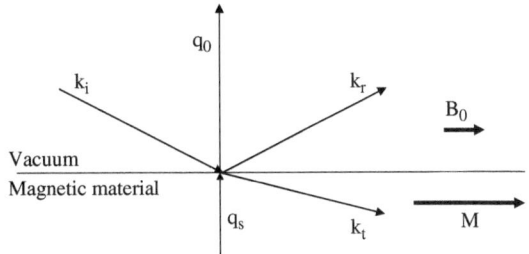

$$r^{+-} = \frac{2\,q_{0z}\cos\frac{\phi}{2}\,\sin\frac{\phi}{2}\,\left(q_{sz}^{+}-q_{sz}^{-}\right)}{\cos^{2}\frac{\phi}{2}\,\left(q_{0z}+q_{sz}^{+}\right)\left(q_{0z}+q_{sz}^{-}\right)+\sin^{2}\frac{\phi}{2}\,\left(q_{0z}+q_{sz}^{-}\right)\left(q_{0z}+q_{sz}^{+}\right)}. \qquad (5.69)$$

The measured intensities are given by

$$R^{++} = \left|r^{++}\right|^{2} \quad \text{and} \quad R^{+-} = \left|r^{+-}\right|^{2}. \qquad (5.70)$$

### 5.4.5.1  Case of a Non-Magnetic Substrate

In this case, corresponding to a zero magnetization ($b_m = 0$), the scattering vectors $q_{sz}^{+}$ and $q_{sz}^{-}$ are equal (Eq. 5.67). The reflection coefficients simplify and can be written in the form of classical Fresnel coefficients:

$$r^{++} = \frac{q_{0z}-q_{sz}}{q_{0z}+q_{sz}} \quad \text{and} \quad r^{+-} = 0. \qquad (5.71)$$

The reflected intensity is given by

$$R^{++} = \left|\frac{q_{0z}-q_{sz}}{q_{0z}+q_{sz}}\right|^{2} = \left|\frac{q_{0z}-\sqrt{q_{0z}^{2}-q_{c}^{2}}}{q_{0z}+\sqrt{q_{0z}^{2}-q_{c}^{2}}}\right|^{2}, \qquad (5.72)$$

where the critical wave vector $q_c$ is equal to $\sqrt{16\pi\rho b_n}$. When $q_{0z} < q_c$, $q_{sz}$ is a pure imaginary number and one finds a reflected intensity equal to 1. When $q_{0z}$ is very large, one can show that the intensity decreases as $1/q_{0z}^{4}$.

### 5.4.5.2  Case of a Magnetic Substrate in a Magnetic Field B₀ Aligned with the Magnetization M ($\phi = 0$)

In this simple case, the expressions of the reflection coefficients simplify and can be written as

$$r^{++} = \frac{q_{0z}-q_{sz}^{+}}{q_{0z}+q_{sz}^{+}}, \quad r^{--} = \frac{q_{0z}-q_{sz}^{-}}{q_{0z}+q_{sz}^{-}} \quad \text{and} \quad r^{+-} = 0. \qquad (5.73)$$

These expressions still correspond to Fresnel reflectivities. The only modification introduced by the magnetism is a difference in the critical angle. The critical angles for the reflectivity curves "up–up" and "down–down" are given by

$$q_c^\pm \approx \sqrt{16\pi\rho \left(b_n \pm b_m\right)}. \tag{5.74}$$

The spin-flip signals ($R^\pm$ and $R^\mp$) are zero.

*N.B.:* the coefficients $r^{++}$ and $r^{--}$ can be deduced one from the other by a $180°$ $\phi$ rotation.

The non-spin-flip signals are plotted as solid lines in Fig. 5.7. We find classical shapes for the reflectivity curves, with a total reflectivity plateau followed by a sharp decrease. The main difference between the "up–up" and "down–down" curves is the extension of the total reflectivity plateau.

The case of the magnetization parallel to the applied field is the most usual case. In this situation, there is no spin-flip cross section and the interaction can again be described using scalar optical indices. A magnetic "optical index" can be derived from the Schrödinger equation whose expression is given by

$$n^\pm \approx 1 - \delta \pm \delta_M = 1 - \frac{\lambda^2}{2\pi}\rho \pm \frac{m\lambda^2}{h^2}\mu.B, \tag{5.75}$$

where $\delta$ is the nuclear contribution to the optical index, and $\delta_M$ is the magnetic contribution to the optical index, the sign of the magnetic contribution depends on the relative orientation of the neutron spin with respect to the magnetization (parallel or anti-parallel). Table 5.2 gives values of optical indexes for some typical materials. One should notice that the magnetic optical index is of the same order of magnitude as the nuclear optical index.

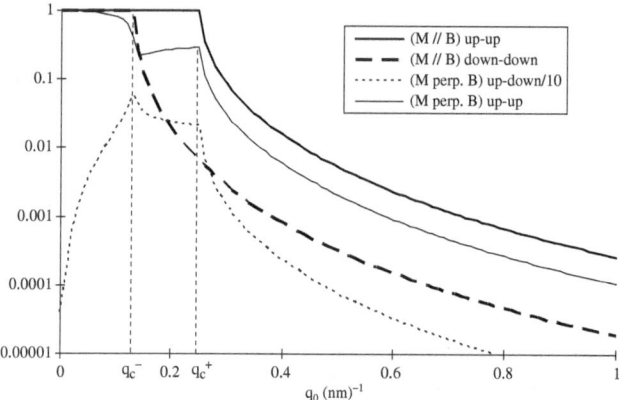

**Fig. 5.7** Reflectivity curves in the case of a magnetization parallel and perpendicular to the magnetic field $\mathbf{B}_0$. When the magnetization is parallel to the field, the non-spin-flip curves "up–up" and "down–down" are distinct (*solid lines*), the spin-flip signal is zero. When the magnetization is perpendicular to the field $\mathbf{B}_0$, the spin-flip curves superimpose (*squares*), a very large spin-flip signal appears (*lozenges*) (the spin-flip signal has been divided by a factor 10 for clarity)

**Table 5.2** Nuclear and magnetic optical index $n = 1 - \delta \pm \delta_M$ for some materials at $\lambda = 0.4$ nm

| Element | $\delta\,(\times 10^{-6})$ | $\delta_M\,(\times 10^{-6})$ | $\sigma_a$ (barns) |
|---|---|---|---|
| Fe | 20.45 | 11.7 | 2.56 |
| Co | 5.7 | 10.3 | 37.2 |
| Ni | 24 | 3.7 | 4.49 |
| Gd | 5.0 | 14.5 | 49,700 |

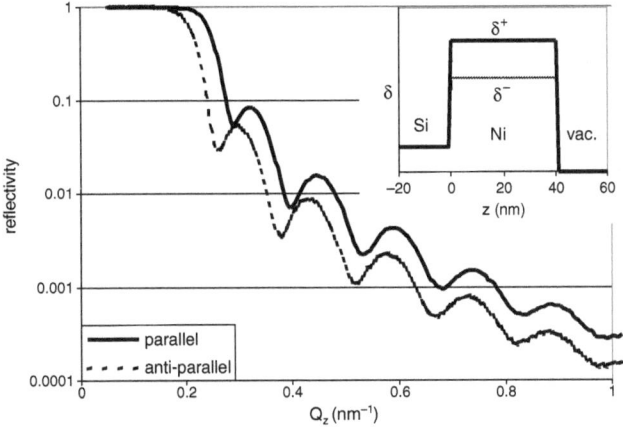

**Fig. 5.8** Reflectivity of a magnetic film Si//Ni (40 nm). The reflectivity depends on the relative orientation of the neutron spin with respect to the magnetization. (Insert) Optical index profile for both neutron polarizations (parallel and anti-parallel)

Figure 5.8 shows the situation of a magnetic thin film on a substrate. In this case, the optical index depends on the relative orientation of the neutron spin with respect to the thin film magnetization. The measured reflectivity is very different for neutron incident with a spin parallel to the magnetization (optical index $n^+ = 1 - \delta - \delta_M$) and for neutrons incident with a spin anti-parallel to the magnetization (optical index $n^- = 1 - \delta + \delta_M$). Note that the neutron spin and magnetic moment have opposite signs so that when the neutron spin is anti-parallel to the magnetization, its magnetic moment is parallel.

### 5.4.5.3  Case of a Magnetic Field Perpendicular to the Substrate Magnetization ($\phi = 90°$)

In this case, the reflection coefficients become

$$r = \frac{\left(q_{0z} - q_{sz}^+\right)\left(q_{0z} + q_{sz}^-\right) + \left(q_{0z} - q_{sz}^-\right)\left(q_{0z} + q_{sz}^+\right)}{2\left(q_{0z} + q_{sz}^+\right)\left(q_{0z} + q_{sz}^-\right)} = \frac{1}{2}\left(r^{++} + r^{--}\right), \quad (5.76)$$

$$r^{+-} = \frac{q_{0z}\left(q_{sz}^{+} - q_{sz}^{-}\right)}{2\left(q_{0z} + q_{sz}^{+}\right)\left(q_{0z} + q_{sz}^{-}\right)}. \tag{5.77}$$

The reflected intensities are given by

$$R^{++} = R^{--} = |r|^2 = \frac{1}{4}\left|r^{++}\right|^2 + \frac{1}{4}\left|r^{--}\right|^2 + \frac{1}{2}Re\left(r^{++} \times r^{--}\right). \tag{5.78}$$

One can notice that the up–up and down–down intensities are the sum of three terms. The first two correspond to the intensities of the non-spin-flip signals in the case of a magnetization aligned with the external magnetic field; they are weighted by a 1/4 coefficient. These terms introduce two discontinuities at the positions $q_c^+$ and $q_c^-$ in the reflectivity curve (see Fig. 5.7, white square curve). To these two terms, an " interference " term adds $\frac{1}{2}Re\left(r^{++} \times r^{--}\right)$ whose analytical expression is not simple. Its variations are plotted in Fig. 5.9a. For $q_{0z} = 0$, this term is equal to 1/2 and the intensity is totally reflected. Its value decreases as soon as $q_{0z}$ increases and becomes negative around $q_c^-$. It becomes positive again around $q_c^+$, then decreases very quickly. However, this contribution does not modify qualitatively the form of the non-spin-flip curve except that there is no total reflectivity plateau.

The spin-flip intensity is given by

$$R^{+-} = \left|r^{+-}\right|^2 = \frac{1}{4}\left|\frac{q_{0z}\left(q_{sz}^{+} - q_{sz}^{-}\right)}{\left(q_{0z} + q_{sz}^{+}\right)\left(q_{0z} + q_{sz}^{-}\right)}\right|^2. \tag{5.79}$$

The characteristic form of the spin-flip signal (see Fig. 5.6) is given by the term $\left|q_{sz}^{+} - q_{sz}^{-}\right|^2$. The variations of this term are plotted in Fig. 5.9b (thick lines). Two successive regime changes appear at the points $q_c^-$ and $q_c^+$. They correspond to the points where $q_{sz}^-$ and $q_{sz}^+$ successively change from pure imaginary to real values. This signal is slightly modulated by the factor $q_{0z}$ which gives a linear increase. The factor $1/\left|\left(q_{0z} + q_{sz}^+\right)\left(q_{0z} + q_{sz}^-\right)\right|^2$ gives a very fast decrease at large $q_z$. Its variations are plotted in Fig. 5.9b (thin lines).

In the case where the magnetization is not fully perpendicular to the applied magnetic field, the three terms in the $R^{++}$ intensity are weighted by $\cos^4\frac{\phi}{2}$, $\sin^4\frac{\phi}{2}$ and

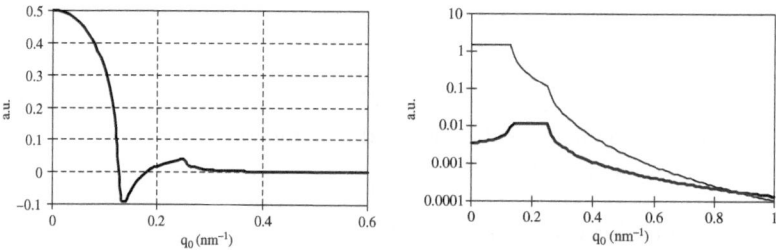

**Fig. 5.9** (**a**) Contribution of the interference term $Re(r^{++} \times r^{--})$ between the $r^{++}$ and $r^{--}$ amplitudes in the non-spin-flip intensities for a reflection on a substrate whose magnetization is perpendicular to the applied magnetic field $\mathbf{B}_0$. (**b**) Variations of two factors of the spin-flip intensity: (*bold lines*) factor $\left|q_{sz}^{+} - q_{sz}^{-}\right|^2$; (*thin lines*) factor $1/\left|\left(q_{0z} + q_{sz}^+\right)\left(q_{0z} + q_{sz}^-\right)\right|^2$

$2\cos^2\frac{\phi}{2}\sin^2\frac{\phi}{2}$ factors, $\phi$ being the angle between the field and the magnetization. In the case of a magnetic layer deposited on a non-magnetic substrate, the above considerations are not qualitatively modified. The main difference is that Kiessig fringes appear after the plateau of total reflection.

### 5.4.5.4 Zeeman effects

In this section we describe an effect which can break the symmetry between the $R^{++}$ and $R^{--}$ signals in a reflectivity experiment. It is related to Zeeman energy changes which can take place when the neutron flips during the reflection on a surface. If a sufficiently high magnetic field is applied on the sample and if the neutrons experience a spin-flip during the reflection, they will either gain or loose magnetic energy. Since the reflection process is an elastic one, the energy is fully transferred

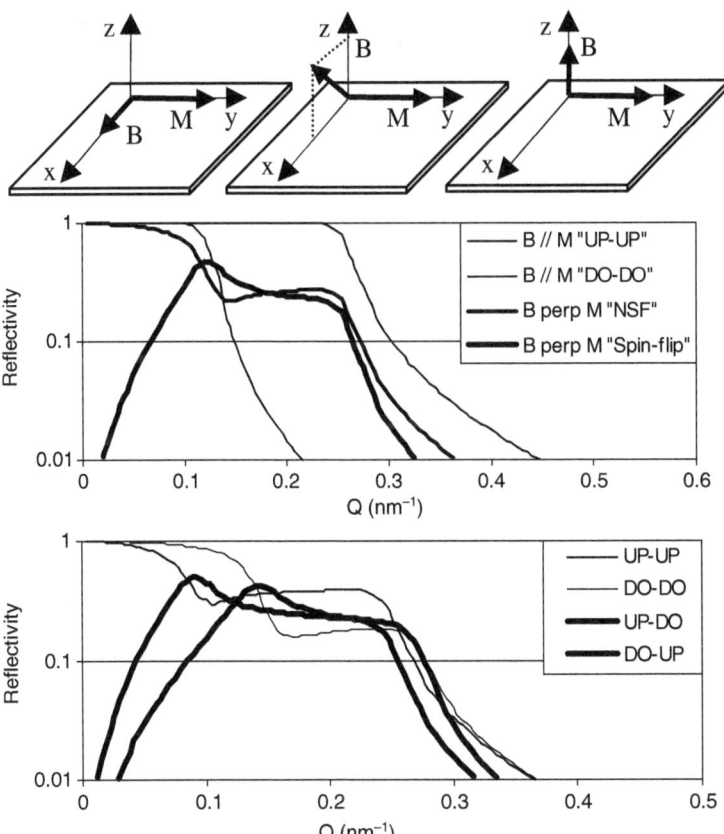

**Fig. 5.10** (**a**) Relative orientations of B and M giving the same results. (**b**) Spin-flip reflectivity with M $\perp$ B in low fields. Both configurations lead to the same spin-flip signals. The reflectivity with B//M is plotted as a reference. (**c**) Reflectivity cross sections in high fields (0.5 T). All top configurations give the same results. Note the very large splitting of the two spin-flip signals up–do and do–up

as a gain or loss in kinetic energy. The conditions required to observe such effects are that (i) a sufficiently high field of a fraction of a tesla is applied on the sample and (ii) there is a sufficiently high spin-flip cross section. Both requirements are opposite since in usual case, the magnetization will align with the applied field and the spin-flip scattering cross section will be zero. In practice, these effects are observed when the magnetic field is applied perpendicular to the sample and the demagnetizing field prevents the magnetization to rotate out of the thin film plane. These are the conditions under which the effect was quantified for the first time [20, 21]. If we consider the situation of an in-plane magnetization, if the guide field B is low (tens of mT), the spin-flip cross section is very large as soon B is non-collinear with M (see Fig 5.10b). However, both spin-flip signals $R^{+-}$ and $R^{-+}$ are equal. The reflectivity does not depend on the fact that B is or not in the film plane. When the applied field is large (fraction of a tesla), significant asymmetry effects are observed in the spin-flip cross sections (see Fig. 5.10c).

### 5.4.6 Conclusion

One should note that polarized reflectivity is sensitive to the induction in the thin films: no difference is made between the spin and orbital magnetic moments. In practice, it is possible to measure four different signals in a polarized reflectivity experiment : two non-spin-flip reflectivities, $R^{++}$ (resp. $R^{--}$), corresponding to the number of incoming "up" (resp. "down") neutrons reflected with an "up" (resp. "down") polarization; two spin-flip reflectivities, $R^{+-} = R^{-+}$, corresponding to the number of neutrons experiencing a spin-flip during the reflection on the sample. In a first approximation, the non-spin-flip reflectivities probe the components of the magnetization which are parallel to the applied field; the spin-flip cross-reflectivities are sensitive to the component of the magnetization perpendicular to the applied field. Combining this information it is possible to reconstruct the magnetization direction and amplitude along the depth of the film. The depth resolution is of the order of 2–3 nm in simple systems. Polarized reflectivity is a surface technique and thus is not sensitive to paramagnetic or diamagnetic contribution from the substrate. There is no absorption. There are no phenomenological parameters. The data are "naturally" normalized. All these characteristics make neutron reflectivity data easy to model and interpret.

## 5.5 Non-Perfect Layers, Practical Problems and Experimental Limits

### 5.5.1 Interface Roughness

Most of the studied systems show imperfect interfaces depending on the deposition process of the layer. We will consider three roughness scales: interface roughness, atomic interdiffusion and homogeneity of the layer thickness. Let $\xi$ represent the characteristic lateral length scale for the roughness. A perfect knowledge of the surface would correspond to the knowledge of $z(x,y)$ for all in-plane length scales. The treatment of the roughness is very similar to that described in Chaps. 2 and 3. According to the resolution of a typical neutron reflectivity experiment, one can (somewhat arbitrarily) distinguish three typical types of roughness which have different origins.

- *Interdiffusion of the species between two successive layers.* This happens during the deposition of a top layer which is miscible with the bottom material. This process is strongly temperature dependent. It corresponds to a typical length scale of $\xi < 0.5\,\mu m$.
- *A roughness induced by rough edges on the substrate or by grains in the case of two successive layers.* This roughness usually occurs during thin film growth. It is also the type of roughness which is difficult to take into account in models. It corresponds to $1\,\mu m < \xi < 100\,\mu m$.
- *Flatness of the sample.* Depending on the deposition process, the atomic flux may have an angular dependence which can lead to an uneven thickness over the sample surface. It corresponds to $\xi > 100\,\mu m$.

These three roughness scales can be modeled in three different ways to account for their effects on the measured reflectivity curves. They induce very different effects on the experimental signals. One has to keep in mind the following limitation: if the lateral fluctuations are not small compared to the layers thicknesses the following treatments are inadequate.

### 5.5.1.1  Thickness Inhomogeneity of the Sample

Thickness variation in a thin film sample (usually between the middle and the sample edges) is a "large" lateral scale problem (a few mm). The experimental measured curve can be treated as the superposition of reflectivity curves calculated for the thicknesses spectrum weighted by the corresponding area. The resulting effect is a blurring of the coherent oscillations for large $\mathbf{q}$.

*N.B.:* Since the Kiessig fringes period is inversely proportional to the wavelength of the incident beam, a thickness fluctuation (which reflects in the Kiessig oscillations period) can be taken into account as an incident wavelength spread $\delta\lambda$. Figure 5.12 (thin line, $\delta\lambda = 10\%$) illustrates the effect of a wavelength spread; it also corresponds to what would be observed for a 10% sample thickness fluctuation.

### 5.5.1.2  Roughness and Interdiffusion

Specular reflectivity cannot distinguish between these two types of roughness. The measurement of the coherent scattering length density $\rho b$ probes a large planar scale compared to the size of the roughness: for a given $z$ depth, one measures a mean value of $\rho b$ averaged over a large surface.

First solution: Nevot–Croce factors

If one assumes a flat distribution of $x$, the two types of interface can be treated by a single model where the step function is replaced by the following error function:

$$erf\left(\frac{z-z_j}{\sigma_j}\right) = \frac{2}{\sqrt{\pi}} \int_0^{(z-z_j)/\sigma_j} e^{-t^2} dt. \qquad (5.80)$$

This curve shows an inflexion point at $z_j$. The value $\sigma_j$ is given by the inverse of the curve slope at $z_j$. The thickness is given by $2\sigma_j$. The effect of a smooth interface surface described by (5.80) is to multiply the reflectivity $\mathscr{R}$ of a perfect flat interface by a Debye–Waller (or better Nevot–Croce, see Chap. 3, Appendix 3.A) factor [22]:

$$\exp(-2k_{z,j}k_{z,j+1}\sigma_j^2). \qquad (5.81)$$

In the case of a stack of multilayers each having a specific roughness, the Nevot–Croce factor is applied to each transfer matrix. Unfortunately, this cannot be applied in the magnetic case, the formalism preventing an easy calculation of the reflectivity $R$ at each interface. However, one can introduce a global factor and then apply this factor to each diagonal element at each interface. In practice, this works quite well except in the case of rather strong magnetic roughness like domains. The main effect of this factor is to decrease the reflectivity at high $\mathbf{q}$.

Second solution: discretization

This technique is efficient to model atomic interdiffusion. The interface is replaced by a finite number of discrete layers describing the concentration index. Either an error function or a linear function profile can be used. For real systems of thin solid films, one layer with an average $\rho b$ usually works well.

If the interdiffusion profile follows an erf function, the result is identical to the Nevot–Croce factor.

### 5.5.1.3 Intermediate Roughness

In the case of intermediate roughness, the previous methods are not completely satisfactory. Actually, this type of roughness not only decreases the specular reflectivity but also creates a non-specular diffuse background which can modify the results. In this case, the diffuse scattering should be measured and the specular reflectivity should be corrected accordingly. This treatment is quite complex and will not be detailed here.

### 5.5.1.4 Magnetic Roughness

This problem is very complex. A typical example where magnetic roughness appears is the case of a demagnetized sample. In this case, each domain has an effective

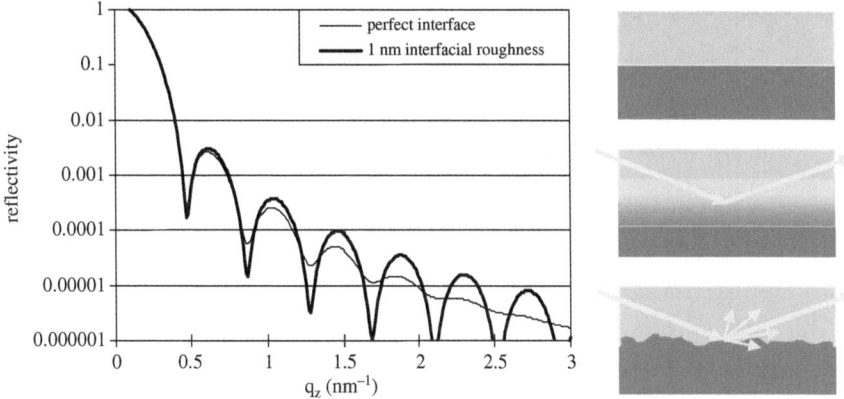

**Fig. 5.11** Roughness effects: comparison between a perfect interface, a 2 nm interfacial roughness, 2 nm Ni–Si interdiffusion. If at the interfacial roughness, the in-plane average scattering length density follows an erf profile, the last two cases provide the same results for the specular reflectivity

scattering length very different from its neighbor. This appears for neutrons as a giant roughness. This situation can be modeled using a DWBA approach [23, 24]. Such a situation leads to magnetic off-specular scattering (see Sect. 9) (Fig. 5.11).

## 5.5.2 Angular Resolution

The different expressions given above are valid for a perfect incident beam. For the fit of experimental data, it is important to have a good knowledge of the beam divergence and homogeneity. The beam angular divergence and wavelength dispersion must be taken into account in the simulations. The divergence of the incident beam, $\delta\theta$, is usually determined by two slits if the beam is smaller than the effective width of the sample seen by the neutron beam, or by the first slit and the sample itself if the sample is small enough to be totally illuminated by the neutron beam. Usually, $\delta\theta$ is fixed during the experiment. We have then to convolute the calculated reflectivity with a function which is the experimental shape of the beam divergence. However, a square function gives in most cases a good approximation of that function. In the case of curved samples, $\delta\theta$ can be slightly adjusted during the treatment. $\delta\theta$ has two effects: a decrease of the amplitude of the oscillations and a rounding of the discontinuity at the critical angle. Figure 5.12 gives an example of this effect. Wavelength dispersion is strongly dependent on the monochromator or on the time resolution in the case of time-of-flight spectrometers (see below). The effect of that dispersion is different from an angular divergence: the oscillations disappear at high angles (see Fig. 5.12). We remind that if a sample has a non-homogeneous thickness, the effect is very similar (see above). A wavelength dispersion can be used to model thickness variations over the sample surface.

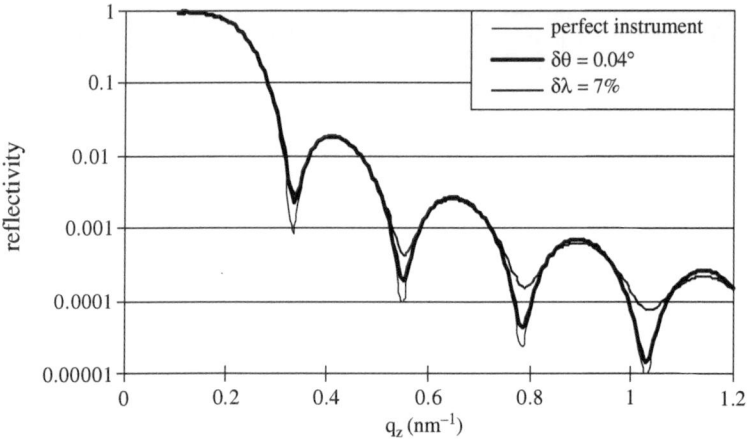

**Fig. 5.12** Effect of $\delta\theta$ and $\delta\lambda$. Comparison between a perfect instrument, an instrumental $\delta\theta$ and a $\delta\lambda$ for a measurement on a single 25-nm-thick Ni layer on a Si substrate

### 5.5.3 Analysis of Experimental Data

Reflectivity curves cannot be directly inverted. For a non-magnetic system, it is even possible to build a family of scattering length density profiles which give the same reflectivity curve. This is due to the fact that we measure only the intensity and loose the phase of the reflectivity [25]. For magnetic systems, the problem of the signal phase is less critical. However, the main source of uncertainty on the result is in general due to the lack of intensity at high angles. The analysis of experimental data is done by adjusting the different parameters involved in the problem until a good fit is obtained. The main source of uncertainty on the result is in general due to the lack of intensity at high $\mathbf{Q}$ and the roughness of the sample. In the case of magnetic systems, we usually know rather well the composition of the different layers. We have then to adjust the roughness, the thicknesses and the magnetic moments magnitude. It is in general very useful to have some external information like x-rays reflectometry and magnetic hysteresis measurements.

### 5.5.4 Sample Environment

The absorption of neutrons is negligible in most materials. The typical penetration depth for materials such as silicon or aluminum is of the order of 50 mm (depending of the wavelength). This makes it easy to set up complex sample environments on neutron spectrometers. The available ancillary equipments include cryomagnets (temperature range 1.6–300 K, magnetic field range 0–7 T), furnaces (temperature up to 800°C) and closed liquid cells (made of quartz or silicon).

## 5.5.5 Sample Sizes – Measurements Time

One has to keep in mind that most existing reflectometers are optimized to perform experiments on samples which have a surface of the order of 10 cm². In the case of optimized reflectometers with focussing systems (such as PRISM at the LLB), it is possible to perform experiments on samples which have a size of the order of 1 cm². These limitations in size imply that the studied samples need to have a very good homogeneity over a very large surface: the thicknesses of the layers need to be homogeneous and the substrate needs to be flat over the whole sample surface. If this is not the case, only averages over the sample surface will be measured and the information that can be obtained about the sample will be limited.

The reflectivity signal drops very quickly with the scattering wave vector value. For a perfect interface, at large $\mathbf{Q}$ values, the reflectivity is proportional to $1/Q^4$. $Q$ values of the order of 2 to $-3\,\mathrm{nm}^{-1}$ typically correspond to reflectivity of the order to $10^{-6}$ and require measurements of the order of 2–6 h.

One has to keep in mind that most existing reflectometers are designed to perform experiments on samples which have a surface of the order of 10 cm². In the case of optimized reflectometers with focusing systems (such as PRISM at the LLB), it is possible to perform experiments on samples which have a size of the order of 1 cm². These limitations in size imply that the studied samples need to have a very good homogeneity over a very large surface: the thickness of the layers needs to be homogeneous and the substrate needs to be flat over the whole sample surface. If this is not the case, only averages over the sample surface will be measured and the information that can be obtained about the sample will be limited.

## 5.6 The Spectrometers

The spectrometers can be divided into two different groups: time-of-flight reflectometers like EROS at the Laboratoire Léon Brillouin (LLB), CRISP and SURF at ISIS, D17 at the ILL and monochromatic reflectometers like PRISM at LLB and ADAM at the Institut Laue Langevin (ILL). Time-of-flight spectrometers are necessary for reflectometry studies on liquids. A list of existing reflectometers can be found in [26].

## 5.6.1 Monochromatic Reflectometers

Monochromatic reflectometers are basically two axes spectrometers. The wavelength is fixed (0.43 nm for PRISM) and the reflectivity curve is obtained by changing the incident angle $\theta$. In this case, the sample is usually vertical. On this type of reflectometers it is easy to put a polarizer and an analyzer in order to select the spin states of the incident and reflected neutrons. The flippers can be of Mezei type (two orthogonal coils) [27]. They allow to flip the neutron spin state from "up" to "down". An example of two-axis spectrometer is presented in Fig. 5.13.

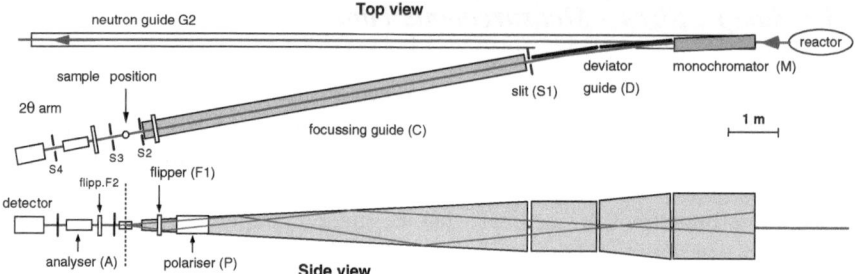

**Fig. 5.13** The two-axis reflectometer PRISM at the LLB

## 5.6.2 Time-of-Flight Reflectometers

The second possibility is to work at a fixed incidence angle and to scan the incident wavelength. Cold neutrons have a typical slow travel speed of the order of 1000 m/s which depends on the wavelength $v = h/m\lambda$. Thus if one sends a neutron pulse (defined with a chopper and of typical duration 0.2 ms), a spatial spread of the neutrons of different wavelengths takes place between the chopper and the detection systems. The neutron wavelength can then simply be measured by the travel time between the chopper and the detector. This technique is called time of flight (ToF). Figure 5.14 shows the typical layout of a ToF reflectometer. One advantage of a ToF's setup is that it is very easy to change the resolution by changing the chopper and slits parameters. Another advantage is that the sample does not need to be moved during an experiment and thus is easier to measure free liquid surfaces.

The time-of-flight technique consists in sending a pulsed white beam on the sample. Since the speed of the neutron varies as the inverse of the wavelength, the latter is directly related to the time taken by the neutron to travel from the pulsed source to the detector (over the distance $L$) by

$$\lambda = \frac{h}{mL}t. \tag{5.82}$$

This relation is also written as

$$\lambda\,(\text{nm}) = \frac{t(\mu s)}{2527L(\text{m})}. \tag{5.83}$$

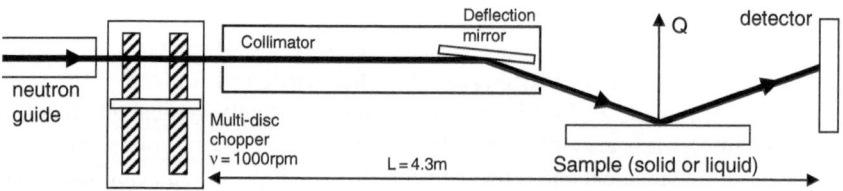

**Fig. 5.14** Description of the time-of-flight reflectometer EROS at the LLB [28]

On a spallation source, the neutron beam is "naturally" pulsed and the time-of-flight technique is used. On a reactor, pulsed neutrons are produced by a chopper.

For a reflectivity measurement, the angle is fixed and the reflectivity curve is obtained by measuring the reflectivity signal for each wavelength of the available spectrum, each wavelength corresponding to a different scattering wave vector magnitude. In practice, the wavelength spectrum (typ. 0.2–2 nm) is not wide enough to cover a very large $Q$-range in the reciprocal space. Thus, usually two or more different incidence angles are used to cover a wider $Q$-range. An example of time-of-flight spectrometer is presented in Fig. 5.14.

## 5.7 Non-Magnetic Reflectivity

### 5.7.1 Isotopic Labeling

The substitution of hydrogen by deuterium in organic materials allows to strongly change the neutron optical index of the material without changing its physical or chemical properties. A very interesting possibility offered by neutron scattering is to do selective labeling by deuteration which leads to a very large contrast between deuterated ($b_D = 6.67$ fm) and protonated ($b_H = -3.7$ fm) systems [29, 30]. Such a labeling is used in two ways: (i) the measurement of the conformation of polymeric chains at the interface in good solvent by using hydrogenated polymers in deuterated solvents (for example, adsorption profiles of polymers at interfaces have been measured by neutron reflectivity [31–34]) and (ii) the determination of the structure of "complex" systems involving two polymers by mixing hydrogenated and deuterated polymers. This can be achieved with deuterated and hydrogenated chains of the same polymer (to study the interdiffusion of chains at the interface of two molten polymers for example [35]) or of different polymers (multilayers of polyelectrolyte of opposite charges for example [36]). Combining these two advantages to determine the structure of a mixture of two different polymers in good solvent is possible by using the variation contrast method: measurements are performed in successive mixtures of hydrogenated and deuterated solvent that either match the neutron optical index of the first polymer, or match the neutron optical index of the second one. It allows to resolve the whole structure of the system [37].

#### 5.7.1.1 Interdiffusion Between Diblock Copolymer Multilayers

We shall illustrate here the use of selective deuteration to study the polymer interdiffusion. By spin coating, it is possible to deposit polymer layers on glass or silicon with a roughness below 1 nm. It is then possible to deposit a second layer on the first one. If one of the layers is deuterated, it is possible to study the interdiffusion as a function of time and annealing temperature. The diffusion will appear as a smearing of the interface between the two layers and thus a decrease of the Kiessig fringes

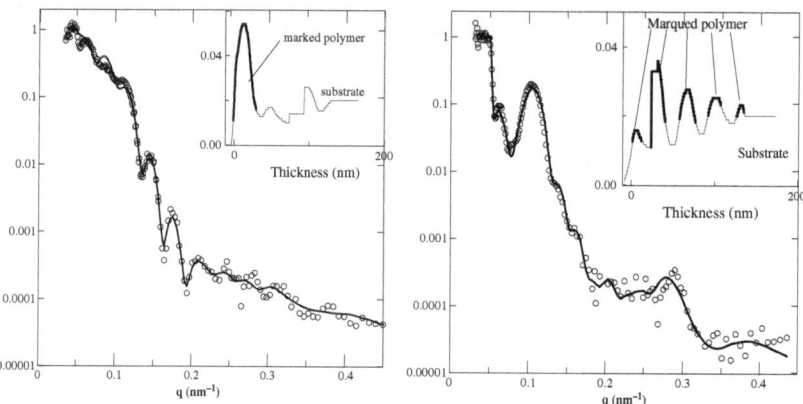

**Fig. 5.15** *Left*: Reflectivity of a quadrilayer consisting in a partially deuterated PS-PBMA copolymer layer deposited on a trilayer of totally hydrogenated polymer (measured on the EROS reflectometer at the Laboratoire L*éon* Brillouin). *Right*: Reflectivity of the quadrilayer after annealing for 1 h at 115°C

amplitude. Diblock copolymers are made of two chains A and B linked together (*A–B*). These systems present a large variety of interesting properties. For example, if A and B are not miscible, they can form self-organized multilayers of a fixed thickness parallel to the surface where the solution is deposited. The observed structure is of the type (*substrate*; *A–B*; *B–A*; *A–B*; *B–A* . . .). We have studied diblock copolymers of the type (polystyrene-polybutylmetacrylate: PS-PBMA). The initial system consisted in a layer of partially deuterated PS-PBMA copolymer deposited on a trilayer of totally hydrogenated copolymers. The reflectivity of this system is shown in Fig. 5.15. The numerical fit shows a large index at the top of the system corresponding to the deuterated copolymer. The system has then been annealed for 12 h at 400 K and then remeasured (Fig. 5.15). On this reflectivity curve, one can observe a clear "Bragg" peak at the position $q = 0.1$ nm$^{-1}$. This indicates the diffusion of the deuterated polymer to the inner layers. Since the diblock copolymers are ordered in multilayers, a periodic variation of the index appears (see insert in Fig. 5.15).

### 5.7.1.2 Interdiffusion Between Polymer Layers

To illustrate these two aspects we shall present a recent study on the conformation of dense grafted brushes of polystyrene (PS) on silicon [38–40]. Such macromolecular architecture is designed to answer the technological demand of controlled and reproducible thin polymer films. It is based on recent *grafting from* techniques that allow to graft polymers onto a surface in an efficient way. Classically, the most common method for polymer grafting is the *grafting onto* where end functionalized polymers react with appropriate surface sites. In this more promising *grafting from* method, the chains grow in situ from preformed surface-grafted initiators [41]. This latter approach is thus a suitable way for building high-density polymer brushes because it is not limited by polymer diffusion. It also allows a fine control of the polymer layer. This strategy has been applied by Devaux et al. [38] to realize grafted brushes of PS on silicon which have been studied by neutron reflectivity.

In order to test the homogeneity of the chains growth *during* the polymerization process, a specific chain designed for neutron reflectivity measurement has been fabricated with a two-step process: the first part of the chains has been grown using deuterated monomers and the second part using hydrogenated monomers (Fig. 5.16a). Chemically, the polymer chains behave as a single physical unit. However, NR allows to easily distinguish between the two parts of the chains as shown in Fig. 5.16c that presents the reflectivity of the polymer layer at the polymer/air

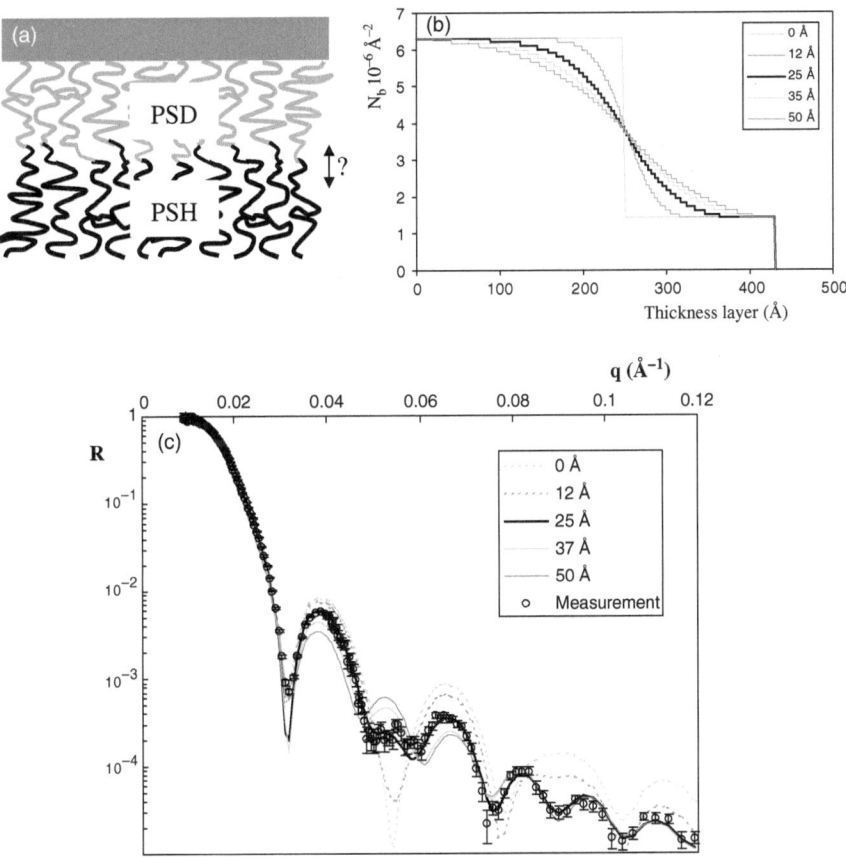

**Fig. 5.16** Reflectivity on a polymer grown using a "grafting from" method. (**a**) A polymer layer grafted on a silicon substrate, half of the layer is hydrogenated, the second half of the layer is deuterated; (**b**) optical index along the thickness of the layer assuming different interface thickness (the profile is assumed to vary as an erf function); and (**c**) reflectivity of the system and theoretical curves for the different thickness of the interfacial layer (zone separating the H- and D-polymer layers). The best agreement is obtained for an interfacial layer of thickness 25 Å

interface: Kiessig fringes arise from the deuterated layer, the hydrogenated layer and the whole layer. It allows a very accurate determination of the width of the interface thickness between the deuterated and the hydrogenated parts of the polymers: if the width is null we would get large oscillations from the deuterated layer and if it is too large we only get small oscillations from the whole layer as other oscillations vanishes (Fig. 5.16c presents the simulated profiles of Fig. 5.16b). Best fit shows that the interface width is limited to 2.5 nm for a brush of thickness 43 nm. This proves that this "grafting from" technique allows to built very well-ordered polymer brushes and that the growth of the brushes is very homogeneous.

### 5.7.1.3 Solid–Liquid Interfaces

The swelling capacities of the PS brushes in good solvent have been measured in a second part of the study: an hydrogenated brush of a dry thickness of 22 nm was placed into a good solvent of PS (deuterated toluene). As it has been said before, the absorption of neutrons is very low and this allows to use the silicon substrate as the incident medium (even though the travel in the silicon is larger than 50 mm) (Fig. 5.17a). Figure 5.17b presents the monomeric concentration profile as a function of the depth deduced from the fit of experimental reflected curves. It shows that the swelling of the layer is limited. The volume fraction $\phi$ of the polymer remains as high as 0.8 of the density of the dry polymer showing that the solvent hardly penetrates the layer. At the polymer–solvent interface, one can observe a parabolic variation of the polymer density. Three fitting methods have been tested and provide very similar results. The detailed information about the very top of the polymer layer is limited because the $Q$-range of the measurement was limited. The maximal stretch of the layer can nevertheless be evaluated as it roughly corresponds to the maximal extension of the profile (30 nm). It shows that the chains were already strongly stretched in their dry state ($\phi \sim 0.7$) as the layer width was 22 nm. Such initial strong stretching is due to a very high density of grafting of polymeric chains that explains the unusual low swelling capacities of the brushes. This example illustrates some of the unique possibilities offered by neutron reflectivity for the study of solid–liquid interfaces.

## 5.7.2 Oxide Layers

Neutron reflectivity may be used to probe oxide layers since the neutron optical index of oxides is usually very different from non-oxidized materials [42]. This makes neutron reflectivity much more sensitive to details in an oxide structure than x-ray reflectivity (Fig. 5.18).

For example, the preparation of $SiO_2$ films on silicon substrates by three different methods (thermal, chemical and electrochemical oxidation) have been compared by Bertagna et al. [43, Fig. 4]. Depending on the preparation method, the obtained films

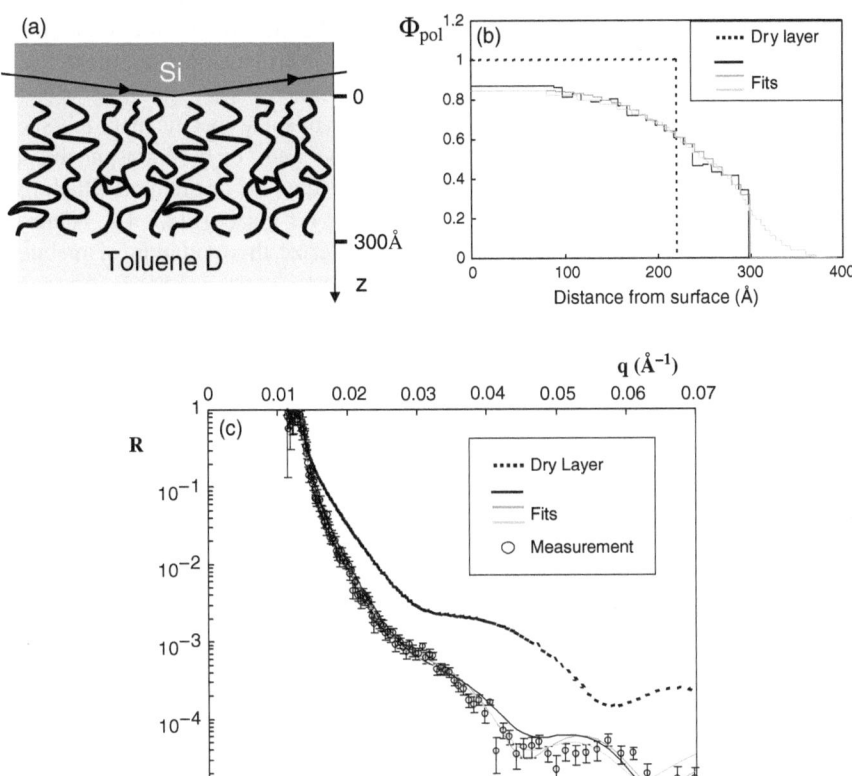

**Fig. 5.17** PS layer in a good solvent (toluene). (**a**) The measurement setup, with the neutron beam incident on the system through the silicon substrate. (**b**) Fit of the polymer density profile $\phi$ in deuterated toluene (three fitting methods of the NR data have been tested and give very similar results). The polymer density $\alpha$ is normalized to 1 for the dry layer. (**c**) Reflectivity measurement and numerical modeling curves [38, 39]

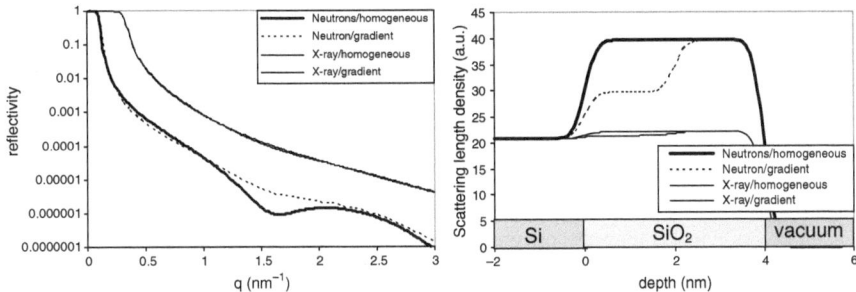

**Fig. 5.18** Comparison of the neutron reflectivity and the x-ray reflectivity on a $SiO_2 \parallel Si$ structure. The contrast between the layers is much larger in the case of neutrons, leading to higher amplitude oscillations, which on the contrary are very small in the case of x-rays. Sub-structures in the oxide layer, due for example to an oxidation gradient, give measurable effects in neutron reflectivity

give very different reflectivity results. Anodic and chemical oxides are found to be not very dense (60–75% of the theoretical density). Thermal oxides are the densest (95%).

Neutron reflectivity may also be useful in the case of some specific materials such as boron which strongly absorbs neutrons (e.g., the study of borophosphosilicate glass thin films used in microelectronic circuit devices) [44] or titanium which has a negative scattering length (such as $TiOx$ coatings for glazing) [45–47]. Neutron reflectometry has sometimes been used to characterize the oxidation of metallic thin films [48–50].

Another key advantage of neutrons is their high sensitivity to $D_2O$ (compared to x-rays). Neutron reflectivity has, for example, been used to characterize the moisture transport through $Al_2O_3/polymer$ multilayered barrier films for flexible displays [51]. It has also been used to characterize the adsorption of water on hydrophobic/hydrophilic $TiO_2$ surfaces under UV illumination [52].

Another advantage which must be mentioned is that in neutron reflectometry it is easy to set up complex sample environments. This is especially true in the case of solid/liquid interfaces where the neutron beam can be sent through the substrate and probe the solid–liquid interface with negligible absorption [53–55].

### 5.7.3 Biological Systems

During the last decade, neutron reflectivity has been used for biological systems, mainly for the study of biophysical problems at solid–liquid interfaces. For the readers interested in this field, they shall refer to the recent reviews of G. Fragneto-Cusani [56] and S. Krueger [57]. The following references give good examples of what can be achieved using neutron reflectometry [58–65].

## 5.8 Polarized Neutron Reflectometry on Magnetic Systems

During the early 1980s, advanced techniques for the deposition of ultra-thin metal films were developed. This led to the fabrication of new artificial materials comprising the stacking of different materials in thin sandwiches (heterostructures). The combination of different types of materials gave rise to new physical phenomena. The first new phenomenon to be probed was the magnetic exchange coupling in superlattices (Fig. 5.19a). It appeared that magnetic layers separated by non-magnetic spacer layers can be magnetically coupled. The coupling can be either ferromagnetic, anti-ferromagnetic or more complex (quadratic or even helicoidal). The coupling can also change sign (from ferro to anti-ferro) as a function of the spacer layer thickness. Such phenomena were observed in rare-earth superlattices (Gd/Y, Dy/Y, Gd/Dy, Ho/Y [66]), transition metal superlattices (Fe/Cr [67,68], Co/Cu [69], Fe/V [70], Co/Ru [71]) and mixing of semiconductors and metals (Fe/Si [72],

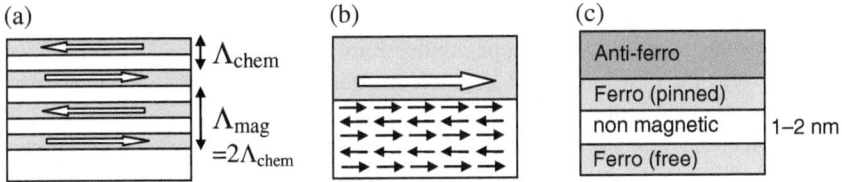

**Fig. 5.19** (**a**) Exchange coupled superlattice with an anti-ferromagnetic order. (**b**) Exchange bias between a ferromagnet and an anti-ferromagnet. (**c**) GMR system or magnetic tunnel junction

Fe/Ge [74]). The field is still open and new systems are still being synthesized, especially with magnetic semiconducting materials (GaMnAs [75,76], EuS/PbS [77]).

These magnetic coupling phenomena are strongly connected to the giant magneto-resistance effect [8]: Depending on the orientation of the magnetization of the different layers in the heterostructure, the resistivity of the system varies significantly. This has opened a new field of study which is now referred to as spintronics.

In the early 1990s, the phenomenon of exchange bias was revived. A ferromagnetic layer in contact with an anti-ferromagnetic material can be magnetically strongly coupled (Fig. 5.19b) [78–82]. The soft magnetic layer is thus strongly pinned along a well-defined direction. This is presently used in most of the spintronics systems (Fig. 5.19c). The phenomenon is used in commercial devices but is still not fully understood from a theoretical point of view. The origin of the coupling depends on the type of materials, their crystallinity, and the fabrication process, etc. [83–85].

In the late 1990s, it appeared that the performances of giant magneto-resistive systems could be enhanced by combining tunnel barriers and magnetic materials (using materials such as $Fe_2O_3$, $Fe_3O_4$, $CoFe_2O_4$, $MgO$, $Al_2O_3$). This field is still very active and a number of phenomena still need to be understood. Electronic devices using magnetic tunnel junctions (such as Magnetic Random Access Memories) are about to be commercialized but significant progress can still be made.

Besides the combination of well-known materials, during the late 1990s, a wealth of new materials were synthesized (typically perovskites of the type $ABMnO_3$). The growth of these materials as epitaxial thin films was quickly mastered following the experience acquired previously on oxide superconductors. These materials have properties ranging from colossal magneto-resistance to magneto-electric effects.

More recently, a new field has developed which is the search for new magnetic semiconductors. After the early studies of Eu-based magnetic semiconductors (EuO and EuS) in the 1970s, the field was dormant until GaMnAs magnetic semiconductors were synthesized in the middle of the 1990s. Since then, a number of new systems have been synthesized in order to find room temperature magnetic semiconductors. The discovery of a suitable material could boost the field of spintronics. These new materials range from diluted semiconductors to magnetically doped insulating oxide materials.

We can mention other types of studies, such as the penetration of the magnetic flux in superconductor thin films [86], the exchange spring effect between soft and hard magnetic layers [87], the magnetism of ultra-thin films [88, 89], proximity effects between magnetism and superconductivity [90, 91], induced magnetism at interfaces (e.g., the magnetism induced in V in contact with Gd [92]) and the super-anti-ferromagnetism (edge effects in Fe/Cr superlattices [93]).

In this section, we shall give examples of specular polarized reflectivity on various types of magnetic systems in order to highlight the information that can be obtained by polarized neutron reflectometry. All the experiments shown here have been performed on the reflectometer PRISM.

## 5.8.1 Superlattices

### 5.8.1.1 Periodic Multilayers

A superlattice consists of a periodic repetition ($n$ times) of a bilayer system $[A/B]_n$ (see Fig. 5.19a). If the material $\mathbf{A}$ is magnetic, then depending on the thickness of the intermediate layer $\mathbf{B}$ (from 0.5 to 3 nm) and the type of the $\mathbf{B}$ material (Cr, Mn, Cu, etc.) a magnetic coupling can be mediated through this non-magnetic $\mathbf{B}$ layer. The coupling energy can be described by using an energy of the form

$$E_{coupling} = -J_1\,\mathbf{S}_1 \cdot \mathbf{S}_2 - J_2\,(\mathbf{S}_1 \cdot \mathbf{S}_2)^2.$$

Depending on the sign and magnitude of the coupling constants $J_1$ and $J_2$, a variety of magnetic orderings can be observed. Usually the coupling constant oscillates between positive and negative values as a function of the thickness of the $\mathbf{B}$ spacer, thus the magnetic order between the $\mathbf{A}$ layers changes from ferromagnetic to anti-ferromagnetic. In some structures, it is even possible to observe non-collinear coupling between the different magnetic layers.

In the case of periodic multilayers, we can observe Bragg peaks corresponding to the period of the multilayer. In the case of anti-ferromagnetic coupling or variable angle coupling, it is possible to obtain directly a mean angle between the different magnetic layers. With polarized neutrons, it is possible to measure very rapidly a precise value of the average magnetic moments. If high-order Bragg peaks are observed, a good estimate of the chemical and magnetic interface can be obtained. In the literature, there is a large amount of results on magnetic multilayers [25, 71, 94]. The most thoroughly studied system is the metallic system Fe/Cr. The pioneering polarized neutron reflectometry studies have been performed on this system [95, 96]. Though the origin of the magnetic coupling is well understood in metallic heterostructures [97], the exact origin of the ordering in structures combining semi-conductors or even insulators is still unclear.

### 5.8.1.2 Metal Superlattice

Figure 5.20 shows an example of PNR on a system $[Fe(2.5\ nm)/Si(1.2\ nm)]_n$ [44]. The reflectivity was measured at 20 K in a planar field of 20 mT. At the position

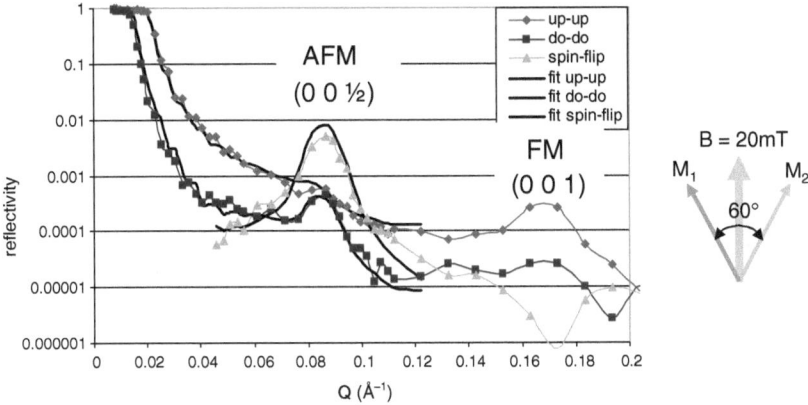

**Fig. 5.20** (a) Reflectivity of a $[Fe\ (2.4\ nm)/Si\ (1.2\ nm)]_n$ multilayer measured at 5 K. (b) Config-uration of the magnetic moments in two adjacent Fe layers

$q = 0.17\ Å^{-1}$, the peak is indicative of the period of the superlattice defined by the thickness 3.7 nm of the $[Fe(2.5\ nm)/Si(1.2\ nm)]$ bilayer. It corresponds to the [001] peak of the superlattice. A magnetic contrast between the UP and DO reflectivities exists corresponding to a net magnetization component along the applied field. The position $q = 0.085\ Å^{-1}$, that is (0 0 1/2), a strong diffraction peak is observed. It indicates an anti-ferromagnetic component. But the existence of a very strong spin-flip peak at (0 0 1/2) indicates that a non-collinear magnetic order has set up in the structure. Numerical modeling suggests that the Fe layers are arranged so that the magnetizations of alternating Fe layers make an angle of 30° with respect to the applied magnetic field. The magnetic moment of the iron layer is however reduced to 1.4 $\mu_B$ per Fe atom because of the Si interdiffusion and of the fact that the Fe layers are very thin. The question of the origin of the coupling remains unclear.

The studies of the magnetic coupling in magnetic superlattices are still numerous: in "all metal" superlattices we can mention Pd/Fe [98], Heussler alloys [99, 100], U/Fe [101]; in semi-conductors Fe/Ge [102]; in rare-earths $DyFe_2/YFe_2$ [103], Ho/Y [104]; in metal oxide layers $Co/Al_2O_3$ [105].

In such multilayer systems, neutron reflectivity is sensitive to very small mag-netic moments. In [GaAs/GaMnAs]_n superlattices, magnetizations as small as 27 kA/m (0.03 T) can be determined [73].

### 5.8.1.3 Magnetic Oxide Superlattice

This example illustrates the use of the magnetic contrast to measure the chem-ical segregation in manganite heterostructures: $[(LaMnO_3)_a/(SrMnO_3)_b]_n$ (with $8 < a < 12\ ;\ 4 < b < 8$). These superlattices are deposited layer by layer in or-der to enforce a cationic order between La and Sr and a cationic segregation be-tween $Mn^{3+}$ and $Mn^{4+}$. The first material is anti-ferromagnetic in its bulk form, the

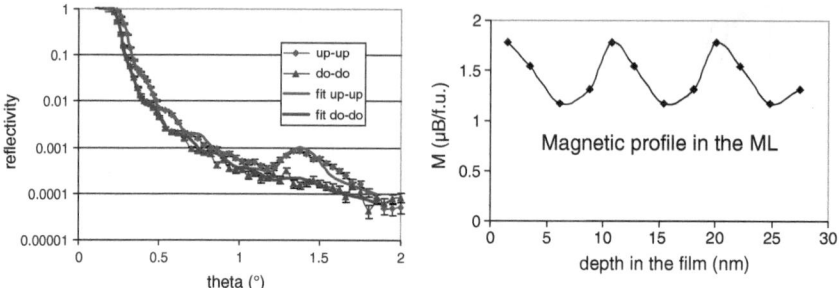

**Fig. 5.21** (**a**) Reflectivity on a superlattice $[(LaMnO_3)_a/(SrMnO_3)_b]_n$. (**b**) Magnetic profile in the superlattice

second is ferromagnetic. The objective of the measurement was to check whether the cationic segregation (La/Sr) effectively induced a (AF/F) stacking. The reflectivity on one of these systems is presented in Fig. 5.21a. Around the angle $\theta=1.3°$, a superstructure peak corresponding to the system's periodicity can be observed. The contrast between the two reflectivity curves up and down is characteristic of the in-depth magnetization profile. In order to model the reflectivity curves, it is only necessary to introduce a small modulation of the magnetization in the system (Fig. 5.21b): the cationic segregation does not lead to a clear magnetic segregation. The magnetization modulation is only 25% between the two types of layers.

### 5.8.1.4 Supermirrors

For technical purposes it is interesting to build systems exhibiting an artificially large optical index. One can build such a structure by stacking periodic multilayers with an almost continuous variation of the period [106]. In such a system, if the periodicity range is well chosen, a large number of Bragg peaks follow the total reflectivity plateau. Since the periodicity of the multilayer is varying continuously, all these Bragg peaks add constructively. Using this technique it is possible to enhance the length of the total reflection plateau by a factor 3–4 (up to 6 in technological demonstrators). Such mirrors are now widely used for neutron guides and for polarization devices. Figure 5.22 gives an example of a polarizing mirror.

## 5.8.2 Magnetic Single Layers

Even though most of the studies are performed on superlattices (usually for scattering intensity reasons), the magnetization of very thin systems can also be probed. The advantage of studying simple systems is that much more detailed information can be obtained since the signal is not blurred by roughness or thickness fluctuations.

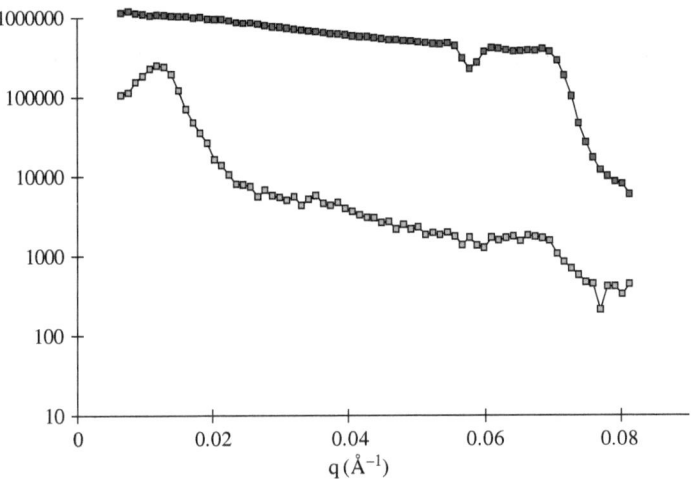

**Fig. 5.22** Polarizing supermirror

### 5.8.2.1 Metal Trilayer

We present here the example of the study of a coupled FeCo/Mn/FeCo tri-layer system [107]. The structure of the sample is shown in Fig. 5.23a. The "active" region is formed by the layers FeCo/Mn/FeCo. The Ag layer is used to promote an epitaxial growth of the system. The Au layer is a simple protective capping. The presented system is $Fe_{0.5}Co_{0.5}/Mn(8\,\text{Å})/Fe_{0.5}Co_{0.5}$. The specificity of this system is that the magnetic couplings between Fe and Mn, and Co and Mn are of opposite sign. Ab initio calculation predicted that in such a system, contrary to a pure Fe/Mn interface, a complex magnetic behavior of the Mn layer arises. A first measurement was performed in a saturating field (not shown). A numerical modeling of the data shows that the magnetic moment in the $Fe_{0.5}Co_{0.5}$ layers is $2.4\mu_B/atom$ (as in bulk materials). A net magnetic moment of $0.8\mu_B/atom$ in Mn is also observed. This induced

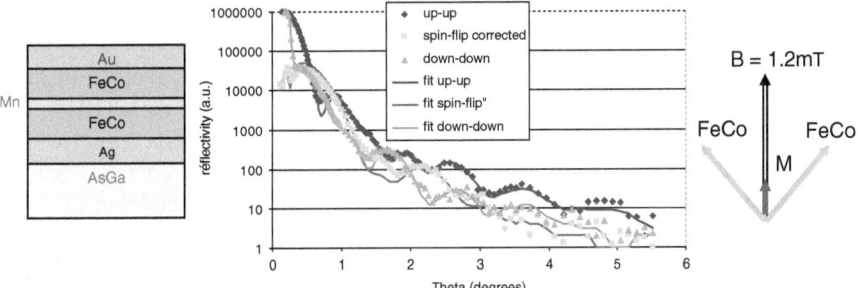

**Fig. 5.23** (**a**) Trilayer system. (**b**) Reflectivity in the remanent state. (**c**) Magnetic configuration as deduced from the fit

magnetization in the Mn layer was theoretically predicted for FeCo alloys by the ab initio calculations. In similar systems without Co, no magnetic moment is observed in the Mn layer.

The applied field was then decreased down to 1.2 mT. The reflectivity was re-measured. In these conditions a large spin-flip signal is observed (Fig. 5.23b). The reflectivity data were fitted by letting the magnetization directions vary. The best adjustment was obtained when the magnetization of the layers makes an angle of $4°$ with respect to the applied field. When the two magnetic layers make an angle of $90°$ we have a quadratic coupling.

### 5.8.2.2 Exchange Bias–Spin Valves

The magnetic thin film system which has enjoyed the most popularity until now is the spin valve. It consists of a stack of two magnetic layers separated by a non-magnetic spacer. The electrical resistance of the system depends on the relative orientation of the magnetizations. In industrial systems, one of the magnetic layers is pinned by a coupling with an anti-ferromagnetic material through the so-called exchange bias mechanism. The materials that are used in such structures are numerous: Co, Fe, Ni, NiFe, $Fe_3O_4, CoFe_2O_4, LaSrMnO_3$, etc., for ferromagnetic layers; Cu, Cr, V, $Al_2O_3$, $HfO_2$, $SrTiO_3$, etc., for the spacer layers; FeMn, IrMn, CoO, NiO, $Fe_2O_3$, $BiFeO_3$, $Co/Ru/Co$, etc., for the anti-ferromagnetic exchange bias layer.

Such spin-valve systems have been extensively characterized [111–114] and are now well understood. However, the microscopic understanding of exchange bias has been a long-standing problem for decades now. A wealth of literature is being produced on numerous and very varied systems [115–121]. It appears that the exchange bias mechanism combines very subtle effects. The reversal process of the coupled magnetic layer has been studied in detail. Since the origin of the phenomenon is often linked to micromagnetic problems, reflectivity studies are often complemented with off-specular scattering which probes the underlying micromagnetic structures. This technique is described in the following.

### 5.8.2.3 Magnetic Oxides

Polarized neutron reflectivity has also been used to probe the magnetism of individual thin films such as oxide layers (manganites [122, 123] or $Fe_3O_4$ [124]). For example, the hysteresis cycle of $La_{0.7}Sr_{0.3}MnO_3$ thin films shows a region with a low coercivity on which is superimposed a contribution which requires 0.3 T to be saturated. This suggests that the films are not homogeneous and that they are composed of several phases having different coercivities. Neutron reflectivity measurements were performed on single $La_{0.7}Sr_{0.3}MnO_3$ thin films in order to probe the magnetization profiles through the depth of the films as a function of the temperature. Figure 5.24 shows the reflectivity on a 16 nm $La_{0.7}Sr_{0.3}MnO_3$. Modeling using a homogeneous magnetic layer does not provide satisfactory fits. In order to

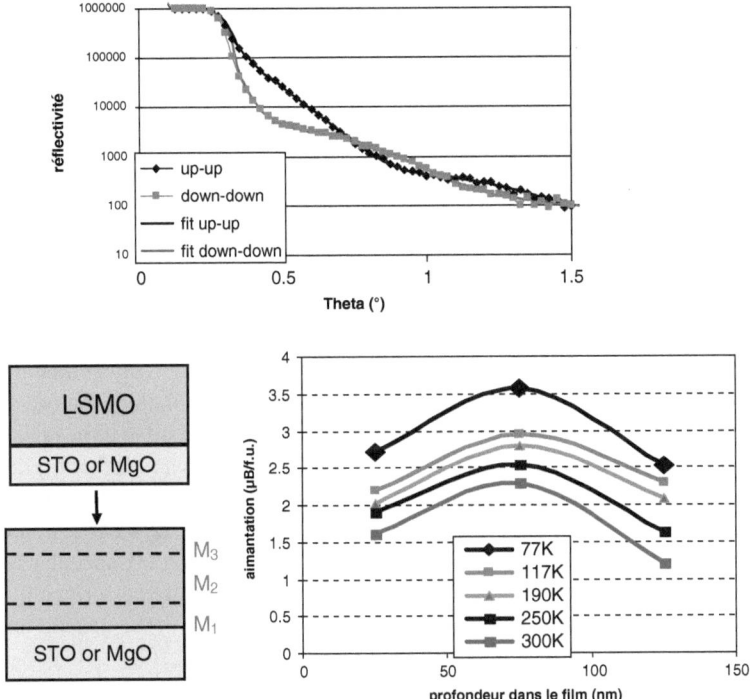

**Fig. 5.24 (a)** Reflectivity of a $La_{0.7}Sr_{0.3}MnO_3$ (16 nm) film deposited on $SrTiO_3$. **(a)** Modeling of the system: (*top*) perfect system, (*bottom*) more realistic model. **(b)** Magnetization profiles as a function of the temperature for the system LSMO (16 nm)//STO

quantitatively model the data, it has been necessary to introduce a model taking into account different magnetizations at the interfaces. We considered a three-layer model with magnetizations $M_1$, $M_2$ and $M_3$ in the depth of the films. Figure 5.24 shows the variations of the magnetizations $M_1$, $M_2$ and $M_3$ as a function of the temperature. One can note that the interface magnetization is reduced by 25–30%.

## 5.9 Off-Specular Scattering

In the recent years, the new developments of polarized reflectivity have been connected to the study of micro and nanostructures especially micromagnetic structures in multilayers. These structures correspond to the formation of magnetic domains in the size range from 100 nm to 10 μm. This is motivated by the fact that the micromagnetic structure plays a key role in the magnetic behavior of superlattices. It is also connected to the present trend which consists in patterning thin films into small structures so as to obtain a confinement not only in one direction but also in the thin film plane. Off-specular scattering which is a technique derived from the specular has been developed for the study of the roughness or the micromagnetism at a micrometric scale. For the study of nanometric structures (in the range below 100 nm), grazing incidence small angle scattering is being considered and is discussed in the next section.

In the case of specular reflectivity, the scattering vector $\mathbf{q}$ is perpendicular to the sample surface and thus one probes the structure of the sample along its depth only. All the structures in the thin film plane are averaged out. This hypothesis is correct as long as there is no formation of magnetic domains in the structure and that it can be assumed that the magnetization is homogeneous in each layer of the system. If this is not the case, by slightly modifying the scattering geometry, that is by introducing a small in-plane component of the scattering wave vector (Fig. 5.25) it is possible to probe in-plane structures. The specificity of the reflectivity geometry is that the in-plane component $q_x$ of the scattering vector is very small, of the order of $0.1–10\,\mu m^{-1}$. In this scattering geometry, one will be mostly sensitive to in-plane lateral structure with a characteristic size ranging from $50\,\mu m$ down to $0.5\,\mu m$. The upper limit is set by the resolution of the spectrometer and the size of the direct beam. The lower limit is set by the available neutron flux. These sizes correspond to typical sizes of micromagnetic domain structures. Thus magnetic off-specular is mostly used to probe such problems. These measurements are usually performed by using a position-sensitive detector after the sample and measuring the scattering on the detector as a function of the incidence angle.

The pioneering work in the field of off-specular scattering was presented in the early 1990s [125]. For flux reasons, until now, most of these studies have been performed on multilayer systems. Figure 5.26 presents an example of the off-specular scattering from a $[Co/Cu]_{50}$ multilayer.

The diffuse signal has been measured as a function of $q_x$ and $q_z$. In Fig. 5.26a and b, one observes the structural correlation peak [001] corresponding to the chemical periodicity. At remanence, a strong diffuse scattering peak is observed at the position [0 0 1/2]. Since the magnetic diffuse scattering is localized around the position [0 0 1/2], it is possible to say that the Co layers are globally anti-ferromagnetically coupled along the thickness of the layer. However, since there is a strong diffuse scattering, it is also possible to say that there exists a significant magnetic disorder in the plane of the Co layers. The width of the diffuse scattering peak around the position [0 0 1/2] (Fig. 5.26c) is inversely proportional to the magnetic domain size and gives an estimate of the mean magnetic domain size which ranges from $1\,\mu m$ at remanence (30 G) and grows to $6\,\mu m$ at 250 G.

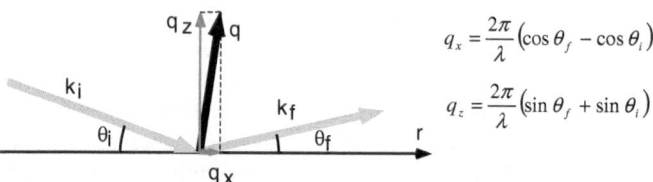

$$q_x = \frac{2\pi}{\lambda}\left(\cos\theta_f - \cos\theta_i\right)$$

$$q_z = \frac{2\pi}{\lambda}\left(\sin\theta_f + \sin\theta_i\right)$$

**Fig. 5.25** Off-specular scattering geometry. The scattering vector $\mathbf{q}$ is not perpendicular to the thin film plane anymore. There is a small component in the thin film plane

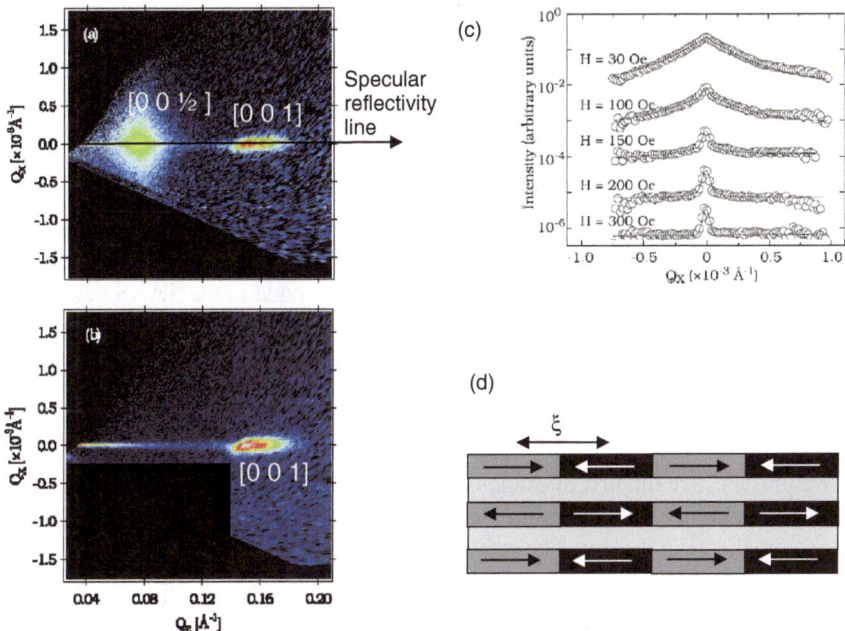

**Fig. 5.26** [Co (2 nm)/Cu (2 nm)]$_{50}$ multilayers (adapted from Langridge et al. [69]). (**a**) Diffuse scattering at $H = 0$. One observes a strong diffuse signal at the AF position. (**b**) Diffuse scattering in a saturating field. The AF peak has disappeared. (**c**) Evolution of the AF peak as a function of the applied field (cut along $Q_z = 0.75\,\text{nm}^{-1}$). (**d**) Magnetic coupling between the layers. $x$ is the lateral correlation length between magnetic domains. The Co layers are locally coupled AF but there is a strong disorder within each Co layer

Magnetic off-specular scattering has been mostly used to probe the magnetic domain sizes in multilayers. Detailed quantitative analysis of the magnetic off-specular scattering can be performed [93]. The effect of the micromagnetic structure can then be correlated with other properties such as the magneto-crystalline anisotropy (in Fe/Cr superlattice [126]) or the magneto-resistive effect (in Fe/Cr [127, 128] or Co/Cu [129] superlattices). The formation of micromagnetic structures is very important with respect to the transport properties in magnetic sensors. The signal-to-noise ratio of giant magneto-resistive systems is very sensitive to the micromagnetic structure [111]. Off-specular studies are also used to complement studies on exchange bias systems: Co/CoO [130] and Ir20Mn80/Co80Fe20 [131]. Off-specular scattering has also been used to study the problem of the reversal process in neutron polarizing supermirrors [132]. In some special cases, it has been shown that it is also possible to probe single interfaces (Fe/Cr/Fe trilayer [133] or waveguide structures [134]).

The trend in nanosciences is shifting from continuous thin films to in-plane nanostructures. These nanostructures can be obtained by patterning or by self-organization [135–139, 143]. In a number of studies, the influence of patterning on the exchange bias has been probed [140–142, 144]. These studies are of interest when the magnetic heterostructures are to be integrated in large-scale microcircuits (typically for magnetic RAMs.)

## 5.10 Grazing Incidence Scattering

Since nanosciences are aiming at smaller scales (well below $1 \mu m$), off-specular will reach its limits since it is limited to probing rather large correlation lengths ($\xi >$ 500 nm). This is why surface scattering has been extended to the SANS geometry. In this case, one looks at the scattering in the plane perpendicular to the incidence plane (Fig. 5.26a, hashed plane). The scattering wave vector is given by $q_x = k_0.\Delta q_y$ and is in a range comparable to the scattering wave vectors in SANS experiments: $10^{-4} < q_y < 3 \text{nm}^{-1}$. This corresponds to correlation lengths $\xi$ ranging from 3 to 100 nm.

We present here the first example of a grazing incidence SANS experiment on a magnetic thin film [145]. FePt thin film layers self-organize themselves in magnetic stripe domains (Fig. 5.27a). The stripes are almost perfectly ordered as a periodic

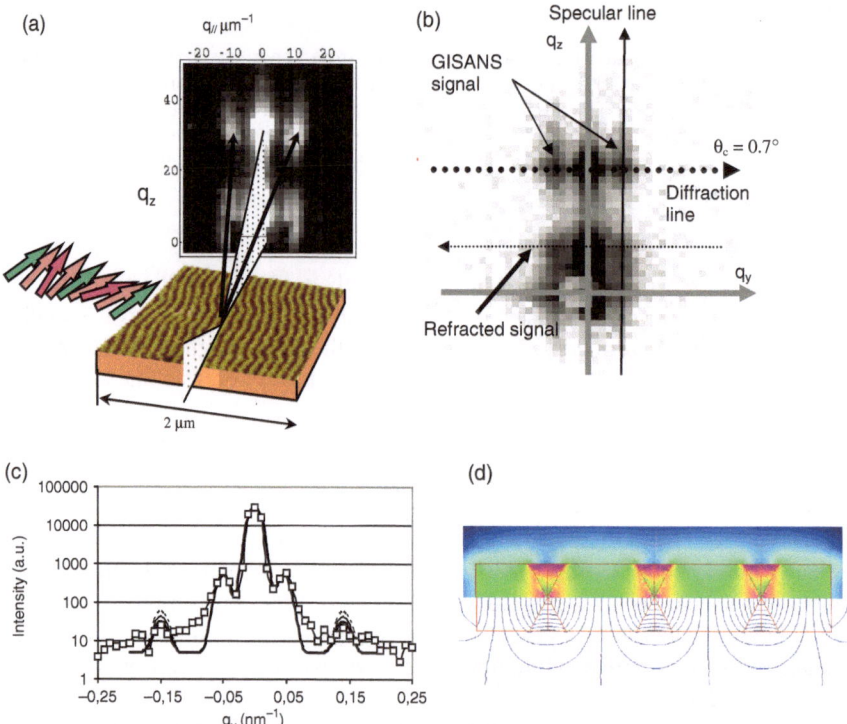

**Fig. 5.27** GISANS signal from a magnetic domain nanostructure. (**a**) Magnetic force microscopy image of the magnetic domain and scattering geometry. (**b**) GISANS signal on the detector for $\theta_{in} = 0.7°$. (**c**) GISANS signal at constant $q_z$. (**d**) Distribution of the magnetic induction in the thin films

pattern with a period of about 100 nm. In order to study in-depth this magnetic pattern, a grazing incidence SANS experiment was performed on the spectrometer PAPYRUS at the LLB. The neutron beam was sent at grazing incidence ($\theta_{in} = 0.7°$) on the layer, the magnetic domains being parallel to the incidence plane (Fig. 5.27a). Diffraction from the magnetic domains can be observed. Figure 5.27b details the different contributions of the grazing incidence SANS signal. An integration at fixed $q_z$ has been performed and is presented in Fig. 5.27c. Three diffraction orders can be observed (the second order being extinct). In order to model the system, it is necessary to take into account the Néel caps between the magnetic stripes as well as the magnetic stray fields (Fig. 5.27d) [146].

Other systems with stripe domains have been studied (e.g., Fe/FeN [147]). Systems of magnetic Fe nanodots have also been observed using grazing incidence SANS [148].

Compared to magnetic force microscopy, which is a direct space probe, the technique permits to probe buried layers and to obtain quantitative information about the magnetization. Force microscopy only gives surface information and no quantitative information. The other advantage is that it is also possible to set up complex sample environments (furnace – cryostat – high magnetic fields). The strong limitation is however that presently the neutron flux is rather low and long counting times are required. As dedicated instruments will be developed the situation will improve.

# 5.11  Conclusion on Neutron Reflectometry

## 5.11.1  Neutron–X-Ray Comparison

In the last decade, great efforts have been made to apply x-ray scattering to the study of the magnetism of thin films. The high flux available on the synchrotron sources compensates for the weak magnetic interaction of x-rays. In this paragraph, we want to underline the strengths and weaknesses of the different scattering techniques. Neutron reflectivity has the following characteristics:

+ It is a direct quantitative probe of the magnetization. The data processing is very simple and quantitative. It is straightforward to obtain the magnetization profile (amplitude and direction) in a thin film system.

+ Complex sample environments are available (very low temperatures, high temperatures, high magnetic fields).

+ It is possible to probe buried layers. Protective capping can be used. The corollary is that it is possible to probe complex systems consisting of several layers. It is not necessary to design the system specifically for the scattering experiment.

+ The flux is low and several hours of measurements are required for each sample and experimental conditions. Dynamics can be probed only down to $\sim 10\,\mu s$ in stroboscopic mode.

− Neutrons have a weak chemical sensitivity and resonant techniques or spectroscopic techniques do not exist.

− It is not possible to distinguish the spin and orbital moments.

The techniques of magnetic x-ray scattering (x-ray dichroism; resonant x-ray reflectivity; x-ray imaging) have the following advantages/disadvantages:

+ High flux.

+ Chemical sensitivity.

+ High speed dynamics.

+ Imaging possibilities (sub-μm).

− The data processing is very complex because the magnetic interaction is tensorial. Quantitative data are difficult to extract on complex materials.

− It is difficult to set up complex sample environments.

− It is difficult to probe buried layers.

− No vector magnetometry.


## 5.11.2 Future Evolutions

This chapter has given an overview of the neutron reflectometry as a tool for the investigation of surfaces. We have presented a matrix formalism which makes it possible to describe the specular reflectivity on non-magnetic and magnetic systems. Neutron reflectometry offers several specificities which makes it very useful for the study of polymer and magnetic thin film systems. In the field of soft matter, the possibility of deuteration and selective labeling makes neutron reflectivity an invaluable tool. In the field of magnetic thin films, the main advantages are that it is a direct probe of the magnetization in a material. It can easily be used to measure AFM, ferro or helical ordering in superlattices, probe complex magnetic ordering in multilayers, give detailed insights in problems such as the magnetism of ultra-thin films or the exchange bias mechanism. This has been illustrated with a few typical examples.

The problem of phase determination in neutron reflectometry is also an active field of research [149–151]. If not only the intensity but also the phase of the reflectivity could be measured a direct inversion of the reflectivity profile would be possible. Nevertheless, during these last 2 decades, polarized neutron reflectometry has proved to be a useful tool for the topics discussed above. In the early studies of magnetic superlattices, new types of magnetic orders were directly and unambiguously probed. Since then it has systematically been used for the study of magnetic thin film heterostructures. It is even used to characterize industrial systems. It is now complemented by new tools using the magnetic sensitivity of x-rays: magnetic x-ray diffraction or reflectivity and x-ray dichroism.

A wide set of surface scattering techniques have become available during the last decade (Fig. 5.28): specular neutron reflectivity which is operated routinely, off-specular scattering which is easily performed but requires complex data processing, grazing incidence SANS which is still in development. A very large range of correlation lengths in thin film systems can now be probed using these different scattering techniques.

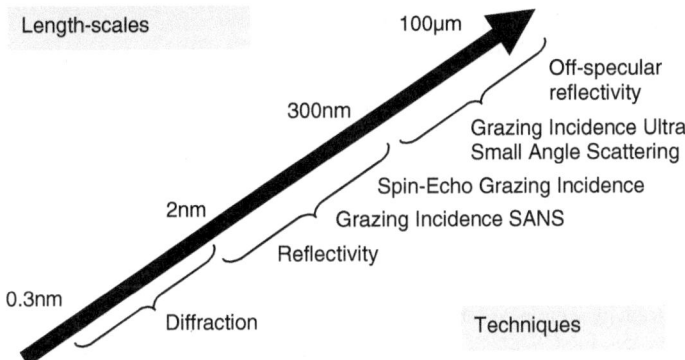

**Fig. 5.28** Correlation lengths and suitable scattering techniques

Presently, a big effort is made in order to increase the flux on neutron reflectometers. Flux gains ranging from 10 to 100 can reasonably be expected in the next decade through the implementation of new types of neutron reflectometers. Quantitative gains in the measuring time and in the minimum sample size will be achieved. Other opportunities may appear when new neutron spallation sources will come into operation in the next 5–10 years. However it is not yet clear if qualitative gains, that is new types of measurements besides the ones presented in this communication will be achieved. For example, the use of neutrons to probe inelastic processes has been barely scratched in the field of thin films.

# References

1. Felcher, G.P., Hilleke, R.O., Crawford, R.K., Haumann, J., Kleb R., Ostrowski, G.: Rev. Sci. Instr. **58**, 609 (1987)
2. Majkrzak, C.F., Cable, J.W., Kwo, J., Hong, M., McWhan, D.B., Yafet Y., Waszcak, J.: Phys. Rev. Lett. **56**, 2700 (1986)
3. Penfold, J., Thomas, R.K.: J. Phys. Condens. Matter **2**, 1369–1412 (1990)
4. Russel, T.P.: Mat. Sci. Rep. **5**, 171–271 (1990)
5. Russel, T.P.: Physica B **221**, 267–283 (1996)
6. Lee, L.T., Langevin, D., Farnoux, B.: Phys. Rev. Lett. **67**, 2678–81 (1991)
7. Penfold, J., Lee, E.M., Thomas, R.K.: Molecular Physics **68**, 33–47 (1989).
8. Baibich, M.N., Broto, J.M., Fert, A., Nguyen Van Dau, F., Petroff, F., Etienne, P., Creuzet, G., Friedrich, A., Chazelas, J.: Phys. Rev. Lett. **61**, 2472 (1988)
9. Sears, V.F.: Physics Report **141**, 281 (1986)
10. Zhou, X.L., Chen, S.H.: Physics Reports **257**, 223–348 (1995).
11. Glättli, H., Goldman, M.: Methods of Experimental Physics, vol. 23C, Neutron Scattering, Academic Press, Orlando, (1987)
12. Dietrich, S., Haase, A.: Physics Reports **260**, 1–138 (1995)
13. Sears, V.F.: Methods of Experimental Physics, vol. 23A, Neutron Scattering, Academic Press, Orlando, (1987)
14. Sears, V.F.: Neutron News **3**, 26 (1992)
15. Lekner, J.: Theory of reflection of electromagnetic and particle waves, Martinus Nijhoff, Dordrecht, (1987)

16. Blundell, S.J., Bland, J.A.C.: Phys. Rev. B **46**, 3391 (1992)
17. Pleshanov, N.K.: Z. Phys. B **94**, 233–243 (1994)
18. Fermon, C., Physica B **213-214**, 910–913 (1995)
19. Fermon, C., Miramond, C., Ott, F., Saux, G.: J. Neutron Res. **4**, 251 (1996)
20. Felcher, G.P., Adenwalla, S., De Haan, V.O., Van Well, A.A.: Nature **377**, 409 (1995)
21. Felcher, G.P., Adenwalla, S., de Haan, V.O., van Well, A.A.: Physica B **221**, 494–499 (1996)
22. Nevot, L., Croce, P.: Revue de Physique Appliquée **15**, 761 (1980)
23. Toperverg, B., Physica B **297**, 160 (2001)
24. Kentzinger, E., Rücher, U., Toperverg, B., OH, F. and Brückel, T.: Phys. Rev. B. **77**, 104435 (2008)
25. Majkrzak, C.F.: Phys. B **221**, 342–356 (1996)
26. https://neutronreflectivity.neutron-eu.net/main/Reflectometers
27. Mezei, F.: Z. Phys. **255**, 146 (1972)
28. Farnoux, B.: Neutron Scattering in the 90', Conf. Proc. IAEA in Jülich, 14–18 january 1985, 205–209, Vienna, 1985, X.D.
29. Russel, T.P., Menelle, A., Hamilton, W.A., Smith, G.S., Satija, S.K., Majkrzak, C.F.: Macromolecules **24**, 5721–5726 (1991)
30. Zhao, X., Zhao, W., Zheng, X., Rafailovich, M.H., Sokolov, J., Schwarz, S.A., Pudensi, M.A.A., Russel, T.P., Kumar, S.K., Fetters, L.J.: Phys. Rev. Lett. **69**, 776 (1992)
31. Lee, L.T., Guiselin, O., Lapp, A., Farnoux, B., Penfold, J., Phys. Rev. Lett. **67**, 2838–2841 (1991)
32. Guiselin, O., Lee, L.T., Farnoux, B., Lapp, A., J. Chem. Phys. **95**, 4632–4640 (1991)
33. Lee, L.T., Jean et B., Menelle, A.: Langmuir **15**, 3267–3272 (1999)
34. Marzolin, C., Auroy, P., Deruelle, M., Folkers, J-P., Leger L., Menelle, A.: Macromolecules **34** 8694–8700 (2001)
35. Geoghegan, M., Boué, F., Bacri, G., Menelle A., Bucknall, D.G.: Eur. Phys. J. B **3**, 83–96 (1998)
36. Lösche, M., Schmitt, J., Decher, G., Bouwman, W.G., Kjaer, K., Macromolecules **31**, 8893 (1998)
37. Cathala, B., Cousin, F., to be published
38. Devaux, C., Beyou, E., PChaumont, h., Chapel, J.P.: Eur. Phys.J. E: Soft Matter E **7**, 345–352 (2002)
39. Devaux, C., Chapel, J.P.: Eur. Phys.J. E: Soft Matter **10**, 77–81 (2003)
40. Devaux, C., Cousin, F., Beyou, E., Chapel, J.-P.: Macromolecules **38**, 4296 (2005)
41. Prucker, O., Rühe, J.: Macromolecules **31**, 592 (1998)
42. Dura, J.A., Richter, C.A., Majkrzak, C.F., Nguyen, N.V., Appl. Phys. Lett. **73**, 2131 (1998)
43. Bertagna, V., Erre, R., Saboungi, M.L., et al.: Appl. Phys. Lett. **84**, 3816–3818 (2004)
44. Chen-Mayer, H.H., Lamaze, G.P., Coakley, K.J., Satija, S.K.: Nucl. Instrum. Methods Phys. Res. A **505**, 531 (2003)
45. Caccavale, F., Coppola, R., Menelle, A., Montecchi, M., Polato, P., Principi, G.: J. Non-Cryst. Solids **218**, 291–295 (1997)
46. Battaglin, C., Caccavale, F., Menelle, A., Montecchi, M., Nichelatti, E., Nicoletti, F., Polato, P.: Thin Solid Films **351**, 176 (1999)
47. Battaglin, G., Menelle, A., Montecchi, M., Nichelatti, E., Polato, P.: Glass Technol. **43**, 203 (2002)
48. Wiesler, D.G., Majkrzak, C.F., Phys. B **198**, 181 (1994)
49. Pusenkov, V.M., Moskalev, K., Pleshanov, N., Schebetov, A., Syromyatnikov, V., Ul'yanov, V., Kobzev, A., Nikonov, O.: Phys. B **276**, 654 (2000)
50. Metelev, S.V., Pleshanov, N.K., Menelle, A., Pusenkov, V.M., Schebetov, A.F., Soroko, Z.N., Ul'yanov, V.A.: Phys. B **297**, 122 (2001)
51. Vogt, B.D., Lee, H.J., Prabhu, V.M., DeLongchamp, D.M., Lin, E.K., Wu, W.L., Satija, S.K.: J. Appl. Phys. **97**, 7 (2005)
52. Jribi, R., to be published
53. Tun, Z., Noel, J.J., Shoesmith, D.W.: Phys. B 241, 1107 (1997)

54. Tun, Z., Noel, J.J., Shoesmith, D.W.: J. Electrochem. Soc. **146**, 988 (1999)
55. Noel, J.J., Jensen, H.L., Tun, Z., Shoesmith, D.W.: Electrochem. Solid State Lett. **3**, 473 (2000)
56. Fragneto-Cusani, G.: J. Phys.: Condens. Matter **13**, 4973–4989 (2001)
57. Krueger, S.: Curr. Opin. Colloid Interface Sci. **6**, 111 (2001)
58. Majewski, J., Kuhl, T.L., Gerstenberg, M.C., Israelachvili, J.N., Smith, G.S.: J. Phys. Chem. B **101**, 3122 (1997)
59. Charitat, T., Bellet-Amalric, E., Fragneto, G., Graner, F.: Eur. Phys. J. B **8**, 583 (1999)
60. Fragneto, G., Graner, F., Charitat, T., Dubos, P., Bellet-Amalric, E.: Langmuir **16**, 4581 (2000)
61. Maierhofer, A.P., Bucknall, D.G., Bayer, T.M.: Biophys. J. **79**, 1428 (2000)
62. Fragneto, G., Charitat, T., Graner, F., Mecke, K., Perino-Gallice, L., Bellet-Amalric, E.: Eur. Lett. **53**, 100 (2001)
63. Krueger, S., Meuse, C.W., Majkrzak, C.F., Dura, J.A., Berk, N.F., Tarek, M., Plant, A.L.: Langmuir **17**, 511 (2001)
64. Schwendel, D., Hayashi, T., Dahint, R., Pertsin, A., Grunze, M., Steitz, R., Schreiber, F.: Langmuir **19**, 2284 (2003)
65. Burgess, I., Li, M., Horswell, S.L., Szymanski, G., Lipkowski, J., Majewski, J., Satija, S.: Biophys. J. **86**, 1763 (2004)
66. Majkrzak, C.F. et al.: J. Appl. Phys. **63**, 3447–3452 (1988)
67. Grünberg, P.: et al.: Phys. Rev. Lett. **57**, 2442 (1986)
68. Schreyer, A., et al.: Phys. Rev. Lett. **79**, 4914–4917 (1997)
69. Langridge, S., et al.: Phys. Rev. Lett. **85**, 4964–4967 (2000)
70. Hjörvasson, B., et al.: Phys. Rev. Lett. **79**, 901 (1997)
71. YHuang, Y., Felcher, G.P., Parkin, S.S.P.: J. Magn. Magn. Mater. **99**, 31–38 (1991)
72. Szuszkiewicz, W., et al.: J. Superconductivity **16**, 205–8 (2003)
73. Szuszkiewicz, W., Dynowska, E., Ott, F., Hennion, B., Jouanne, M., Morhange, J.F., Sadowski, J.: J. Superconductivity: Incorporating Novel Magnetism, **16** 209–12 (2003)
74. Singh, S., et al.: J. Appl. Phys. **101**, Art. No. 033913 (2007)
75. Kepa, H., et al.: Phys. Rev. B **64**, 121302 (2001)
76. Szuszkiewicz, W., et al.: J. Superconductivity **16**, 209–12 (2003)
77. Kepa, H., et al.: EuroPhysics Lett. **56**, 54–60 (2001)
78. te Velthuis, S.G.E., et al.: Phys. Rev. Lett. **89**, 127203 (2002)
79. te Velthuis, S.G.E., et al.: J. Appl. Phys. **87**, 5046 (2000)
80. te Velthuis, S.G.E., et al.: Appl. Phys. Lett. **75**, 4174–6 (1999)
81. Radu, F., et al.: J. Magn. Magn. Mater. **240**, 251–253 (2002)
82. Radu, F., et al.: Phys. Rev. B **67**, 134409 (2003)
83. Nogués, J. et al.: J. Magn. Magn. Mater. **192**, 203–232 (1999)
84. Fitzsimmons, M.R., et al.: Phys. Rev. Lett. **84**, 3986–89 (2000)
85. Gierlings, M., et al.: Phys. Rev. B **65**, 092407/1–4 (2002)
86. Lauter-Pasyuk, V. et al., Phys. B **248**, 166–170 (1998)
87. O'Donovan, K.V., Borchers, J.A., Majkrzak, C.F., Hellwig, O., Fullerton, E.E.: Phys. Rev. Lett. **88**, 067201/1–4 (2002)
88. Haskel, D., et al.: Phys. Rev. Lett. **87**, 207201/1–4 (2001)
89. Leiner, V., et al.: Phys. B **283**, 167–70 (2000)
90. Stahn, J.: et al.: Phys. Rev. B **71**, 140509 (2005)
91. Chakhalian, J., et al.: Nature Phys. **2**, 244–248 APR 2006
92. Baczewski, L.T., et al.: Phys. Rev. B **74**, 075417 (2006)
93. Lauter-Pasyuk, V., et al.: Phys. Rev. Lett. **89**, 167203 (2002)
94. Schreyer, A., Aukner, J.F., Zeidler, T., Zabel, H., Majkrzak, C.F., Schaefer, M., Gruenberg, P.: Euro. Phys. Lett., 595–600 (1995)
95. Bland, J.A.C., et al.: J. Magn. Magn. Mater. **93**, 513–522 (1991)
96. Schreyer, A., et al., Euro. Phys. Lett. **32**, 595–600 (1995)

97. Bürgler, D.E., Grünberg, P., Demokritov, S.O., Johnson, M.T.: Interlayer exchange coupling in layered magnetic structures. In: Buschow, K.H.J. (ed.) Handbook of Magnetic Materials, vol. 13, p. 1. Elsevier, CityplaceAmsterdam (2001)
98. Cheng, L., et al.: Phys. Rev. B **69**, 5 (2004)
99. Bergmann, A., et al.: Phys. Rev. B **72**, 12 (2005)
100. Vadala, M., et al.: J Phys. D-Appl. Phys. **40**, 1289 (2007)
101. Beesley, A.M., et al.: J. Phys-Condens Mat. **16**, 8507 (2004)
102. Singh, S., et al.: Electrochem. Solid State Lett. **9**, J5 (2006)
103. Fitzsimmons, M.R., et al.: Phys. Rev. B **73**, 134413 (2006)
104. Leiner, V., Ay, M., Zabel, H.: Phys. Rev. B **70**, 1 (2004)
105. Bedanta, S., et al.: Phys. Rev. B **74**, 5 (2004)
106. Böni, P.: Phys. B **234–236**, 1038–1043 (1997)
107. Nerger, S., et al.: Phys. B **297**, 185-188 (2001)
108. Fritzsche, H., et al.: Phys. Rev. B **70**, 214406 (2004)
109. Laloe, J.B., et al.: IEEE T. Magn. **42** (2006)
110. Schmitz, D., et al.: J. Magn. Magn. Mater. **269**, 89–94 (2004)
111. Pannetier, M., Doan, T.D., Ott, F., Berger, S., Persat, N., Fermon, C., Europhys. Lett. **64**, 524–528 (2003)
112. Moyerman, S., et al.: J. Appl. Phys. **99**, (2006)
113. Schanzer, C., et al.: Physica B **356**, 46–50(2004)
114. Zhao, Z.Y., et al.: Phys. Rev. B **71**, 1098–0121(2005)
115. Fitzsimmons, M.R., et al.: Phys. Rev. Lett. **84**, 3986–89 (2000)
116. Leighton, C., et al.: Phys. Rev. B **65**, 064403/1–7 (2002)
117. Fitzsimmons, M.R., et al.: Phys. Rev. B **65**, 134436 (2002)
118. Gierlings, M., et al.: Phys. Rev. B **65**, 092407/1–4 (2002)
119. Blomqvist, P., et al.: J Appl. Phys. **96**, 6523–6526 (2004)
120. Paul, A., et al.: Phys. Rev. B 1098–0121 (2006)
121. Roy, S., et al.: Phys. Rev. Lett. **95**, 047201 (2005)
122. Ott, F., et al.: J. Mag. Mag. Mat. **211**, 200–205 (2000)
123. Borges, R.P., et al.: J. Appl. Phys. **89**, 3868-3873 (2001)
124. Moussy, J.-B., et al.: Phys. Rev. B **70**, 174448 (2004)
125. Felcher, G.P., et al.: Neutron News **5**, 18–22 (1994)
126. Nagy, D.L., et al.: Phys. Rev. Lett. **88**, 4 (2002)
127. Lauter, H., et al.: J. Magnet. Magnetic Mater. **258**, 338 (2003)
128. Lauter-Pasyuk, V., et al.: J. Magnet. Magnetic Mater. **226**, 1694 (2001)
129. Paul, A., et al.: Physica B **356**, 31 (2005)
130. Gierlings, M., et al.: Physica B **356**, 36–40 (2004)
131. Paul, A., et al.: Physica B **356**, 26–30 (2005)
132. Rücker, U., Kentzinger, E., Toperverg, B., Ott, F., Brückel, Th.: Appl. Phys. A **74**, S607–S609 (2002)
133. Kentzinger, E., Rucker, U., Toperverg, B., et al.: Physica B **335**, 89–94 (2004)
134. Kozhevnikov, S.V., Ott, F., Kentzinger, E., Paul, A.: Physica B
135. Ott, F., Menelle, A., Fermon, C., Humbert, P.: Physica B **283**, 418–421 (2000)
136. Ott, F., Humbert, P., Fermon, C., Menelle, A., Physica B **297**, 189–193 (2001)
137. Theis-Brohl, K., Physica B **345**, 161–168 (2004)
138. Theis-Brohl, K., et al.: Phys. Rev. B **71**, 4 (2004)
139. Theis-Brohl, K., et al.: Physica B **356**, 14 (2005)
140. Popova, E., et al.: Eur. Phys. J. B **44** 491–500 (2005)
141. Temst, K., et al.: Eur. Phys. J. B **45**, 261–266 (2005)
142. Temst, K., et al.: J Magnet. Magnetic Mater. **304**, 14–18 (2006)
143. Langridge, S., et al.: Phys. Rev. B **74**, 6 (2006)
144. Theis-Brohl, K., et al.: Phys. Rev. B **73**, 14 (2006)
145. Fermon, C., et al.: Physica B **267-268**, 162–7 (1999)
146. Pannetier, M., Ott, F., Fermon, C., Samson, Y.: Physica B **335**, 54–8 (2003)
147. Szuszkiewicz, W., et al.: J Alloys Compd 423, 172–175 (2006)
148. Li, C.-P., et al.: J. Appl. Phys. **100**, 074318–074325 (2006)
149. Majkrzak, C.F., Berk, N.F.: Phys. Rev. B **52**, 10827 (1995)
150. Majkrzak, C.F., Berk, N.F.: Phys. Rev. B **58**, 15416 (1998)
151. Kasper, J., Leeb, H., Lipperheide, R.: Phys. Rev. Lett. **80**, 2614–2617 (1998)

# Chapter 6
# X-Ray Reflectivity by Rough Multilayers

T. Baumbach and P. Mikulík

## 6.1 Introduction

One tendency in present material research is the increasing ability to structure solids in one, two and three dimensions at a sub-micrometer scale. Based on various material systems artificial *mesoscopic-layered superstructures* such as multilayers, superlattices, layered gratings, quantum wires and dots have been fabricated successfully. This has opened new perspectives for manifold technological applications (e.g. for anticorrosion coating and hard coating, micro and optoelectronic devices, neutron and x-ray optical elements, magnetooptical recording).

The perfection of mesoscopic-layered superstructures is characterised by

1. the perfection of the superstructure (grating shape, periodicity, layer thickness, etc.),
2. the interface quality (roughness, graduated heterotransition, interdiffusion, etc.),
3. crystalline properties (strain, defects, mosaicity, etc.).

*Roughness* is of crucial importance for the physical behaviour of interfaces. Roughness reduces the specular reflectivity of mirrors and waveguides for x-ray and neutron optics. Moreover it creates unintentional diffuse scattering. In magnetic layers it changes the interface magnetisation. Roughness promotes corrosion and influences the hardness of materials. It disturbs the electronic band structure in semiconductor devices. Interface roughness supports the generation of crystalline defects in layered structures. In multilayers already the roughness of the substrate or the buffer layer influences the quality of all subsequent layers. Depending on the growth process the roughness profile can be partially replicated from interface to interface.

Interface roughness is a random deviation of the layer shape from an ideally smooth plane. We consider here roughness with correlation properties of *mesoscopic* (sub-micrometer) scale. Irradiating a macroscopic area of the sample,

T. Baumbach (✉)
Institut für Synchrotronstrahlung (ISS/ANKA), Forschungzentrum Karlsruhe,
Hermann von Helmholtz Platz 1, D-76344 Eggenstein-Leopoldshafen, Germany

Baumbach, T., Mikulík, P.: *X-Ray Reflectivity by Rough Multilayers*. Lect. Notes Phys. **770**, 235–282 (2009)
DOI 10.1007/978-3-540-88588-7_6      © Springer-Verlag Berlin Heidelberg 2009

surface-sensitive x-ray scattering allows the investigation of the statistical behaviour of the roughness profile.

Interface roughness in multilayers can be studied by all surface-sensitive x-ray scattering methods (x-ray reflection (XRR), grazing incidence diffraction (GID), strongly asymmetric x-ray diffraction (SAXRD)) employing physical principles similar to the case of simple surfaces. They are based on

1. the reduction of the information depth at grazing angles of incidence and exit,
2. reflection of x-rays by the individual interfaces of a multilayer (ML) at small angles of incidence,
3. interference of the waves reflected by different interfaces,
4. diffuse scattering of x-rays by interface disturbances.

Specular x-ray reflection (SXR) as the most frequently used method studies the depth profile of the electron density. It detects the density gradient at the interface between two layers, where from we conclude on the r.m.s. roughness. Grazing incidence diffraction and strongly asymmetric x-ray diffraction detect interface roughness via the strain and the depth profile of the Fourier components of the electron density. The measurement of diffuse x-ray scattering (DXS) gives a clear evidence of interface roughness, distinguishing between roughness and graduated interfaces due to transition layers, inter-diffusion or graduated heterotransitions. Up to now DXS has frequently been observed in the XRR mode [1–10]. First measurements of DXS in the diffraction mode have been reported recently [11, 12]. DXS by multilayers enables one to characterise the lateral correlation properties of interfaces similar to DXS by surfaces. Moreover it allows to detect vertical roughness replication from interface to interface. DXS at grazing incidence occurs under condition of simultaneous intense specular reflection. This gives rise to strong effects of multiple scattering [5, 7, 8, 10, 12–14]. That is why semi-dynamical methods such as the distorted wave Born approximation (DWBA) are more appropriate to explain the DXS features than kinematical treatments.

This chapter intends to give an introduction into theoretical and experimental aspects of x-ray reflection by solid multilayers with rough interfaces, illustrated by various examples. We start in Sect. 6.2 with a short presentation of rough multilayers and of the notations used in this chapter.

In Sect. 6.3 we will introduce the experimental set-up and usual experimental scans and in the following sections we apply the results of the Chaps. 3 and 4 on multilayered samples with different types of interface correlation properties. There we discuss typical features of the reflection curves and reciprocal space maps by various experimental examples. Afterwards, we mention the investigation of roughness by *surface-sensitive diffraction methods* and at the end we study the reflectivity by intentionally laterally structured multilayers (*gratings*).

Throughout the chapter the *reciprocal space representation* of the optical potential and the scattering processes allows us to outline the scattering principles in a geometrical way. The basic principles of it are summarised in the appendix.

## 6.2 Description of Rough Multilayers

The scattering potential of a sample can be represented by the polarisability $\chi(r)$, by the refractive index $n(r)$ as well as by the dielectric function $\epsilon(r)$. In classical optics it is common to use $n(r)$ or $\epsilon(r)$, x-ray optics uses also $\delta(r) = 1 - n(r)$. In order to pronounce similarities in the procedures and expressions for all x-ray scattering methods, thus *reflection* and *diffraction*, we preferred to use in this chapter the polarisability $\chi(r)$. We recall the relation between $\chi$ and $\delta$

$$\chi(r) = -2\delta(r) . \tag{6.1}$$

Furthermore, we will make use of the *optical potential* defined by

$$V(r) = -k_0^2 \chi(r) , \tag{6.2}$$

where $k_0 = 2\pi/\lambda$ is the vacuum wave vector and $\lambda$ is the wavelength of the scattered radiation

$$\chi(r) = \sum_{j=1}^{N} \chi_j(r) . \tag{6.3}$$

We represent $\chi(r)$ of the multilayer by the *polarisability of the individual layers* (see Fig. 6.1).

In order to distinguish between the interface properties of the layers and their volume properties, each layer is presented by the product of the volume polarisability $\chi_{\infty j}(r)$ and the *layer size (shape) function* $\Omega_j(r)$

$$\chi_j(r) = \chi_{\infty j}(r) \Omega_j(r) . \tag{6.4}$$

X-ray reflection methods measure the scattered intensity in the region near the origin of reciprocal space (000). There, only the mean polarisability plays a role and we can replace $\chi_{\infty j}(r)$ by the zero-order Fourier component $\chi_{0 j}(r)$ which is not sensitive to crystalline properties.

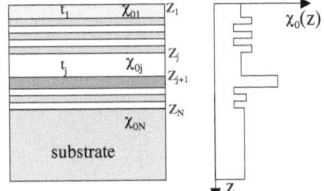

**Fig. 6.1** The schematic set-up of an "ideal" planar multilayer (*left*). Its optical potential is characterised by the polarisability depth profile (*right*)

### 6.2.1 Ideal Planar Multilayers

Let us first deal with a laterally extended *"ideal" multilayer with sharp and smooth interfaces*. Then $\chi_{0j}^{id}(r)$ will be constant within each "ideal" layer. The *layer size (shape) function* of a smooth layer with sharp interfaces is the difference of two Heaviside functions corresponding to the upper and lower interfaces,

$$\Omega_j^{\mathrm{id}}(r) = H(z - Z_j) - H(z - Z_{j+1}) \ . \tag{6.5}$$

Sharp interfaces do not allow any overlapping of neighbouring layers; thus $\Omega_j^{\mathrm{id}}(r) = 1$ predicts $\Omega_k^{\mathrm{id}}(r) = 0$ for all other layers $k \neq j$.

## 6.2.2 Multilayers with Rough Interfaces

Similar to the smooth multilayer we express the polarisability by the sum of the individual layer contributions

$$\chi(r) = \sum_{j=1}^{N} \chi_{0j}\,\Omega_j(r) \ . \tag{6.6}$$

We will further consider vertically layered structures with a *random defect structure*, which we assume to be *laterally statistically homogeneous*. We concentrate on defects, which vary *the layer shape and interface sharpness* $\Omega_j(r)$ (interdiffusion and roughness) in contrast to those influencing the *layer volume properties* $\chi_{\infty j}$ (porosity, inclusions).

Interdiffusion and graduated heterotransition between neighbouring layers produce *vertically graduated interfaces*. Then $\Omega_j(r)$ can have all values between 1 and 0. The layer is defined within the region $\Omega_j(r) \neq 0$. We allow an intermixing of neighbouring layers only, in order to keep the layer sequence. We define here by *interface roughness* the random profile of locally sharp interfaces. The vertical shift of the actual interface position with respect to its mean position is characterised by the displacement function $z_j(r_\parallel) = Z_j(r_\parallel) - Z_j$ of each interface, Fig. 6.2(a), modifying the actual layer size function,

$$\Omega_j(r) = H\left(z - [Z_j + z_j(r_\parallel)]\right) - H\left(z - [Z_{j+1} + z_{j+1}(r_\parallel)]\right) \tag{6.7}$$

(a)                                                    (b)

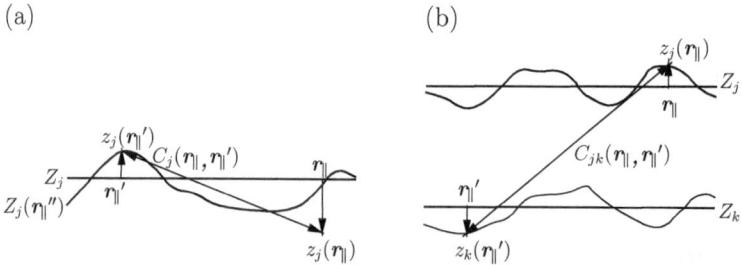

**Fig. 6.2** Notation of the interface displacements and schematical representation of the correlation function of one (**a**) and of two interfaces (**b**)

and the actual layer thickness is $t_j(r_\parallel) = t_j + z_{j+1}(r_\parallel) - z_j(r_\parallel)$, where the "ideal" thickness is $t_j = Z_{j+1} - Z_j$.

## 6.2.3 Correlation Properties of Different Interfaces

Correlation properties of single rough interfaces have been studied in Chap. 2. There were introduced the probability density of heights $p_1(z)$ and lateral height–height correlation function $C_{zz'}(r_\parallel, r_\parallel') = \langle z(r_\parallel)z(r_\parallel')\rangle$ for *one* interface, i.e. for a substrate. In this section, we will treat the correlation properties between different interfaces of a multilayer. We introduce the two-dimensional probability density of *two* interfaces, Fig. 6.2(b),

$$p_2(z_j, z_k') = p\big(z_j(r_\parallel), z_k(r_\parallel')\big) \tag{6.8}$$

and height–height correlation function

$$C_{jk}(r_\parallel - r_\parallel') = \langle z_j(r_\parallel)z_k(r_\parallel')\rangle . \tag{6.9}$$

Usually the perfection of interfaces in multilayers is essentially influenced by the quality of the substrate or buffer surface. The surface defects can be replicated in growth direction. Different replication behaviours have been observed, depending on the material system, layer set-up and the growth conditions. The following replication model has been proposed in [15]: (1) during the growth of the $j$th layer, the roughness profile $z_{j+1}(r_\parallel)$ of the lower interface is *partially replicated* and (2) other defects, an *intrinsic roughness* $\Delta_j(r_\parallel)$, are induced by imperfections of the growth process

$$
\begin{aligned}
z_j(r_\parallel) &= \Delta_j(r_\parallel) + \int dr_\parallel' \, z_{j+1}(r_\parallel') a_j(r_\parallel - r_\parallel') \\
&= \Delta_j(r_\parallel) + z_{j+1}(r_\parallel) \otimes a_j(r_\parallel),
\end{aligned}
\tag{6.10}
$$

where $\otimes$ denotes a convolution product. Here a non-random replication function $a_j(r_\parallel)$ has been introduced, determining the "degree of memory" of the interface at the top for the roughness profile at the bottom interface. If the replication function is zero, the upper interface of a layer "forgets" the interface profile at the layer bottom and its profile is entirely determined by the intrinsic roughness (*no replication*). *Identical profile replication* is achieved for zero intrinsic roughness and full replication ($a_j(r_\parallel)$ equals the delta function). Other cases are discussed in detail in [15] and will win our interest within the discussion of the experimental results.

In later sections we will use the Fourier transformation of the interface correlation functions

$$\tilde{C}_{jk}(q_\parallel) = \int dR_\parallel \, C_{jk}(R_\parallel) e^{iq_\parallel R_\parallel} = \langle \tilde{z}_j(q_\parallel)\tilde{z}_k^*(q_\parallel)\rangle \tag{6.11}$$

with

$$\tilde{z}_j(q_\parallel) = \tilde{\Delta}_j(q_\parallel) + \tilde{z}_{j+1}(q_\parallel)\tilde{a}_j(q_\parallel) . \tag{6.12}$$

In the following we neglect any statistical influence of the interface profile $z_{j+1}(r_\parallel)$ on the intrinsic roughness $\Delta_j(r_\parallel)$. Also the intrinsic roughness of different interfaces shall be statistically independent. Then we find the recursion formula for the Fourier transform of the correlation function

$$\tilde{C}_{jk}(q_\parallel) = \tilde{C}_{j+1,k+1}(q_\parallel)\, \tilde{a}_j(q_\parallel)\, \tilde{a}_k(q_\parallel) + \delta_{jk}\tilde{K}_j(q_\parallel) \,, \qquad (6.13)$$

where $\tilde{K}_j(q_\parallel)$ is the Fourier transform of the correlation function of the intrinsic roughness

$$K_j(r_\parallel - r_\parallel') = \langle \Delta_j(r_\parallel)\Delta_j(r_\parallel') \rangle \,. \qquad (6.14)$$

If we assume for all layers the same replication function $a(r_\parallel)$ and the same intrinsic roughness $\Delta(r_\parallel)$ (replicated substrate roughness $z_N(r_\parallel)$, for instance) we get the explicit expressions for the Fourier transforms of the *in-plane correlation function*

$$\tilde{C}_{jj}(q_\parallel) = \tilde{C}_{NN}(q_\parallel)\, [\tilde{a}(q_\parallel)]^{2(N-j)} + \tilde{K}(q_\parallel)\frac{[\tilde{a}(q_\parallel)]^{2(N-j-1)} - 1}{[\tilde{a}(q_\parallel)]^2 - 1} \qquad (6.15)$$

($\tilde{C}_{NN}(Q_\parallel)$ is the correlation function of the substrate) and of the *inter-plane correlation function*

$$\tilde{C}_{k \geq j}(q_\parallel) = \tilde{C}_{kk}(q_\parallel)\, [\tilde{a}(q_\parallel)]^{(k-j)} \,. \qquad (6.16)$$

The physical meaning of the particular terms in (6.15) is obvious. The first term on the right-hand side represents the influence of the substrate surface modified by the replication function, the second term is due to the intrinsic roughness of the layers beneath the layer $j$.

Knowing $\tilde{C}_{jj}(q_\parallel)$ we can calculate the mean square roughness $\sigma_j^2$ of the $j$th interface:

$$\sigma_j^2 = \langle z_j^2(r_\parallel) \rangle = \int dq_\parallel \tilde{C}_{jj}(q_\parallel) \,. \qquad (6.17)$$

## 6.3 Set-Up of X-Ray Reflectivity Experiments

In this section we outline the experimental set-up to investigate the fine structure of the reflected intensity pattern in vicinity of the origin of reciprocal space (000) under conditions of small angles of incidence and exit with respect to the sample surface.

### 6.3.1 Experimental Set-Up

A conventional x-ray reflectometer is drawn in Fig. 6.3. The x-ray source (a conventional x-ray tube or a synchrotron) emits a more or less divergent and polychromatic

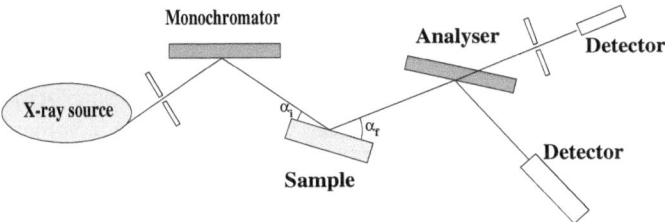

**Fig. 6.3** Schematic set-up of an x-ray reflectometer (source, monochromator, sample, slits and detector) and of a triple-crystal-like diffractometer (source, monochromator, sample, analyser and detector)

beam. The *monochromator* (a crystal or a multilayer mirror) and entrance slits produce a sufficiently monochromatic and parallel beam, hitting the sample surface under the incident angle $\theta_{in}$. Its angular divergence is characterised by the spatial angle $\Delta\Omega_{in}$. The sample is mounted on a goniometer, which allows one to change the incident angle $\theta_{in}$ by the rotation $\omega$. The x-rays are reflected (scattered) by the sample. The coherently reflected beam leaves the sample in specular direction (under the exit (final) angle $\theta_{sc} = \theta_{sc}$ in the plane of incidence). Due to roughness there occurs diffuse scattering into the upper half-space of the sample. A detector rotates around the sample and measures the *flux* of photons (in units of counts per second) through the detector window, which defines the spatial angle interval $\Delta\Omega_{det}$ around a certain spatial angle $\Omega_{sc}$ (sufficiently defined by $\theta_{sc}$ in the coplanar case). If we suppose a perfectly monochromatic and parallel incident beam of intensity $I_0$ then the idealised flux through the detector window is related with the differential scattering cross section by

$$J = I_0 \int d\sigma = I_0 \int_{\Omega_{sc}-\Delta\Omega_{det}/2}^{\Omega_{sc}+\Delta\Omega_{det}/2} \left(\frac{d\sigma}{d\Omega}\right) d\Omega \;. \tag{6.18}$$

Taking the divergence and the intensity profile of the incident beam into account, we obtain

$$J = \int_{\Delta\Omega_{in}} d\Delta\Omega_{in} \, I_0(\Delta\Omega_{in}) \int_{\Omega_{sc}-\Delta\Omega_{det}/2}^{\Omega_{sc}+\Delta\Omega_{det}/2} \frac{d\sigma(\Omega_{in}+\Delta\Omega_{in},\Omega)}{d\Omega} d\Omega \;. \tag{6.19}$$

Actually, in the case of a large sample, the detector slits select another angular interval for each point on the illuminated sample area. That can be overcome replacing the detector slits by an analyser (also a perfect crystal or a multilayer mirror) in front of the detector similar to a triple-crystal diffractometer (TCD). The monochromator is the "first crystal", the sample the "second crystal" and the analyser the "third crystal". The flux measured by the TCD is

$$J = \int_{\Delta\Omega_{in}} d\Delta\Omega_{in} \, I_0(\Delta\Omega_{in}) \int d\Omega \, \frac{d\sigma(\Omega_{in}+\Delta\Omega_{in},\Omega)}{d\Omega} \, \mathscr{D}(\Omega-\Omega_{sc}) \;, \tag{6.20}$$

where $\mathscr{D}(\Delta\Omega)$ is the reflectivity of the analyser.

## 6.3.2 Experimental Scans

Mapping the measured flux for different angles of incidence and exit, we can plot the measured scattering pattern in angular space, $\mathcal{J}(\Omega_{\text{in}}, \Omega_{\text{sc}})$, or by three reciprocal space coordinates and one angular coordinate of the sample, e.g. $\mathcal{J}(k_{\text{sc}} - k_{\text{in}}, \theta_{\text{in}})$. Restricting ourselves on coplanar reflection ($k_{\text{sc}}, k_{\text{in}}$ and the surface normal are in the same plane), the angular representation $\mathcal{J}(\theta_{\text{in}}, \theta_{\text{sc}})$ and the reciprocal space representation $\mathcal{J}(q)$ with the *scattering vector* $q = k_{\text{sc}} - k_{\text{in}}$ are equivalent.

The principal rotations of a (coplanar) TCD are as follows:

1. The rotation $2\theta$ of the detector arrangement in the coplanar scattering plane around the sample: $2\theta$ measures the *scattering angle* ($2\theta = \theta_{\text{in}} + \theta_{\text{sc}}$), the variation of $2\theta$ changes $\theta_{\text{sc}}$ ($\Delta 2\theta = \Delta \theta_{\text{sc}}$).
2. The rotation $\omega$ of the sample around the same axis: $\theta_{\text{in}} = \omega$, $\theta_{\text{sc}} = 2\theta - \omega$, a variation of $\omega$ changes simultaneously $\theta_{\text{sc}}$ and $\theta_{\text{in}}$ ($\Delta \omega = \Delta \theta_{\text{in}} = -\Delta \theta_{\text{sc}}$).

Different experimental scans can be performed by coupling both rotations. In Fig. 6.4 the most usual scans are illustrated in real and reciprocal space. They are given as follows:

**Detector scan or $2\theta$-scan.** The incident wave vector $k_{\text{in}}$ opens out the Ewald sphere $\epsilon$. If we keep the angle of incidence fixed ($\omega = $ const) and rotate the detector arrangement, we move in reciprocal space along the Ewald sphere $\epsilon$.

**$\omega$-Scan or constant $q$-scan.** The $\omega$-scan rotates the Ewald sphere around the origin of reciprocal space. Fixing the scattering angle $2\theta$, we fix the modulus of the scattering vector. Then the $\omega$-scan represents a *constant q-scan* since we move in reciprocal space on a circle of radius $q = |q|$ around the origin.

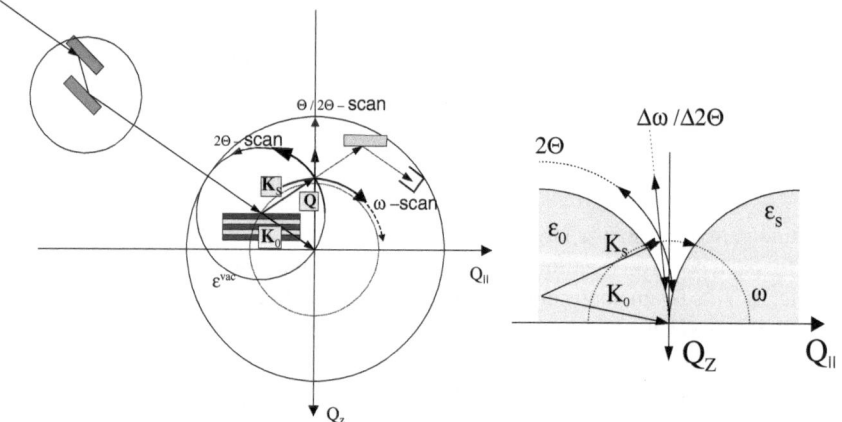

**Fig. 6.4** Illustration of the experimental scans in the reciprocal space. *Right* figure shows the enlargement around its origin, where x-ray reflection takes place. The $2\theta$-scan (detector scan) follows the Ewald circle of the incident wave. The $\omega$-scan represents rocking scan, which is transversal for XRR. For $2\omega = 2\theta$ it is a $q_z$-scan with $q_{||} = 0$ (specular scan)

**$\Delta\theta/\Delta2\theta$-Scan or radial scans.** Rotating the sample and the detector arrangement in a ratio $\Delta\omega/\Delta2\theta = 1/2$, we drive the TCD in reciprocal space in *radial direction* from the origin of reciprocal space.

**$\theta/2\theta$-Scan on the $q_z$ axis or specular scan.** This special radial scan with $\omega/2\theta = 1/2$ keeps the condition $\theta_{in} = \theta_{sc}$ and performs a $q_z$-scan at $q_x = 0$. This experimental mode is also called *specular scan*, since the detector selects always the specularly reflected beam.

**$q_x$-Scan and $q_z$-scan.** These scans go parallel to the $q_x$ and $q_z$ axes at fixed $q_z$ and $q_x$ positions, respectively.

Sometimes it is useful to measure a **reciprocal space map**, i.e. to measure the map of the scattered intensity by combining different scans, e.g. measuring a series of $\omega$-scans (rocking scans) in the interval from $\omega=0$ to $\omega=2\theta$ for varying $2\theta$. Using a position-sensitive detector (PSD), one would detect PSD spectra for different omega positions.

The angular region investigated by a reflection experiment is limited by the horizon of the sample. The limiting cases for grazing incidence ($\theta_{in} = 0$) and grazing exit ($\theta_{sc} = 0$) are illustrated in Fig. 6.5. The situation in reciprocal space is represented by the two limiting half-spheres $\epsilon_0$ and $\epsilon_s$.

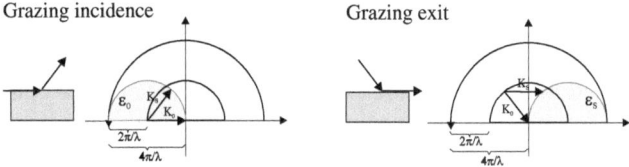

**Fig. 6.5** Situation of grazing incidence (*left*) and grazing exit (*right*) in reciprocal and real space

X-ray reflection experiments are usually realised at very small scattering angles. In Fig. 6.4(*right*) we show the introduced experimental scans in the x-ray reflection mode and their restrictions due to the sample horizon. Especially the $\omega$-scans are narrowed down. In the accessible region of reflection, i.e. near the origin of the reciprocal space, they perform approximately a transversal scan ($q_{\parallel}$-scan).

## 6.4 Specular X-Ray Reflection

In this section we discuss some theoretical and experimental examples of coherent specular x-ray reflection by layered structures with the aim to show *typical features* created by *different surface roughness point properties*. The coherent scattering intensity is concentrated along the specular rod. That means, the appropriate experimental scan is the specular or $\theta/2\theta$-scan.

## 6.4.1 Roughness with a Gaussian Interface Distribution Function

### 6.4.1.1 Single Surface

The predominant number of samples have been successfully characterised assuming a Gaussian probability density of the interface roughness profile (see (2.19))

$$p_1(z) = \frac{1}{\sigma\sqrt{2\pi}} \, e^{-z^2/2\sigma^2} \,. \tag{6.21}$$

In this case, as shown in Chap. 3, Eq. (3.115), we obtain for a single surface (e.g. a substrate) the *amplitude ratio* of *dynamic reflection* [16, 17]

$$r_{\text{dyn}}^{\text{coh}} = r_{\text{dyn}}^{\text{flat}} \, e^{-2k_{z,0}k_{z,1}\sigma^2} \tag{6.22}$$

with the amplitude ratio of the flat substrate being the dynamical Fresnel reflection coefficient of the substrate surface $r_{\text{dyn}}^{\text{flat}} = (k_{z,0}-k_{z,1})/(k_{z,0}+k_{z,1})$, see Eq. (3.80).

The ratio of *kinematical reflection* coefficients is (Eq. (3.116))

$$r_{\text{kin}}^{\text{coh}} = r_{\text{kin}}^{\text{flat}} \, e^{-2k_{z,0}^2\sigma^2} \tag{6.23}$$

with the kinematical Fresnel reflection coefficient of the surface $r_{\text{kin}}^{\text{flat}} = q_c^2/4q_z^2$, Eq. (3.103).

Both the kinematical and the dynamical Fresnel reflection coefficients are multiplied with a diminution factor containing the r.m.s. roughness $\sigma$ in the exponent. The kinematical diminution factor decreases with the square of the scattering vector $q_z$, which is proportional to the angle of incidence. Its form resembles the static Debye–Waller factor. The dynamical diminution factor contains the product of the scattering vector in vacuum $q_{z,0}$ and that in the medium $q_{z,1}$. The angular dependence of the diminution factors in the dynamical and the kinematical theory differs substantially for small angles near the critical angle of total external reflection $\theta_c$, see Fig. 6.6. Neglecting absorption, the scattering vector $q_{z,1}$ becomes purely imaginary below $\theta_c$. Consequently there is no influence of roughness on the reflectivity in this angular range within the dynamical description. At large incident angles both diminution factors coincide.

A more detailed discussion of both formulae (6.22) and (6.23) is given in [13]. There the contribution of the incoherent scattering to the specular direction has been studied by means of second-order DWBA, showing its dependence on the lateral correlation length $\Lambda$. Concluding therefrom, the specularly reflected intensity can be described by the "dynamical" equation (6.22) for short $\Lambda$ below 1 µm. For larger $\Lambda$ the kinematical formula (6.23) with $r_{\text{kin}}^{\text{flat}}$ becomes more appropriate.

Surface roughness of numberless samples of amorphous, polycrystalline and monocrystalline material systems has been studied by SXR. In Fig. 6.7 we plotted one experimental example, the reflectivity of a rough GaAs substrate.

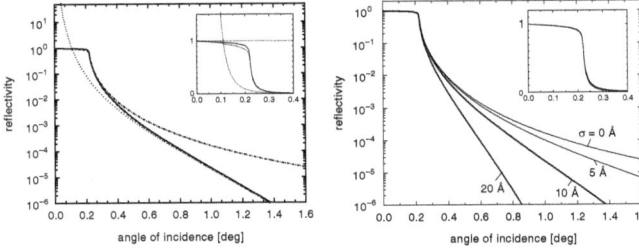

**Fig. 6.6** The coherent reflectivity of a rough Si surface. In the *left* panel the reflectivity of a flat surface (*dashed*) is compared with that for the roughness $\sigma = 1$ nm, calculated by the "dynamical" theory (6.22) (*full*) and the kinematical theory (6.23) (*dotted*). The kinematical reflectivity diverges at grazing incidence. The "dynamical" curve coincides nearly with that of the flat surface below the critical angle $\theta_c$. In the subfigure, the *dashed* line represents the coherent reflectivity of a rough surface calculated with *dynamical* Fresnel reflection coefficient and *kinematical* diminution factor. Thus the reflectivity decreases also below $\theta_c$. In the *right* figure, influence of different roughness, calculated by dynamical formulae, is demonstrated. Close to $\theta_c$ (see subfigure), no essential change is observed

**Fig. 6.7** Measured (*points*) and calculated (*line*) reflectivity curves of a GaAs substrate, $\sigma = 12$ Å [18]. In the inset the mean coverage of the surface is plotted

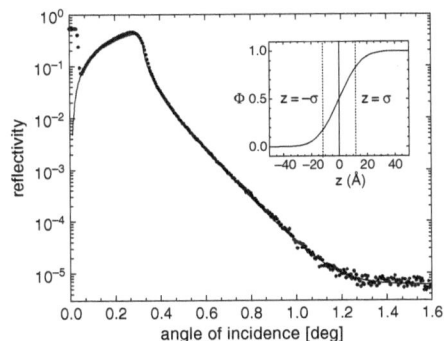

### 6.4.1.2 Multilayer

Conventional SXR simulation and fit programs are today based on a multilayer model with *independent r.m.s. roughness profiles* of each interface supposing a Gaussian probability density. This leads to effective Fresnel reflection and transmission coefficients (Eq. 3.115):

$$r_{j,j+1} = r_{j,j+1}^{\text{flat}} e^{-2k_{z,j}k_{z,j+1}\sigma_{j+1}^2} \quad \text{and} \quad t_{j,j+1} = t_{j,j+1}^{\text{flat}} e^{(k_{z,j}-k_{z,j+1})^2\sigma_{j+1}^2/2} \quad (6.24)$$

for each interface. The influence on the transmission function is rather small according to the small difference in the vertical scattering vector components of the layers. However, the interface reflection is exponentially diminished by roughness, creating a strong change in the interference pattern. The effect of interface roughness versus surface roughness is shown in Fig. 6.8. The surface roughness mainly decreases the specular intensity of the whole curve progressively with $q_z$, where

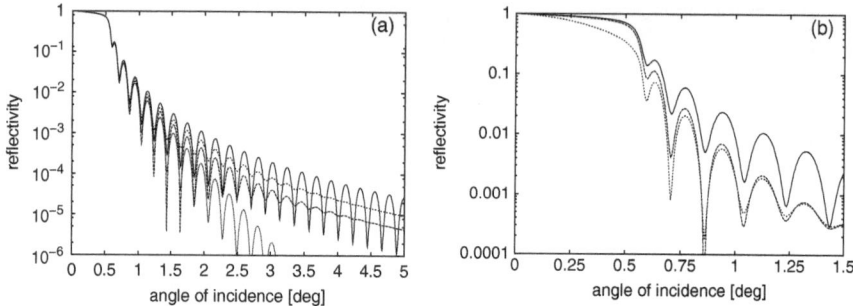

**Fig. 6.8** Calculation of the specular reflectivity of a single layer (20 nm tungsten) on a substrate (sapphire) for different r.m.s. roughness and diminution factors. (**a**) Dynamical diminution factor. From the *upper* to the *lower* curve: without roughness, interface roughness 0.5 nm, surface roughness 0.5 nm, both surface and interface roughnesses 0.5 nm. Surface roughness yields a faster decay of the reflectivity, while interface roughness attenuates the peaks. (**b**) Different diminution factors. Surface roughness 1.2 nm and interface roughness 0.3 nm calculated for the kinematical "slow" roughness (*lower curve*), dynamical "rapid" roughness (*middle curve*) and without roughness (*upper curve*)

the interface roughness gives rise to a progressive dampening of the interference fringes (thickness oscillations). However, locally the variation in the Fresnel coefficients can cause more pronounced oscillations, too. In Fig. 6.9 we plotted the experimental and simulated curves of a magnetic rare earth/transition metal multilayer (Cr/TbFe$_2$/W on sapphire Al$_2$O$_3$), grown by laser ablation deposition. It shows a quite complicated non-regular interference pattern. A good agreement with the simulation was realised by considering a thin oxide film at the sample surface.

**Fig. 6.9** Measurement (*points*) and the fit (*full curve*) of the specular reflectivity of a Cr/TbFe$_2$/W multilayer [19]. We determined the thicknesses (34.6 nm W, 4.8 nm TbFe$_2$, 50.5 nm Cr, 3 nm oxidised Cr) and the roughnesses (0.2 nm above sapphire, 2.0 nm W, 0.9 nm TbFe$_2$, 2.2 nm Cr)

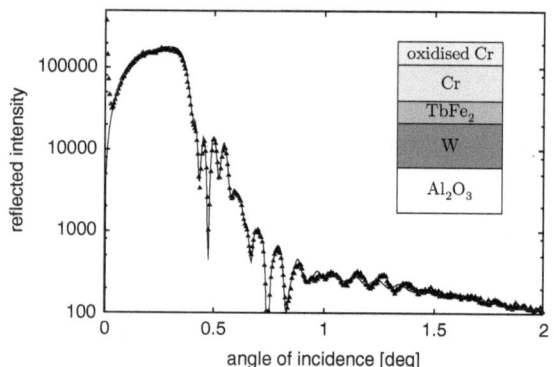

### 6.4.1.3 Periodic Multilayer

The main feature of the specular scans of a periodic multilayer is the multilayer Bragg peaks, giving evidence for the vertical periodicity, see Fig. 6.10 and Sect. 6.A.2.

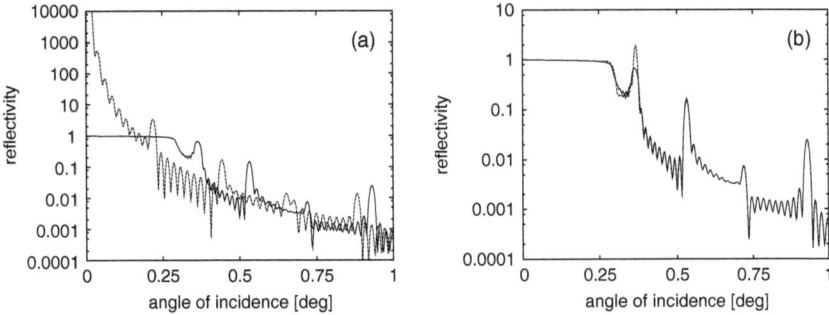

**Fig. 6.10** Specular reflection by an "ideal" periodic multilayer—calculated curves for a [GaAs (13 nm)/AlAs (7 nm)] superlattice with 10 periods on a GaAs substrate, flat interfaces (no roughness). (**a**) Comparison of the dynamical theory (*full curve*) with the kinematical theory. The kinematical multilayer Bragg peaks correspond to the positions of the satellites of the (000) RLP. The curve diverges at low incident angles. The dynamical calculation shows the plateau of total external reflection below the critical angle. Due to refraction the multilayer Bragg peaks are shifted to larger angles. The first multilayer Bragg peak broadening is caused by multiple reflection (extinction effect). (**b**) Comparison of the dynamical theory with the semi-dynamical approximation (single-reflection approximation [18]). The satellite positions of all Bragg peaks coincide, also the shape and intensities except for the intense Bragg peaks

The intensity ratio of the Bragg peaks depends on the layer set-up within the multilayer period. The difference in the electron density determines the Fresnel coefficients, and the thickness ratio of the layers characterises the phase relations of the reflected waves of different interfaces. The laterally averaged gradual interface profile caused by interdiffusion or interface roughness leads to a damping mainly of the multilayer Bragg peaks progressively with $q_z$, whereas the roughness of the sample surface reduces the intensity of the whole curve. This is demonstrated in Fig. 6.11.

In Fig. 6.12 we plotted the measured SXR curves of an epitaxial CdTe/CdMnTe superlattice on a CdZnTe substrate. Due to the low contrast of the electron density of

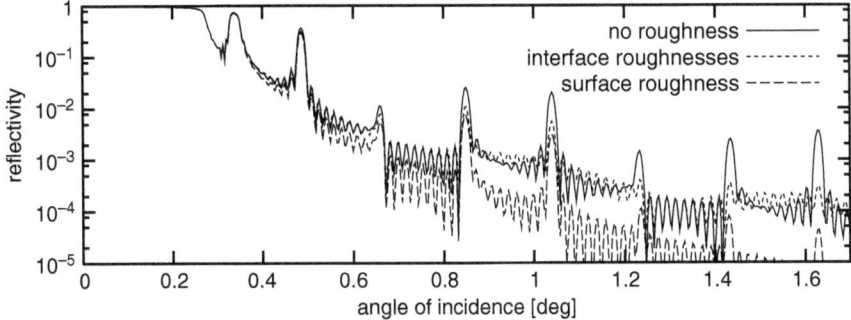

**Fig. 6.11** Simulation of coherent reflectivity of a [GaAs (7 nm)/AlAs (15 nm)]10× periodic multilayer with no roughness (*full curve*) or 1 nm roughness of surface (*dashed lower curve*) or of all interfaces (*dotted*)

**Fig. 6.12** Measured and calculated specular reflectivity of a [CdTe (14.2 nm)/CdMnTe (2.5 nm)]20× superlattice on CdZnTe [20]. In the subfigure, the roughness is represented by an effective MnTe concentration depth profile

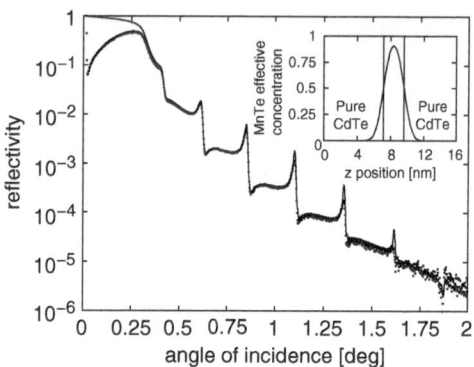

both layer materials the first-order Bragg peak appears only as a very weak hump on the slope of the surface. The other Bragg peaks have a shape similar to a resonance line. From the best fit we obtain the mean compositional profile.

### 6.4.1.4 Increasing and Decreasing Roughness in Multilayers

The influence of roughness increasing or decreasing during the growth from the substrate towards the surface can be described by the use of the roughness replication model introduced in Sect. 6.2.

We start the layer growth from a substrate with a Gaussian surface roughness profile,

$$C_{NN}(r_\| - r_\|') = \sigma_N^2 \, e^{-\left(\frac{|r_\| - r_\|'|}{\Lambda_N}\right)^2} . \tag{6.25}$$

For the non-random replication function in (6.10) we choose for all layers a Gaussian function

$$a(r_\| - r_\|') = \frac{1}{2\pi L^2} \, e^{-\frac{|r_\| - r_\|'|^2}{2L^2}} . \tag{6.26}$$

The factor $L$ determines the loss of memory from interface to interface. This choice arises from the aim to explain the different limiting cases of roughness replication models by one class of functions. It is not supported by any physical reason. However, the model allowed to describe measured curves of SXR and NSXR showing good agreement [8, 15].

We assume the intrinsic correlation function (6.13) of all interfaces

$$K(r_\| - r_\|') = (\Delta\sigma)^2 e^{-\left(\frac{|r_\| - r_\|'|}{\Delta\Lambda}\right)^2} . \tag{6.27}$$

Now we continue as in Sect. 6.2. The Fourier transform of the *in-plane correlation function* is under these assumptions

$$\tilde{C}_{jj}(q) = \frac{1}{2}(\sigma_N \Lambda_N)^2 e^{-\frac{(q\Lambda'_j)^2}{2}} + \frac{1}{2}(\Delta\sigma\Delta\Lambda)^2 \sum_{k=j}^{N-1} e^{-\frac{(q\Lambda'_k)^2}{4}}, \qquad (6.28)$$

where we have denoted

$$\Lambda'_j = \sqrt{\Lambda_N^2 + 4L^2(N-j)}. \qquad (6.29)$$

The *inter-plane correlation* is then simply given by

$$\tilde{C}_{j\geq k}(q) = \tilde{C}_{jj}(q) e^{-\frac{(qL)^2}{2}(j-k)}. \qquad (6.30)$$

We obtain the mean square roughness of the $j$th interface

$$\sigma_j^2 = \int dq\, \tilde{C}_{jj}(q) = \sigma_N^2 \frac{\Lambda_N^2}{\Lambda_j'^2} + (\Delta\sigma\Delta\Lambda)^2 \sum_{k=j}^{N-1} \frac{1}{\Lambda_k'^2}. \qquad (6.31)$$

Let us see what does it give for some limiting cases of the model:

1. *Identical interface roughness* is achieved with maximum replication and no intrinsic roughness: $L = 0$ and $\Delta\sigma = 0$. Consequently $\sigma_j = \sigma_N$, and all interfaces reproduce the profile of the substrate surface, $z_j(x) = z_N(x)$.
2. *Increasing roughness* towards the free surface is obtained by maximum replication and a non-zero intrinsic roughness ($L = 0$ and $\Delta\sigma > 0$). From (6.28) and (6.31) we find

$$\sigma_j^2 = \sigma_N^2 + (\Delta\sigma)^2 (N-j), \qquad (6.32)$$

describing the *roughening* during the growth.
3. *Partial replication and no intrinsic roughness* ($L > 0$ and $\Delta\sigma = 0$) leads to decreasing r.m.s. roughness towards the free surface (*smoothing* of the multilayer during growth), described by

$$\sigma_j^2 = \frac{\sigma_N^2}{1 + 4\left(\frac{L}{\Lambda_N}\right)^2 (N-j)}. \qquad (6.33)$$

4. *No replication* occurs for diverging $L$, where $a(r_\parallel - r_\parallel')$ goes to zero. The roughness profile of each interface is independent.

We compare here the experimental example of two *periodic Si/Nb multilayers*, grown by magnetosputtering for superconductivity studies. The multilayer is deposited on a Si substrate with a thick $SiO_2$ layer and an Al buffer layer. The roughness of the buffer layer depends on its thickness and influences the quality of the interfaces. Two samples of different Al thickness have been investigated and the results are shown in Fig. 6.13. The multilayer periodicity generates the multilayer Bragg peaks or reflection satellites, which are dampened by interface roughness. The roughness of the substrate and the buffer layers has less influence on the

**Fig. 6.13** Measurement
(*points*) and simulation (*full
curve*) of the specular
reflectivity of a periodic
Nb/Si multilayer of 10
periods [19]. (a) Sample A,
fitted by the model of
constant roughness,
(b) sample B, fitted by the
model of increasing
roughness

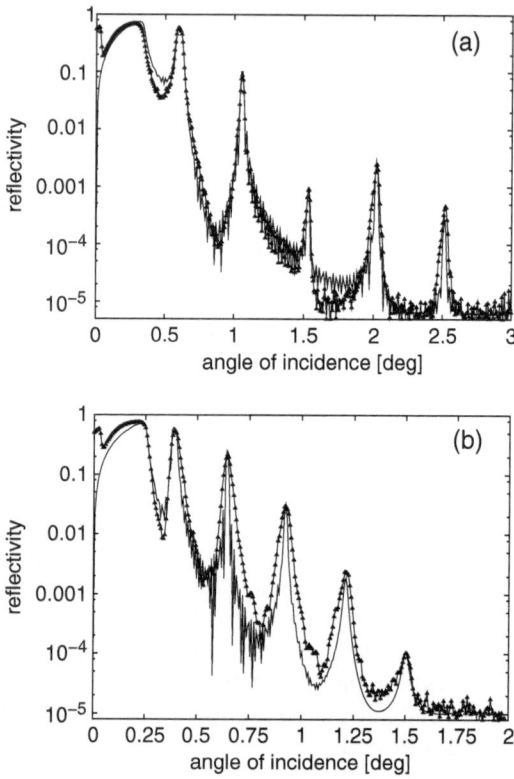

reflection pattern. Sample A can be fitted by a roughness model of constant r.m.s. roughness for all interfaces. The peak widths of the first intense Bragg peak is broadened by extinction due to dynamical multiple scattering. For all higher order Bragg peaks we observe a narrower (kinematical) peak width. The satellite reflections of sample B are also rapidly damped, indicating a large interface roughness. Besides the widths of the peaks increases with $q_z$. That cannot be explained by model 1. The satellite intensities and shape can be successfully reproduced by supposing increasing roughness according to (6.32). Due to their increased roughness, the upper layers near the surface contribute with decreasing effective Fresnel coefficients to the reflected wave. Within the Bragg position the contributions of all interfaces are still in phase; however, slightly away from the Bragg condition the contribution of interfaces near the substrate and those near the sample surface do not cancel completely, giving rise to the peak broadening.

## 6.4.2 Stepped Surfaces

The surface morphology of monocrystalline samples can also be described by a *discrete* surface probability distribution following the concept of terraces or small

**Fig. 6.14** Multilayer with
random two-level islands

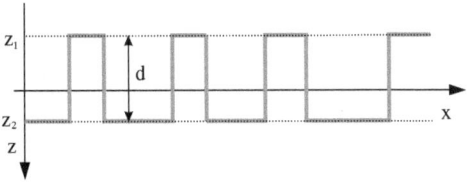

separated islands. In the simplest case, the *two-level surface* consists of randomly
placed *islands* of uniform height $d$, so that the displacement $z(r_{\parallel})$ has two possible
values $z_1$ and $z_2 = d + z_1$ with the corresponding probabilities $p_1$ and $p_2 = 1 - p_1$,
see Fig. 6.14 [21]. The surface probability distribution function $p(z)$ for this case
writes

$$p(z) = p_1\delta(z_1) + p_2\delta(z_2) . \tag{6.34}$$

Since $\langle z(r_{\parallel})\rangle = 0$, then $Z_1 = -p_2 d$ and $Z_2 = p_1 d$. The mean square roughness is

$$\sigma^2 = p_1 Z_1^2 + p_2 Z_2^2 = p_1 p_2 d^2 \tag{6.35}$$

and the characteristic function (2.10) is

$$\chi(q_z) = e^{-iq_z d p_2}\left(p_1 + p_2 e^{iq_z d}\right) . \tag{6.36}$$

Putting this in the formulae for the reflected amplitude ratio of rough surfaces,
we get the *amplitude ratio* of *kinematical specular reflection*

$$r_{\text{kin}}^{\text{coh}} = e^{-iq_z d p_2}\left(p_1 r_{0,1} + p_2 r_{1,2} e^{iq_z d}\right) . \tag{6.37}$$

A surface region perturbed in this way acts as a thin, homogeneous layer forming an upper and a lower interface with the Fresnel reflection coefficients $p_1 r_{0,1}$
and $p_2 r_{1,2}$. They give rise to interference fringes which represent the height $d$
(Fig. 6.15).

The example of a thin surface layer of porous silicon fits approximately this simple model, if its thickness is smaller than the vertical correlation lengths of the crys-
tallites (Fig. 6.16(a)) [22]. Since the surface "layer" density is quite different from
that of the substrate, we can observe two critical angles $\theta_1$ and $\theta_2$. The second one,
$\theta_2$, corresponds to silicon, the first one, $\theta_1$, to the averaged surface region. Above
$\theta_1$ the wave can penetrate into the perturbed surface region; however, total external reflection occurs at the "interface" with the non-perturbed region. That is why
very intense fringes appear in this region between $\theta_1$ and $\theta_2$, which drop rapidly
above $\theta_2$. The whole curve is similar to that of a homogeneous layer of much less
density or to that of a surface grating. In the fitted curves a small Gaussian deviation of the actual displacement around the $z_1$ and $z_2$ has been supposed, which
leads to roughness diminution factors of the Fresnel reflection coefficients similar to
(6.24).

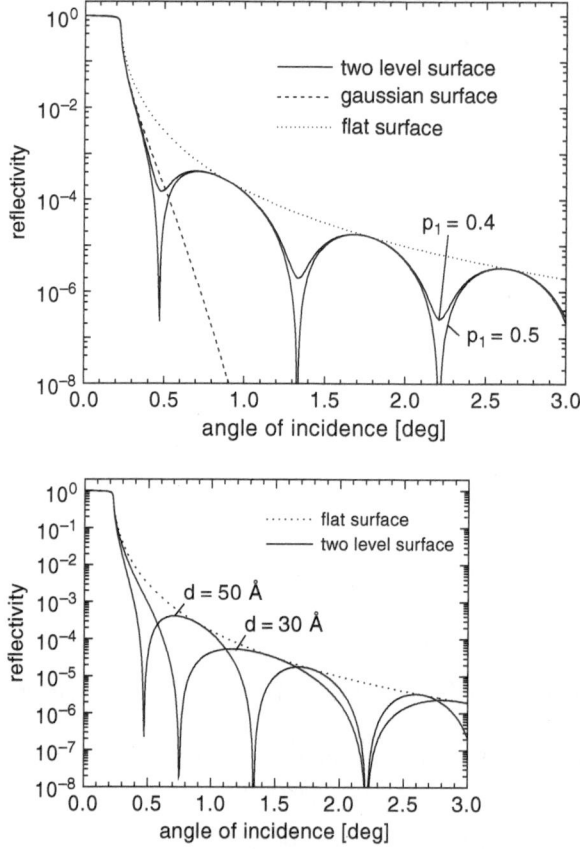

**Fig. 6.15** Coherent reflectivity of a two-level surface calculated within the kinematical theory for two values of the probability $p_1$ and the step height $d = 5$ nm (*left*) and for two values of $d$ and a symmetrical probability distribution $p_1 = p_2 = 0.5$ (*right*)

### 6.4.3 Reflection by "Virtual Interfaces" Between Porous Layers

Porous silicon layers are fabricated by electrochemical etching in a mono-crystalline silicon wafer. By a variation of the anode voltage, multilayers of modulated porosity can be produced. Following our division of the layer polarisability we can distinguish between the porous layer volume and the size of the layer of equal porosity. The interface between two layers of different porosities is not a microscopic laterally continuous and sharp interface between two media of different densities, but an interface of two degrees of porosity. According to the coherent approach (used also in Sect. 3.4) we take for the coherent reflection an effective averaged refractive index into account. Layers of statistically homogeneous porosity are assumed. We treat the slow "roughness" of the transition between two layers of different porosities by

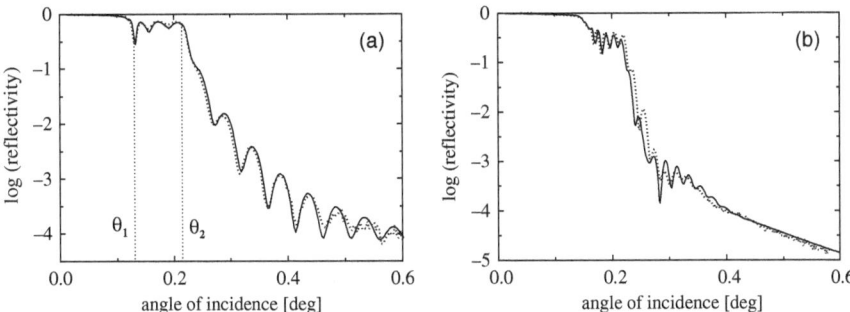

**Fig. 6.16** Measured (*full*) and fitted (*dashed*) reflectivity curves of a thin porous silicon surface layer (**a**) and of a porous silicon double layer (**b**) on silicon substrate [22]. Positions $\theta_1, \theta_2$ are the critical angles of the porous layer and the substrate, respectively

a Gaussian probability function. Same results are obtained by introducing a graduated transition of porosity from layer to layer. An experimental example is given in Fig. 6.16(b) for a double-layer sample [22]. The thickness of the surface layer is much smaller than that of the buried layer. The fast oscillating fringes represent the total thickness. The fringe amplitude is modulated by a period, which corresponds approximately to the thickness of the surface layer. It has been found from the simulation that the interface between the two layers of different porosities is much sharper than the interface with the substrate (which is the end front of the etching process).

The occurrence of the modulation of thickness oscillations in Fig. 6.16(b) is a direct proof for the validity of the coherent scattering approach. Between the two porous layers there is nowhere a real roughly smooth lateral interface between two media. Nevertheless the x-rays are specularly reflected at this "microscopically non-existent interface" showing all features of the continuum theory of dynamical reflection by multilayers.

## 6.5 Non-Specular X-Ray Reflection

In this section we use the incoherent scattering approach (2) within the DWBA (Chap. 4) and derive some explicit expressions for the incoherent scattering cross section for x-ray reflection by rough multilayers. We discuss the main features of the scattering patterns illustrated by experimental examples. The representation of the scattering in reciprocal space allows a simple interpretation of the findings by the various scattering processes. We will treat samples with interfaces having a Gaussian roughness profile, diffuse scattering from terraced interfaces and finally non-coplanar diffuse scattering.

## 6.5.1 Interfaces with a Gaussian Roughness Profile

We will deal with interfaces having a Gaussian roughness profile. We start with the
scattering from a single surface. Then we continue with a multilayer showing the
effects of different roughness replication as well as dynamical scattering effects on
reciprocal space maps.

### 6.5.1.1 Single Surface

First we will deal with *surfaces* of a Gaussian probability distribution. The *pair
probability distribution function* is in the stationary case (see [1, 23–25], for instance)

$$p_2(z,z') = \frac{1}{2\pi\sqrt{\sigma^4 - C_{zz}^2(r_\parallel - r_\parallel')}} \exp\left\{ -\frac{z^2 + z'^2 - \frac{2zz'}{\sigma^2}C_{zz}(r_\parallel - r_\parallel')}{2\sigma^2[1 - \frac{1}{\sigma^4}C_{zz}^2(r_\parallel - r_\parallel')]} \right\} \tag{6.38}$$

with the two-dimensional characteristic function

$$\chi_{zz'}(q,q') = \langle e^{i(qz - q'z')} \rangle = e^{-\sigma^2(q^2+q'^2)/2} e^{qq'C_{zz}(r_\parallel - r_\parallel')}. \tag{6.39}$$

One correlation function, which has been successfully applied to interpret the experimental findings, follows from similarities between the description of interfaces
with fractal roughness properties and the Brownian motion, if we replace the lateral
position by time. Supposing a behaviour like [1, 24]

$$\left\langle [z(r_\parallel) - z(r_\parallel')]^2 \right\rangle = A|r_\parallel - r_\parallel'|^{2h}, \qquad 0 < h \le 1, \tag{6.40}$$

leads together with

$$\left\langle [z(r_\parallel) - z(r_\parallel')]^2 \right\rangle = 2\sigma^2 - 2C_{zz}(r_\parallel - r_\parallel') \tag{6.41}$$

to a correlation function, which only depends on the distance $|r_\parallel - r_\parallel'|$. The so-
called *Hurst factor h* describes the jagged shape of the interface, determining the
*fractal dimension D* of the interface, $D = 3 - h$. For $h = 1$ the fractal dimension is
2 and corresponds to the topological dimension of an interface (without a fractal
structure). This function diverges for large distance $|r_\parallel - r_\parallel'|$. Thus it is suitable to
introduce a *cut-off radius* $\xi$. Below $\xi$ the correlation function shall approximately
behave like (6.40), but above it should converge to zero. A function with such a
behaviour is

$$C_{zz}(r_\parallel, r_\parallel') = \langle z(r_\parallel) \cdot z(r_\parallel') \rangle = \sigma^2 e^{-(|r_\parallel - r_\parallel'|/\xi)^{2h}}. \tag{6.42}$$

The cut-off radius represents the *lateral correlation length of the interface*. Let
us now determine the incoherent cross section for a surface with such properties.

Using Eq. (4.47) we find for the incoherent scattering cross section of a single rough surface within the *full DWBA*

$$d\sigma_{\text{incoh}} = d\Omega \, \frac{k_0^4}{16\pi^2} \, \left|t_{0,1}^{\text{in}}\right|^2 \left|t_{0,1}^{\text{sc}}\right|^2 \tilde{Q}_1 \tag{6.43}$$

with the covariance function (4.D28)

$$\tilde{Q}_1 = A\left|n_1^2 - n_0^2\right|^2 \frac{e^{-\frac{1}{2}\sigma^2(q_{z,1}^2 + q_{z,1}^{*2})}}{|q_{z,1}|^2} \tag{6.44}$$
$$\times \int_A d(r_\| - r_\|') \, e^{iq_\|(r_\| - r_\|')} \left[ e^{|q_{1z}|^2 C_{zz}(r_\| - r_\|')} - 1 \right],$$

where $A$ is the area of integration, which means the illuminated surface of the sample. The result can be interpreted as follows: the incident wave transmits through the surface considered by the Fresnel transmission coefficients. This "distorted wave" is diffusely scattered by the surface disturbance. Thus the non-specularly reflected intensity depends on the r.m.s. roughness and is proportional to the Fourier transform of

$$\left[ e^{|q_{1z}|^2 C_{zz}(r_\| - r_\|')} - 1 \right].$$

Taking the correlation function (6.42), we have $C_{zz}(r_\| - r_\|') < \sigma^2$. For small roughness or small $q_z$ fulfilling $(\sigma q_z)^2 \ll 1$, we can approximate (6.42) by the first two terms of its Taylor series and obtain finally

$$d\sigma_{\text{incoh}} = d\Omega \, \frac{k_0^4 A |n_1^2 - n_0^2|^2}{16\pi^2} \, |t_{01}^{\text{in}}|^2 |t_{01}^{\text{sc}}|^2 \, e^{-\frac{1}{2}\sigma^2\left(q_{z,1}^2 + q_{z,1}^{*2}\right)} \tilde{C}(q_\|), \tag{6.45}$$

i.e. an expression, which is proportional to the Fourier transform of the correlation function.

The according *kinematical* expressions are found by setting the transmission coefficients equal to 1 and substituting the scattering vectors in the medium $q_{z,1}$ by the scattering vectors in vacuum, $q_z$.

### 6.5.1.2 Multilayer with No Vertical Roughness Replication

In the case of *independent roughness profiles* of all different interfaces we have the replication function $a_m(r_\|) = 0$ ($L \to \infty$ in (6.26)). There is no inter-plane correlation, that is why only the in-plane correlation functions have to be considered. We can proceed for each interface like in the case of a single surface described above. However, now we take four scattering processes (corresponding to downwards and upwards propagating incident and scattered waves), see (4.D27), into account instead of one in (6.43). Consequently, we consider $4 \times 4$ covariance functions for each interface. The incoherent scattering cross section adds up the contribution of all *single interfaces*

$$\left(\frac{d\sigma}{d\Omega}\right)_{\text{incoh}} = \frac{k_0^4}{16\pi^2} \sum_{j=0}^{N} \sum_{\pm} \sum_{\pm} \sum_{\pm} \sum_{\pm} \tag{6.46}$$

$$U_j(\pm k_{\text{in}z,j}, Z_j) U_j(\pm k_{\text{sc}z,j}, Z_j) U_j(\pm k_{\text{in}z,j}^*, Z_j) U_j(\pm k_{\text{sc}z,j}^*, Z_j)$$

$$\tilde{Q}_{jj}(\pm k_{\text{in}z,j} \pm k_{\text{sc}z,j}, \pm k_{\text{in}z,j} \pm k_{\text{sc}z,j})$$

with

$$\tilde{Q}_{jj}(q_z, q_z') = \frac{A |\chi_{0\,j+1} - \chi_{0\,j}|^2}{q_z(q_z')^*} e^{-\frac{1}{2}\sigma_j^2 \left[q_z^2 + (q_z')^{*2}\right]} \tag{6.47}$$

$$\times \int_A d(r_\| - r_\|') e^{iq_\|(r_\| - r_\|')} \left(e^{q_z(q_z')^* C_{jj}(r_\| - r_\|')} - 1\right),$$

where we have used the polarisabilities $\chi_{0\,j+1} - \chi_{0,j} = n_{j+1}^2 - n_j^2$ instead of the optical indices. Assuming the same in-plane correlation functions for all interfaces the $\tilde{Q}_{jj}$ of different interfaces differ only by the scattering vectors and the differences of polarisability.

Figure 6.17 shows a measurement and fit of an $\omega$-scan from a single-layer sample.

**Fig. 6.17** Measurement (*points*) and fit (*full line*, shifted down 2×) of an $\omega$-scan at $2\Theta = 2.63°$ single layer W (11.1 nm) on Si substrate

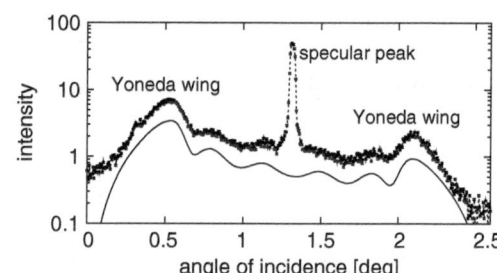

### 6.5.1.3 Multilayer with Partial Vertical Roughness Replication

In the case of partial vertical roughness replication also the covariance functions of scattering at *different interfaces* have to be included. We get (cf. (4.55), (4.56) and (4.D28), (4.D29))

$$\left(\frac{d\sigma}{d\Omega}\right)_{\text{incoh}} = \frac{k_0^4}{16\pi^2} \sum_{j=0}^{N} \sum_{k=0}^{N} \sum_{\pm} \sum_{\pm} \sum_{\pm} \sum_{\pm} \tag{6.48}$$

$$U_j(\pm k_{\text{in}z,j}, Z_j) U_j(\pm k_{\text{sc}z,j}, Z_j) U_k(\pm k_{\text{in}z,k}^*, Z_k) U_k(\pm k_{\text{sc}z,k}^*, Z_k)$$

$$\tilde{Q}_{jk}(\pm k_{\text{in}z,j} \pm k_{\text{sc}z,j}, \pm k_{\text{in}z,k} \pm k_{\text{sc}z,k})$$

with the covariance function (see Fig. 6.18)

**Fig. 6.18** Illustration considering the covariance function $\tilde{Q}_{jk}(q_z, q_z')$ of one scattering process $q_z$ at the interface $j$ and another scattering process $q_z'$ at the interface $k$

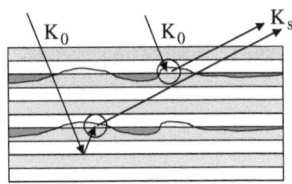

$$\tilde{Q}_{jk}(q_z, q_z') = \frac{A\,(\chi_{0\,j+1} - \chi_{0\,j})(\chi_{0\,k+1} - \chi_{0\,k})^*}{q_z(q_z')^*}\, e^{-\frac{1}{2}[\sigma_j^2 q_z^2 + \sigma_k^2 (q_z')^{*2}]}$$

$$\times \int_A d(r_\parallel - r_\parallel')\, e^{iq_\parallel(r_\parallel - r_\parallel')} \left( e^{q_z(q_z')^* C_{jk}(r_\parallel - r_\parallel')} - 1 \right). \qquad (6.49)$$

Here $\sigma_j$ and $\sigma_k$ are the r.m.s. roughnesses of the corresponding interfaces determined by (6.31), $C_{jk}$ are their inter-plane correlation functions. Restricting ourselves on small roughness $(\sigma q_z)^2 \ll 1$, we can make approximations similar to (6.45) using the Fourier transform of the correlation functions $\tilde{C}_{jk}(Q_\parallel)$ obtained in Sect. 6.4, Eqs. (6.28) and (6.30).

The treatment of the corresponding expressions of the *simpler DWBA* for multilayers (p. 180) is straightforward. It neglects the influence of specular interface reflection on the diffuse scattering. Only the primary scattering processes are taken into account.

## 6.5.2 The Main Scattering Features of Non-Specular Reflection by Rough Multilayers

Let us give an overview of the main features in the non-specular reflected intensities and discuss their physical origin. The diffuse x-ray scattering (DXS) pattern is characterised by the *transmitted/reflected wave amplitudes* $U_j(\pm k_z)$ of the incident and final wave fields in the layers and by the 16 *covariances of the scattering processes*, $\tilde{Q}_{jk}(q_z, q_z')$ for each pair of interfaces $j, k$. We want to study the features of the DXS pattern under the aspect whether they are particularities of scattering by the *roughness profiles*, caused by the correlation properties, or of the *excited non-perturbed wave amplitudes*. In other words, we want to distinguish between effects of the random disturbance potential and the non-perturbed potential. The latter effects do not depend on the statistical roughness properties, we call them *dynamical scattering effects*.

### 6.5.2.1 Resonant Diffuse Scattering

First we investigate the influence of the interface roughness correlation. One essential characteristic caused by the inter-plane correlation is the so-called *resonant diffuse scattering* (RDS). We simplify the discussion of this phenomenon by introducing a simpler model of vertical roughness correlation [26], where the inter-plane correlation function $C_{jk}$ depends on the in-plane correlation function $C_{ll}$, $l = \max(j, k)$ of the lower interface, by

$$C_{jk}(r_\parallel - r_\parallel{}') = C_{ll}(r_\parallel - r_\parallel{}')\,e^{-|z_j - z_k|/\Lambda_\perp} . \qquad (6.50)$$

In this phenomenological model the vertical correlation of the roughness profiles is limited by a *vertical correlation length* $\Lambda_\perp$. The model does not explain the effects of smoothening and roughening studied in Sect. 6.4, since it neglects the interdependence of the r.m.s. roughness and the lateral correlation length (6.30). However, it makes the calculation and the discussion simpler. In Fig. 6.19 we see some calculated reciprocal space maps of the diffusely scattered intensity for a GaAs/AlAs

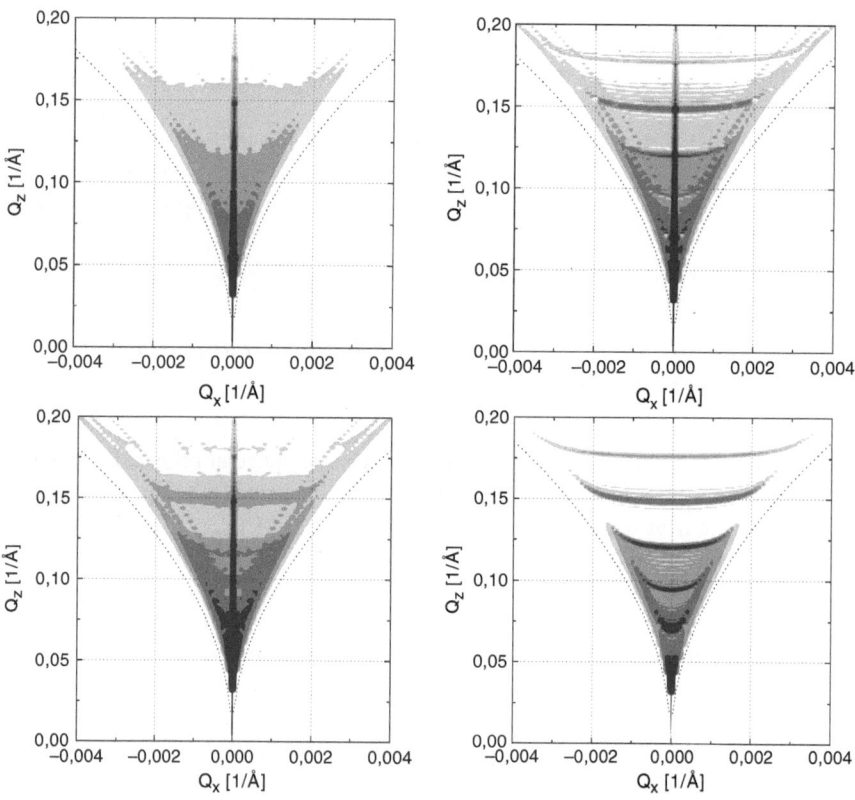

**Fig. 6.19** Reciprocal space maps of the diffusely scattered intensity calculated for a [GaAs (7 nm)/AlAs (15 nm)]10× multilayer using the DWBA method and the simpler replication model (6.50) [18]. All the interfaces have the same r.m.s. roughness 1 nm, the correlation lengths 50 nm and different vertical correlation lengths $\Lambda_\perp$. *Upper left panel*: no replication, $\Lambda_\perp = 0$. *Upper right panel*: full replication, $\Lambda_\perp = \infty$. *Bottom left panel*: $\Lambda_\perp = 100$ nm. *Bottom right panel*: full replication, $\Lambda_\perp = \infty$, calculated by the simpler DWBA. The full lines represent the arcs of the Ewald spheres for the limiting cases of $\theta_{\mathrm{in}} = 0$ and $\theta_{\mathrm{sc}} = 0$. The RDS disappear, if the roughness profiles are not replicated (*upper left panel*). Bragg-like resonance lines are visible in all maps calculated by the full DWBA. They are not reproduced by the simpler DWBA (*bottom right panel*)

superlattice assuming this vertical replication model. All the interfaces have the same r.m.s. roughness $\sigma = 1\,\text{nm}$, and the *lateral* correlation length $\Lambda = 50\,\text{nm}$. It shows the cases of *no replication, partial replication* and *full replication*. In the first case all interfaces scatter independently, the diffuse intensities of all individual interfaces superpose. The other two cases give rise to scattering with partial coherence, the resonant diffuse scattering. It occurs due to the vertical replication of the roughness profiles of different interfaces. The partial phase coherence of the waves diffusely scattered from different interfaces leads to a concentration of the scattered intensity in narrow sheets. These sheets of resonant diffuse scattering intersect the specular rod in the multilayer Bragg peaks. Neglecting refraction the sheets would be horizontally oriented with the centre fulfilling the one-dimensional Bragg conditions

$$q_z = k_0 \left( \sin \theta_{\text{in}} + \sin \theta_{\text{sc}} \right) = \frac{2\pi m}{D_{\text{ML}}}, \tag{6.51}$$

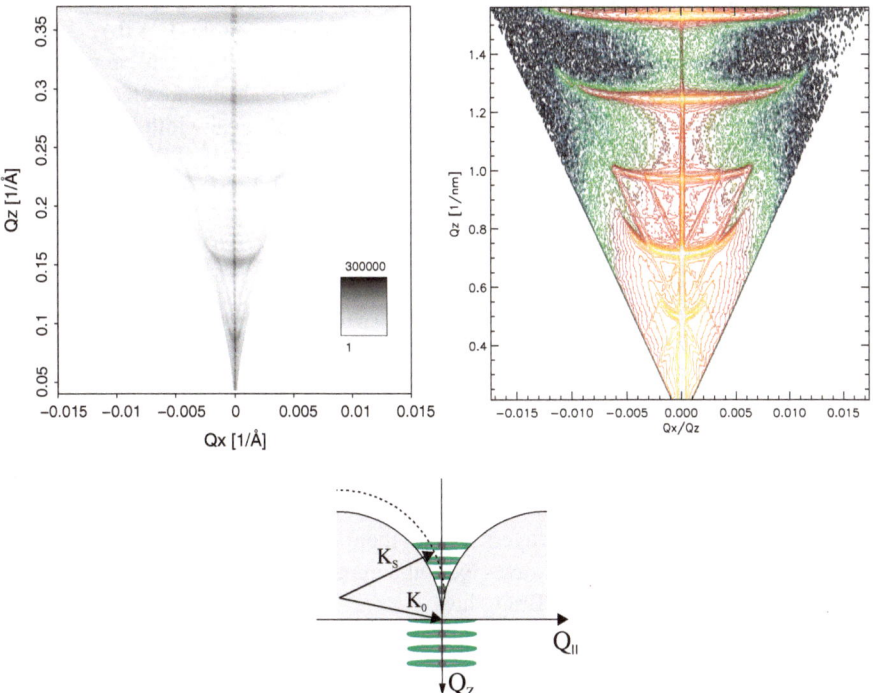

**Fig. 6.20** Measured reciprocal space maps (*top*). *Left map*: periodic multilayer [Si (3.0 nm)/Nb (5.8 nm)] 10× starting from a rough Si substrate of $\sigma = 0.46\,\text{nm}$ and with interface roughness *decreasing* towards the free surface [19]. *Right map*: periodic multilayer with the set-up corresponding to that of Fig. 6.19 with interface roughness *increasing* towards the free surface [8]. *Left schema*: the reciprocal space representation of diffuse scattering by a multilayer with interface roughness replication. The essential features are (1) the multilayer truncation rod through the RLP (000) with the multilayer satellite peaks and (2) horizontal sheets crossing the TR in the satellite positions

schematised in Fig. 6.20, which is the case in a kinematical treatment. Due to the angle-dependent refraction of x-rays the sheets are curved forming "RDS bananas" following the modified Bragg law

$$\langle q_z \rangle_{ML} = k_0 \left( \sqrt{\sin^2 \theta_{in} + \langle \chi_0 \rangle_{ML}} + \sqrt{\sin^2 \theta_{sc} + \langle \chi_0 \rangle_{ML}} \right) = \frac{2\pi m}{D_{ML}}, \qquad (6.52)$$

where $\langle \chi_0 \rangle_{ML} = \Sigma_{j=1}^N \chi_{0j} / D_{ML}$ is the mean polarisability of the multilayer period and $\langle q_z \rangle_{ML} = q_z(\theta_{in}, \theta_{sc}, \langle \chi_0 \rangle_{ML})$ the mean scattering vector in the medium. The length of the RDS bananas in $q_x$ direction is inversely proportional to some effective correlation length $\Lambda_{eff}$ depending on the correlation length $\Lambda_j$ of the interfaces. If all interfaces have the same correlation length, $\Lambda_{eff}$ would equal $\Lambda_j$. The widths of the RDS bananas in $q_z$ direction represent the degree of replication. In the simple model it depends inversely on $\Lambda_\perp$ and for large $\Lambda_\perp$ on the total thickness of the multilayer. The sheets disappear if there is no vertical replication, $\Lambda_\perp = 0$, turning into a broad vertical maximum similar to that for a single surface. The RDS bananas have no dynamical nature, their existence is not related with any kind of multiple scattering. They are also produced by the kinematical theory and the simpler DWBA.

RDS has been experimentally observed at amorphous, polycrystalline as well as epitaxial multilayers as it is shown in Fig. 6.20. The RDS sheets are clearly visible, bent is due to the refraction. Their existence and narrow vertical width give evidence for full roughness replication in both samples.

### 6.5.2.2 Dynamical Scattering Effects

One typical dynamical feature is known from NSXR by rough surfaces. The so-called *Yoneda wings* arise if the incident or the exit angle equals the critical angle, $\theta_{in/sc} = \theta_c$. The wings are generated by the enhancement of the transmitted wave amplitude at the inner sample surface, Fig. 6.17. In the case of a single-layer structure interference fringes can also be created due to the waveguide behaviour of the two interfaces in the layer structure. In general, this behaviour can produce *dynamical fringes* in $\omega$-scans as well as in $2\theta$-scans.

In the case of periodic multilayers we call them *Bragg-like resonance lines*, since the amplitudes of the reflected waves exhibit a maximum if the incident or exit wave fulfils the refraction-corrected Bragg law

$$k_0 \sqrt{\sin^2 \theta_{in/sc} + \langle \chi_0 \rangle} = \frac{\pi m_{in/sc}}{D_{ML}}, \qquad (6.53)$$

where $m_{in}, m_{sc}$ are integers. It is easy to prove that the zero-order Bragg-like resonances are identical with the Yoneda wings. The resonance lines have a particular maximum, the so-called *Bragg-like peak (BL)*, where the incident and exit waves are simultaneously in Bragg condition and the Bragg-like resonances intersect, that is at the positions

$$Q_{z,m_{in}n_{sc}} = \sqrt{(m_{in}2\pi/D)^2 - k_0^2 \langle \chi_0 \rangle} + \sqrt{(m_{sc}2\pi/D)^2 - k_0^2 \langle \chi_0 \rangle},$$

$$Q_{\parallel,m_{in}m_{sc}} = \sqrt{k_0^2 - (m_{in}2\pi/D)^2 + k_0^2 \langle \chi_0 \rangle} + \sqrt{k_0^2 - (m_{sc}2\pi/D)^2 + k_0^2 \langle \chi_0 \rangle}. \tag{6.54}$$

The existence of the Yoneda wings, dynamic fringes and Bragg-like peaks is of completely dynamical origin. They occur independent of the actual interface correlation function. However, their form and intensity are influenced by the interface correlation.

In the case of vertically replicated roughness we see with (6.52), (6.53) and (6.54) that all Bragg-like peaks of an even number $m_{in}+m_{sc}$ are situated on RDS sheets, Fig. 6.21. These Bragg-like peaks are very pronounced with respect to the others. That can be interpreted by the concept of *Umweganregung* (excitation of a reflection by another reflection), well known from x-ray diffraction and outlined in Fig. 6.22.

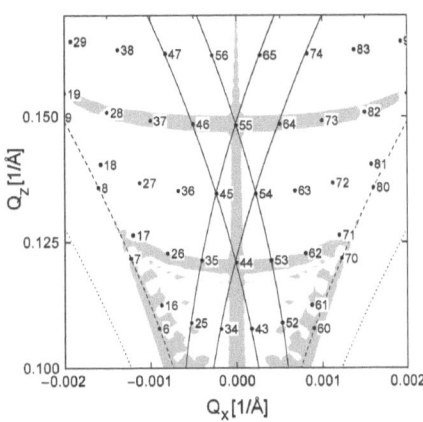

**Fig. 6.21** The schema of the positions of the Bragg-like peaks (*points*) and the RDS bananas (*grey areas*). The numbers denote the orders $m_{in}$ and $m_{sc}$ of the Bragg-like peaks according to (6.54). The *dotted lines* denote the positions of the Yoneda wings. The *full lines* are Bragg-like resonance lines, corresponding to $m_{in/sc} = 4$ and 5

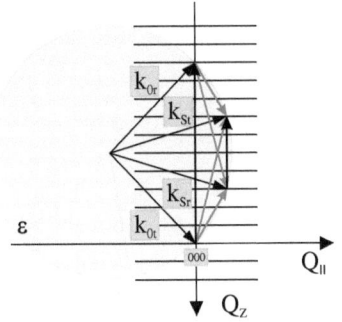 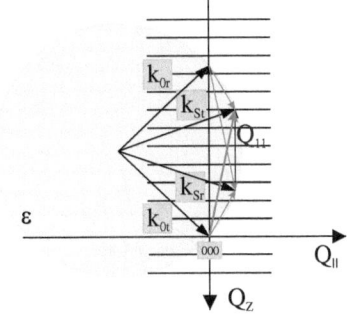

**Fig. 6.22** Generation of Bragg-like peaks on the RDS sheets and interpretation by the concept of Umweganregung. On the *left side*, both the incident and final non-perturbed states fulfil the Bragg condition (6.53). Simultaneously all four diffuse scattering processes are in the situation of resonant diffuse scattering (6.52). On the *right side*, the situation of RDS (6.52) is fulfilled for the primary scattering process. The incident wave is out of Bragg condition, consequently also the final state is out of Bragg condition. Additionally all three secondary diffuse scattering processes are out of resonance

In our experimental map of Fig. 6.20 the Yoneda wings and the Bragg-like resonances are well resolved. Along the RDS sheets we observe intense Bragg-like peaks. All the features are reproduced by the calculation using the full DWBA treatment for multilayers.

Not always it is possible and necessary to measure a full well-resolved map. In general $\omega$-scans at different $q_z$ and offset-scans or $2\theta$-scans are employed. Already one offset-scan or $2\theta$-scan is sufficient to give evidence for vertical replication.

### 6.5.3 Stepped Surfaces and Interfaces

The model of islands of nearly uniform height discussed in Sect. 6.4.2 is the simplest case for a discrete stepped $n$-level surface. An infinite number of levels exist at a *terraced surface*, see Fig. 6.23, which is mostly the case of multilayers grown on slightly miscut substrates [27–29]. The miscut angle $\alpha$ equals the mean ratio of the step height $\langle h \rangle$ and the terrace widths $\langle L \rangle$: $\alpha = \langle h \rangle / \langle L \rangle$. The lateral correlation properties of such a stepped surface are determined by the conditional probability $p(\Delta x, z)$ giving the probability of displacement $z$ for two surface points with the distance $\Delta x$. The two-dimensional characteristic function $\chi_{zz'}$ of such a stair-like surface can be described based on the approach of stationary random processes [21]. Using (6.47) one can calculate the covariance function $\tilde{Q}$ and with (6.46) the differential scattering cross section for the diffuse scattering by the stair-like surface. In [27] the gamma distribution of order $M$ has been supposed for the distribution of the terrace widths $L$

$$p(L) = \frac{1}{\Gamma(M)} \left( \frac{M}{\langle L \rangle} \right)^M e^{-\frac{ML}{\langle L \rangle}} L^{M-1}, \tag{6.55}$$

with the dispersion of the distribution

$$\sigma_L^2 = \frac{\langle L \rangle^2}{M} . \tag{6.56}$$

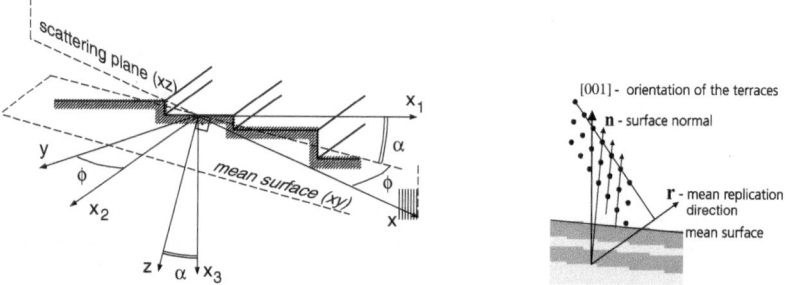

**Fig. 6.23** (*left*) Model of a step-like surface. (*right*) Illustration of the stair-like interface pattern in the superlattice and the corresponding fine structure in the reciprocal space

The terrace length was described by a similar distribution. The step height between the terraces $h$ was assumed to be normally distributed with the dispersion $\sigma_h$. For such a model the correlation function and the two-dimensional characteristic function have been calculated [27] and implemented in the expressions of the DWBA. The terrace size and its statistical distribution can be determined by transversal scans in reciprocal space or by $\omega$-scans. In Fig. 6.24 the DXS intensity has been calculated for a terraced surface of GaAs with a slight miscut of $0.3°$. Between the Yoneda wings there occur maxima, which are equidistant in reciprocal space and their distance is inversely proportional to the mean terrace size. The positions of these maxima correspond to the grating satellites of a mean surface grating with the lateral grating period $D_G$. The DXS peaks are broadened with increasing dispersion of the terrace lengths and the step height.

Growing an epitaxial layer on a miscut substrate, the staircase profile can be replicated from the substrate/layer interface to the sample surface. In a superlattice on off-oriented substrates, the staircase profile can be replicated from interface to interface [28, 29]. The direction of the replication may be inclined with respect to the growth direction (see Fig. 6.23). For simplicity we suppose first laterally uniform terrace lengths and perfect interface replication, giving the recursion formulae for the layer size functions

$$\Omega_j(r) \approx \Omega_{j-2}(r + D_{SL}\hat{z} + D_\parallel \hat{x}) , \qquad (6.57)$$

where $D_\parallel$ is the lateral shift of the stair-like pattern during the growth of one superlattice period (here we assume a bilayer superlattice period). Such a two-dimensionally periodic morphological superstructure creates a two-dimensional fine structure, similar to later discussed multilayer surface gratings. In this case the whole reflected intensity would be concentrated along so-called grating truncation rods perpendicular to the sample surface, representing the lateral periodicity. Each truncation rod would contain the multilayer Bragg peaks due to the multilayer

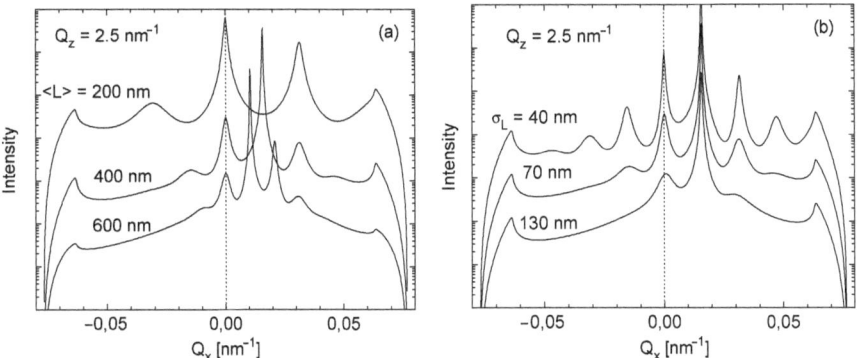

**Fig. 6.24** $\omega$-Scans of a $3°$ miscut GaAs surface. (**a**) Calculation for different terrace sizes and (**b**) for different dispersions of the terrace size [27]

periodicity. An inclined replication direction of the interface profile creates inclined branches of multilayer Bragg peaks. All are shown schematically in Fig. 6.23. In reality there will be a rather partial interface replication, characterised by an effective replication length $\Lambda_\perp$. In the Gaussian roughness model (discussed in Sect. 6.5.1) the vertical replication in the periodic multilayer caused horizontal bananas of resonant diffuse scattering, crossing the multilayer Bragg peaks in the specular scan. In the present case of the lateral correlation of the interface steps similar horizontal sheets appear. However, they are, in addition, horizontally structured by lateral DXS maxima, which indicate the laterally and vertically correlated stair-like interfaces, see Fig. 6.25.

As a result a two-dimensionally structured pattern of resonant diffuse scattering is obtained with longitudinal DXS satellites due to the superlattice periodicity and transversal DXS satellites

$$\delta q_x \approx \frac{2\pi}{\langle L \rangle_{\mathrm{av}}} \, , \qquad (6.58)$$

which represent the more or less periodic lateral morphological order of the interfaces. Both together form longitudinal stripes perpendicular to the mean sample surface, which remind us of the grating truncation rods of multilayer surface gratings (see Sect. 6.7). Considering the $q_z$-dependence of the diffuse intensity one observes that the envelope of the intensity follows with its maximum the direction of the terrace orientation. However, the simultaneous existence of large terraces formed by *step bunching* and atomic scale micro-terraces can modify the DXS pattern (see Fig. 6.26) [27–29].

The investigation of step-like interface morphology by interface-sensitive *diffraction* methods is briefly discussed in Sect. 6.6.

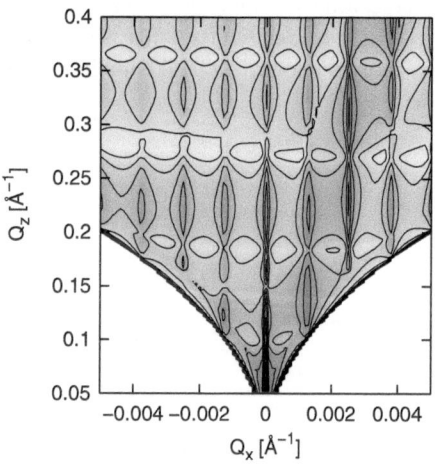

**Fig. 6.25** Calculated map for a (7 nm GaAs/15 nm AlAs)10× superlattice grown on a 0.5° miscut GaAs substrate. Averaged terrace distance is $\langle L \rangle$ =500 nm and interface steps are fully replicated at 40°.

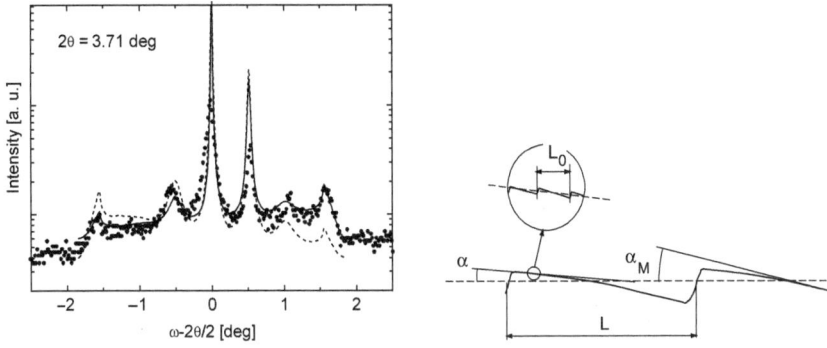

**Fig. 6.26** Measured $\omega$-scan of a GaInAs/GaAs/GaAsP/GaAs multilayer (*dots*) and its fit by the theory using a single type of steps (*full*), and two sets of the steps (*dashed*) [27]. The left-hand figure shows a possible microscopic structure of terraces

### 6.5.4 Non-Coplanar NSXR

XRR in *coplanar* geometry is most common and simple to realise with conventional diffractometers and reflectometers. The intensity distribution is resolved in the $q_x/q_z$-plane which contains the surface normal. The region in the $q_x/q_z$-plane accessible by coplanar reflection geometry is restricted by the Ewald spheres for the limiting cases of grazing incidence and grazing exit, which represent the horizon of the sample surface. Especially for small values of $q_z$ the measurable lateral momentum transfer decreases and consequently the information is cut about roughness with small lateral dimensions of nanoscopic scale.

By the use of a *non-coplanar* scattering geometry this limitation has been overcome [10, 30, 31]. The equipment requires monochromatic beam collimated in two directions, which can be provided by synchrotron radiation sources. First experiments used the set-up of a small angle scattering instrument with a well-collimated beam and a two-dimensional position-sensitive detector. Other set-ups are based on surface diffraction instruments, working usually in a strongly non-coplanar (grazing incidence diffraction) geometry, see Fig. 6.27. The detection of the diffusely scattered intensity up to a parallel momentum transfer of $1\,\text{\AA}^{-1}$ enables to study the correlation properties up to a few Å. The diffusely scattered intensity is usually drawn in a double logarithmic scale. Fitting the asymptotic intensity decay with increasing $Q_\parallel$ by a power law, the Hurst factor introduced in Sect. 6.5.1 can be determined with good precision, wherefrom one can conclude on the validity of different growth models.

Figure 6.28(**a**) shows measured $\theta_{sc}$-scans of an amorphous W/Si superlattice for different $q_\parallel$. They cross the RDS sheets indicated by roman numbers. For increasing $q_\parallel$ the width of the RDS sheets increases and finally the resonant diffuse scattering disappears, indicating a reduction of the vertical replication length $L$ for the higher frequencies of the roughness profile. In Fig. 6.28(**b**), the decrease of the intensity of

**Fig. 6.27** Schema of non-coplanar x-ray reflectivity set-up [30]

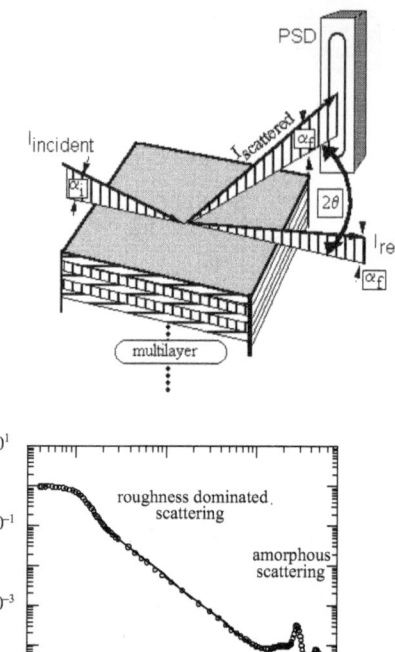

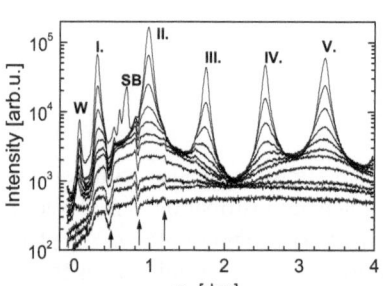

**Fig. 6.28** Measurement of non-specular x-ray reflectivity of an amorphous W/Si superlattice. (**a**) $\theta_{sc}$-Scans for different $q_{\parallel}$ [10]. Intersections with the RDS sheets are indicated by roman numbers. (**b**) Intensity profile of the first RDS sheet [30]

the first RDS sheet is plotted. The measurements prove the validity of a logarithmic scaling behaviour as predicted by the Edward–Wilkinson equation [32].

## 6.6 Interface Roughness in Surface-Sensitive Diffraction Methods

In the case of epitaxial multilayers, surface and interface roughness can also be studied by surface-sensitive x-ray *diffraction* methods such as *grazing incidence diffraction* (GID) and *strongly asymmetric x-ray diffraction* (SAXRD). Besides reflection at the interfaces there occurs diffraction by the layer lattices. The principles of diffraction by rough multilayers are similar to those described in more detail for x-ray reflection. All used theoretical treatments can be extended.

The polarisability of each layer can be developed in a Fourier series after its reciprocal lattice vectors

$$\chi_j^{\text{layer}}(r) = \sum_g \tilde{\chi}_{g,j}(r) e^{-igr} . \tag{6.59}$$

Measuring the intensity pattern of a Bragg reflection with the reciprocal lattice vector $\boldsymbol{h}$ in a conventional diffraction geometry (so-called *two-beam case*) only the Fourier components with the indices $\boldsymbol{h}$, $-\boldsymbol{h}$ and 0 are of importance.

Crystal truncation rods through each reciprocal lattice point characterise the structure amplitude of a crystalline layer. All truncation rods of a periodic multilayer contain the fine structure of equidistant superlattice satellites similar to the schema in Fig. 6.20.

The non-perturbed wave field of diffraction by a planar epitaxial multilayer under conditions of grazing incidence consists in each plane layer of eight plane waves for each polarisation

$$E_j^{\text{pl}}(r) = \sum_{n=1}^4 \left[ T_j^n e^{-ik_{0\|}r_\|} e^{-ik_{0z,j}^n(z-Z_j)} + R_j^n e^{-ik_{h\|}r_\|} e^{-ik_{hz,j}^n(z-Z_j)} \right] \Omega_j^{\text{pl}}(z) \tag{6.60}$$

with the rough interface shape function $\Omega_j^{\text{pl}}(z) = H(z - [Z_j + z_j]) - H(z - Z_j)$. For superlattices with rough interfaces the layer disturbance includes the variation of the Fourier components of the polarisability and the lattice displacement $\Delta u(r)$ due to the lattice deformation created by the interface roughness profile. In layer $j$,

$$\Delta\chi_j^{B\text{layer}}(r) = \sum_{g=0,-h,h} \Delta\chi_{g,j}(r) e^{-igr} \qquad \text{with} \tag{6.61}$$

$$\Delta\chi_{g,j}(r) = \left[ \chi_{g,j}\left( e^{ig\Delta u(r)} - 1 \right) + \Delta\chi_{g,j} e^{ig\Delta u(r)} \right] e^{igu_0(z)} \Omega_j^{\text{pl}}(z) .$$

Similar to x-ray reflection by the rough interfaces the disturbances give rise to diffuse scattering. The number of possible diffuse scattering processes between two non-perturbed states at one interface, see (4.D23), increases up to 64. Fortunately a certain number of them is almost negligible. If the roughness profile is replicated, the diffusely scattered intensity is concentrated in horizontal sheets of resonant diffuse scattering crossing the crystal truncation rods in the position of the diffraction satellites. Their origin arises now from partially coherent *diffraction and reflection* by the interface disturbances. For weak strain the covariance functions are formally quite similar to (6.49) found for x-ray reflection

$$\tilde{Q}_{jk}^{mnop} = \frac{A\Delta\chi_{g,j}(\Delta\chi_{g',k})^*}{\delta q_{z,j}^{mn}(\delta q_{z,k}^{op})^*} \int_S d(r_\| - r_\|') e^{iq_\|(r_\| - r_\|')} \tag{6.62}$$

$$\times \left[ \chi_{z_j,z_k}(\delta q_{z,j}^{mn}, (\delta q_{z,k}^{op})^*) - \chi_{z_j}(\delta q_{z,j}^{mn}) \chi_{z_k}((\delta q_{z,k}^{op})^*) \right] ,$$

however now with the *reduced* scattering vectors of the corresponding scattering process in the layers, which depend on the *local* reciprocal lattice vectors in the layers by $\delta q_{z,j} = q_{z,j} - g_{z,j}$.

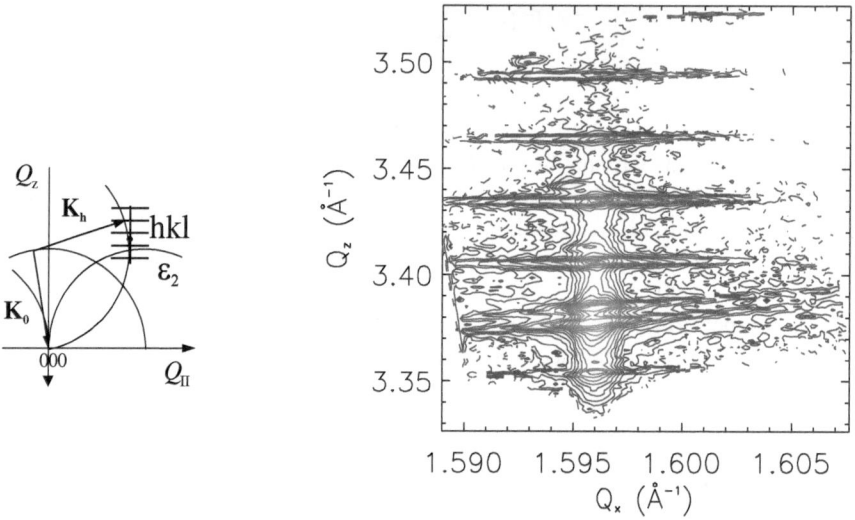

**Fig. 6.29** Diffuse x-ray scattering by rough interfaces in the strongly asymmetric diffraction mode. *Left*: schematic situation in reciprocal space. *Right*: reciprocal space map of (113) diffraction of a GaAs/AlAs superlattice for $\lambda = 1.47$ Å. The coherent crystal truncation rod (CTR) is crossed by horizontal RDS sheets, indicating correlated roughness. The sheets are laterally not limited by the experimental geometry

In Fig. 6.29 the scattering geometry in reciprocal space and the corresponding experimental results of strongly asymmetric diffraction by a GaAs/AlAs superlattice are shown. The measured sheets of resonant diffuse scattering (RDS) of the diffraction mode are clearly visible. It is an advantage of the AXRD measurements that the RDS sheets are not limited by the sample horizon, in contrast to coplanar XRR. So the full range of momentum transfer can be detected in a coplanar scattering geometry.

The application of x-ray diffraction methods is limited on epitaxial structures. On the other hand, x-ray reflection experiments are less successful for many semiconductor systems due to the missing contrast in the electron density modulation. Thus the choice of suitable Bragg reflections allows increasing the contrast between the layers in the diffraction mode.

GID, a non-coplanar surface-sensitive diffraction method, was successfully applied for the measurement of RDS by rough multilayers in [12].

Beside Gaussian roughness correlation behaviour, the *step-like interface morphology* was also investigated by various diffraction methods. In Fig. 6.30 we show the measured 200-reciprocal space map of a GaInAs/GaAs/GaAsP/GaAs-superlattice on a 2° off-oriented GaAs substrate, measured by grazing incidence diffraction. This reflection is highly sensitive for the morphological ordering, since the scattering contrast of the corresponding Fourier components of the susceptibility is much larger than that in the above-discussed reflection mode. Similar to Figs. 6.23 and 6.25, the diffuse scattering is concentrated in stripes, resonant diffuse scattering

**Fig. 6.30** Diffuse x-ray scattering by stepped interfaces in the grazing incidence diffraction mode. Reciprocal space map of (0$\bar{2}$0) diffraction of a GaInAs/GaAs/GaAsP/GaAs multilayer on a 2° off-oriented [001]-substrate. The axes are normalised in crystallographic units (*HKL*)

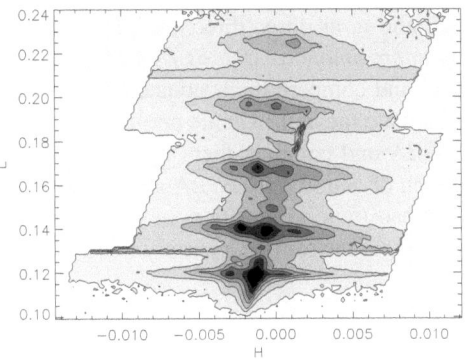

along so-called grating truncation rods, which are *perpendicular to the averaged surface*. The grating rods are therefore inclined with respect to the *crystallographic* orientation, which is simultaneously the orientation of the terraces. Each grating rod contains multilayer Bragg peaks. The Bragg peaks of the same vertical order but of different grating rods form branches which are inclined with respect to the sample surface according to the inclination of the morphological interface replication via the surface normal. The envelope maximum of the diffuse scattering follows the 001-direction, which is the orientation of the terraces.

## 6.7 X-Ray Reflection from Multilayer Gratings

In this section we discuss the calculation of the x-ray reflection from *multilayer gratings* (MLGs), Fig. 6.31. Gratings are etched into planar multilayers so that their lateral structure is formed by wires distributed equidistantly with period $d$ along the surface. We focus the present study mainly on the short-period gratings with the periodicity at about micrometers, which are of most interest in semiconductor physics.

The part etched out (dips between wires) can be several hundred nanometers deep. Thus these structures can be considered as a special case of huge deterministic roughness or as an artificial lateral one-dimensional crystals contrary to the crystals

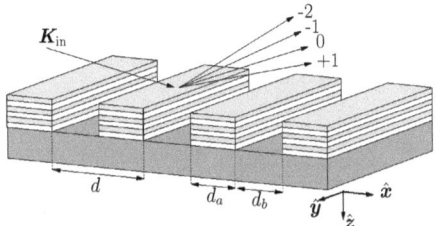

**Fig. 6.31** A sketch of a multilayer grating with a fan consisting of four diffracted–reflected waves

periodic in all three directions. Thus the reflectivity from gratings can be treated by *approximate as well as rigorous methods* [19, 33–39], thus making possible to treat and compare the adequateness of various approximations. In this section, we formulate the approximate perturbative treatment by the kinematical theory and by DWBA and compare them to the exact dynamical calculation. We determine the region of validity of DWBA and we show that the correct choice of the eigenstates can lead to good results even when the perturbed potential is present in the most volume of the sample, contrary to the small roughness of interfaces.

### *6.7.1 Theoretical Treatments*

MLG possesses the translation symmetry so that it is fully sufficient to determine its susceptibility $\chi(r)$ in one period $\left(-\frac{d}{2} \leq x \leq \frac{d}{2}\right)$ only. Therefore we first describe it for any of the layer $j$. The period consists of two parts (wires) namely $\underline{a}_j$ and $\underline{b}_j$ (for the case of an etched grating, one of the parts is the air). We denote their susceptibilities $\chi_j^a, \chi_j^b$ and their widths $d_j^a = \Gamma_j d,\ d_j^b = (1-\Gamma_j)d$ with $0 \leq \Gamma_j \leq 1$. We introduce the *shape function* $\Omega_j^{a1}(r)$ of the material $\underline{a}_j$ in the period. It equals unity inside the volume occupied by the material $\underline{a}_j$ and it is zero elsewhere, see Fig. 6.32. Then the susceptibility of one period is

$$\chi_j^1(r) = \chi_j^a \Omega_j^{a1}(x,z) + \chi_j^b\left(1 - \Omega_j^{a1}(x,z)\right) . \tag{6.63}$$

By $\tilde{\Omega}_j^{a1,h}(q_z)$ we denote the two-dimensional Fourier transform of the shape function of one period.

Because of the presence of two types of interfaces, horizontal and vertical ones, different theories treat the respective reflectivities using different approximations. Further, we mean by the single-scattering approaches those related to the lateral diffraction case. We treat separately the perturbative (single-scattering) and rigorous dynamical (multiple-scattering) theories.

**Fig. 6.32** Notation of the variables describing a laterally structured layer

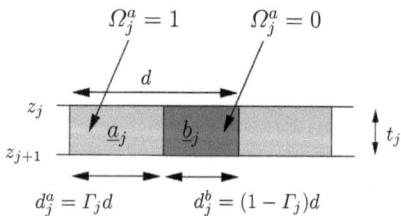

#### 6.7.1.1 Perturbative Treatments

MLG is periodic along the axis $\hat{x}$ with the lateral periodicity $d$, Fig. 6.31. Then the scattering potential $V(r)$ of the sample, as defined in Sect. 6.2, can be given as a convolution of the scattering potential of one period $V^1(r)$ (defined in the interval $-\frac{d}{2} \leq x \leq \frac{d}{2}$) with a periodic arrangement of $\delta$-functions

$$V(r) = V^1(r) \otimes \sum_n \delta(x - nd) \,. \tag{6.64}$$

Its Fourier transform is a product of two terms ($A$ denotes the sample area)

$$\tilde{V}(q) = \int dr\, V(r)\, e^{iqr} = \frac{A}{d}\, \tilde{V}^1(q_x, q_z) \cdot \sum_{h = \frac{2\pi}{d} m} \delta_{q_x,h}\, \delta_{q_y,0} \,. \tag{6.65}$$

The second (summation) term expresses the reciprocal lattice of the grating, which are *grating truncation rods (GTRs)* in $Q_z$ direction positioned equidistantly along the axis $q_x$ at points $q_x = h_m = \frac{2\pi}{d} m$, where $m$ is an integer (see Fig. 6.33). The first term $\tilde{V}^1$ is the Fourier transform of the potential in one period, behaving like an envelop function for the wave fields associated with the GTRs.

In a multilayer grating, the potential of one period $V^1(r)$ is the sum of the potentials of individual layers $V_j^1(r)$, and similarly for their two-dimensional Fourier transforms $\tilde{V}_j^1(h, q_z)$. The latter separates into the zeroth component proportional to the laterally averaged susceptibility and the discrete Fourier components proportional to the susceptibility contrast

$$\tilde{V}_j^1(h, q_z) = \begin{cases} -K^2 d \int dz\, \chi_{0\,j}(z)\, e^{iq_z z} & \text{for } h = 0 \,, \\ -K^2 (\chi_j^a - \chi_j^b)\, \tilde{\Omega}_j^{a1,h}(q_z) & \text{for } h \neq 0 \,. \end{cases} \tag{6.66}$$

Now we will consider the scattering from the sample we characterised generally above. We first determine the directions $K_h$ of scattered waves. We use the principles for the Ewald construction, discussed in Appendix 6.A, which state that the wave vector end-points lay at the intersection of the sample reciprocal lattice and the Ewald sphere of the incident wave. Thus the incident wave is scattered into the fan of reflected and transmitted waves associated with each GTR, see Figs. 6.31 and 6.35.

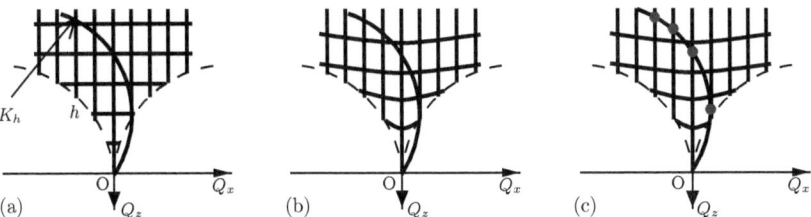

**Fig. 6.33** Schematical drawing of the reciprocal space maxima of a laterally periodic grating etched into a periodic multilayer. The "Bragg" sheets are parallel to the $q_x$ axis in the kinematical treatment (**a**), whereas they are curved and shifted upwards in the DWBA (**b**) and dynamical (**c**) calculations due to refraction. In addition, subfigure (**c**) illustrates the multiple-scattering interaction among wave fields of the simultaneously excited GTRs which is taken into account within the dynamical theory

Further, we calculate the reflection amplitudes. Scattering potential of MLG is *deterministic* and thus the reflection amplitude of all GTRs comes from the *coherent scattering* only (even though into non-specular directions). It is expressed similar to the coherent specular reflection amplitude of rough MLs calculated by DWBA. Using the formalism from Appendix 4.D, the amplitude at the sample surface is $R^h(K_h) = T_{0h}/2iK_{hz}A$. The scattering matrix element $T_{0h} = \langle E_h|V(r)|E_0\rangle$ can be decomposed into the sum over the individual layer contributions $\tau_j^h$. The sample reflectivity along GTR $h$ is finally $|R^h|^2 K_{hz}/K_z$.

The reflection amplitude then depends on the approximation used in the evaluation of the scattering matrix element. We discuss briefly the calculation by the kinematical theory and by the first-order DWBA applying the approach of Sects. 4.D.2 and 4.D.3, respectively.

### 6.7.1.2 Kinematical Calculation

Kinematical theory is equivalent to the first Born approximation [36,37], thus calculating the scattering process as the single-scattering transition of the incident vacuum plane wave $|E_0\rangle = e^{-iK_0r}$ into the diffracted vacuum plane wave $|E_h\rangle = e^{-iK_hr}$, see Fig. 6.35(a). The scattering matrix element for one period and one layer is proportional to the Fourier transform of the layer potential in one period (with the scattering vector $Q_h = K_h - K_0$)

$$\tau_j^h = \langle e^{-iK_hr}|V_j^1(r)|e^{-iK_0r}\rangle = \tilde{V}_j^1(h, Q_{hz}). \tag{6.67}$$

According to (6.66), we can see that the Fourier transform for $h=0$ is determined by the profile of laterally averaged susceptibility. Thus the *specular reflectivity* profile coincides with a kinematical reflection from laterally averaged planar multilayer and the specular reflectivity curve exhibits the same features as those calculated in the framework of the kinematical theory and the stationary phase method (SPM) [19, 39]. (SPM helps to avoid the Fraunhofer approximation which is not suitable for laterally extended samples).

Considering the intensity of the *non-specular truncation rods* ($h \neq 0$), the scattering matrix contribution is

$$\tau_j^h = -k_0^2(\chi_j^a - \chi_j^b) \cdot \tilde{\Omega}_j^{a1,h}(Q_{hz}) . \tag{6.68}$$

By calculating the kinematical scattering integral by the stationary phase method we generalise the *kinematical Fresnel reflection coefficient for lateral diffraction case*

$$r_{j,j+1}^{h,\text{kin}} = \frac{k_0^2\left(\tilde{\chi}_j^h - \tilde{\chi}_{j+1}^h\right)}{2K_{hz}Q_{hz}}. \tag{6.69}$$

For specular reflection it perfectly coincides with the kinematical Fresnel reflection coefficient for planar multilayers $r_{j,j+1}^{0,\text{kin}} = k_0^2\left(\chi_{0,j} - \chi_{0,j+1}\right)/Q_z^2$, cf. (3.103), as we said above.

As all the kinematical theories, also in the present case the effects of absorption and refraction are not comprised. Thus the kinematical intensity is much larger than unity below the critical angle and it diverges for the specular scan at the origin of the reciprocal space. Further, the kinematical period of oscillations of an MLG converges slowly to that calculated by a theory including the refraction.

Let us figure out the positions of maxima of a periodic multilayer grating using a reciprocal space schema, Fig. 6.33. They lay on the intersections of the grating truncation rods (reciprocal lattice of the grating represents the lateral periodicity) and the sheets passing through the ML maxima on the specular truncation rod (which represents the vertical periodicity).

### 6.7.1.3  Calculation by DWBA

We follow the basis of the DWBA as treated for the roughness and we split the MLG potential $V(r)$ into two parts, see Fig. 6.34. We choose the *ideal (unperturbed) potential* $V^A(r)$ as that of a planar laterally averaged multilayer and thus calculate the eigenstates $|E_K^A\rangle$, see (4.D19), according to (3.59). For the simplicity of further treatment we restrict ourselves to the rectangular gratings only [40]. From (6.65) and (6.66) it follows that the ideal potential $V_j^A$ is constant in each etched layer, $V_j^A(r) = \tilde{V}_j^1(0,0)/dt_j$, while the *perturbed potential* $V^B(r) = V(r) - V^A(r)$ is the sum of non-zero Fourier components, $V_j^B(r) = \sum_{h \neq 0} \tilde{V}_j^1(h,0)e^{ihx}/dt_j$.

Consequently the scattering element of the perturbed potential does not intervene into the *specular* term

$$\tau_j^{h=0} = \langle E_0^A|V_j^A|E_0\rangle + \langle E_0^A|V_j^B|E_0^A\rangle \, . \tag{6.70}$$

The specular reflection amplitude from the whole MLG then equals the (dynamically) calculated reflection from the laterally averaged multilayer. From this it clearly follows that this DWBA considers multiple scattering between the horizontal interfaces of averaged layers by using the dynamical Fresnel reflection coefficients, but neglects the influence of multiple scattering by the vertical side walls.

The amplitude of the wave scattered into a *non-specular* GTR $h \neq 0$ is

$$\tau_j^h = \langle E_h^A|V_j^B|E_0^A\rangle \, . \tag{6.71}$$

The contribution of each laterally structured layer consists of four terms

$$\tau_j^h = -K^2\left( T_{k_{h,j}}S_j^{11}T_{k_{0,j}} + T_{k_{h,j}}S_j^{12}R_{k_{0,j}} + R_{k_{h,j}}S_j^{21}T_{k_{0,j}} + R_{k_{h,j}}S_j^{22}R_{k_{0,j}} \right), \tag{6.72}$$

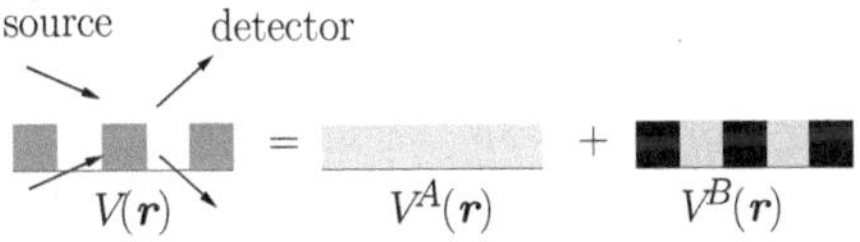

**Fig. 6.34** Splitting of the MLG potential $V(r)$ into an ideal and perturbed part

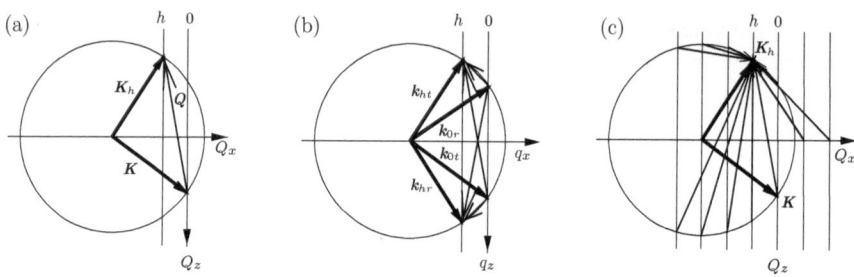

**Fig. 6.35** Single-scattering approaches, i.e. kinematical (**a**) and DWBA (**b**), calculate the diffracted field as a single-scattering process from GTR 0 to a GTR $h$, while the multiple-scattering approaches (**c**) take the contributions from all the GTRs into account

where the amplitudes $T_{k_j}, R_{k_j}$ are equal to $U_j(\pm k_{z,j})$ in (3.60) and the layer structure factor (4.D13) is $S_j^{mn} = S_j(q_j^{mn}) = \left(\chi_j^a - \chi_j^b\right)\tilde{\Omega}_{q_{x,j}^{mn}}^{a1}(-q_{z,j}^{mn})$. The four scattering wave vectors $q_j^{11}, \ldots, q_j^{22}$ are defined as in the case of diffuse scattering, see (4.D23) and Fig. 6.40. We draw them in the reciprocal space schema in Fig. 6.35(b) while demonstrating there the single-scattering character of the diffraction from the incident to the diffracted wave fields.

Because the eigenstates of the ideal potential are calculated using the usual dynamical matrix formalism for specular reflectivity from a planar multilayer, thus the effects of absorption and refraction are taken into account. Then the maxima of a periodic multilayer grating, Fig. 6.33(b), lay on the intersection of the truncation rods and the refraction-curved sheets passing through the maxima on the specular truncation rod.

#### 6.7.1.4 Multiple-Scattering Treatment by the Dynamical Theory

The dynamical theory treats the reflection from a multilayer grating by rigorously solving the wave equation under the condition of a one-dimensional periodicity

$$\chi(r) = \chi_0(z) + \sum_h \tilde{\chi}_h(z)\, e^{-ihx} . \tag{6.73}$$

There are miscellaneous approaches found in the literature, reviewed, e.g. in [19, 39]. Their formulation comes from the optics of visible light [33], while they have been applied in XRR only for surface gratings [38] using integral Rayleigh–Maystre formulae. XRR from multilayered gratings is studied deeply in [19,39] using matrix modal method. Dynamical theory takes into account the multiple scattering among the wave fields (each consisting of pair of a transmitted and reflected wave), which are associated with all truncation rods, including the real as well as evanescent GTRs as shown in Fig. 6.35(c).

Using a convenient matrix formalism similar to that for planar multilayers, the *generalisation of the Fresnel coefficients for lateral diffraction case*, compare (3.80) and (3.82), has been found [19,39]

$$r_{j,j+1}^{hg} = \frac{k_{z,j}^h - k_{z,j+1}^g}{k_{z,j}^h + k_{z,j+1}^g} \quad \text{and} \quad t_{j,j+1}^{hg} = \frac{2k_{z,j}^h}{k_{z,j+1}^h + k_{z,j+1}^g} . \tag{6.74}$$

Here, the indices $h$ and $g$ relate the transmission and reflection processes to simultaneous diffraction between wave fields of two GTRs $h$ and $g$. Wave vectors $k_h$ of scattered waves do not point to a *spherical* Ewald sphere, but to a so-called *dispersion surface* like in dynamical theory of x-ray diffraction.

In the dynamical theory the energy is conserved. Therefore a strong wave field corresponding to a certain GTR can influence significantly the intensity profile of another GTR. This may be the case, for instance, in the angular region where the wave field of the first GTR changes from evanescent to real (near the intersection of the Ewald sphere with the GTR +1, see Fig. 6.33(c)). There the specular intensity can be enhanced with respect to the specular intensity of an averaged planar multilayer.

### 6.7.2 Discussion

For the following discussion we will consider short period rectangular gratings (period around $1\,\mu m$) and the wavelength at about $1\,\text{Å}$. Since we already mentioned that the kinematical theory does not involve the refraction, which is of crucial importance in XRR, we will further devote our discussion to the comparison of DWBA to the dynamical theory. We choose the ratio $\Gamma$ of the wire width with respect to the period one half. Then we can find truncation rods of three types:

*Specular truncation rod* ($h = 0$). Here, the DWBA and dynamical theory give the same profiles, except for the known angular region of the enhanced interaction with GTR +1 as discussed earlier.

*Weak, kinematically forbidden truncation rods* ($h$ is even). The associated Fourier coefficients are zero, and therefore single-scattering theories, including the

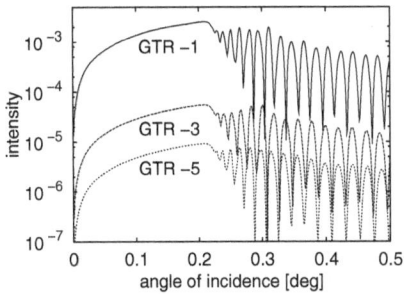
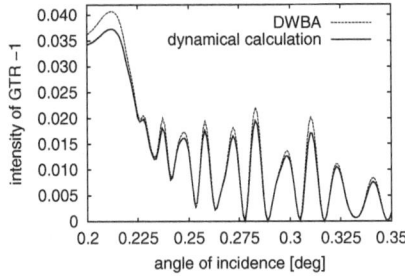

**Fig. 6.36** Calculation of the odd-order GTRs for a GaAs surface grating (thickness 300 nm) for period of (**a**) $0.8\,\mu m$ and (**b**) $5\,\mu m$ for wavelength $1.54\,\text{Å}$. In the former case, DWBA gives the same results as the dynamical theory. In the latter case the multiple scattering starts to be important and DWBA of the first order gives only approximative result

kinematical one and DWBA, predict zero intensity for them. Thus these GTRs are excited by multiple scattering in the etched layers and consequently their profiles can be calculated by the dynamical theory or by higher order DWBA.

*Strong truncation rods* (*h* is odd). Here, both DWBA and dynamical theory coincide, see Fig. 6.36(a). The good coincidence depends on the force of the dynamical interaction between diffracted wave fields. There are more GTRs excited in the Ewald sphere of the incident wave for large periods or small wavelengths, thus the dynamical effects will be enhanced and DWBA starts to be only approximative, see Fig. 6.36(b). We found possible to formulate a condition separating the two cases using a two-beam approximation of the dynamical theory [19].

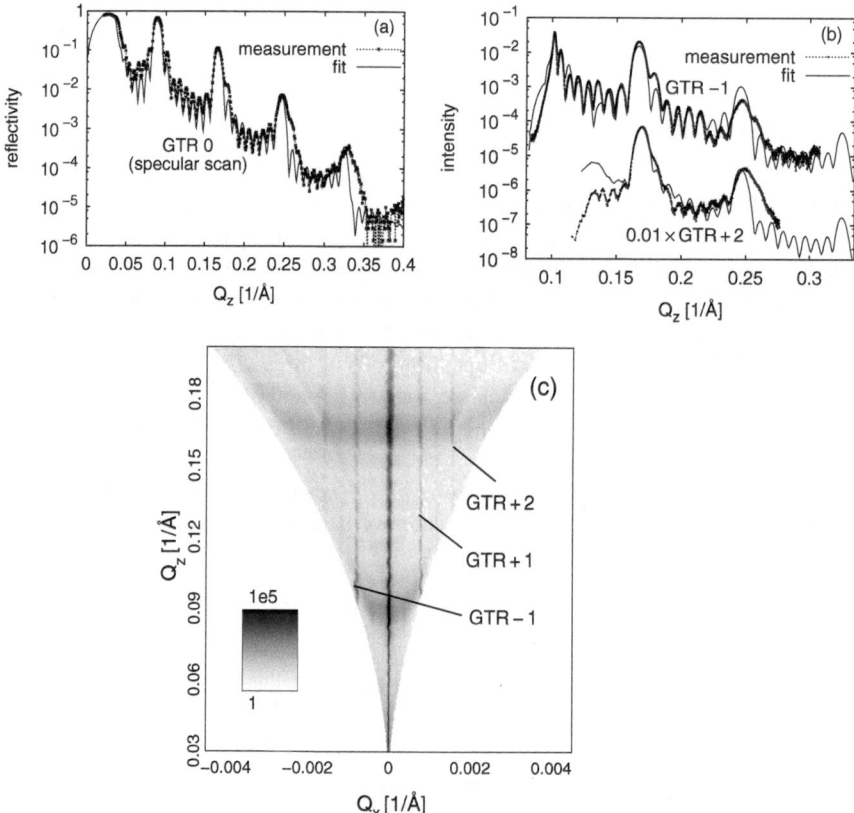

**Fig. 6.37** (**a**), (**b**) Measurement and fit of GTRs −1, 0 and +2 from a [W (1.5 nm)/Si (6.23 nm)]10× multilayer grating (lateral period 780 nm, wire width to period ratio 0.7) [41]. The measured reciprocal space map (**c**) shows the coherent intensity scattered into GTRs and the diffuse (incoherent) intensity concentrated in sheets of resonant diffuse scattering, which indicates vertically correlated roughness

### 6.7.3 Reflectivity from Rough Multilayer Gratings

The influence of interface roughness on grating reflectivity can be studied within all three theoretical treatments discussed earlier. Within the matrix approach of the dynamical theory [19, 39], the generalised Fresnel coefficients (6.74) corrected for roughness were found formally similar to those for rough planar multilayers (6.24)

$$r^{hg}_{j,j+1} = r^{hg,\text{flat}}_{j,j+1} e^{-2k^h_{z,j}k^g_{z,j+1}\sigma^2_{j+1}} \text{ and } t^{hg}_{j,j+1} = t^{hg,\text{flat}}_{j,j+1} e^{(k^h_{z,j}-k^g_{z,j+1})^2\sigma^2_{j+1}/2} . \qquad (6.75)$$

Roughness in gratings decreases the scattered intensity for the incidence angles even *below* the critical angle. Furthermore, there is different sensitivity to the surface and interface roughnesses for *weak and strong GTRs*, respectively. Finally we can notice that the kinematical reflection coefficients (6.69) are attenuated by $e^{-Q^2_{hz}\sigma^2_{j+1}/2}$ similar to (3.116).

In Fig. 6.37 we show XRR results of a periodic W/Si multilayer grating. Structural parameters (lateral periodicity and wires width, layer thicknesses and interface roughnesses) of the sample were obtained by fitting the measured GTR profiles employing the dynamical theory for rough gratings.

Finally, the calculation (by DWBA) of the diffuse scattering from MLGs, such as simulation of the map in Fig. 6.37(c), is even more tricky procedure which requires the preliminary calculation of the eigenstates either using the DWBA for perfect MLG or the dynamical theory.

**Acknowledgments** The work was supported by Deutsche Forschungsgemeinschaft (grant BA1642/1-1), Lise Meitner Fellowship of FWF, Austria (project M428-PHY) and the grant VS 96102 of the Ministry of Education of the Czech Republic.

## 6.A  Appendix: Reciprocal Space Constructions for Reflectivity

In some of the previous chapters in the book, the *reciprocal space* representation was used for drawing the experimental scattering geometry: experimental scans and inaccessible regions for coplanar reflectivity (Figs. 6.4 and 6.5). In addition, throughout this chapter we use the reciprocal space to describe graphically the scattering events of x-ray reflection. Since this approach may not be common to the reader who is not accustomed to that representation, we give here some schematic interpretations of the reflection by multilayers in reciprocal space, which help in finding the intuition for an easy understanding of the scattering features in a simple geometrical way. We start by the interpretation of fundamental laws of reflection and refraction at interfaces. We relate the reflection curves of thin films and periodic multilayers to their particular reciprocal space features and discuss multiple scattering as it is considered within the treatment by a DWBA.

The idea to represent x-ray scattering by reciprocal space constructions has been introduced by P.P. Ewald in the early stage of the dynamical theory of x-ray diffraction. The goal is to relate the directions of the scattered waves and the symmetry of the sample represented by the Fourier transform of the crystal lattice and/or the shape function of the scatterers. *Ewald (reciprocal space) construction* visualises two basic physical principles:

1. Energy conservation. X-ray reflection is an elastic scattering process, conserving the wave vector length. Then the end-points of all scattered waves can lay only on the *Ewald sphere* of the radius of the wave vector length, Fig. 6.38(a).
2. Momentum conservation except of a reciprocal lattice vector if the diffraction condition is fulfilled. This reflects the symmetry properties of the sample.

In this book we use Ewald construction for the illustration of the reflection by layers and multilayers, including the wave vectors in the vacuum and medium.

### 6.A.1 Reflection from Planar Surfaces and Interfaces

Let us discuss the reflection and refraction laws in reciprocal space, Fig. 6.38, by using Ewald construction. The wave propagation in the vacuum and media is determined by the different lengths of the wave vectors. In case of a homogeneous half-space of a slightly absorbing medium with a flat surface or planar layers with smooth interfaces their reciprocal space structure is defined by a so-called *truncation rod* passing through the origin and normal to the surface (i.e. it usually coincides with the axis $q_z$). We call the truncation rod through the origin of the reciprocal space here as *specular rod*, since it defines the conditions for specular reflection. It intersects the vacuum Ewald sphere $\epsilon^{\mathrm{vac}}$ of the incident wave $k_0$ in two points, which pin down the wave vectors of the reflected wave $k_r$ and the transmitted wave in the vacuum, see subfigure (a). Therefrom we obtain the *law of reflection*—the reflected wave makes the same angle with the surface as the incident one. The Ewald construction with the specular rod represents the symmetry of the sample and the scattering process, which permits a momentum transfer only along the $q_z$ direction (along the surface normal).

Inside a layer $j$ of a multilayer (or in a substrate) the wave vectors are determined

1. by the *dispersion relation* $k_j = n_j k_0$ giving the radius of the Ewald sphere within the medium $\epsilon_j$,
2. by the continuity of the lateral wave vector components at the interface.

These two conditions lead to the *Snell's law* (also refraction law) for the transmitted wave as outlined in subfigures (b), (c) and (d). The tie points $T_j$ and $R_j$ of the transmitted and reflected waves in the layer $j$, respectively, are located at the intersections of the specular rod and the "inner" Ewald sphere $\epsilon_j$. For x-rays $n < 1$ ($\chi < 0$), thus three distinct cases may happen in each layer. Case (b) marks the refraction law above the *critical angle*: two waves, reflected $k_{rj}$ and transmitted $k_{tj}$,

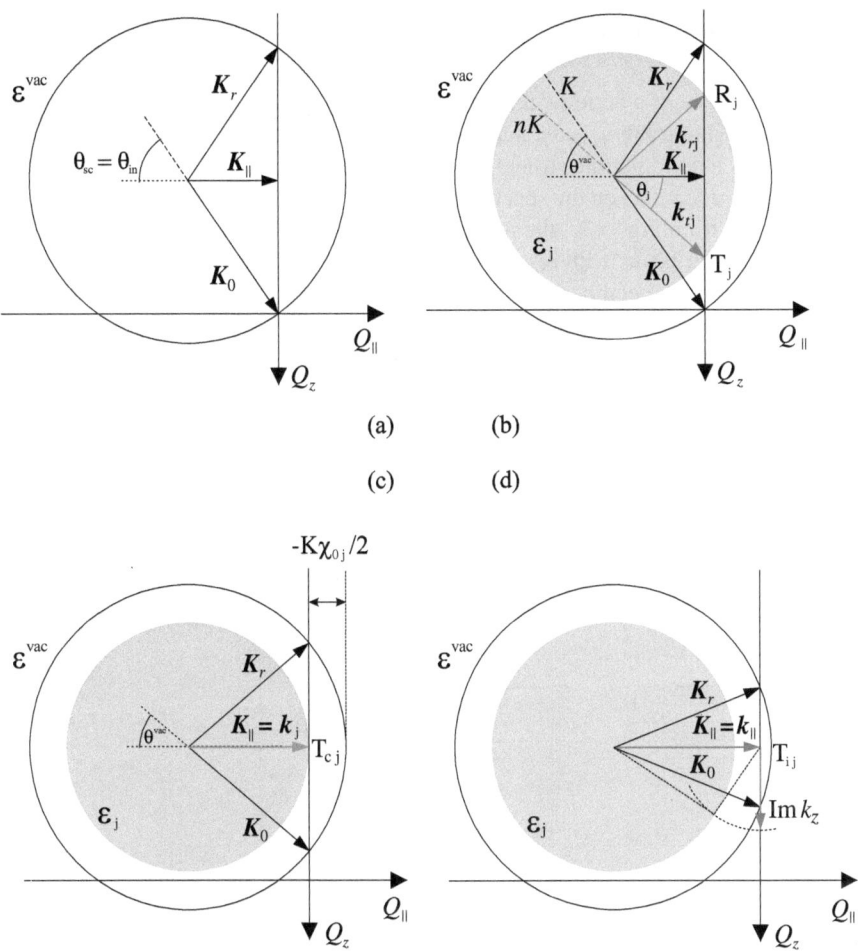

**Fig. 6.38** Graphical representation of the laws of reflection and refraction by an interface by means of the Ewald construction. (**a**) The law of reflection, (**b**)–(**d**) Snell's law: (**b**) above, (**c**) at and (**d**) below the critical angle. Below the critical angle the lateral component of $k$ is larger than the radius of the Ewald sphere of the medium $j$, thus it has purely imaginary $k_z$ component (neglecting absorption) and the wave is called evanescent

propagate in the layer. Case (c) visualises the situation at the critical angle for total external reflection in the layer. There is one tie point $T_{cj}$ only and the wave in the layer propagates parallel to the interface, $k_j = k_{\parallel}$. Case (d) interprets the generation of the *evanescent wave* in the layer, propagating parallel to the interface and exponentially damped perpendicular to it.

According to the Fresnel formulae, see (3.80) and (3.82), the reflected and transmitted wave amplitudes depend exclusively on the complex wave vectors of the media bordering the interface.

## 6.A.2 Periodic Multilayer

Reciprocal lattice of a periodic multilayer, Fig. 6.39(a), is a set of points positioned equidistantly along the $q_z$ axis, subfigure (c). Thus the "super-periodicity" in real space causes a periodic fine structure along the specular rod, and we find so-called *multilayer Bragg peaks* on the specular reflectivity curve, see Fig. 6.11, for instance.

Following from Fig. 6.38, the refraction in the layers causes a shift of the actual multilayer Bragg peaks with respect to the position of the reciprocal lattice points. This is shown by the comparison between the kinematical the dynamical reflection curve of a smooth multilayer in Fig. 6.10. The position of the kinematical Bragg peaks coincides exactly with the reciprocal lattice points.

The finite total multilayer thickness gives rise to additional side maxima, so-called *Kiessig fringes* between the multilayer Bragg peaks (not shown in the figure). There are $p-2$ maxima in between two Bragg peaks for a flat multilayer with $p$ periods.

Reciprocal lattice of a laterally periodic multilayer grating etched into a planar periodic multilayer is shown in Fig. 6.33. The lateral periodicity gives rise to a grating rod pattern. The grating rods are equidistantly positioned along the direction of patterning with the specular rod in the centre.

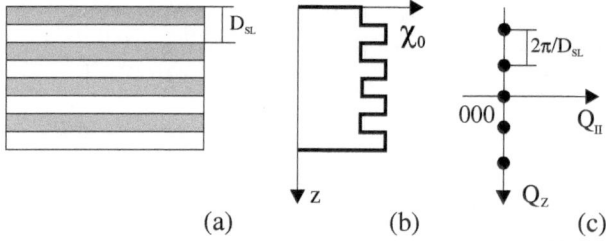

**Fig. 6.39** Schematic set-up of a periodic multilayer: (**a**) in real space, (**b**) the polarisability profile, and (**c**) in reciprocal space

## 6.A.3 Reciprocal Space Representation of DWBA

The formulae for the calculation of the first-order DWBA have been derived in Chap. 4. Here, we show the graphical representation of the corresponding scattering events. Each of the two eigenstates of the unperturbed potential $V^A$ consists of a transmitted and reflected wave $T = U(+k_z)$, $R = U(-k_z)$. The four wave vector transfers $q^{11}, \ldots, q^{22}$ defined by (4.D23) and corresponding to $(k_{sc\parallel} - k_{in\parallel}, \pm k_{sc,z} \pm k_{in,z})$ in (4.47), (4.56) or (6.48) are represented in the reciprocal space by the four intervening scattering processes. They are schematically drawn in Fig. 6.40. We call the first (transmission–transmission) term the *primary* scattering process $q_j^{11}$, since it is directly excited by the incident wave and it corresponds to the measured scattering vector in vacuum $q = k_{sc} - k_{in}$. The other three terms are *secondary* scattering processes. They are of purely dynamical nature, called *Umweganregung* (detour or non-direct excitation), which occurs exclusively due to multiple scattering (direct or non-direct excitations).

The division of the perturbed potential $V^B$ into the layer disturbances $V_j^B$ allowed to represent the scattering in terms of structure factors $S_j$, Eq. (4.D13), an advantage usually reserved for the kinematical theory. The contribution of one scattering process in a single layer to the amplitude reflected by the whole sample depends on the structure factor of the layer disturbance and on the amplitudes of the participating waves.

Reciprocal space representation of the scattering processes in the Born approximation, DWBA and dynamical theory for reflection by gratings is shown in Fig. 6.35.

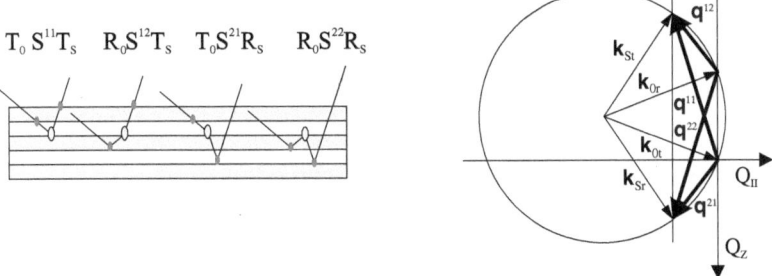

**Fig. 6.40** Schematic representation of the four x-ray reflection processes in real space (*left*) and in the reciprocal space (*right*) of the first-order DWBA. The *full circles* denote the dynamical reflection and transmission in the ideal multilayer, *open circles* indicate the diffuse scattering due to the interface roughness. The process with the indices 11 is the primary scattering process, described also by the kinematical approximation. The other three are processes of Umweganregung

# References

1.  Sinha, S.K., Sirota, E.B., Garoff, S., Stanley, H.B.: Phys. Rev. B **38**, 2297 (1988)
2.  Stearns, D.G.: J. Appl. Phys. **65**, 491 (1989)
3.  Kortright, J.B.: J. Appl. Phys. **70**, 3620 (1991)
4.  Savage, D.E., Kleiner, J., Schimke, H., Phang, Y.H., Jankowski, T., Jacobs, J., Kariotis, R., Lagally, M.G.: J. Appl. Phys. **69**, 1411 (1991)
5.  de Boer, D.K.G.: Phys. Rev. B **44**, 498 (1991)
6.  Daillant, J., Bélorgey, O.: J. Chem. Phys. **97**, 5824 (1992)
7.  Holý, V., Kuběna, J., Ohlídal, I., Lischka, K., Plotz, W.: Phys. Rev. B **47**, 15896, (1993)
8.  Holý, V., Baumbach, T.: Phys. Rev. B **49**, 10668 (1994)
9.  Schlomka, J.-P., Tolan, M., Schwalowsky, L., Seeck, O.H., Stettner, J., Press, W.: Phys. Rev. B **51**, 2311 (1995)
10. Salditt, T., Lott, D., Metzger, T.H., Peisl, J., Vignaud, G., Høghøj, P., Schärpf, J.O., Hinze, P., Lauer, R.: Phys. Rev. B **54**, 5860 (1996)
11. Holý, V., Baumbach, G.T., Bessière. M.: J. Phys. D **28**, A220, (1995)

12. Stepanov, S.A., Kondrashkina, E.A., Schmidbauer, M., Köhler, R., Pfeiffer, J.-U.: Phys. Rev. B **54** 8150 (1996)
13. de Boer, D.K.G.: Phys. Rev. B **49** 5817 (1994)
14. Baumbach, G.T., Holý, V., Pietsch, U., Gailhanou, M.: Physica B **198**, 249, (1994)
15. Spiller, E., Stearns, D., Krumrey, M.: J. Appl. Phys. **74**, 107 (1993)
16. Croce, P., Névot, L.: Revue Phys. Appl. **11**, 113 (1976)
17. Névot, L., Croce, P.: Revue Phys. Appl. **15**, 761 (1980)
18. Holý, V., Pietsch, U., Baumbach, G.T.: High-Resolution X-Ray Scattering from Thin Films and Multilayers, Springer-Verlag, Berlin (1999)
19. Mikulík, P.: PhD thesis, Université Joseph Fourier, Grenoble and Masaryk University, Brno (1997)
20. Eymery, J., Hartmann, J.M., Baumbach, G.T.: J. Cryst. Growth **184**, 109 (1998)
21. Pukite, P.R., Lent, C.S., Cohen, P.I.: Surf. Sci. **161** 39 (1985)
22. Buttard, D., Bellet, D., Dolino, G., Baumbach, T.: J. Appl. Phys. **83**, 5814 (1998)
23. Krim, J., Palasantzas, G.: Int. J. Mod. Phys. **9**, 599 (1995)
24. Palasantzas, G., Krim, J.: Phys. Rev. B, **48**, 2873 (1993)
25. Palasantzas, G.: Phys. Rev. B, **48**, 14472 (1993)
26. Ming, Z.H., Krol, A., Soo, Y.L., Kao, Y.H., Park, J.S., Wang, K.L.: Phys. Rev. B **47**, 16373 (1993)
27. Holý, V., Giannini, C., Tapfer, L., Marschner, T., Stolz, W.: Phys. Rev. B **55**, 55 (1997)
28. Holý, V., Darhuber, A.A., Stangl, J., Bauer, G., Nützel, J.F., Abstreiter, G.: Il Nuovo Cimento **19D**, 419 (1997)
29. Holý, V., Darhuber, A.A., Stangl, J., Bauer, G., Nützel, J.F., Abstreiter, G.: Semicond. Sci. & Technol. **13**, 590 (1998)
30. Salditt, T., Metzger, T.H., Peisl, J.: Phys. Rev. Lett. **73**, 2228 (1994)
31. Salditt, T., Metzger, T.H., Brandt, Ch., Klemradt, U., Peisl, J.: Phys. Rev. B **51**, 5617 (1995)
32. Edwards, S.F., Wilkinson, D.R.: Proc. R. Soc. London, Ser. A **381**, 17 (1982)
33. Maystre, D.: Rigorous vector theories of diffraction gratings. In: Wolf, E. (ed.) Progress in Optics XXI, North-Holland, Amsterdam (1984)
34. Nevière, M.: J. Opt. Soc. Am. A **11**, 1835 (1994)
35. Sammar, A., André, J.-M., Pardo, B.: Opt. Commun. **86**, 245 (1991)
36. Sammar, A., André, J.-M.: J. Opt. Soc. Am. A **10**, 2324 (1993)
37. Bac, S., Troussel, P., Sammar, A., Guerin, P., Ladan, F.-R., André, J.-M., Schirmann, D., Barchewitz, R.: X-ray Sci. Technol. **5**, 161 (1995)
38. Tolan, M., Press, W., Brinkop, F., Kotthaus, J.P.: Phys. Rev. B **51**, 2239 (1995)
39. Mikulík, P., Baumbach, T.: Phys. Rev. B **59**, (1999)
40. Mikulík, P. Baumbach, T.: Physica B **248**, 381 (1998)
41. Jergel, M., Mikulík, P., Majková, E., Luby, Š., Senderák, R., Pinčík, E., Brunel, M., Hudek, P., Kostic, I., Konecnikova, A.: J. Appl. Phys. **85**, 1225 (1999)

# Chapter 7
# Grazing Incidence Small-Angle X-Ray Scattering from Nanostructures

R. Lazzari

## 7.1 Introduction

Since its first use in the late 1930s, bulk small-angle scattering of X-rays or neutrons [1, 2] has become a well-established technique to probe density inhomogeneities at length scales greater than the interatomic distances. This tool is particularly used in the soft condensed matter community where brittleness of the sample made of organic molecules such as polymers or liquid crystals hampers the use of electron microscopy techniques. However, for a long time, its extension to the study of surfaces or interfaces suffered from the lack of surface sensitivity in transmission geometry as the signal is directly proportional to the scattering volume. More dramatically, the surface signal may be reduced to an unmeasurable quantity owing to (i) the small cross section for surface scattering, (ii) photoelectric absorption by the substrate and (iii) bulk scattering from defects. All these problems can be overcome by using the grazing incidence geometry rather than the transmission one and by taking advantage of the brilliance and coherence of synchrotron radiation. By selecting an incidence angle on the sample surface close and even below the angle of total external reflection of X-rays (see Sect. 3.1.1), the wavefield penetration depth is considerably decreased down to a few nanometers thus enhancing the surface or subsurface signal compared to the volume one. Any discontinuity in the local electronic density (surface roughness, islands, inclusions, particles, clusters, etc.) scatters either the transmitted or the reflected beam. For incidence below the angle of total external reflection, the evanescent refracted wave is confined to the top layer and provides a considerable enhancement of surface sensitivity. Owing to the involved distances of the nanometer order, the grazing incidence small-angle X-ray scattering (GISAXS) is collected close to the origin of the reciprocal space around the specularly reflected beam at small angles.

For morphological studies, GISAXS offers practical advantages over conventional microscopies:

R. Lazzari (✉)
Institut des NanoSciences de Paris, Université Pierre et Marie Curie, Paris 6 CNRS UMR 7588, Campus Boucicaut, 140 Rue de Lourmel, 75015 Paris, France

Lazzari, R.: *Grazing Incidence Small-Angle X-Ray Scattering from Nanostructures.* Lect. Notes Phys. **770**, 283–342 (2009)
DOI 10.1007/978-3-540-88588-7_7

- As a non-invasive technique, the beam-induced damage and complex sample preparations are avoided at variance to transmission electron microscopy (TEM)
- By varying the angle of incidence, depth sensitivity from surface to buried interfaces up to a few hundreds of nanometers can be achieved contrary to near-field microscopies
- The statistical average over the illuminated surface is intrinsic to X-rays, thus giving a picture of the mean scattering particle that is, in other respects, characterized by many ensemble measurements
- The use of photons does not suffer from charge build-up for insulating samples
- With synchrotron radiation, multi-wavelength measurements close to an absorption edge allow to enhance the signal coming from a specific element through the anomalous component of its atomic form factor (see Sect. 1.4.6)
- The technique can be used in situ to monitor time-dependent phenomena (at roughly the second time scale) in various environments from ultra-high vacuum to gas or vapor atmospheres.

To weight up the pro and the cons, the main technique drawbacks are

- The flatness and low roughness of the sample are prerequisite for valuable measurements
- The preferable use of synchrotron radiation to have reasonable counting time
- Like for all scattering techniques, extracting morphological parameters implies data analysis in reciprocal space with an account of dynamical effects.

The first GISAXS experiments were carried out 15 years ago during the Ph.D. thesis of I. Levine in the group of J.B. Cohen [3, 4]. This came naturally after the birth of surface X-ray crystallography by Eisenberg and Marra [5,6] 10 years before and with the advent of dedicated synchrotron sources. However, the tour de force of this pioneer work was the use of a rotating anode connected to a vacuum deposition chamber to study the growth of gold aggregates on glass as function of coverage, annealing time and temperature. Data collection for one GISAXS pattern took 1 or 2 hours with a position-sensitive detector! Only scattering patterns parallel to the sample surface were analyzed in terms of correlation distance between islands and of Porod radius. It was observed that the island spacing exhibits a linear dependence on island size during deposition. The main conclusion of this first *in situ* study was that the island mobility dominates mass transfer during the annealing process and that usual coarsening models are not adequate. Concomitantly, the group of Naudon [7] started the study of Guinier–Preston zone of Al–Ag alloy with a similar experimental set-up. Their initial goal was to examine clustering near surface as compared to bulk using as a probe the evanescent X-ray refracted wave.

Thereafter, it was realized that the full potential of this technique can be achieved when a synchrotron source with its inherent properties of high flux, collimation and multi-wavelength availability is combined with 2D detectors [8]. Thus, there was a surge of available beamlines with set-ups dedicated to GISAXS studies. It was quickly understood that this surface morphology-sensitive technique can be combined efficiently with grazing incidence diffraction, a technique sensitive to the atomic scale arrangement. The research topics covered a wide range of systems:

- Islands on surfaces either metals aggregates (Ag/MgO(100) [9]) or semiconductors quantum dots (Ge/Si [10], InAs/GaAs(001) [11])
- Discontinuous semiconductors superlattices (quantum dots and wires of SiGe/Si [12–15] or Ge in C [16]) or metallic multilayers in insulating matrices (Co–Al$_2$O$_3$ [17], Au–C [18], Fe–BN [19])
- Nanocermets or buried aggregates in matrices obtained by sputtering (C–Ag [20], Pt–Al$_2$O$_3$ [21])
- Soft condensed matters topics from polymer dewetting [22, 23], diblock copolymers [24] to surfactant mesophase templates [25–28].

However, since the work of Levine, except annealing, only a few surveys were performed during the sample elaboration process [3, 9, 29, 30]. The latest experimental developments are directed toward ultra-small-angle scattering (GIUSAXS) [31], the use of neutrons for light organic elements (GISANS) [23], micro-focalization [32, 33] to probe inhomogeneous samples, in situ measurements [28, 30, 34, 35] and anomalous GISAXS [36].

As a subset of diffuse scattering, GISAXS theoretical treatment benefited from the development of the distorted wave Born approximation (DWBA) (see Chap. 4). Even if the theoretical background developed in bulk small-angle scattering is of great use, the specificities due to the grazing incidence have to be accounted for; it is not only the incident beam that is scattered as in transmission geometry but also the reflected and transmitted waves. Even though the DWBA treatment of scattering was established for a long time in quantum mechanics [37], two seminal works of Vineyard [38] and Sinha [39] introduced this concept in the field of X-ray scattering from rough surface. DWBA was thereafter generalized to multilayers [40, 41] or multilayer gratings [42], to x-ray fluorescence at grazing incidence [43, 44], to magnetic diffuse scattering of x-ray [45, 46], to diffuse and small-angle scattering [47, 48].

As each different sample morphology leads to a different GISAXS patterns, an on-purpose analysis methodology and model have to be developed for analyzing data. In a very restrictive way, this chapter will be mainly devoted to the use and the theory of GISAXS technique, as it was historically developed, for probing the size, shape and spatial organization of nanostructures (islands, inclusions, particles, clusters, etc.). The goal of this chapter is to give a feeling of the most used approximation. An emphasis will be put on (i) the calculation of the scattering cross section from density inhomogeneities accounting for the surface-induced multiple scattering effects and (ii) the interplay between coherent and diffuse scattering for size-distributed samples. A few selected examples, without any claim of exhaustiveness, will be given in the field of hard and soft condensed matter in order to illustrate the potentiality of the GISAXS technique. A free GISAXS analysis software [50] encompassing most of the material developed in this chapter is available from http://www.insp.jussieu.fr.

## 7.2 The GISAXS Scattering Geometry

A typical GISAXS experiment is illustrated in Fig. 7.1; it consists in measuring the diffuse scattering around the specularly reflected beam at fixed incidence angle $\theta_{in}$

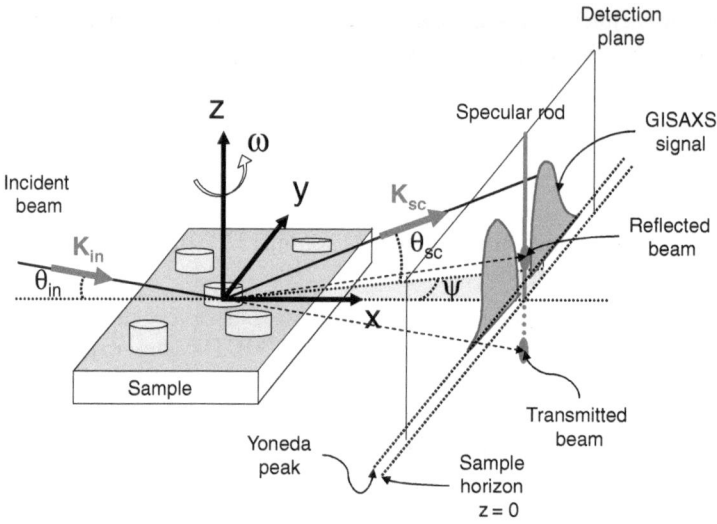

**Fig. 7.1** Principle of a GISAXS experiment. An x-ray beam of wavevector $\mathbf{k}_{in}$ impinges on the sample surface under a grazing incidence $\theta_{in}$ and is reflected and transmitted by the smooth surface but also scattered along $\mathbf{k}_{sc}$ in the directions $(\psi, \theta_{sc})$ by the surface or subsurface roughness or density heterogeneities. The specular rod is often hidden by a beam-stop in experimental conditions

upon sample rotation $\omega$ around its normal. The scattering angles $(\psi, \theta_{sc})$ are related to the wavevector transfer $\mathbf{q} = \mathbf{k}_{sc} - \mathbf{k}_{in}$ through

$$q_x = k_0[\cos(\theta_{sc})\cos(\psi) - \cos(\theta_{in})]$$
$$q_y = k_0[\cos(\theta_{sc})\sin(\psi)]$$
$$q_z = k_0[\sin(\theta_{sc}) + \sin(\theta_{in})]. \tag{7.1}$$

In a conventional GISAXS experiment, the recorded data as function of $(\psi, \theta_{sc})$ are labeled and analyzed as a map of the $(q_\parallel = q_y, q_\perp = q_z)$ reciprocal plane. This is affordable if the forward wavevector transfer $q_x$ is negligible or, in other words, if the curvature of the Ewald sphere can be reasonably neglected which is not obvious for long-range ordered systems. Putting some figures in Eq. (7.1) leads to $\lambda = 0.1$ nm, $\theta_{in} = \theta_{sc} = 0.1°$ and $\Psi = 0.5°$ to $q_x = 2.4\,10^{-3}$ nm$^{-1}$, $q_y = 0.54$ nm$^{-1}$, $q_z = 0.21$ nm$^{-1}$. A typical GISAXS pattern is made of two sharp peaks due to the transmitted and specularly reflected beams, of an intense specular rod and of diffuse scattering. The $z = 0$ sample surface is called the horizon while an enhancement of scattering, the so-called Yoneda peak (see Sect. 4.3.1) [52], is found at an exit angle $\theta_{sc}$ close to the angle of total external reflection $\theta_c$.

In conventional X-ray reflectivity (see Fig. 7.2), the intensity of the reflected beam is recorded as function of the incident angle by integrating a small angular slice around $\psi = 0$ ($\theta, 2\theta$ scan). As only the perpendicular wavevector transfer $q_z$ is involved, the measurement is sensitive to the perpendicular profile of index of refraction (see Chap. 3). An off-specular scan, obtained for example by rocking the

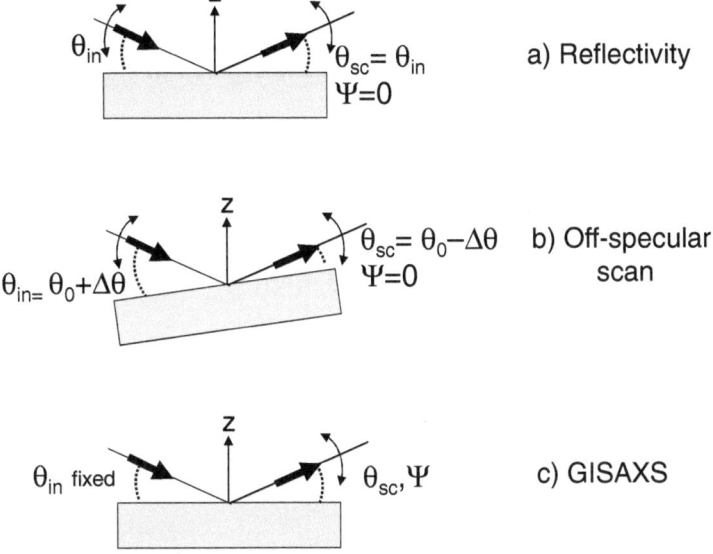

**Fig. 7.2** Experimental scans performed to probe surface roughness: (**a**) reflectivity measurement, (**b**) off-specular x-ray reflectivity, (**c**) GISAXS

sample,[1] gives access to a wavevector transfer in the surface plane $q_x$ and allows to probe length scale parallel to the surface. Most often, a GISAXS experiment does not consider the specular reflection or the specular rod and focuses only on the diffuse scattering. One advantage of GISAXS over off-specular reflectivity in the plane of incidence is to probe in-plane roughness along $\Psi$ at a fixed scattering depth [51] (see Sect. 4.4.1), i.e., $\theta_{in}, \theta_{sc} = Cte$.

## 7.3 Scattering from Density Inhomogeneities in DWBA: The Case of Isolated Particles

The exact Green function formalism of diffuse scattering was developed in Chap. 4 as well as the several approximations, i.e., Born and distorted wave Born approximations which allow to obtain tractable expressions of the scattering cross section. The scattering from roughness, either of a single surface or multilayers, was previously treated in Chap. 6. The goal of this part is to give practical applications of DWBA in the case of surface or subsurface nanostructures like densities inhomogeneities made of particles. The generic term of "particle" encompasses inclusions in a matrix, islands on surfaces or holes in a substrate with a well-defined geometrical shape.

---

[1] The limitation due to the sample surface with off-specular reflectivity measurements disappears in the GISAXS geometry.

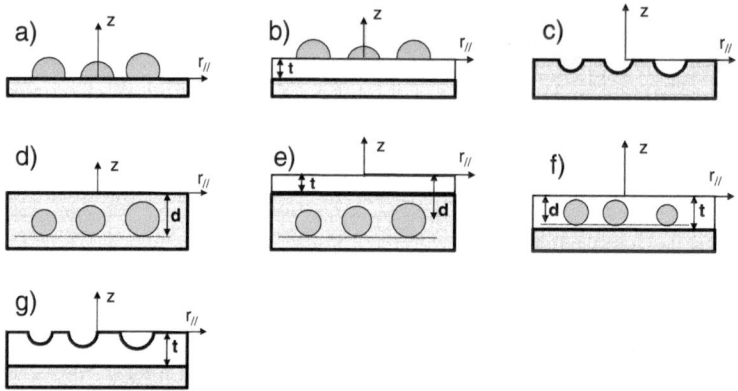

**Fig. 7.3** Schematic layout of the particle morphologies: (**a**) islands on a substrate; (**b**) islands on an overlayer; (**c**) holes in a substrate; (**d**) inclusions in a matrix; (**e**) inclusions below an overlayer; (**f**) inclusions in an overlayer; (**e**) holes in an overlayer. $t$ is the thickness of the overlayer and $d$ the burying depth of the inclusions ($t, d \geq$)

It will be shown that the effective form factor of a particle involves a four event-scattering process including the reflection and refraction of waves at interfaces. The problem of particle assemblies is the topic of the next section. The morphological cases sketched in Fig. 7.3 will be handled:

- Islands supported on a substrate
- Layer of inclusions in a matrix or holes with defined shapes in a surface
- Inclusions or holes in an overlayer on substrate

These morphologies of densities inhomogeneities are quite restrictive but give an overview of the treatment of small-angle scattering from particles in terms of form factor.

In the first Born approximation, the cross section of a particle is, apart from prefactors, nothing else than the Fourier transform of its shape. However, if the incoming $\theta_{in}$ or outgoing $\theta_{sc}$ angles are close to the critical angle of total external reflection $\theta_c$ as it is the case in a GISAXS experiment, better results are obtained with the semi-dynamical treatment of scattering of the DWBA. This first-order perturbation theory [37, 39] needs to define a reference state which is chosen according to the sample morphology and a perturbation potential which is the particle; of course, advantage is taken by calculating the reference wavefields for flat interfaces (see Chap. 4). The problem is restricted to one single scattering particle, collective effects being treated in the next section. Polarization effects will be dropped out as they induce minor correction of the fourth order in the scattering angles.

## 7.3.1 Islands on a Substrate

The problem is simplified by considering only one island on an infinite smooth surface [48]. To calculate the scattering, the starting point is the reflection–refraction of a plane wave on a perfectly smooth surface. The Helmholtz propagation equation is reduced to a scalar one for the amplitude of the electric field $E(\mathbf{r})$ in $s$-polarization:

$$\left[\nabla^2 + k_0^2 n^2(\mathbf{r})\right] E(\mathbf{r}) = 0. \tag{7.2}$$

$k_0 = 2\pi/\lambda$ is the vacuum wavevector. To work out the perturbation formalism, the dielectric constant is decomposed according to

$$n^2(\mathbf{r}) = n_0^2(z) + (n_i^2 - 1)S(\mathbf{r}). \tag{7.3}$$

$n_0(z)$ is the index of refraction of the unperturbed system, i.e., for $z > 0$, $n_0(z) = 1$ and for $z < 0$, $n_0(z) = n_s = 1 - \delta_s - i\beta_s$ is the index of the substrate. $\delta n^2(\mathbf{r}) = (n_i^2 - 1)S(\mathbf{r})$ is the scattering perturbation induced by the island of index $n_i = 1 - \delta_i - i\beta_i$ and of shape $S(\mathbf{r})$. $S(\mathbf{r})$ is one inside the island and zero outside. For the bare substrate, the wavefield in reciprocal space is given by the Fresnel functions:

$$E_{\text{ref}}(\mathbf{r}, \mathbf{k}_{in}) = E_{in} E_1^{PW}(k_{inz,0}, z) e^{-i\mathbf{k}_\parallel \cdot \mathbf{r}_\parallel},$$

$$\text{with} \quad E_1^{PW}(k_{inz,0}, z) = \begin{cases} e^{-ik_{inz,0}z} + r_{0,1}^{in} e^{ik_{inz,0}z} & \text{for} \quad z > 0 \\ t_{0,1}^{in} e^{-ik_{inz,1}z} & \text{for} \quad z < 0 \end{cases} \tag{7.4}$$

$\mathbf{k}_\parallel$ is the component of the wavevector parallel to the surface while $k_{inz,0} = -\sqrt{k_0^2 - k_\parallel^2}$ and $k_{inz,1} = -\sqrt{k_0^2 n_s^2 - k_\parallel^2}$ are the components perpendicular to the surface, respectively, in vacuum and in the substrate, the second one resulting from the Snell–Descartes second law. The introduced Fresnel coefficients in reflection $r_{0,1}$ or in transmission $t_{0,1}$ (see Sect. 3.1) are given by

$$r_{0,1}^{in} = \frac{k_{inz,0} - k_{inz,1}}{k_{inz,0} + k_{inz,1}}, \quad t_{0,1}^{in} = \frac{2k_{inz,0}}{k_{inz,0} + k_{inz,1}}. \tag{7.5}$$

As explained in Chap. 4, the far-field Green function $E_{\text{det}}(\mathbf{R}, \mathbf{r}')$ needed for the DWBA perturbation treatment is evaluated in vacuum while the unperturbed state is the reflected–refracted waves $E_{\text{ref}}(\mathbf{r}, \mathbf{k}_{in})$, Eq. (4.17):

$$E_{\text{det}}(\mathbf{R}, \mathbf{r}') = \frac{k_0^2 \exp(-ik_0 R)}{4\pi\epsilon_0 R} E_1^{PW}(-k_{scz}, z') e^{i\mathbf{k}_{sc\parallel} \cdot \mathbf{r}'_\parallel}. \tag{7.6}$$

$\mathbf{n}_{sc} = \mathbf{k}_{sc}/k_0$ gives the direction of the detector. Putting Eqs. (7.4, 7.5 and 7.6) in the equation giving the amplitude of the scattered field and using the expression of the perturbation induced by the island Eq. (7.3), one obtains the total field to first order:

$$E_{sc}(\mathbf{r}, \mathbf{k}_{in}, \mathbf{k}_{sc}) = E_{\text{ref}}(\mathbf{r}, \mathbf{k}_{in}) +$$
$$\frac{k_0^2 \exp(-ik_0 R)}{4\pi R} E_{in} (n_i^2 - 1) \int d\mathbf{r} e^{i\mathbf{q}_\parallel \cdot \mathbf{r}_\parallel} E_1^{PW}(-k_{scz}, 0, z) S(\mathbf{r}) E_1^{PW}(k_{inz,0}, z). \tag{7.7}$$

Using the expression of $E_1^{PW}$ (Eq. (7.4)), the differential scattering cross section per particle and per unit area in an out-specular direction obtained through the flux of Poynting vector reads

$$\left(\frac{d\sigma}{d\Omega}\right)_{incoh} = \frac{k_0^4}{16\pi^2}\left|n_i^2 - 1\right|^2 \left|\mathscr{F}(\mathbf{q}_{\parallel}, k_{inz,0}, k_{scz,0})\right|^2. \tag{7.8}$$

The DWBA island form factor $\mathscr{F}(\mathbf{q}_{\parallel}, k_{inz,0}, k_{scz,0})$ is defined through

$$\mathscr{F}(\mathbf{q}_{\parallel}, k_{inz,0}, k_{scz,0}) = F(\mathbf{q}_{\parallel}, k_{scz,0} - k_{inz,0}) + r_{0,1}^{sc} F(\mathbf{q}_{\parallel}, -k_{scz,0} - k_{inz,0})$$
$$+ r_{0,1}^{in} F(\mathbf{q}_{\parallel}, k_{scz,0} + k_{inz,0}) + r_{0,1}^{in} r_{0,1}^{sc} F(\mathbf{q}_{\parallel}, -k_{scz,0} + k_{inz,0}). \tag{7.9}$$

$\mathbf{q} = \mathbf{k}_{sc} - \mathbf{k}_{in}$ is the wavevector transfer and $F(\mathbf{q})$ the Fourier transform of the particle shape known as the form factor:

$$F(\mathbf{q}) = \int_{S(\mathbf{r})} d\mathbf{r} e^{i\mathbf{q}\cdot\mathbf{r}}. \tag{7.10}$$

The diagrammatic interpretation of the DWBA calculation is depicted on Fig. 7.4. Four scattering events with different effective wavevector transfers $q_z^{eff} = \pm k_{scz} \pm k_{inz}$ on the island interfere coherently. The first one, the classical Born term, involves a direct scattering by the island. The other ones introduce the possibility of a reflection of either the incident or the scattered waves on the flat substrate surface. The island form factors calculated with the appropriated vertical wavevector transfers are weighted by the Fresnel reflection coefficients, either in incidence $r_{0,1}^{in}(k_{zin})$ or in emergence $r_{0,1}^{sc}(k_{zsc})$. The sharp variation of these coefficients close to the substrate critical angle of total external refection $\theta_c$ (see Fig. 3.3) leads to an enhancement of the scattered intensity known as the Yoneda peak [52] around $\theta_{sc} = \theta_c$. The maximum of diffuse scattering is observed close to the critical angle $\theta_c$. A noticeable point is that the classical wavevector transfer $\mathbf{q}$ is not enough to fully describe the scattering contrary to the first Born approximation. Indeed, in this case, the cross section is limited to the first term of Eq. (7.9) as the Green function and the electric field are simply evaluated in vacuum.

A typical example of DWBA form factor is given in Fig. 7.5 for a cylinder. At $\theta_{sc} = \theta_c$, an enhancement of intensity, known as the Yoneda peak [52], is found whose shape is driven by $\theta_{in}$ and the index of refraction of the substrate. In DWBA, the interference fringes[2] of the Born form factor $F(\mathbf{q}_{\parallel}, q_z)$ are smeared out by the

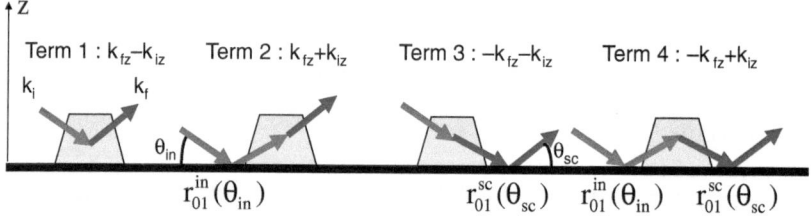

**Fig. 7.4** The four scattering events and their associated vertical wavevector transfers in the case of the DWBA form factor of an island. From [48]

---

[2] These fringes are analogous the Kiessig fringes observed in reflectivity.

**Fig. 7.5** Typical DWBA form factor of a cylindrical island Eq. (7.9) as function of the exit angle $\theta_{sc}$ for various incident angles $\theta_{in}$. Also added the classical Born form factor for two incident angles $\theta_{in} = \theta_c, 2\theta_c$. The parameters are $\delta_s = 5.10^{-6}, \beta_s = 2.10^{-8}$. The intensities are normalized by the volume of the scattering particle

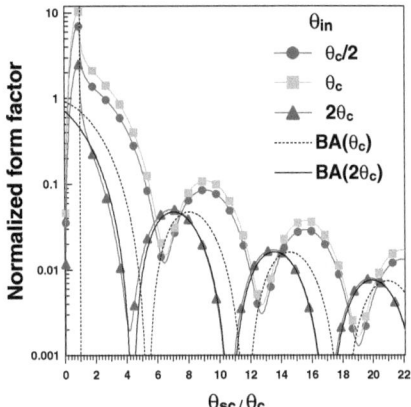

coherent interference between the four scattering events. Furthermore, the $\theta_{sc}$ location of the minima becomes dependant on the incident angle $\theta_{in}$ and no more simply related to the height of the islands as it would be expected in the Born approximation. However, the fringes spacing is still simply related to the particle height. Except close to the critical angle $\theta_{sc} = \theta_c$, because of the steep decrease of $r_{0,1}^{sc}$ with $\theta_{sc}$, two terms dominate the DWBA form factor (see Fig. 7.6): the Born term and that involving a reflection of the incident wave before scattering. As expected, the Born approximation prevails in the range $\theta_{in}, \theta_{sc} \gg \theta_c$.

Some morphological modifications of the substrate can be accounted for very simply in Eq. (7.8) by modifying the Fresnel reflectivities. If the substrate is covered by a continuous overlayer of thickness $t$ (case b of Fig. 7.3), the reflection coefficient of Eqs. (7.9) has to be modified accordingly (Sect. 3.2.4). Also for substrate roughness of wavelength higher than that induced by the islands (to avoid interferences between the scattered waves), the Croce–Névot factor (Sect. 3.A.3) can be introduced to reduce the substrate reflectivity [49].

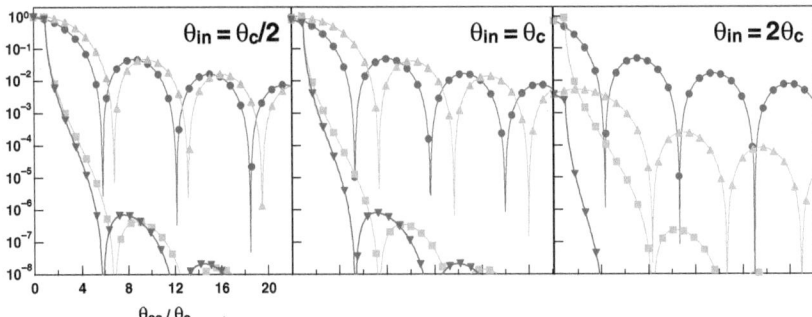

**Fig. 7.6** The modulus squared of the various components involved in the cylinder DWBA form factor shown in Fig. 7.5. The symbols (*circles, uptriangles, downtriangles, squares*) correspond, respectively, to the four scattering events of Fig. 7.4 from left to right

## 7.3.2 Particles Buried in a Substrate or Holes with Well-Defined Shapes in Substrate

The case of a particle (or more generally a buried density fluctuation) buried at a distance $d$ under the smooth surface of a substrate (see Fig. 7.3-d) is handled in a very similar way as the case of an island. The perturbation potential is now given by the difference of dielectric constant between the inclusion of shape $S(\mathbf{r})$ and the substrate, i.e., $\delta n^2(\mathbf{r}) = (n_i^2 - n_s^2)S(\mathbf{r})$. The eigenstates of scattering are still the Fresnel solutions of the propagation equation, Eq. (7.4), and the Green function that is in vacuum, Eq. (7.6). The difference with the previous case comes from the integration in the scattering cross section that is limited to $z < 0$ thus involving the transmitted part of the Fresnel field. The final result is given by

$$\left(\frac{d\sigma}{d\Omega}\right)_{incoh} = \frac{k_0^4}{16\pi^2}\left|n_i^2 - n_s^2\right|^2 \left|\mathscr{F}(\mathbf{q}_\parallel, k_{inz,0}, k_{scz,0})\right|^2, \qquad (7.11)$$

$$\mathscr{F}(\mathbf{q}_\parallel, k_{inz,0}, k_{scz,0}) = t_{0,1}^{in} t_{0,1}^{sc} F(\mathbf{q}_\parallel, k_{scz,1} - k_{inz,1})e^{id(k_{scz,1} - k_{inz,1})}.$$

The cross section is weighted by the transmission coefficients, either in reflection or in transmission (see Sect. 4.3.1) which lead to the Yoneda peak. These latter are a signature of the source–observer reciprocity theorem of wave propagation; they would be absent in the first Born approximation. The particle form factor Eq. (7.11) is calculated for the wavevector transfer evaluated inside the substrate and multiplied by a phase factor which accounts for the propagation and attenuation of the waves along their paths to the particle or to the substrate. Some morphological characteristics of the substrate like a covering homogeneous layer (see Fig. 7.3-e) or an uncorrelated roughness can also be accounted for in the transmission prefactor coefficients.

This treatment is also relevant for the case of holes with well-defined shapes characterized by a function $S(\mathbf{r})$. The contrast of dielectric constant is given by $\delta n^2(\mathbf{r}) = (n_s^2 - 1)S(\mathbf{r})$; the phase factor is useless ($d = 0$) if the particle form factor Eq. (7.10) as calculated in the forthcoming section Sect. 7.3.5 (see Fig. 7.9) is replaced by $F(q_x, -q_y, -q_z)$, i.e., if the trihedron linked to the particle has its $z$-axis inverted. Nothing allows to distinguish between scattering from holes or true inclusions (except the phase factor in Eq. (7.11)) as it is expected from the Babinet principle in optics.

## 7.3.3 Holes or Particles Encapsulated in a Layer on Substrate

The morphology under consideration is made of particles or holes in an overlayer of thickness $t$ and index of refraction $n_l = 1 - \delta_l - i\beta_l$ on a bulk substrate (see Fig. 7.3-f,g). The inclusions are buried at a distance $d$ from the top interface. The solution of the propagation equation Eq. (7.2) with the following index of refraction:

$$n_0(z) = \begin{cases} 1 & \text{if} \quad z > 0 \\ n_l & \text{if} \quad -t < z < 0 \\ n_s & \text{if} \quad z < -t \end{cases} \tag{7.12}$$

is given by a set of upward and downward propagating waves (see Sect. 3.2.1):

$$E_{\text{ref}}(\mathbf{r}, \mathbf{k}_{in}) = E_{in} E_1^{PW}(k_{inz,0}, z) e^{-ik_\parallel \cdot r_\parallel}, \quad \text{with} \tag{7.13}$$

$$E_1^{PW}(k_{inz,0}, z) = \begin{cases} A_0^+ e^{ik_{inz,0}z} + A_0^- e^{-ik_{inz,0}z} & \text{for} \quad z > 0 \\ A_1^+ e^{ik_{inz,1}z} + A_1^- e^{-ik_{inz,1}z} & \text{for} \quad -t < z < 0, \\ A_2^+ e^{ik_{inz,2}z} & \text{for} \quad z < -t \end{cases} \tag{7.14}$$

where $k_{inz,j}(j = 0, 1, 2) = -\sqrt{n_j^2 k_0^2 - k_\parallel^2}$ is the perpendicular component of the wavevector in each media (vacuum, layer, substrate). Matching the boundaries conditions (continuity of the field and its first derivative) leads to the expressions of the coefficients $A_j^\pm$ as function of the reflection $r_{i,j}^{in}$ and transmission $t_{i,j}^{in}$ of each interface (Eq. (7.5)). The only useful ones for the DWBA scattering cross section are those inside the layer normalized by the incident wave coefficient $A_0^+$:

$$\frac{A_1^-}{A_0^+} = \tilde{A}_1^- = \frac{t_{0,1}^{in}}{1 + r_{0,1}^{in} r_{1,2}^{in} e^{2ik_{inz,1}t}}, \tag{7.15}$$

$$\frac{A_1^+}{A_0^+} = \tilde{A}_1^+ = \frac{t_{0,1}^{in} r_{1,2}^{in} e^{2ik_{inz,1}t}}{1 + r_{0,1}^{in} r_{1,2}^{in} e^{2ik_{inz,1}t}}. \tag{7.16}$$

The perturbation in the case of inclusions (index $n_i$ and shape $S(\mathbf{r})$) is given by $\delta n^2(\mathbf{r}) = (n_i^2 - n_l^2)S(\mathbf{r})$ and exists only inside the layer. The Fresnel unperturbed state solution in the far-field Green function leads to the scattering cross section per unit area:

$$\left( \frac{d\sigma}{d\Omega} \right)_{\text{incoh}} = \frac{k_0^4}{16\pi^2} \left| n_i^2 - n_l^2 \right|^2 \left| \mathscr{F}(\mathbf{q}_\parallel, k_{inz}, k_{scz}) \right|^2. \tag{7.17}$$

Once again an effective form factor has been introduced:

$$\mathscr{F}(\mathbf{q}_\parallel, k_{inz,0}, k_{scz,0})$$

$$= \tilde{A}_1^-(k_{inz,1}) \tilde{A}_1^-(-k_{scz,1}) e^{i(+k_{scz,1} - k_{inz,1})d} F(\mathbf{q}_\parallel, +k_{scz,1} - k_{inz,1})$$

$$+ \tilde{A}_1^+(k_{inz,1}) \tilde{A}_1^-(-k_{scz,1}) e^{i(+k_{scz,1} + k_{inz,1})d} F(\mathbf{q}_\parallel, +k_{scz,1} + k_{inz,1})$$

$$+ \tilde{A}_1^-(k_{inz,1}) \tilde{A}_1^+(-k_{scz,1}) e^{i(-k_{scz,1} - k_{inz,1})d} F(\mathbf{q}_\parallel, -k_{scz,1} - k_{inz,1})$$

$$+ \tilde{A}_1^+(k_{inz,1}) \tilde{A}_1^+(-k_{scz,1}) e^{i(-k_{scz,1} + k_{inz,1})d} F(\mathbf{q}_\parallel, -k_{scz,1} + k_{inz,1}). \tag{7.18}$$

The diagrammatic picture of Eq. (7.18) is depicted in Fig. 7.7. The inclusion gives rise to four different scattering events with different wavevector transfers:

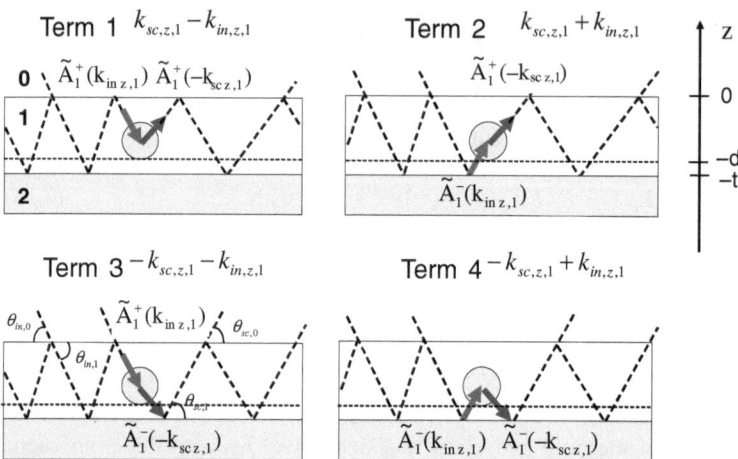

**Fig. 7.7** The four terms involved in the DWBA form factor of an inclusion in a layer on a substrate

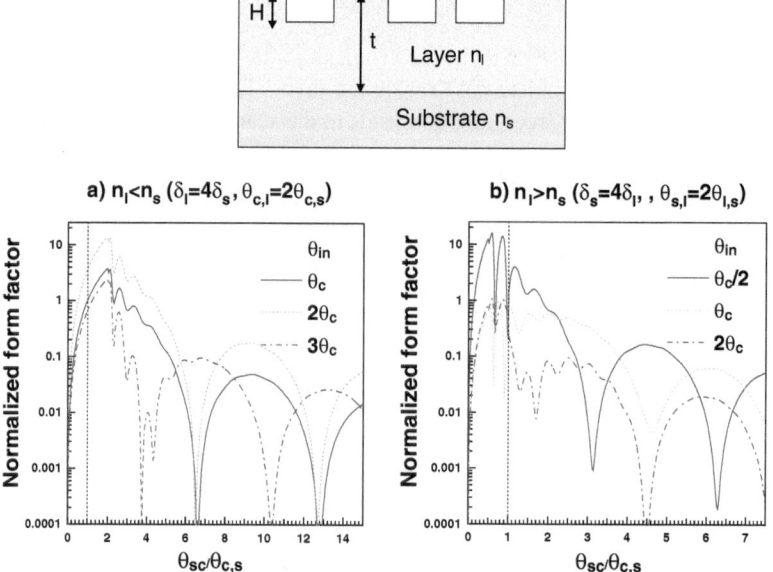

**Fig. 7.8** The DWBA form factor of cylindrical holes of height $H$ in a layer of thickness $t = 3.4H$ which is (**a**) less or more (**b**) refringent than the substrate. The curves are plotted against the exit angle $\theta_{sc}$ for various incident angle $\theta_{in}$. The intensities are normalized by the hole volume and the angle by the critical angle of the substrate $\theta_{c,s}$

$q_z^{eff} = \pm k_{scz,1} \pm k_{inz,1}$ which correspond to a scattering from an upward or downward to an upward or downward propagating waves. Once again, it appears that the wavevector transfer $q_z$ alone is insufficient to fully characterize the scattering process. Contrary to the case of an island, the involved wavevectors are those inside the layer. Each term is multiplied by a phase factor accounting from the path of these waves to the interfaces (vacuum-layer or layer-substrate) and by a reflection $\widetilde{A}_1^-$ or transmission $\widetilde{A}_1^+$ coefficients for the incident or scattered waves. $\widetilde{A}_1^-, \widetilde{A}_1^+$ (Eq. 7.16) include the reflection at the bottom interface $r_{1,2}$ and the transmission at the top interface $t_{0,1}$, plus what is equivalent to Fabry–Pérot interference term $1 + r_{0,1}^{in} r_{1,2}^{in} e^{2ik_{inz,1}t}$ leading to layer thickness interference fringes.

Holes (Fig. 7.3-g) in a layer is only a special case of inclusions if the right contrast of dielectric constant $n_l^2 - 1$ is used. The previous expression of the DWBA form factor is usable if the burying depth $d = 0$ and the particle form factor $F(\mathbf{q}_\parallel, q_z)$ is calculated with the $z$-axis of the trihedron associated to the hole oriented downward (see Fig. 7.9).

Figure 7.8 shows an example of such form factor of holes. The dynamical effects are more complex than for a single interface. The two superimposed frequencies of beating are due to the hole depth $H$ and to the layer thickness $t$. The higher the incident angle $\theta_{in}$ the sharper the fringes of the layer as the probed depth and the sensitivity to the substrate–layer interface are greater. In the case of a layer less refringent than the substrate ($\delta_s > \delta_l$) (Fig. 7.8-b), the incident wave or the time-inverted wave starts to penetrate first in the layer and then in the substrate yielding a kind of double Yoneda peak.

## 7.3.4 General Concluding Remarks

As a general rule at small angles [1], the incoherent scattering cross section is proportional to the contrast of dielectric constant between the scatterer and the embedding medium. Indeed, the obtained prefactors are $|n_i^2 - 1|^2$ for island (Eq. (7.8)), $|n_i^2 - n_s^2|^2$ or $|1 - n_s^2|^2$ for inclusions or holes in a substrate (Eq. (7.11)) and $|n_i^2 - n_l^2|^2$ or $|1 - n_l^2|^2$ for inclusions or holes in a layer.

As introduced in Chap. 1, the above DWBA formalism is also suitable to the case of neutron scattering (GISANS) if the Thomson scattering length is replaced by the Fermi pseudopotential scattering length of neutrons in the expression of the refraction index $n = 1 - 2\pi/k_0^2 \langle \rho \rangle b_{at}$. Moreover, the energy dispersive behavior for X-rays are implicitly included in the refraction index through the anomalous terms $b_{at} = r_e(f + f' + f'')$.

For very shallow angles $\theta_{in}, \theta_{sc} \leq \theta_c$, all the effective perpendicular wavevector transfers $\pm k_{scz,0-1} \pm k_{inz,0-1}$ of Eq. (7.9) for islands Eq. (7.18) or inclusions in a layer Eq. (7.18) are small. Therefore, the form factor is close to $F(\mathbf{q}_\parallel, 0)$ and can be factorized. For islands, one obtains

$$\mathscr{F}(\mathbf{q}_{\|},k_{inz},k_{scz}) \simeq F(\mathbf{q}_{\|},0)(1+r_{0,1}^{sc}+r_{0,1}^{in}+r_{0,1}^{sc}r_{0,1}^{in})$$
$$= F(\mathbf{q}_{\|},0)(1+r_{0,1}^{in})(1+r_{0,1}^{sc})$$
$$= F(\mathbf{q}_{\|},0)t_{0,1}^{in}t_{0,1}^{sc}. \tag{7.19}$$

A similar result is obtained for inclusions in a layer. Whatever the morphology is, at $\theta_{in},\theta_{sc} \le \theta_c$, the dynamical scattering effects in the perpendicular direction are driven by the transmission functions inside the substrate as in the case of inclusions in a substrate Eq. (7.11).

The semi-dynamical treatment of scattering of DWBA is necessary only if $\theta_{in}$ or $\theta_{sc}$ are close the critical angle $\theta_c$ of the materials. Well above this value, all the reflection coefficients that decrease at least as $1/\theta_{in,sc}^4$ are negligible while the transmission ones reach their final value 1. Therefore, within this limit, the Born approximation or the simple kinematical treatment of scattering is valid. The analysis is thus considerably simplified as the particle form factor is given only by the Fourier transform of its shape. However, working at grazing incidence $\theta_{in}$ and exit $\theta_{sc}$ angles not only reduces the bulk background but also enhances the surface signal in absolute value (see Fig. 7.5). For inclusions in a non-absorbing substrate, at $\theta_{in}=\theta_{sc}=\theta_c$, $t_{0,1}^{in}(\theta_c)=t_{0,1}^{sc}(\theta_c)=2$ and $k_{scz,1}=k_{inz,1}=0$:

$$\mathscr{F}(\mathbf{q}_{\|},k_{inz}^c,k_{scz}^c) = 4F(\mathbf{q}_{\|},0). \tag{7.20}$$

The waves are exactly traveling parallel to the surface. For islands, as $r_{0,1}^{in}=r_{0,1}^{sc}=1$,

$$\mathscr{F}(\mathbf{q}_{\|},k_{inz}^c,k_{scz}^c) = F(\mathbf{q}_{\|},2k_z^c)+F(\mathbf{q}_{\|},0)+F(\mathbf{q}_{\|},0)+F(\mathbf{q}_{\|},-2k_z^c). \tag{7.21}$$

The enhancement is slightly lower than for inclusions.

## 7.3.5 Form Factor Expressions of Simple Geometrical Shapes

### 7.3.5.1 The Form Factor Expressions

When particles have a well-defined shape, it is possible to calculate analytically their form factor $F(\mathbf{q})$ (Eq. (7.10)) by using symmetry.

The cartesian frame linked to each particle is defined with its origin at the bottom of the object,[3] its $z$-axis oriented upward and it $x$-axis along one side of the particle. The herein library of shapes as well as the geometrical parameters are all defined in Fig. 7.9. The form factors $F(\mathbf{q})$, the volume $V$, the surface seen from top $S$ and the radius $R_{pa}$ obtained by rotating the object around its $z$-axis are the following:

---

[3] This convention find its meaning when evaluating the mean form factors.

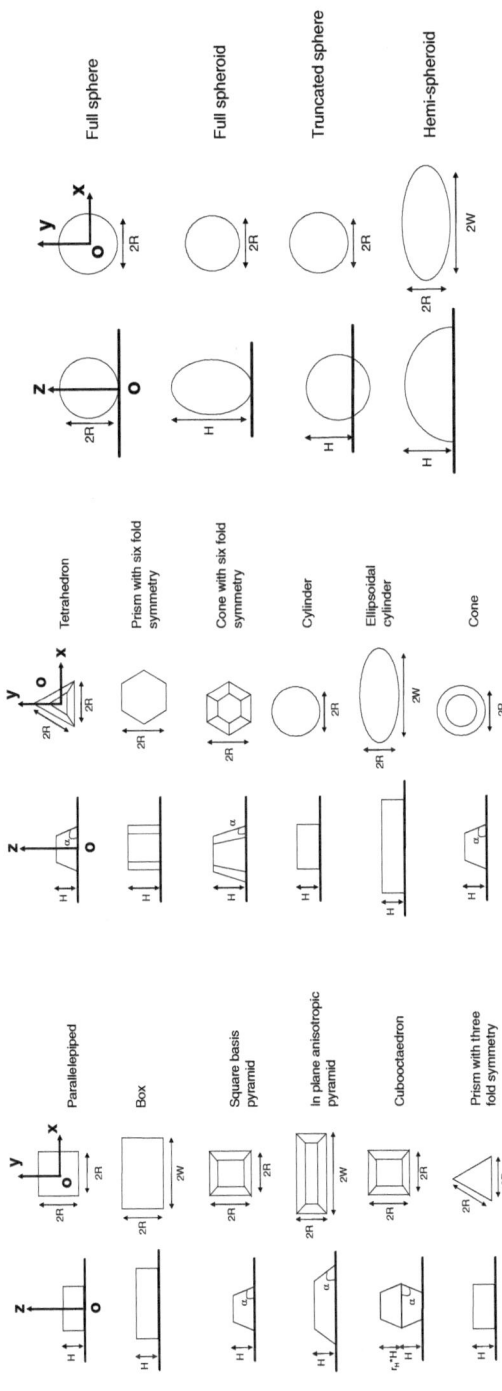

**Fig. 7.9** Particle shapes which form factors are given in Sect. 7.3.5. *Left panel: side view, right panel: top view*

- *parallelepiped*:

$$F_{pa}(\mathbf{q},R,H) = 4R^2 H \mathrm{sinc}(q_x R)\mathrm{sinc}(q_y R)\mathrm{sinc}(q_z H/2)e^{iq_z H/2},$$

$$V_{pa} = 4R^2 H,\ S_{pa} = 4R^2,\ R_{pa} = \sqrt{2}R. \tag{7.22}$$

- *box*:

$$F_{box}(\mathbf{q},R,W,H) = 4RWH \mathrm{sinc}(q_x R)\mathrm{sinc}(q_y W)\mathrm{sinc}(q_z H/2)e^{iq_z H/2},$$

$$V_{box} = 4RWH,\ S_{box} = 4RW,\ R_{box} = \sqrt{R^2 + W^2}. \tag{7.23}$$

- *square basis pyramid*:

$$F_{py}(\mathbf{q},R,H,\alpha) = \int_0^H 4R_z^2 \mathrm{sinc}(q_x R_z)\mathrm{sinc}(q_y R_z)e^{iq_z z}\,dz,$$

$$R_z = R - z/\tan(\alpha),\ H/R < \tan(\alpha),$$

$$V_{py} = \frac{4}{3}\tan(\alpha)\left[R^3 - \left(R - \frac{H}{\tan(\alpha)}\right)^3\right],\ S_{py} = 4R^2,\ R_{py} = \sqrt{2}R;$$

$$F_{py}(\mathbf{q},R,H,\alpha) = \frac{H}{q_x q_y}$$

$$\times\left\{\cos[(q_x - q_y)R]K_1 + \sin[(q_x - q_y)R]K_2\right.$$

$$\left. - \cos[(q_x + q_y)R]K_3 - \sin[(q_x + q_y)R]K_4\right\},$$

$$K_1 = \quad \mathrm{sinc}(q_1 H)e^{iq_1 H} + \mathrm{sinc}(q_2 H)e^{-iq_2 H},$$

$$K_2 = -i\mathrm{sinc}(q_1 H)e^{iq_1 H} + i\mathrm{sinc}(q_2 H)e^{-iq_2 H},$$

$$K_3 = \quad \mathrm{sinc}(q_3 H)e^{iq_3 H} + \mathrm{sinc}(q_4 H)e^{-iq_4 H},$$

$$K_4 = -i\mathrm{sinc}(q_3 H)e^{iq_3 H} + i\mathrm{sinc}(q_4 H)e^{-iq_4 H},$$

$$q_1 = \frac{1}{2}\left[\frac{q_x - q_y}{\tan(\alpha)} + q_z\right],\ q_2 = \frac{1}{2}\left[\frac{q_x - q_y}{\tan(\alpha)} - q_z\right],$$

$$q_3 = \frac{1}{2}\left[\frac{q_x + q_y}{\tan(\alpha)} + q_z\right],\ q_4 = \frac{1}{2}\left[\frac{q_x + q_y}{\tan(\alpha)} - q_z\right]. \tag{7.24}$$

- *in-plane anisotropic pyramid*:

$$F_{anpy}(\mathbf{q},R,W,H,\alpha) = \int_0^H 4R_z W_z \mathrm{sinc}(q_x R_z)\mathrm{sinc}(q_y W_z)e^{iq_z z}\,dz,$$

$$R_z = R - z/\tan(\alpha),\ W_z = W - z/\tan(\alpha),$$

$$H/R < \tan(\alpha),\ W/R < \tan(\alpha), \tag{7.25}$$

$$V_{anpy} = 4\left[WRH - \frac{H^2(R+W)}{2\tan(\alpha)} + \frac{H^3}{3\tan^2(\alpha)}\right],\ S_{anpy} = 4RW,$$

$$R_{anpy} = \sqrt{R^2 + W^2},$$

$$F_{anpy}(\mathbf{q},R,W,H,\alpha) = \frac{H}{q_x q_y}$$

$$\times \left\{\cos[q_x R - q_y W]K_1 + \sin[q_x R - q_y W]K_2\right.$$

$$\left. -\cos[q_x R + q_y W]K_3 - \sin[q_x R + q_y W]K_4\right\},$$

$$K_1 = \ \ \mathrm{sinc}(q_1 H)e^{iq_1 H} + \mathrm{sinc}(q_2 H)e^{-iq_2 H},$$

$$K_2 = -i\mathrm{sinc}(q_1 H)e^{iq_1 H} + i\mathrm{sinc}(q_2 H)e^{-iq_2 H},$$

$$K_3 = \ \ \mathrm{sinc}(q_3 H)e^{iq_3 H} + \mathrm{sinc}(q_4 H)e^{-iq_4 H},$$

$$K_4 = -i\mathrm{sinc}(q_3 H)e^{iq_3 H} + i\mathrm{sinc}(q_4 H)e^{-iq_4 H},$$

$$q_1 = \frac{1}{2}\left[\frac{q_x - q_y}{\tan(\alpha)} + q_z\right],\ q_2 = \frac{1}{2}\left[\frac{q_x - q_y}{\tan(\alpha)} - q_z\right],$$

$$q_3 = \frac{1}{2}\left[\frac{q_x + q_y}{\tan(\alpha)} + q_z\right],\ q_4 = \frac{1}{2}\left[\frac{q_x + q_y}{\tan(\alpha)} - q_z\right]. \tag{7.26}$$

- *cubooctahedron*:

$$F_{cu}(\mathbf{q},R,H_1,H_2,\alpha) = e^{iq_z H}\left[F_{py}(q_x,-q_y,-q_z,R,H,\alpha)\right.$$

$$\left. + F_{py}(q_x,q_y,q_z,R,r_H H,\alpha)\right],$$

$$H/R < \tan(\alpha),\ r_H H/R < \tan(\alpha), \tag{7.27}$$

$$V_{cu} = \frac{4}{3}\tan(\alpha)\left[2R^3 - \left(R - \frac{H}{\tan(\alpha)}\right)^3 - \left(R - \frac{Hr_H}{\tan(\alpha)}\right)^3\right],$$

$$S_{cu} = 4R^2,\ R_{cu} = \sqrt{2}R.$$

- *prism with threefold symmetry*:

$$F_{pr3}(\mathbf{q},R,H) = \frac{2\sqrt{3}e^{-iq_yR/\sqrt{3}}}{q_x(q_x^2 - 3q_y^2)} \left[ q_x e^{iq_yR\sqrt{3}} - q_x\cos(q_xR) \right.$$
$$\left. - i\sqrt{3}q_y\sin(q_xR) \right] \times \ \mathrm{sinc}(q_zH/2)e^{iq_zH/2},$$
$$V_{pr3} = \sqrt{3}R^2H, \ S_{pr3} = \sqrt{3}R^2, \ R_{pr3} = \frac{2}{\sqrt{3}}R. \qquad (7.28)$$

- *tetrahedron*:

$$F_{te}(\mathbf{q},R,H,\alpha) = \frac{2\sqrt{3}}{q_x(q_x^2 - 3q_y^2)} \int_0^H e^{-iq_yR_z/\sqrt{3}} \left[ q_x e^{iq_yR_z\sqrt{3}} - q_x\cos(q_xR_z) \right.$$
$$\left. \times -i\sqrt{3}q_y\sin(q_xR_z) \right] e^{iq_zz}dz,$$

$$R_z = R - \sqrt{3}/\tan(\alpha)z, \ H/R < \tan(\alpha)/\sqrt{3},$$

$$V_{te} = \frac{1}{3}\tan(\alpha)\left[ R^3 - \left( R - \frac{\sqrt{3}H}{\tan(\alpha)} \right)^3 \right], \ S_{te} = \sqrt{3}R^2, \ R_{te} = \frac{2}{\sqrt{3}}R;$$

$$F_{te}(\mathbf{q},R,H,\alpha) = \frac{H}{\sqrt{3}q_x(q_x^2 - 3q_y^2)}e^{iq_zR\tan(\alpha)/\sqrt{3}}\left\{ -(q_x + \sqrt{3}q_y) \right.$$
$$\left. \mathrm{sinc}(q_1H)e^{iq_1L} + (-q_x + \sqrt{3}q_y)\mathrm{sinc}(q_2H)e^{-iq_2L} + 2q_x\mathrm{sinc}(q_3H)e^{iq_3L} \right\},$$

$$q_1 = \frac{1}{2}\left[ \frac{\sqrt{3}q_x - q_y}{\tan(\alpha)} - q_z \right], \ q_2 = \frac{1}{2}\left[ \frac{\sqrt{3}q_x + q_y}{\tan(\alpha)} + q_z \right],$$

$$q_3 = \frac{1}{2}\left[ \frac{2q_y}{\tan(\alpha)} - q_z \right], \ L = \frac{2\tan(\alpha)R}{\sqrt{3}} - H. \qquad (7.29)$$

- *prism with sixfold symmetry*:

$$F_{pr6}(\mathbf{q},R,H) = \frac{4\sqrt{3}}{3q_y^2 - q_x^2} \left[ q_y^2R^2\mathrm{sinc}(q_xR/\sqrt{3})\mathrm{sinc}(q_yR) + \cos(2q_xR/\sqrt{3}) \right.$$
$$\left. - \cos(q_yR)\cos(q_xR/\sqrt{3}) \right] \mathrm{sinc}(q_zH/2)e^{iq_zH/2},$$
$$V_{pr6} = 2\sqrt{3}R^2H, \ S_{pr6} = 2\sqrt{3}R^2, \ R_{pr6} = \frac{2}{\sqrt{3}}R. \qquad (7.30)$$

- *cone with sixfold symmetry:*

$$F_{co6}(\mathbf{q}, R, H) = \frac{4\sqrt{3}}{3q_y^2 - q_x^2} \times \int_0^H \left[ q_y^2 R_z^2 \operatorname{sinc}(q_x R_z / \sqrt{3}) \operatorname{sinc}(q_y R_z) \right.$$
$$\left. + \cos(2q_x R_z / \sqrt{3}) - \cos(q_y R_z) \cos(q_x R_z / \sqrt{3}) \right] e^{iq_z z} dz,$$

$$R_z = R - z / \tan(\alpha), \ H / R < \tan(\alpha),$$

$$V_{co6} = \frac{2\tan(\alpha)}{\sqrt{3}} \left[ R^3 - \left( R - \frac{H}{\tan(\alpha)} \right)^3 \right], \ S_{co6} = 2\sqrt{3} R^2, \ R_{co6} = \frac{2}{\sqrt{3}} R.$$

$$(7.31)$$

- *cylinder:*

$$F_{cy}(\mathbf{q}, R, H) = 2\pi R^2 H \frac{J_1(q_\parallel R)}{q_\parallel R} \operatorname{sinc}(q_z H / 2) e^{iq_z H / 2},$$

$$q_\parallel = \sqrt{q_x^2 + q_y^2},$$

$$V_{cy} = \pi R^2 H, \ S_{cy} = \pi R^2, \ R_{cy} = R.$$

$$(7.32)$$

- *ellipsoidal cylinder:*

$$F_{ell}(\mathbf{q}, R, W, H, \alpha) = 2\pi R W H \frac{J_1(\gamma)}{\gamma} \operatorname{sinc}(q_z H / 2) e^{iq_z H / 2},$$

$$\gamma = \sqrt{(q_x R)^2 + (q_y W)^2},$$

$$V_{ell} = \pi R W H, \ S_{anpy} = \pi R W, \ R_{anpy} = \sqrt{R^2 + W^2}.$$

$$(7.33)$$

- *cone:*

$$F_{co}(\mathbf{q}, R, H, \alpha) = \int_0^H 2\pi R_z^2 \frac{J_1(q_\parallel R_z)}{q_\parallel R_z} e^{iq_z z} dz,$$

$$q_\parallel = \sqrt{q_x^2 + q_y^2}, \ R_z = R - z / \tan(\alpha), \ H / R < \tan(\alpha),$$

$$(7.34)$$

$$V_{co} = \frac{\pi}{3} \tan(\alpha) \left[ R^3 - \left( R - \frac{H}{\tan(\alpha)} \right)^3 \right], \ S_{co} = \pi R^2, \ R_{co} = R.$$

- *full sphere:*

$$F_{fsp}(\mathbf{q}, R) = 4\pi R^3 \frac{\sin(qR) - qR\cos(qR)}{(qR)^3} e^{iq_z R},$$

$$V_{fsp} = \frac{4}{3} \pi R^3, \ S_{fsp} = \pi R^2, \ R_{fsp} = R.$$

$$(7.35)$$

- *full spheroid*:

$$F_{fsph}(\mathbf{q},R,H) = e^{iq_z H/2} \int_0^{H/2} 4\pi R_z^2 \frac{J_1(q_\| R_z)}{q_\| R_z} \cos(q_z z)\,dz,$$

$$q_\| = \sqrt{q_x^2 + q_y^2},\, R_z = R\sqrt{1 - 4\frac{z^2}{H^2}},$$

$$V_{fsph} = \frac{4}{3}\pi R^2 H,\, S_{fsph} = \pi R^2,\, R_{fsph} = R. \tag{7.36}$$

- *truncated sphere*:

$$F_{sp}(\mathbf{q},R,H) = e^{iq_z(H-R)} \int_{R-H}^{H} 2\pi R_z^2 \frac{J_1(q_\| R_z)}{q_\| R_z} e^{iq_z z}\,dz,$$

$$q_\| = \sqrt{q_x^2 + q_y^2},\, R_z = \sqrt{R^2 - z^2},\, 0 < H/R < 2,$$

$$V_{sp} = \pi R^3 \left[ \frac{2}{3} + \frac{H-R}{R} - \frac{1}{3}\left(\frac{H-R}{R}\right)^3 \right],$$

$$S_{sp} = \begin{cases} \pi R^2 & \text{if } H > R \\ \pi(2RH - H^2) & \text{if } H < R \end{cases},\, R_{sp} = \begin{cases} R & \text{if } H > R \\ \sqrt{2RH - H^2} & \text{if } H < R \end{cases}$$

$$\tag{7.37}$$

- *hemi-spheroid*:

$$F_{hsphe}(\mathbf{q},R,W,H) = 2\pi \int_0^H R_z W_z \frac{J_1(\gamma)}{\gamma} e^{iq_z z}\,dz,$$

$$R_z = R\sqrt{1 - \left(\frac{z}{H}\right)^2},\, W_z = W\sqrt{1 - \left(\frac{z}{H}\right)^2},$$

$$\gamma = \sqrt{(q_x R_z)^2 + (q_y W_z)^2},$$

$$V_{hsphe} = \frac{2}{3}\pi RWH,\, S_{hsphe} = \pi RW,\, R_{hsphe} = \sqrt{R^2 + W^2}. \tag{7.38}$$

$\sin_c(x) = \sin(x)/x$ is the cardinal sine and $J_1(x)$ is the Bessel function of first order [53].

If the island frame makes an angle $\zeta$ with the impinging beam, it is necessary to apply a rotation matrix $\mathscr{R}(\zeta)$ to the wavevector transfer $\mathbf{q}$ before applying the above formulae:

$$\mathscr{R}(\zeta)\mathbf{q} = \begin{pmatrix} \cos(\zeta) & \sin(\zeta) & 0 \\ -\sin(\zeta) & \cos(\zeta) & 0 \\ 0 & 0 & 1 \end{pmatrix} \begin{pmatrix} q_x \\ q_y \\ q_z \end{pmatrix}. \tag{7.39}$$

### 7.3.5.2 Scattering Anisotropy of the Form Factor

Rod of Scattering by Facets

Since it is the Fourier transform of the particle shape, the form factor $F(\mathbf{q})$ contains the symmetry of the object and displays interference fringes that are characteristics of the particle sizes. For polyhedrons, it is shown in [54] that the volume integral in $F(\mathbf{q})$ can be reduced to a surface integral over the facets or even a line integral along the edges. In particular, rods of scattering appear in the direction perpendicular to facets, if any. The widths of these rods are inversely proportional to the facet areas while their tilt with respect to the specular rod gives the angle between the substrate plane and the facet normals. As shown in Fig. 7.10 for a tetrahedron, these rods appear when, in the plane of the substrate, the impinging beam is parallel to the particle edge. At the extreme limit, these rods are parallel to the substrate plane for particle shapes which are invariant by translation along z-axis like parallelepiped or cylinder.

Experimental scattering patterns (Fig. 7.11) from Ge quantum dots grown on Si(100) are displayed in Fig. 7.11. The data analysis showed that the particles have a size of 50 nm and are separated by 135 nm. An annealing allowed to enlarge the size of the {311} lateral facets. Rod of scattering are clearly observed with a tilt of $26 \pm 1°$ from the substrate (100) orientation; this corresponds to {311} planes in cubic system. The shape of this rod can be modulated with the incident angle. Below the critical angle of the substrate ($\theta_{in} = 0.1° < \theta_c = 0.2°$), they point at the specularly reflected beam just below the Yoneda peak; above $\theta_c$, i.e., $\theta_{in} = 0.4°$, they point at the direct beam position while for $\theta_{in} = \theta_c = 0.2°$, the rods are splitted. This can be rationalized if the various scattering terms in the DWBA island form factor (Fig. 7.4) are considered. As shown in Fig. 7.6, for $\theta_{sc} > \theta_c$, two terms dominate the form factor, the Born term with $q_z^{eff} = k_{scz} - k_{inz}$ and the term involving the reflection of the incident beam on the substrate with $q_z^{eff} = k_{scz} + k_{inz}$. They point respectively toward $k_{scz} = k_{inz} < 0$, i.e., the reciprocal space origin and toward $k_{scz} = -k_{inz} > 0$, i.e., the specular position (see Fig. 7.1). The relative weight of these two scattering terms is given by $r_{0,1}^{in}(\theta_{in})$; therefore the second one is strongly damped for $\theta_{in} > \theta_c$.

The Friedel Rule in Grazing Incident Geometry [48]

In a scattering experiment, the phase factor is lost and only the modulus square form factor is recorded. As a result, the scattering is centrosymmetrical $d\sigma/d\Omega(\mathbf{q}) = d\sigma/d\Omega(-\mathbf{q})$, a fact known as the Friedel rule in X-ray diffraction [55, 56]. This means that, in a bulk small-angle scattering experiment, a rotation of the particle by 180° around the perpendicular to the incoming beam does not change the scattering pattern. The sixfold symmetry of the in-plane ($q_z = 0$) BA form factor of a tetrahedron shown in Fig. 7.12-a demonstrates that, with conventional bulk scattering, distinguishing between three- and sixfold particles is impossible. This phase problem

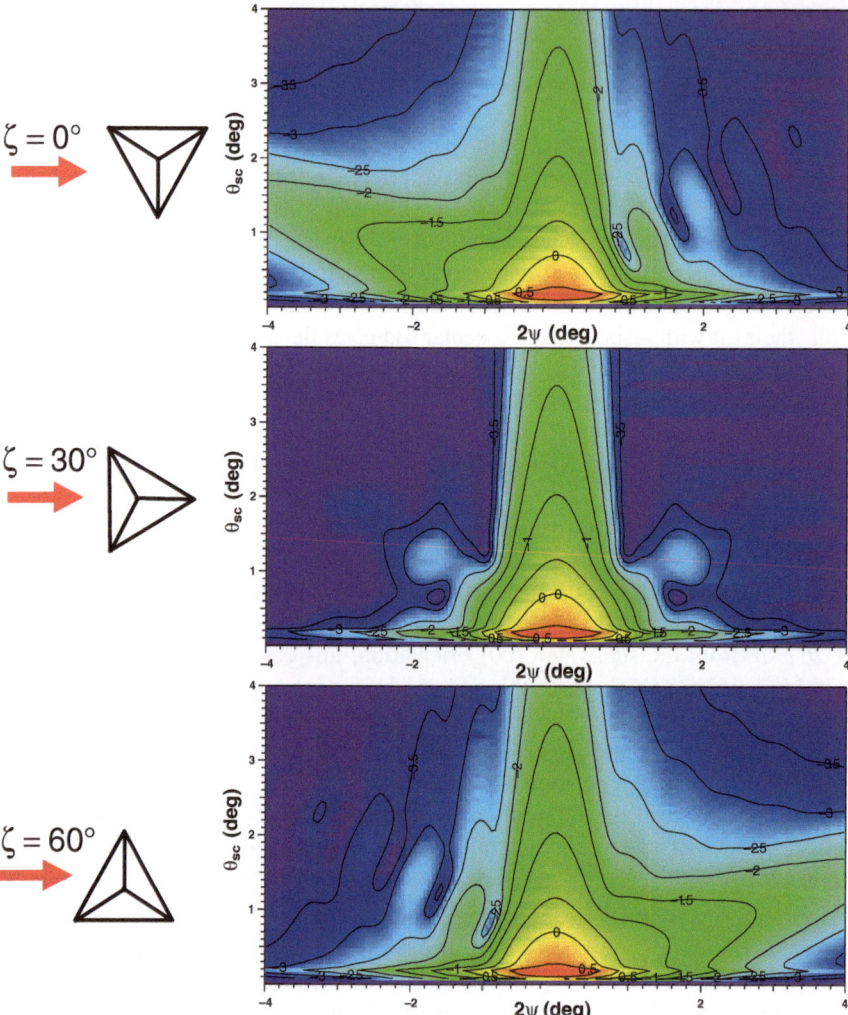

**Fig. 7.10** DWBA form factor of a full tetrahedral island for several azimuthal $\zeta$ orientations of the beam. The angle $\alpha$ is those between (111) planes in cubic system ($\alpha = 70.53°$). The simulation parameters are $\lambda = 0.1$ nm, $R = 5$ nm, $\theta_{in} = \theta_c$, $\delta = 5.10^{-6}$, $\beta = 2.10^{-8}$. The signal is normalized by the particle volume and displayed on a logarithmic scale

is due to the lack of wavevector transfer $q_z$ in the direction parallel to the threefold symmetry. In a grazing incidence geometry, this is not the case and an azimuthal rotation by $180°$ of the sample leads to a different scattering pattern as, even within the Born approximation, $d\sigma/d\Omega(\mathbf{q}_\parallel, q_z) \neq d\sigma/d\Omega(-\mathbf{q}_\parallel, q_z)$. The Friedel rule will be checked only through a combined rotation and reflection at the $x - y$ plane. The threefold symmetry of a tetrahedron is thus recovered as shown in Fig. 7.12-b when a small wavevector transfer is added in the perpendicular direction. However, it is

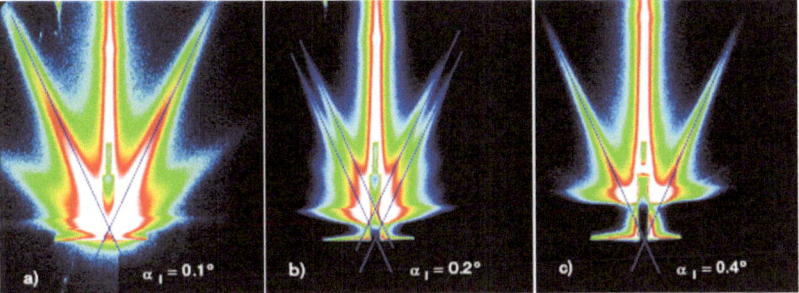

**Fig. 7.11** GISAXS patterns from Ge islands grown on Si(001) (equivalent thickness 1 nm) and annealed at 900 K for 15 min. The photon energy was set to 10 keV. The beam is aligned along the $\langle 110 \rangle$ direction of silicon. The angle of incidence $\theta_{in}$ is given in figure while $\theta_c \sim 0.2°$. Adapted from [80]

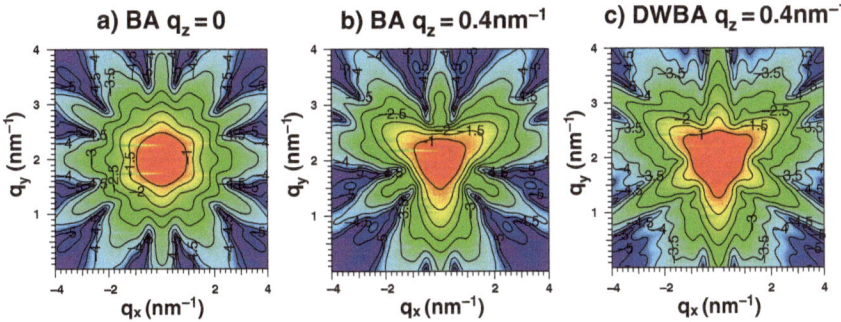

**Fig. 7.12** $(q_x, q_y)$ maps of the form factor of a full tetrahedron within (**a**) BA at $q_z = 0$, $\theta_{in} = \theta_{sc} = 0$; (**b**) BA at $q_z = 0.4\,\text{nm}^{-1}$, $\theta_{in} = \theta_{sc} = \theta_c$; (**c**) DWBA at $q_z = 0.4\,\text{nm}^{-1}$, $\theta_{in} = \theta_{sc} = \theta_c$. Same simulation parameters as in Fig. 7.10

not possible to determine the symmetry of any kind of particle. For instance, the up-down symmetry of a threefold prism results in a sixfold scattering pattern. Within the DWBA, the situation for an tetrahedral island is complicated by the superposition of the four scattering terms but the conclusions are close to those obtained in BA as four different wavevector transfers are involved $q_z^{eff} = \pm k_{scz} \pm k_{inz}$. However, if $\theta_{in}, \theta_{sc}$ are close to zero, the weighting reflectivities involved in Eq. (7.9) are close to one; this limit corresponds to a direct transmission of the beam through the island and leads again to applicability of Friedel rule upon azimuthal rotation.

### 7.3.5.3 The Core–Shell Particles

The core–shell particle is made of a core of shape $S_{co}(\mathbf{r})$ of index of refraction $n_{co} = 1 - \delta_{co} - i\beta_{co}$ and a shell of shape $S_{sh}(\mathbf{r})$ of thickness $t_{sh}$ and of index of

refraction $n_{sh} = 1 - \delta_{sh} - i\beta_{sh}$. Its form factor in the DWBA formalism is deduced
from the expression of the perturbation potential:

$$\delta n^2(\mathbf{r}) = \delta n_{co}^2 S_{co}(\mathbf{r}) + \delta n_{sh}^2 [S_{sh}(\mathbf{r}) - S_{co}(\mathbf{r})], \qquad (7.40)$$

where

- $\delta n_{co}^2 = n_{co}^2 - 1, \delta n_{sh}^2 = n_{sh}^2 - 1$ for islands

- $\delta n_{co}^2 = n_{co}^2 - n_s^2, \delta n_{sh}^2 = n_{sh}^2 - n_s^2$ for inclusions in a substrate

- $\delta n_{co}^2 = 1 - n_s^2, \delta n_{sh}^2 = n_{sh}^2 - n_s^2$ for holes in a substrate surface

- $\delta n_{co}^2 = n_{co}^2 - n_l^2, \delta n_{sh}^2 = n_{sh}^2 - n_l^2$ for inclusions in layer

- $\delta n_{co}^2 = 1 - n_l^2, \delta n_{sh}^2 = n_{sh}^2 - n_l^2$ for holes in layer

Therefore, the scattering cross sections Eqs. (7.8, 7.11, 7.17) are unmodified if
the following effective form factor is used:

$$\mathscr{F}(\mathbf{q}_\parallel, k_{in\,z}, k_{sc,z}) = \mathscr{F}_{co}(\mathbf{q}_\parallel, k_{in\,z}, k_{sc,z})$$
$$+ \tau \left[ \mathscr{F}_{sh}(\mathbf{q}_\parallel, k_{in\,z}, k_{sc,z}) - \mathscr{F}_{co}(\mathbf{q}_\parallel, k_{in\,z}, k_{sc,z}) \right] \quad (7.41)$$

with $\tau = \delta n_{sh}^2 / \delta n_{co}^2$. $\tau$ measures the contrast induced by the difference of dielectric
constant between the core and the shell.

## 7.4 Scattering from Collections of Particles

Up to now, only single isolated particle has been considered as a source of scattering. Its scattering cross section has been evaluated within the DWBA for various
geometries (islands, inclusions or holes in a substrate or in layer). To proceed
with a collection of particles, the multiple scattering between particles will be neglected and a simple hypothesis of additivity in the scattering potential $\delta n^2(\mathbf{r})$
will be made. In other words, the incident wave and the scattered wave by one
particle are supposed to be unperturbed by the other particles. The limitations of
such hypothesis will be commented in the following. The particles are supposed
to lie in the same plane as the main topic of this chapter is the use of grazing incidence geometry to enhance surface or subsurface out-of-specular scattering. The case of particles distributed in 3D will be dropped out on purpose. The
formalism developed in the following can be found in many standard text books
on crystallography [1, 2, 55, 56] or in the field of non-crystallized materials or
bulk small-angle scattering. The only specificity comes from the 2D treatment
of the problem.

## 7.4.1 The Scattering Cross Section: The Particle–Particle Partial Pair Correlation Functions

### 7.4.1.1  General Formalism: Coherent, Incoherent Scattering and Specular Rod

Consider a set of $N \gg 1$ particles labeled by an index $i$ with shapes $S_i(\mathbf{r})$ located at $\mathbf{r}_{\|,i}$. The scattering potential that comes into play in the DWBA is written as follows:

$$\delta n^2(\mathbf{r}) = \delta n_{ge}^2 \sum_{i=1}^{N} S_i(\mathbf{r}) \otimes \delta(\mathbf{r} - \mathbf{r}_{\|,i}), \qquad (7.42)$$

where $\delta n_{ge}^2(\mathbf{r})$ is the contrast of dielectric constant which depends on the considered morphology (Fig. 7.3). Using the previous analysis in terms of DWBA form factor, the decoupling between the parallel and perpendicular scattering directions and the linearity of the potential allow to write the differential cross section per particle $\left(\dfrac{d\sigma}{d\Omega}\right)_{part}$ as

$$\left(\frac{d\sigma}{d\Omega}\right)_{tot} = \frac{k_0^4}{16\pi^2}|\delta n_{ge}^2|^2 N \left(\frac{d\sigma}{d\Omega}\right)_{part},$$

$$\left(\frac{d\sigma}{d\Omega}\right)_{part} = \frac{1}{N}\left|\sum_{i=1}^{N} \mathscr{F}_i(\mathbf{q}_\|, k_{inz}, k_{scz})e^{i\mathbf{q}_\| \cdot \mathbf{r}_{\|,i}}\right|^2. \qquad (7.43)$$

After having isolated the self term, Eq. (7.43) reads

$$N\left(\frac{d\sigma}{d\Omega}\right)_{part} = \sum_{i=1}^{N}\left|\mathscr{F}_i(\mathbf{q}_\|, k_{inz}, k_{scz})\right|^2$$

$$+ \sum_{i=1}^{N}\sum_{j=1(i\neq j)}^{N} \mathscr{F}_i(\mathbf{q}_\|, k_{inz}, k_{scz})\mathscr{F}_j^*(\mathbf{q}_\|, k_{inz}, k_{scz})e^{i\mathbf{q}_\| \cdot (\mathbf{r}_{\|,i}-\mathbf{r}_{\|,j})}.$$

$$(7.44)$$

The star symbol points out to the complex conjugate.

If the system is ergodic, the ensemble average can be replaced by a configurational average over the "coherent domains" that are size limited either by the incident beam coherence (divergence and wavelength spread) or by the finite aperture of the detector (see Sect. 2.3). The position disorder and the substitution disorder, i.e., shape and morphological parameter disorder, can be separated by sorting out particles, over the typical size of the coherent domain $S_{coh}$, in class of size and shape $\alpha$ whose probability of occurrence is $p_\alpha$. Moreover, the discrete sum over the positions can be replaced by a continuous one thanks to the reduced (or normalized) partial pair correlation functions $g_{\alpha\beta}(\mathbf{r}_{\|,\alpha}, \mathbf{r}_{\|,\beta})$ between particles of class $\alpha$ and $\beta$. Both previously introduced concepts rely on the statistical limit hypothesis. The coherent domain $S_{coh}$ is supposed to be sufficiently large to define the statistical quantities $p_\alpha$ and $g_{\alpha\beta}(\mathbf{r}_{\parallel,\alpha}, \mathbf{r}_{\parallel,\beta})$. However, the illuminated area $A$ is assumed to contain so much coherent domain $S_{coh}$ to average out the speckle behavior. If those hypothesis are reasonably fulfilled (see Sect. 2.4), one ends up with

$$\left(\frac{d\sigma}{d\Omega}\right)_{part} = \sum_\alpha p_\alpha \left|\mathscr{F}_\alpha(\mathbf{q}_\parallel, k_{inz}, k_{scz})\right|^2$$

$$+ \frac{n_S^2}{N} \sum_\alpha \sum_{\beta \neq \alpha} p_\alpha p_\beta \mathscr{F}_\alpha(\mathbf{q}_\parallel, k_{inz}, k_{scz}) \mathscr{F}_\beta^*(\mathbf{q}_\parallel, k_{inz}, k_{scz})$$

$$\iint_A d\mathbf{r}_{\parallel,\alpha} d\mathbf{r}_{\parallel,\beta} g_{\alpha\beta}(\mathbf{r}_{\parallel,\alpha}, \mathbf{r}_{\parallel,\beta}) e^{i\mathbf{q}_\parallel \cdot (\mathbf{r}_{\parallel,\alpha} - \mathbf{r}_{\parallel,\beta})}, \qquad (7.45)$$

where $n_S = N/A$ is the number of particles per surface unit. $n_S^2 p_\alpha p_\beta g_{\alpha\beta}(\mathbf{r}_{\parallel,\alpha}, \mathbf{r}_{\parallel,\beta}) d\mathbf{r}_{\parallel,\alpha} d\mathbf{r}_{\parallel,\beta}$ counts the number per unit surface of couples of particles of kinds $(\alpha, \beta)$ located at $(\mathbf{r}_{\parallel,\alpha}, \mathbf{r}_{\parallel,\beta})$. The constraint $i \neq j$ of Eq. (7.44) is implicitly included in the partial pair correlation functions through a hard core type effect. Usually, the variations of the number of particles around its mean value are singled out by subtracting 1 to $g_{\alpha\beta}(\mathbf{r}_{\parallel,\alpha}, \mathbf{r}_{\parallel,\beta})$.

With the hypothesis of spatial homogeneity, the pair correlation function depends only on the relative position of the scatterers: $g_{\alpha\beta}(\mathbf{r}_{\parallel,\alpha}, \mathbf{r}_{\parallel,\beta}) = g_{\alpha\beta}(\mathbf{r}_{\parallel,\beta} - \mathbf{r}_{\parallel,\alpha})$. Eq. (7.45) can be rewritten as

$$\left(\frac{d\sigma}{d\Omega}\right)_{part} = N\left|\langle \mathscr{F}(q_\parallel = 0, k_{inz}, k_{scz})\rangle\right|^2 \delta(q_\parallel) + \Phi_0(\mathbf{q}_\parallel, k_{inz}, k_{scz})$$

$$+ \sum_{\beta \neq \alpha} p_\alpha p_\beta \mathscr{F}_\alpha(\mathbf{q}_\parallel, k_{inz}, k_{scz}) \mathscr{F}_\beta^*(\mathbf{q}_\parallel, k_{inz}, k_{scz}) S_{\alpha\beta}(\mathbf{q}_\parallel),$$

$$(7.46)$$

$$\Phi_0(\mathbf{q}_\parallel, k_{inz}, k_{scz}) = \left\langle \left|\mathscr{F}(\mathbf{q}_\parallel, k_{inz}, k_{scz})\right|^2 \right\rangle - \left|\langle \mathscr{F}(\mathbf{q}_\parallel, k_{inz}, k_{scz})\rangle\right|^2, \qquad (7.47)$$

$$S_{\alpha\beta}(\mathbf{q}_\parallel) = 1 + n_S \int_A \left(g_{\alpha\beta}(\mathbf{r}_\parallel) - 1\right) e^{i\mathbf{q}_\parallel \cdot \mathbf{r}_\parallel} d\mathbf{r}_\parallel. \qquad (7.48)$$

The brackets $\langle \ldots \rangle$ stand for the average over the distribution of particle kind. $S_{\alpha\beta}(\mathbf{q}_\parallel)$, the Fourier transform of the partial pair correlation functions, are known as the partial interference functions. The Faber–Ziman definition of the $g_{\alpha\beta}$ was used although other definitions are found in the literature [57]. The scattering cross section was decomposed in to three terms :

(i)   The specular rod in $\delta(q_\parallel)$ (see below).
(ii)  An incoherent term $\Phi_0(\mathbf{q}_\parallel, k_{inz}, k_{scz})$ including the effect of size and shape fluctuations along the sample surface. It is worth noting that its counterpart in neutron crystallography is the incoherent scattering due to the isotopic effects on the nucleus scattering lengths.

(iii)  A coherent term (third term of Eq. (7.46)). "Coherent" means that the waves scattered by each particle are allowed to partially interfere. With our 2D geometry of a plane of particles, this interference appears along the substrate surface.

### 7.4.1.2 The Specular Rod

The name of "specular rod" is used in this chapter instead of "coherent scattering" as previously introduced in Chap. 4 to avoid any confusion with the terminology of small-angle scattering, namely the third term of Eq. (7.46). The intensity of the specular rod $N \left\langle \left| \mathscr{F}(q_\| = 0, \ k_{inz}, k_{scz}) \right|^2 \right\rangle \delta(q_\|)$ is directly proportional to the number of scatterers $N$ and to the value of mean form factor at origin. Thus, this term is orders of magnitude more intense than the diffuse scattering and is often not measured experimentally in GISAXS because of limited detector dynamic. The introduction of a Dirac peak $N \, \delta(q_\|)$ as a result of the integral $n_S \int_A e^{i\mathbf{q}_\| \cdot \mathbf{r}_\|} d\mathbf{r}_\|$ is valid only in the limit of infinite coherent domain size. In fact, experimentally, all the sources of loss of coherence (monochromator resolution, beam divergence, detector acceptance, sample macroscopic curvature, etc.) contribute to the finite width of the specular rod. On a formal point of view, the specular rod can be reintroduced in the interference function [50,58] but the simulation of its shape is difficult as it relies on the knowledge of the main sources of coherence loss. That is why this rod is usually integrated with slit aperture for reflectivity measurements.

Within the DWBA, Eq. (7.46) gives the integrated intensity of the specular rod in terms of the particle form factor. Moreover, on this specular rod, the specularly reflected beam should be added. Using the notations of Sect. 3.2.1, its intensity is $A \sin^2(\alpha_f) \left| A_0^+ (k_{inz,0}) \right|^2$ [39], $A_0^+$ being calculated with the formalism of Parratt [59] or of the matrix method for stratified media (see Sect. 3.2.1).

### 7.4.1.3 Decoupling and Local Monodisperse Approximations

Besides the fact that the size–shape distributions can be reduced to a small number of parameters, the practical use of Eq. (7.46) implies the statistical knowledge of the whole system, in particular that of all the partial pair correlation functions. The correlations between the kinds of the scatterers and their relative positions included in $g_{\alpha\beta}(\mathbf{r}_\|)$ are difficult to rationalize. The same problem is found for scattering from correlated roughnesses in multilayers (see Chap. 6). Approximations are thus needed.

Decoupling Approximation

In the decoupling approximation (DA) [1], such correlations are ignored and $g_{\alpha\beta}(\mathbf{r}_\|)$ is supposed to be independent of the scatterer kind: $g_{\alpha\beta}(\mathbf{r}_\|) = g(\mathbf{r}_\|)$. DA is equivalent to an unrestricted disorder. For instance, this hypothesis is justified for isotopic disorder in the field of neutron scattering. For particles, this approximation is restricted to dilute systems. A factorization becomes possible in Eq. (7.46):

$$\left(\frac{d\sigma}{d\Omega}\right)_{part} = \Phi_0(\mathbf{q}_\parallel) + \left|\left\langle \mathscr{F}(\mathbf{q}_\parallel, k_{inz}, k_{scz})\right\rangle\right|^2 S(\mathbf{q}_\parallel), \qquad (7.49)$$

$$S(\mathbf{q}_\parallel) = 1 + n_S \int_A \left(g(\mathbf{r}_\parallel) - 1\right) e^{i\mathbf{q}_\parallel \cdot \mathbf{r}_\parallel} d\mathbf{r}_\parallel. \qquad (7.50)$$

$S(\mathbf{q}_\parallel)$, the so-called total interference function, is the Fourier transform of the particle position autocorrelation function irrespective of the kind of the particle. Expression and properties of the pair correlation and interference functions will be the topic of Sect. 7.4.3. The first numerical applications of DA were carried out by Kotlarchyk and Chen [60].

Local Monodisperse Approximation

This hypothesis widely used in the literature assumes that the system is made of locally monodisperse domains that interfere incoherently. The particle–particle pair correlation function can vary from domain to domain. In other words, the surrounding of each particle is supposed to be made of particles of the same size and shape in such a way that the particle kind varies slowly across the sample but with a spatial wavelength lower than the coherence of the beam. The cross section reads

$$\left(\frac{d\sigma}{d\Omega}\right)_{part} = \left\langle \left|\mathscr{F}_\alpha(\mathbf{q}_\parallel, k_{inz}, k_{scz})\right|^2 S_\alpha(\mathbf{q}_\parallel)\right\rangle. \qquad (7.51)$$

This approximation known as the local monodisperse approximation (LMA) [61, 62] works nicely on a point of view of data analysis because partial correlation between particles can be reintroduced in each domain. But, one have to bear in mind that LMA relies on a unphysical description of most of the experimental systems. Upon a progressive increase of the size correlation between neighboring particles, a continuous transition from DA to LMA is obtained [63].

### 7.4.2 Size and Shape Distribution Effects: The Mean Scattering Form Factors

#### 7.4.2.1 General Case

To perform the average in Eqs. (7.46, 7.47, 7.51), it is compulsory to define all the morphological parameters $v_i, (i = 1, \ldots, n)$ that characterize the particles and their joint probability distribution $p(v_1, \ldots, v_n)$:

$$\left\langle \left| \mathscr{F}(\mathbf{q}_{\|}, k_{inz}, k_{scz}) \right|^2 \right\rangle =$$

$$\int \dots \int p(v_1, \dots, v_n) \left| \mathscr{F}(\mathbf{q}_{\|}, k_{inz}, k_{scz}, v_1, \dots, v_n) \right|^2 dv_1 \dots dv_n,$$

(7.52)

$$\left| \left\langle \mathscr{F}(\mathbf{q}_{\|}, k_{inz}, k_{scz}) \right\rangle \right|^2 =$$

$$\left| \int \dots \int p(v_1, \dots, v_n) \mathscr{F}(\mathbf{q}_{\|}, k_{inz}, k_{scz}, v_1, \dots, v_n) dv_1 \dots dv_n \right|^2.$$

(7.53)

For instance, as defined previously in Fig. 7.9, $v_i = R, H, W, \zeta$. On a practical point of view, the parameters $v_i$ are often assumed to be independent $p(v_1, \dots, v_n) = p(v_1) \dots p(v_n)$.

The general influence of the size distribution is illustrated in the case of cylinders in Fig. 7.13 on the quantities $\left\langle |F(\mathbf{q})|^2 \right\rangle$, $|\langle F(\mathbf{q}) \rangle|^2$ and $\Phi_0(\mathbf{q}) = \left\langle |F(\mathbf{q})|^2 \right\rangle - |\langle F(\mathbf{q}) \rangle|^2$ normalized by the mean volume $\langle V \rangle$ of the size-distributed particles. Sharp fringes of destructive interference are observed for monodisperse particles at values which are specific to the particle shape and size either along the parallel or the perpendicular direction [64]. Their position correspond to the zero of the Bessel cardinal $J_1(x)/x$ or sine cardinal $\sin(x)/x$ functions (see Sect. 7.3.5.1). These fringes are smoothed upon increasing the size distribution for $\left\langle |F(\mathbf{q})|^2 \right\rangle$ through a transfer of intensity from the lobes to the minima. On the contrary, $|\langle F(\mathbf{q}) \rangle|^2$ still shows sharp minima but with a faster damping of the intensity with increasing wavevector transfer. It is worth noticing that the incoherent scattering $\Phi_0(\mathbf{q})$ peaks at $\mathbf{q}_{\|} = 0$.

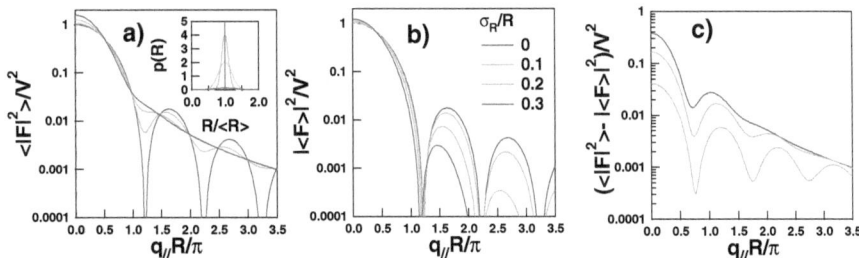

**Fig. 7.13** Influence of the polydispersity on the Born form factor of a cylinder: **(a)** $\left\langle |F(q_{\|}R)|^2 \right\rangle$, **(b)** $|\langle F(q_{\|}R) \rangle|^2$, **(c)** incoherent scattering $\Phi_0(q_{\|}R) = \left\langle |F(q_{\|}R)|^2 \right\rangle - |\langle F(q_{\|}R) \rangle|^2$. The gaussian size distribution has an increasing standard deviation width $\sigma_R/R = 0, 0.1, 0.2, 0.3$ as shown in the inset of **(a)**

### 7.4.2.2 Asymptotic Behavior of the Form Factor: the Porod and Guinier Limits

Guinier Limit

For uncorrelated particles $S_{\alpha\beta}(\mathbf{q}_{\parallel}) = 1$, the scattering cross section Eq. (7.46) is given only by the mean particle form factor $\left|\langle \mathscr{F}(\mathbf{q}_{\parallel}, k_{inz}, k_{scz})\rangle\right|^2$. Within this limit of independent particles, the Guinier law [1,2] deals with the asymptotic behavior of the scattered intensity at small wavevector transfer. It allows to get a characteristic size known as gyration radius $R_g$ along a given direction. Let's consider the case of bulk small-angle scattering from an isolated fixed particle and let's suppose that the wavevector transfer $\mathbf{q}$ tends toward zero:

$$|F(\mathbf{q})|^2 = \int_{S(\mathbf{r})} \int_{S(\mathbf{r}')} e^{i\mathbf{q}(\mathbf{r}-\mathbf{r}')} d\mathbf{r}d\mathbf{r}' \stackrel{q\to 0}{\simeq} V^2 - \frac{1}{2}\int_{S(\mathbf{r})}\int_{S(\mathbf{r}')} \left[\mathbf{q}.(\mathbf{r}-\mathbf{r}')\right]^2 d\mathbf{r}d\mathbf{r}'. \tag{7.54}$$

The center of mass $G$ of the particle defined by

$$\int_{S(\mathbf{r})} (\mathbf{r}-\mathbf{r}_G)d\mathbf{r} = 0 \tag{7.55}$$

allows to recast the scalar product in Eq. (7.54):

$$\left[\mathbf{q}.(\mathbf{r}-\mathbf{r}')\right]^2 = \left[\mathbf{q}.(\mathbf{r}-\mathbf{r}_G)\right]^2 + \left[\mathbf{q}.(\mathbf{r}'-\mathbf{r}_G)\right]^2 - 2\left[\mathbf{q}.(\mathbf{r}-\mathbf{r}_G)\right]\left[\mathbf{q}.(\mathbf{r}'-\mathbf{r}_G)\right]. \tag{7.56}$$

By definition of $\mathbf{r}_G$, the cross-product disappears upon integration. If the wavevector transfer $\mathbf{q}$ tends toward zero along a constant unitary vector $\mathbf{n}$, Eq. (7.56) reads

$$|F(\mathbf{q})|^2 \stackrel{q\to 0}{\simeq} V^2(1-q^2R_g^2) \simeq V^2 e^{-q^2 R_g^2}, \tag{7.57}$$

where $R_g^2$ is the inertia moment of the particle with respect to the plane $\Pi$ perpendicular to $\mathbf{n}$ and going through the gravity center $G$ of the particle. $R_g^2$ is defined by

$$R_g^2 = \frac{1}{V}\int_V d^2[M(\mathbf{r}),\Pi]d\mathbf{r}. \tag{7.58}$$

In size-distributed samples, if the Guinier limit ($q^2 R_g^2 \ll 1$) is valid for each class $\alpha$ of particles, Eq. (7.57) is still valid but with the average inertia moment $\langle R_g^2\rangle$.

A complexity arises in GISAXS compared to standard SAXS because of (i) the anisotropy of particles between the parallel and the perpendicular directions and (ii) the refraction–reflection of the incoming and scattered beams at interfaces. In other words, as the particle form factor in GISAXS is not a simple Fourier transform, the validity of the Guinier law under grazing incidence is questionable; this is all the more relevant than the analysis being hindered along the $z$-direction by the Yoneda peak. For instance, the transmission functions $t_{0,1}^{in} t_{0,1}^{sc}$ used as prefactors in the inclusion particle form factor Eq. (7.11) tends toward zero at small $\theta_{in}$ and $\theta_{sc}$. The validity along the direction parallel to interfaces is justified, stricto sensu, if the particle shape is invariant along the $z$-axis as for cylinders. The DWBA form factor can hence be separated into two independent components $\mathscr{F}(\mathbf{q}_\parallel, k_{inz}, k_{scz}) = F_\parallel(\mathbf{q}_\parallel) \mathscr{F}_\perp(k_{inz}, k_{scz})$; the Guinier law can thus be applied safely to $F_\parallel$. However, one should keep in mind that, in that case, such a parallel gyration radius results from a surface and not a volume integral as in Eq. (7.58).

## Porod Limit

In bulk small-angle scattering, the Porod limit uses the integral of scattered intensity over all the reciprocal space and its asymptotic behavior at high-$\mathbf{q}$ to get the average volume/surface ratio of the particles [1,2].

If the set of particles is characterized by a loss of long range order, the partial pair correlation functions and the partial interference functions tend toward one at long distance and high wavevector transfer. From Eq. (7.46), one sees that

$$\lim_{\psi, \theta_{sc} \to +\infty} \left(\frac{d\sigma}{d\Omega}\right)_{part} (\mathbf{q}_\parallel, k_{inz}, k_{scz}) = \left\langle \left| \mathscr{F}(\mathbf{q}_\parallel, k_{inz}, k_{scz}) \right|^2 \right\rangle. \qquad (7.59)$$

The signal in the high-angle scattering range is dominated by the mean cross section per particle. For size-distributed samples, further insight can be gained by studying, on logarithmic plot, the slope of the form factor in high $q_\parallel$ and $q_z$ range. This quantity is not hindered by the refraction effects as in this wavevector range the Born Approximation is valid. A decrease in power law $q_\parallel^{-n}$ and $q_z^{-m}$ is observed [64], with an exponent $n, m$ that depends on the degree of sharpness of the particle. The case of cylinder shape is easily tractable in an analytical way. As its form factor Eq. (7.32) can be decomposed along the two directions

$$F_{cy}(\mathbf{q}, R, H) = F_{cy}(\mathbf{q}_\parallel, R) F_{cy}(q_z, H),$$

$$F_{cy}(\mathbf{q}_\parallel, R) = \pi R^2 \frac{J_1(q_\parallel R)}{q_\parallel R}; \quad F_{cy}(q_z, H) = H \mathrm{sinc}(q_z H/2), \qquad (7.60)$$

the asymptotic behavior[4] is given by

$$\left\langle \left| F_{cy}(\mathbf{q}_\parallel, R) \right|^2 \right\rangle \overset{q_\parallel \to +\infty}{\simeq} \frac{2}{\pi q_\parallel^3} \left\langle R \cos^2(q_\parallel R - 3\pi/4) \right\rangle$$

$$= \frac{1}{\pi q_\parallel^3} \left( \langle R \rangle - \langle R \sin(2q_\parallel R) \rangle \right), \qquad (7.61)$$

$$\left\langle \left| F_{cy}(q_z, H) \right|^2 \right\rangle = \frac{4}{q_z^2} \left\langle \sin^2(q_z H/2) \right\rangle = \frac{2}{q_z^2} \left( 1 - \langle \cos(q_z H) \rangle \right). \quad (7.62)$$

---

[4] $J_1(x) \overset{x \to +\infty}{\simeq} \sqrt{\frac{2}{\pi x}} \cos(x - \frac{3\pi}{4})$

As the size distribution $p(x)$ are bounded functions of $x$, the Fourier transform over the size distributions that appear in Eqs. (7.61 and 7.62) tend toward zero at high-$q$ value. The exponents are $n = 3, m = 2$. The asymptotic regime is reached when these characteristic function, i.e., $\sim \langle R \sin(2q_{\parallel}R) \rangle$ and $\sim \langle \cos(q_z H) \rangle$ are negligible compared to the constant factor $\langle R \rangle$ or 1. If $\sigma_{R,H}$ is a typical standard deviation of the size distribution, as a rule of thumb, these oscillations of the form factor are sufficiently damped as $\sigma_R q_{\parallel} \gg 1$ and $\sigma_H q_z \gg 1$. Other straightforward cases are the full sphere (Eq. (7.35)) $n = m = 4$ or the parallelepiped (Eq. (7.22)) $n = m = 2$. It is important to keep in mind that $n, m$ depends, of course, on the shape of the particle (for a truncated sphere, the exponents depend on the truncation ratio; $n = 4, m = 2$ for an hemisphere), but also on the coupling between the size parameters $v$ (Eq. (7.52)) and on the in-plane orientational average $\zeta$ (Eq. (7.39)). Randomly oriented parallelepipeds give $n = 3$ like for cylinders!

To conclude, on a practical point of view, as recording data far away in reciprocal space gives a signal that decreases as power law, the Porod analysis is often hampered by the noise level. Furthermore, determining Porod dimensions through the ratio of the Porod constant and the invariants [1, 2] (i.e., the integral of the intensity over reciprocal space) is quite difficult in GISAXS, because of the inherent anisotropy between the two main directions. Only approximate results can be obtained in return with many unjustified hypothesis [65, 66].

## 7.4.3 Some Models of Interference Functions

Providing simple models for partial interference functions Eq. (7.48) is far from being an easy task and requires a considerable knowledge of the structure of the system. However, DA and LMA allow to reduce the problem to that of the total interference function. Three useful cases will be briefly summarized since a full treatment can be found in standard crystallography text books [55, 56]:

(i)   highly disordered systems described by an isotropic pair correlation function
(ii)  ordered lattice with defects
(iii) lattice with loss of long range order in one or two directions for which the paracrystal treatment is relevant

### 7.4.3.1 Highly Disordered Case

Highly disordered system of particles are statistically described by the reduced total pair correlation function $g(\mathbf{r}_{\parallel})$. The hypothesis of homogeneity reduces the dependence of $g(\mathbf{r}_{\parallel})$ to the relative positions of the objects $\mathbf{r}_{\parallel}$. As $n_S g(\mathbf{r}_{\parallel})$ is the number of particles per surface unit located at $\mathbf{r}_{\parallel}$ knowing that there is one object at origin, the autocorrelation function of the particle positions $g_{pp}(\mathbf{r}_{\parallel})$ (irrespective of their sizes) is given in terms of $g(\mathbf{r}_{\parallel})$ by

$$g_{pp}(\mathbf{r}_\parallel) = \frac{1}{N} \left\langle \sum_{i,j} \delta(\mathbf{r}_\parallel - \mathbf{r}_{\parallel,i} + \mathbf{r}_\parallel, j) \right\rangle = \delta(\mathbf{r}_\parallel) + n_S g(\mathbf{r}_\parallel).  \qquad (7.63)$$

The term $i = j$, i.e., the particle at origin is isolated through $\delta(\mathbf{r}_\parallel)$. Since long range order is lacking, $g(\mathbf{r}_\parallel)$ tends toward one when $r_\parallel \rightarrow +\infty$. Therefore, the oscillating part of $g(\mathbf{r}_\parallel)$ around this limit can be singled out as it contains all the structural information:

$$g_{pp}(\mathbf{r}_\parallel) = \delta(\mathbf{r}_\parallel) + n_S + n_S[g(\mathbf{r}_\parallel) - 1].  \qquad (7.64)$$

The divergence found in the Fourier transform of the previous expression (i.e., the interference function $S(\mathbf{q}_\parallel)$) is thus cured:

$$S(\mathbf{q}_\parallel) = 1 + N\delta(\mathbf{q}_\parallel) + n_S \int_A \left[ g(\mathbf{r}_\parallel) - 1 \right] e^{i\mathbf{q}_\parallel \cdot \mathbf{r}_\parallel} d\mathbf{r}_\parallel.  \qquad (7.65)$$

$N\delta(\mathbf{q}_\parallel)$ is broadened upon folding with the limited coherence of the beam. This coherent scattering leading to the specular rod is often dropped out in the total interference function. For isotropic systems, $g(\mathbf{r}_\parallel)$ and $S(\mathbf{q}_\parallel)$ depend only on the modulus $r_\parallel, q_\parallel$ and are linked in two dimensions through an Hankel transform:

$$S(q_\parallel) = 1 + 2\pi n_S \int_0^\infty [g(r_\parallel) - 1]J_0\left(r_\parallel q_\parallel\right) r_\parallel dr_\parallel,$$

$$g(r_\parallel) = 1 + \frac{1}{2\pi n_S} \int_0^\infty [S(q_\parallel) - 1]J_0\left(r_\parallel q_\parallel\right) q_\parallel dq_\parallel.  \qquad (7.66)$$

$g(r_\parallel)$ and $S(q_\parallel)$ are related to the partial pair correlation functions $g_{\alpha\beta}(r_\parallel)$ and the partial interference functions $S_{\alpha\beta}(q_\parallel)$ by substituting in Eq. (7.46) the particle shape by a Dirac peak and the particle form factor $\mathscr{F}_\alpha$:

$$g(r_\parallel) = \sum_{\alpha,\beta} p_\alpha p_\beta g_{\alpha\beta}(r_\parallel), \quad S(q_\parallel) = \sum_{\alpha,\beta} p_\alpha p_\beta S_{\alpha\beta}(q_\parallel).  \qquad (7.67)$$

The Dirac contribution at origin was ignored in the previous equation. The $g(r_\parallel)$ and $S(q_\parallel)$ curves are made of broader and broader peaks whose positions are linked to the mean distance $D$ between scatterers; as they are normalized functions, their limit $r_\parallel, q_\parallel \rightarrow +\infty$ is equal to one. However, any functions with such features are not necessarily valid since $g(r_\parallel)$ should be an autocorrelation function ($S(q_\parallel) > 0$). The analytical build-up of such functions is quite difficult as it relies on a knowledge of the types of "interactions" that are responsible for the observed morphology. Approximations were developed in the field of thermodynamics of gazes or glasses to determine them from the interaction potential between particles [67]. Figure 7.14 illustrates such calculations for monodisperse particles of diameter $\sigma_0$ that do not overlap in two dimensions. Even this simple hard core constraints induces a short-range structuring at the preferential distance $\sigma_0$ upon increasing the coverage. It is worth noticing that, apart from the excluded coherent term, $S(q_\parallel = 0)$ is finite. This value is given by the statistical fluctuations of the number of particles in the probed area [55].

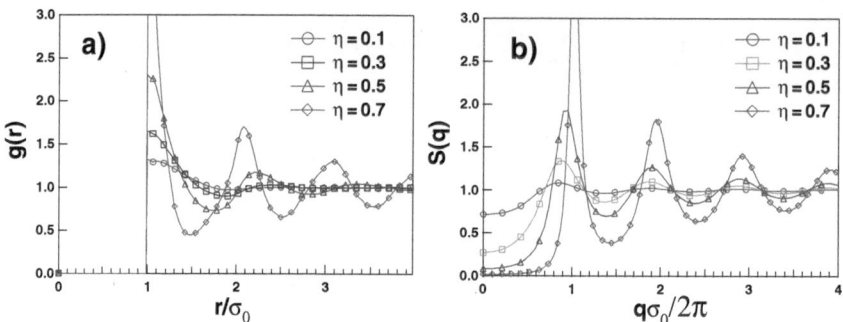

**Fig. 7.14** (a) Pair correlation function and (b) interference function of particles of diameter $\sigma_0$ interacting through an hard core potential as function of the coverage $\eta = n_S \pi \sigma_0^2 / 4$. From [50]

### 7.4.3.2 Ordered Lattices with Defects

The Lattice

A lattice of particles is defined by two basis vectors $\mathbf{a}, \mathbf{b}$ and a pattern made of several particles linked to each lattice site. The interference function Eq. (7.65) is made of sharp Bragg rods extending perpendicular to the surface and located at the nodes of the reciprocal lattice defined through the two vectors:

$$\mathbf{a}^* = 2\pi \frac{\mathbf{b} \times \mathbf{n}}{\mathbf{a}.\,[\mathbf{b} \times \mathbf{n}]}, \quad \mathbf{b}^* = 2\pi \frac{\mathbf{n} \times \mathbf{a}}{\mathbf{b}.\,[\mathbf{n} \times \mathbf{a}]}, \tag{7.68}$$

where $\mathbf{n}$ is the normal to the surface and $\times$ the vector product. Even if scattering is performed at small angles, the curvature of the Ewald sphere finds its importance when diffraction from super lattice is involved.

The Rod Shape in the Interference Function

If the lattice is perfect, the rods are Dirac peaks. Defects of any kinds, finite-size effects and limited coherence length of the beam induce an homogeneous broadening of these rods. All over the reciprocal space, the shapes of the rods depend on the underlying models of disorders. Accounting for this broadening can be done in an efficient and effective way (as it is done for reflections in powder diffraction) by folding the lattice sites with simple rod shapes $\mathscr{S}(\mathbf{q}_{\parallel})$ like gaussian or lorentzian:

$$S(\mathbf{q}_{\parallel}) = \sum_n \sum_m \mathscr{S}(\mathbf{q}_{\parallel} - n\mathbf{a}^* - m\mathbf{b}^*). \tag{7.69}$$

The rod does not follow necessarily the symmetry of the lattice and its width is inversely proportional to the involved coherence lengths [50].

The Unit Cell Structure Factor

The scattering unit in the case of a regular lattice is a pattern made of $N_p$ particles located at $\mathbf{r}_{\|,k}$ within the unit cell $(\mathbf{a},\mathbf{b})$. The associated unit cell structure factor is

$$\mathscr{F}_C(\mathbf{q}_\|,k_{inz},k_{scz}) = \sum_{k=1}^{N_p} \mathscr{F}_k(\mathbf{q}_\|,k_{inz},k_{scz})e^{i\mathbf{q}_\|\cdot\mathbf{r}_{\|,k}}. \tag{7.70}$$

Taking the configuration average of the previous equation leads to the same questioning as in Sect. 7.4.1 about correlations between the particles and their locations inside the unit cell. Neglecting them corresponds to the decoupling approximation. By denoting $\langle\ldots\rangle_S$ and $\langle\ldots\rangle_P$, the average over the particle sizes and over the intracell position, this hypothesis leads to

$$\langle\mathscr{F}_C(\mathbf{q}_\|,k_{inz},k_{scz})\rangle = \left\langle \sum_{k=1}^{N_p} \langle\mathscr{F}_k(\mathbf{q}_\|,k_{inz},k_{scz})\rangle_S e^{i\mathbf{q}_\|\cdot\mathbf{r}_{\|,k}} \right\rangle_P. \tag{7.71}$$

By introducing the fluctuations $\delta\mathbf{r}_{\|,k}$ of the $k$th-particle position around its mean value $\mathbf{r}_{\|,k}^0$ ($\langle\delta\mathbf{r}_{\|,k}\rangle_P = 0$) and after an expansion of the exponential term to second order in displacement, the classical Debye–Waller factor appears:

$$\langle\mathscr{F}_C(\mathbf{q}_\|,k_{inz},k_{scz})\rangle = \sum_{k=1}^{N_p} \langle\mathscr{F}_k(\mathbf{q}_\|,k_{inz},k_{scz})\rangle_S e^{i\mathbf{q}_\|\cdot\mathbf{r}_{\|,k}^0}e^{-W_k(\mathbf{q}_\|)/2}, \tag{7.72}$$

where $W_k(\mathbf{q}_\|) = \langle(\mathbf{q}_\|\cdot\delta\mathbf{r}_{\|,k})^2\rangle_P = \mathbf{q}_\|\mathscr{B}_k\mathbf{q}_\|$. $\mathscr{B}_k$ is the symmetric tensor of the standard deviations of the particle displacements. The other mean form factor involved in Eqs. (7.46 and 7.47) can be expressed as

$$\langle|\mathscr{F}_C(\mathbf{q}_\|,k_{inz},k_{scz})|^2\rangle = \sum_{k=1}^{N_p} \langle|\mathscr{F}_C(\mathbf{q}_\|,k_{inz},k_{scz})|^2\rangle_S$$
$$+ \sum_{k,l=1(k\neq l)}^{N_p} \langle\mathscr{F}_k(\mathbf{q}_\|,k_{inz},k_{scz})\mathscr{F}_l^*(\mathbf{q}_\|,k_{inz},k_{scz})\rangle_S$$
$$\times e^{i\mathbf{q}_\|\cdot(\mathbf{r}_{\|,k}^0-\mathbf{r}_{\|,l}^0)}\left\{e^{-W_k(\mathbf{q}_\|)/2}+e^{-W_l(\mathbf{q}_\|)/2}-1\right\}. \tag{7.73}$$

By gathering Eqs. (7.72 and 7.73) in Eqs. (7.46 and 7.47), the cross section reads

$$\left(\frac{d\sigma}{d\Omega}\right)_{part} = \Phi_0(\mathbf{q}_\|,k_{inz},k_{scz}) + S(\mathbf{q}_\|)S_C(\mathbf{q}_\|,k_{inz},k_{scz}), \tag{7.74}$$

$$\Phi_0(\mathbf{q}_\parallel, k_{inz}, k_{scz}) = \frac{1}{N_p} \sum_{k=1}^{N_p} \left\{ \left\langle \left| \mathscr{F}_k(\mathbf{q}_\parallel, k_{inz}, k_{scz}) \right|^2 \right\rangle_S \right.$$

$$\left. - \left| \left\langle \mathscr{F}_k(\mathbf{q}_\parallel, k_{inz}, k_{scz}) \right\rangle_S \right|^2 e^{-W_k(\mathbf{q}_\parallel)} \right\}, \tag{7.75}$$

$$S_C(\mathbf{q}_\parallel, k_{inz}, k_{scz}) = \frac{1}{N_p} \left| \sum_{k=1}^{N_p} \left\langle \mathscr{F}_k(\mathbf{q}_\parallel, k_{inz}, k_{scz}) \right\rangle_S e^{i\mathbf{q}_\parallel \cdot \mathbf{r}_{\parallel,k}^0} e^{-W_k(\mathbf{q}_\parallel)/2} \right|^2. \tag{7.76}$$

The lattice interference function $S(\mathbf{q}_\parallel)$ is weighted by the average unit cell structure factor $S_C(\mathbf{q}_\parallel, k_{inz}, k_{scz})$ with a decrease of the Bragg rod intensities through a Debye–Waller factor. This intensity lost in the Bragg peaks is recovered spread all over the reciprocal space in the incoherent scattering term $\Phi_0(\mathbf{q}_\parallel, k_{inz}, k_{scz})$.

Within the local monodisperse approximation, the domains that interfere incoherently are made of a lattice with monodisperse particles at fixed positions in the unit cell. This is equivalent to weight the lattice site with a average structure factor that accounts for size and Debye–Waller disorders:

$$\left( \frac{d\sigma}{d\Omega} \right)_{part} = S(\mathbf{q}_\parallel) S_C(\mathbf{q}_\parallel, k_{inz}, k_{scz}), \tag{7.77}$$

$$S_C(\mathbf{q}_\parallel, k_{inz}, k_{scz}) = \frac{1}{N_p} \left| \sum_{k=1}^{N_p} \sqrt{\left\langle \left| \mathscr{F}_k(\mathbf{q}_\parallel, k_{inz}, k_{scz}) \right|^2 \right\rangle_S} e^{i\mathbf{q}_\parallel \cdot \mathbf{r}_{\parallel,k}^0} e^{-W_k(\mathbf{q}_\parallel)/2} \right|^2. \tag{7.78}$$

### 7.4.3.3 The Paracrystal

The paracrystal model was mainly developed in the early 1950s by Hosemann [68] and coworkers before the publication of a comprehensive book [69]. Even if the paracrystal is based on an underlying lattice of nodes, the long range order is progressively destroyed in a cumulative way. This so-called second kind disorder [55][5] induces a broadening of the Bragg reflections with wavevector transfer.

The analytical treatment of the paracrystal is easily handled in 1D. The site–site autocorrelation function $g(x)$ of the 1D paracrystal is calculated by taking advantage on the folding product through the knowledge of the statistics of the distances between neighboring sites $p(x)$. After putting the first site at origin, the second one is put at a distance $x$ with a density probability $p(x)$ that is peaked at a mean value $D$. The third one is added at a distance $y$ from the second site using the same rule. Thus, its probability density for a distance $x$ from origin is the product $p(y) \times p(x-y)$ integrated over all the possible intermediate distances $y$; this is nothing else than the

---

[5] The first kind of disorder, introduced in the previous section, keeps the sharpness of the Bragg peaks everywhere in reciprocal space as long as an average lattice can be defined.

convolution product $p(x) \otimes p(x)$. A straightforward generalization to all the sites yields for $x > 0$

$$g_+(x) = p(x) + p(x) \otimes p(x) + p(x) \otimes p(x) \otimes p(x) + \cdots \qquad (7.79)$$

The total pair correlation function is given by $g(x) = \delta(x) + g_+(x) + g_-(x)$, where $g_-(x) = g_+(-x)$. For the folding products giving simple product in reciprocal space, the total interference function reads

$$S(q) = 1 + \mathcal{P}(q) + \mathcal{P}(q) \cdot \mathcal{P}(q) + \mathcal{P}(q) \cdot \mathcal{P}(q) \cdot \mathcal{P}(q) \cdots + c.c. = \text{Re} \left[ \frac{1 + \mathcal{P}(q)}{1 - \mathcal{P}(q)} \right],$$
$$(7.80)$$

where $\mathcal{P}(q)$ is the 1D Fourier transform of the site–site spacing probability $p(x)$. Finite-size effects can be introduced by truncating the previous sum [70]. A useful example is given by the gaussian paracrystal for which the statistics is normal:

$$p(x) = \frac{1}{\sigma\sqrt{2\pi}} \exp\left[ -\frac{(x-D)^2}{\sigma^2} \right], \ \mathcal{P}(q) = \exp\left[ \pi q^2 \sigma^2 \right] \exp[iqD]. \qquad (7.81)$$

The resulting Hosemann scattering function

$$S(q) = \frac{1 - \phi(q)^2}{1 + \phi(q)^2 - 2\phi(q)\cos(qD)}, \quad \phi(q) = \exp\left[ \pi q^2 \sigma^2 \right] \qquad (7.82)$$

shows broader and broader reflections upon increasing the width $\sigma$ of the site spacing distribution (see Fig. 7.15). At the extreme limit, as a result of the cumulative disorder, only one broad peak remains in the interference function.

The generalization to higher dimensions is far from being simple except for the perfect paracrystal. This latter relies on a set of lattice vectors with fully independent position statistics [69]. However, the perfect paracrystal constraints the unit cells

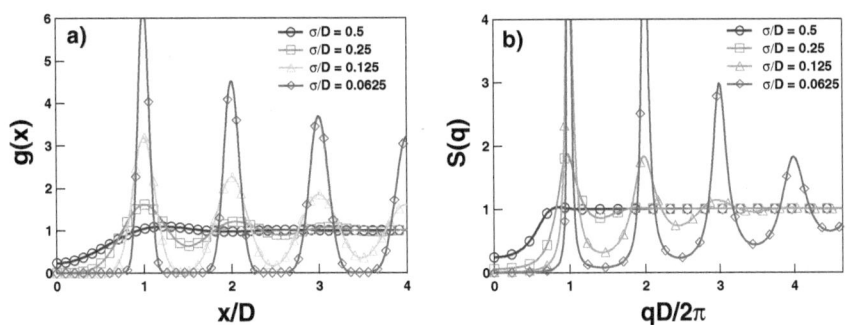

**Fig. 7.15** (**a**) Pair correlation function and (**b**) interference function of a 1D gaussian paracrystal for various values of the variance of site spacing distribution. From [50]

to be parallelograms[6] and the associated Fourier transform generate a too intense scattering at low wavevector transfer that is cured only with finite-size effects.

## 7.4.4 DA, LMA or Beyond?

### 7.4.4.1 Differences Between LMA and DA

Scattering from polydispersed systems of particles are most of the time analyzed within DA and LMA as they offer two tractable expressions for the scattering cross section. However, one has to bear in mind that they give an oversimplified view of the morphology. The DA supposes that the particles are positioned in a way that is completely independent of their kinds (shape, sizes, etc.). The LMA assumes a more stringent correlation, i.e., each particle is surrounded, within the coherence length of the X-ray beam, by objects of the same kind giving rise to an incoherent interference between homogeneous domains.

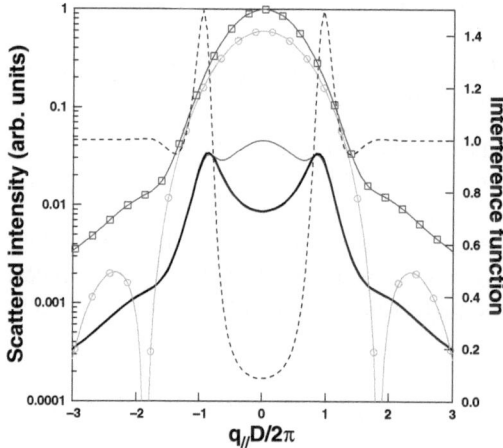

**Fig. 7.16** Scattering from a polydispersed collection of cylinders on a paracrystalline chain as a function of $q_\parallel$ ($q_z = 0$): LMA scattering with the same $S(q_\parallel)$ in each domain (thick line) and DA scattering (thin line) (shifted down by one decade for clarity). The size distribution is normal with $\langle R \rangle / D = \sigma_D / D = 0.29$, $\sigma_R / \langle R \rangle = 0.3$. Also shown are the mean form factor $\langle |\mathscr{F}(q_\parallel)|^2 \rangle$ (*line with square*), the incoherent term $\Phi_0(q_\parallel) = \langle |\mathscr{F}(q_\parallel)|^2 \rangle - |\langle \mathscr{F}(q_\parallel) \rangle|^2$ (*line with circle*) and the interference function (*dotted line – right scale*). The intensity have been normalized by the value $\langle |\mathscr{F}(q_\parallel = 0)|^2 \rangle = \langle |V|^2 \rangle$. The specular rod is not shown

---

[6] For instance, the sixfold symmetry is not recovered with a perfect gaussian paracrystal based on two vectors at 120° [75]

For disordered system, the scattering at high wavevector transfer is dominated only by the particle form factor since the interference function tends toward one. Within this limit, both approximations give similar results.[7] The differences shows up close and below the correlation peak as shown in Fig. 7.16. Close to the origin, the DA signal is dominated by the incoherent term $\Phi_0(q_\parallel) = \left\langle |\mathscr{F}(q_\parallel)|^2 \right\rangle - |\langle \mathscr{F}(q_\parallel) \rangle|^2$ while the LMA signal is lower. It is worth noticing that the location of the maximum of intensity, the so-called correlation peak, is shifted toward low $q_\parallel$ value compared to the maximum of the interference function because of the product with the form factor. Therefore, the correlation peak position gives only a rough estimate of the interparticle distance [50, 64].

### 7.4.4.2 Toward the Account of Correlations Between Particles: the 1D Size-Spacing Correlation Approximation

In concentrated systems, the DA breakdowns because of correlations. For instance, in the case of islands on surface, the bigger the particles the farther apart they are because their capture area scales with their size. The analysis of some transmission electron micrographs for Pd islands on MgO (see Sect. 7.6.1 and [64]) has given clear evidence of a lack of correlation between sizes but pointed out a correlation between the island radius and the distance to its first neighbor. The expected scattering computed from the microscopy results yielded to a reduction of the scattered intensity as compared to the DA just below the correlation peak while LMA reproduced the experimental results only with an ad hoc interference function. The key point is to reintroduce some correlations between particle sizes and distances. This can be done in LMA by linking the size of the particles in each monodisperse domain to the interference function parameters [61, 62]. A more subtle approach called scaling approximation (SA) was developed by Gazzillo and coworkers [76, 77]. The idea is to obtain all the partial pair correlation functions $g_{\alpha\beta}(r_\parallel)$ used in Eq. (7.46) from a suitable scaling of that of a monodisperse system. For high particle concentration, the efficiency of SA was tested with success against the exact scattering solution of the only one tractable polydisperse case, namely the hard core interacting particles in 3D [78, 79]. However, to our knowledge, the method was never really applied to actual data analysis. On a practical point of view, whatever the approximations, the obtained morphological parameters are valid if one takes care to fit properly the high-$q$ range even to the detriment of a good fit of the correlation peak.

Some attempts to rationalize the problem of getting the partial interference functions at least for 3D particles aligned in one dimension have been made in [58, 63, 71]. The main interest of this analytical model is to highlight the influence of such correlations. Basically, the particles are aligned along a chain with a spacing from particle to particle that depends on statistical way on their relative sizes. The idea of this size-spacing correlation approximation (SSCA) is to take advantage of

---

[7] Despite the fall in signal/noise ratio, this points out the importance of reliable measurements far in the reciprocal space where the sensitivity to the form factor is by far the better.

the folding properties of the scattering density autocorrelation function within the cumulative paracrystal model. As the perpendicular $\mathbf{r}_\perp$ (or $\mathbf{z}$) and the parallel $r_\parallel$ (or $y$) directions behave independently, the analysis is obtained only by using the simple Fourier transform, the DWBA scattering cross section being recovered upon replacing the simple particle form factor by the DWBA one (see Sect. 7.3).

The autocorrelation function of the scattering volume $c_{\rho\rho}^+(r_\parallel, \mathbf{r}_\perp)$ (see Sect. 7.4.3) for $r_\parallel \geq 0$ is calculated step by step along the chain from the knowledge of (i) the joint density probability $p(\alpha_0, \ldots, \alpha_n)$ of having a sequence of particles of kind $\alpha_0, \ldots, \alpha_n$ along the chain and (ii) the conditional density probability of having an algebraic distance $d_n$ between the particles $n-1$ and $n$ knowing the sequence $\alpha_0, \ldots, \alpha_n$, i.e., $P_n(d_n/[\alpha_0, \ldots, \alpha_n])$:

$$C_{\rho\rho}^+(r_\parallel, \mathbf{r}_\perp) = c_{\rho\rho}^0(r_\parallel, \mathbf{r}_\perp) + c_{\rho\rho}^+(r_\parallel, \mathbf{r}_\perp), \tag{7.83}$$

$$c_{\rho\rho}^0(r_\parallel, \mathbf{r}_\perp) = \int p(\alpha_0) \{\mathscr{S}_0(-y, -\mathbf{z}, \alpha_0) \otimes \mathscr{S}_0(y, \mathbf{z}, \alpha_0) \otimes \delta(y)\}(r_\parallel, \mathbf{r}_\perp) d\alpha_0, \tag{7.84}$$

$$\begin{aligned}
c_{\rho\rho}^+(r_\parallel, \mathbf{r}_\perp) = &\iint p(\alpha_0, \alpha_1) \{\mathscr{S}_0(-y, -\mathbf{z}, \alpha_0) \\
&\otimes \mathscr{S}_1(y, \mathbf{z}, \alpha_1) \otimes P_1(y/[\alpha_0, \alpha_1])\}(r_\parallel, \mathbf{r}_\perp) d\alpha_0 d\alpha_1 \\
&+ \iiint p(\alpha_0, \alpha_1, \alpha_2) \{\mathscr{S}_0(-y, -\mathbf{z}, \alpha_0) \otimes \mathscr{S}_2(y, \mathbf{z}, \alpha_2) \\
&\otimes P_1(y/[\alpha_0, \alpha_1]) \otimes P_2(y/[\alpha_0, \alpha_1, \alpha_2])\}(r_\parallel, \mathbf{r}_\perp) d\alpha_0 d\alpha_1 d\alpha_2 + \cdots
\end{aligned} \tag{7.85}$$

$\otimes$ is the folding product in space and as previously introduced, $\mathscr{S}(y, \mathbf{z}, \alpha)$ is the shape function of the particle of kind $\alpha$. Indeed, the probability of having a given distance $d$ between the origin and the $n$th particle is the product of all the probabilities of the intermediate distances between particles before the $n$th summed over such distances; the constraint that the sum of such distances is $d$ shows that the required probability is the folding product of all the intermediate distance probabilities. After having introduced $\mathscr{P}_n(q_\parallel/[\alpha_0, \ldots, \alpha_n])$, the Fourier transform along the chain of $P(d_n/[\alpha_0, \ldots, \alpha_n])$, the scattered intensity per particle is obtained from the total autocorrelation function $C_{\rho\rho}(r_\parallel, \mathbf{r}_\perp) = c_{\rho\rho}^0(r_\parallel, \mathbf{r}_\perp) + c_{\rho\rho}^+(r_\parallel, \mathbf{r}_\perp) + c_{\rho\rho}^-(r_\parallel, \mathbf{r}_\perp)$ $(c_{\rho\rho}^-(r_\parallel, \mathbf{r}_\perp) = c_{\rho\rho}^+(-r_\parallel, -\mathbf{r}_\perp))$ by simple Fourier transform:

$$\begin{aligned}
\left(\frac{d\sigma}{d\Omega}\right)_{part} = &\tilde{z}_0(\mathbf{q}_\perp)\delta(q_\parallel) + \int p(\alpha_0) |\mathscr{F}_0(\alpha_0, q_\parallel, \mathbf{q}_\perp)|^2 d\alpha_0 \\
&+ 2\mathrm{Real}\left\{ \iint p(\alpha_0, \alpha_1) \mathscr{F}_0^*(q_\parallel, \mathbf{q}_\perp, \alpha_0) \mathscr{F}_1(q_\parallel, \mathbf{q}_\perp, \alpha_1) \right. \\
&\qquad \mathscr{P}_1(q_\parallel/[\alpha_0, \alpha_1]) d\alpha_0 d\alpha_1 \\
&\qquad + \iiint p(\alpha_0, \alpha_1, \alpha_2) \mathscr{F}_0^*(q_\parallel, \mathbf{q}_\perp, \alpha_0) \mathscr{F}_2(q_\parallel, \mathbf{q}_\perp, \alpha_2) \mathscr{P}_1(q_\parallel/[\alpha_0, \alpha_1]) \\
&\qquad \left. \mathscr{P}_2(q_\parallel/[\alpha_0, \alpha_1, \alpha_2]) d\alpha_0 d\alpha_1 d\alpha_2 + \ldots \right\}.
\end{aligned} \tag{7.86}$$

A divergence proportional to the size of the system appears at $q_\| = 0$ where all the particles scatter exactly in phase. It is proportional to $V^2(\alpha, \mathbf{q}_\perp)$, the $q_\| = 0$ limit of the form factor which reduces to the volume of the particle at $\mathbf{q}_\perp = 0$.

The obtained formula is quite general, whatever the correlations between particles are. However, to go further on, a complete lack of correlation between sizes of neighbors will be assumed: $p(\alpha_0, \ldots, \alpha_n) = p(\alpha_0) \ldots p(\alpha_n)$ (see [63] for the influence of such correlations). But, to account for an excluded volume effect, the distance between two neighbors is supposed to depend linearly on their respective sizes $R_\|(\alpha_i)$ along the chain direction. This size-spacing correlation approximation (SSCA) is included into $P_n(d_n/[\alpha_0, \ldots, \alpha_n])$ through

$$\int_{-\infty}^{+\infty} d_n P_n(d_n/[\alpha_0, \ldots, \alpha_n]) dd_n = D + \kappa \left[ \Delta R_\|(\alpha_{n-1}) + \Delta R_\|(\alpha_n) \right], \qquad (7.87)$$

with $\Delta R_\|(\alpha_i) = R_\|(\alpha_i) - \langle R_\|(\alpha) \rangle$. $D$ and $\langle R_\|(\alpha) \rangle$ are, respectively, the average distance along the chain between particles irrespective of their sizes and the parallel average radius. This gives in reciprocal space:

$$\mathscr{P}_n(q_\|/[\alpha_0, \ldots, \alpha_n]) = \phi(q_\|) e^{iq_\| D} e^{i\kappa q_\| [\Delta R_\|(\alpha_{n-1}) + \Delta R_\|(\alpha_n)]}. \qquad (7.88)$$

$\kappa$ is the size-spacing coupling parameter. $\kappa > 0$ corresponds to repelling particles. It is worth noticing that in the $\kappa = 0$ limit, Eq. (7.88) gives the statistic of the classical paracrystal in reciprocal space $\phi(q_\|) e^{iq_\| D}$. The introduction of these two approximations in Eq. (7.86) yields

$$\left( \frac{d\sigma}{d\Omega} \right)_{part} = \tilde{z}_0(\mathbf{q}_\perp) \delta(q_\|) + |\langle \mathscr{F}(q_\|, \mathbf{q}_\perp) \rangle|^2 + 2 \operatorname{Real} \left\{ \sum_{n=1}^{+\infty} \Gamma_n(q_\|, \mathbf{q}_\perp) \right\},$$
$$(7.89)$$

$$\Gamma_n(q_\|, \mathbf{q}_\perp) = \phi^n(q_\|) \exp(inq_\| D) \int \ldots \int p(\alpha_0) \ldots p(\alpha_n) \mathscr{F}^*(q_\|, \mathbf{q}_\perp, \alpha_0)$$
$$\times \mathscr{F}(q_\|, \mathbf{q}_\perp, \alpha_n) \times \exp \left[ i\kappa q_\| \left( \Delta R_\|(\alpha_0) + 2 \sum_{k=1}^{n-1} \Delta R_\|(\alpha_k) + \Delta R_\|(\alpha_n) \right) \right]$$
$$d\alpha_0 \ldots d\alpha_n.$$

The geometric series allows to carry out the summation in Eq. (7.89):

$$\left( \frac{d\sigma}{d\Omega} \right)_{part} = \tilde{z}_0(\mathbf{q}_\perp) \delta(q_\|) + |\langle \mathscr{F}(q_\|, \mathbf{q}_\perp) \rangle|^2$$
$$+ 2 \operatorname{Real} \left\{ \widetilde{\mathscr{F}}_\kappa(q_\|, \mathbf{q}_\perp) \widetilde{\mathscr{F}}^*_\kappa(q_\|, \mathbf{q}_\perp) \frac{\Omega_\kappa(q_\|)}{\tilde{p}_{2\kappa}(q_\|) [1 - \Omega_\kappa(q_\|)]} \right\}, \qquad (7.90)$$

$$\Omega_\kappa(q_\|) = \tilde{p}_{2\kappa}(q_\|) \phi(q_\|) e^{iq_\| D}. \qquad (7.91)$$

The characteristic function of the particle kind distribution evaluated along the parallel size distribution was introduced in the previous equation:

$$\tilde{p}_\kappa(q_\parallel) = \int p(\alpha) e^{i\kappa q_\parallel \Delta R_\parallel(\alpha)} d\alpha. \tag{7.92}$$

$\widetilde{\mathscr{F}}_\kappa(\mathbf{q}_\parallel)$ defined as

$$\widetilde{\mathscr{F}}_\kappa(q_\parallel, \mathbf{q}_\perp) = \int p(\alpha) \mathscr{F}(q_\parallel, \mathbf{q}_\perp, \alpha) e^{i\kappa q_\parallel \Delta R_\parallel(\alpha)} d\alpha \tag{7.93}$$

is a generalization of the average particle form factor (obtained for $\kappa = 0$). Notice that in Eq. (7.90) the complex conjugate is applied to the particle form factor before the average over $\alpha$ (Eq. (7.93)).

The total interference function $S(q_\parallel)$ follows from Eq. (7.90) upon replacement of the particle shape by a Dirac peak, i.e., the particle form factor $\mathscr{F}(q_\parallel, \mathbf{q}_\perp, \alpha)$ by one:

$$S(q_\parallel) = 1 + 2\,\text{Real}\left\{ \frac{\tilde{p}_\kappa^2(q_\parallel)\Omega_\kappa(q_\parallel)}{\tilde{p}_{2\kappa}(q_\parallel)\left[1 - \Omega_\kappa(q_\parallel)\right]} \right\}. \tag{7.94}$$

If $\kappa = 0$, $\widetilde{\mathscr{F}}_\kappa(q_\parallel, \mathbf{q}_\perp) = \langle \mathscr{F}(q_\parallel, \mathbf{q}_\perp, \alpha) \rangle$ and $\tilde{p}_{2\kappa}(q_\parallel) = 1$. Thus, Eq. (7.90) reduces to the DA (Eq. (7.49)) with an interference function given by the Hosemann 1D paracrystal (Eq. (7.82)). The specular $q_\parallel = 0$ can also be included in Eq. (7.89) through finite-size effects or correlation length limitation [58]. Through an expansion around $q_\parallel = 0$, it is possible to demonstrate [58] that the scattering cross section reaches a minimum for a special $\kappa_0$ value:

$$\kappa_0(\mathbf{q}_\perp) = \frac{1}{2}\frac{D}{\sigma_{R_\parallel}^2} \frac{\langle \mathscr{F}(q_\parallel = 0, \mathbf{q}_\perp)\Delta R_\parallel(\alpha) \rangle}{\langle \mathscr{F}(q_\parallel = 0, \mathbf{q}_\perp) \rangle} \tag{7.95}$$

with $\sigma_{R_\parallel}^2 = \left\langle \left(R_\parallel - \langle R_\parallel \rangle\right)^2 \right\rangle$.

Equation (7.90) is illustrated in Figs. 7.17 and 7.18 for DA, LMA and SSCA with fixed particle aspect ratio or fixed height. As expected for disordered systems, the different approximations give an equivalent result well above the correlation peak where the scattering is given only by the particle form factor. The main differences show up at and below the correlation peak, i.e., the maximum of the scattered intensity. The size-spacing correlation induces a shape modification of the correlation peak and a shift of its position that depend on the coupling parameter $\kappa$. Even if the particle density is constant for all the approximations, at variance to DA or LMA, the SSCA peak position $q_\parallel^p$ does not give a direct measurement of the particle density using $D \simeq 2\pi/q_\parallel^p$ as expected from Eqs. (7.50 and 7.51). At $\kappa = \kappa_0$, a dip of scattering shows up below the correlation peak while at higher or lower $\kappa$ value, the curves are dominated only by the mean particle form factor $\left\langle |\mathscr{F}(\mathbf{q}_\parallel)|^2 \right\rangle$. Within this limit, the particles can be said to be "uncorrelated" and the Guinier analysis

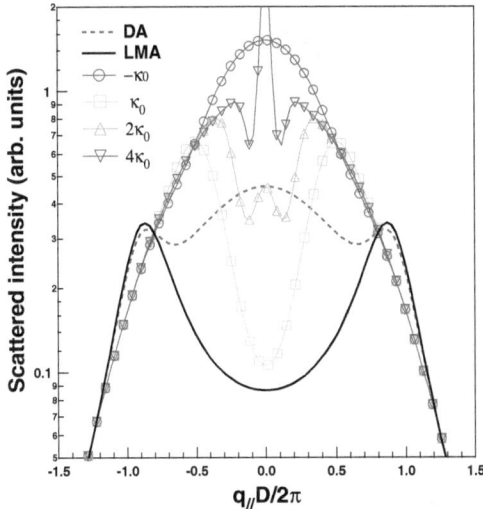

**Fig. 7.17** Scattered intensity within the SSCA from a 1D chain of correlated cylinders at $q_z = 0$. The form factor is calculated in the Born approximation. The set of simulation parameters is the same as that of Fig. 7.16 with $H/R = 1$. Various models of calculations have been used for the sake of comparison: DA, LMA, SSCA. For SSCA, the coupling parameter has been set to multiple of the special $\kappa_0$ value (see text). The left scale has been normalized by $\left\langle |\mathscr{F}(q_\| = 0)|^2 \right\rangle = \left\langle |V|^2 \right\rangle$. Adapted from [71]

is valid (see Sect. 7.4.2). To some extent, this complex behavior can be explained through the expansion of the scattering cross section along the partial interference functions $S_{\alpha\beta}(\mathbf{q}_\|)$ Eq. (7.48) (see [58, 63]). The behavior along $q_\|$ is all the more complicated than it depends on the dimensionality of the scatterer. For instance, the $(q_\|, q_z)$ scattering of a cylinder chain (Fig. 7.18) is different along the chain axis if the particles have an constant aspect ratio $H/R$ or a constant height as the particle shape involved in $\widetilde{\mathscr{F}}_\kappa(q_\|, \mathbf{q}_\perp)$ (Eq. (7.93)) scales either with $R_\|^3$ or with $R_\|^2$. An interesting feature (Fig. 7.18-c) is the tilt of the perpendicular scattering lobes of the cylinder shape. While the perpendicular and the parallel directions behave independently for the cylinder form factor, at variance to DA and LMA, the SSCA leads to a coupling in the scattering pattern between both directions. At constant aspect ratio and for $\kappa > 0$, on a statistical point of view, the bigger particles that scatter closer to the origin in $q_z$ are farther apart and thus scatter closer to the origin along $q_\|$. The reverse is true for the smaller ones thus leading to a tilt of the scattering lobes.

The main drawback of the SSCA is its 1D treatment of the partial interference function as compared to 2D or 3D real systems. It is difficult to weight up the degree of approximation made but SSCA seems to catch the most important physical aspects of scattering by dense collection of particles. An example of analysis applied to islands on a surface can be found in [71]. A generalization to 2D at least using the perfect paracrystal theory [69] is possible.

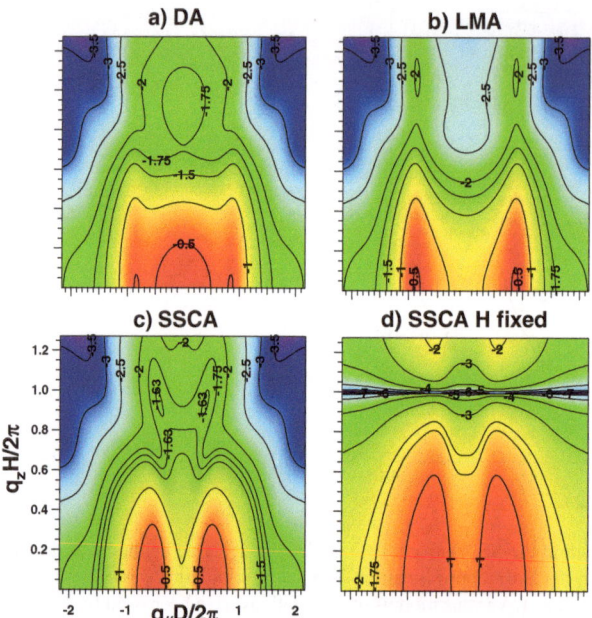

**Fig. 7.18** Same as in Fig. 7.17 but in $(q_\parallel, q_z)$ space. The color scale is logarithmic as given by the contour lines. The particle aspect ratio is constant $H/R = 1$ except for case(d) where $H/\langle R \rangle = 1$. Adapted from [71]

## 7.4.5 Dynamical Scattering and DWBA Limitations: The Graded Interface

To calculate the incoherent scattering cross section in Sects. 7.3 and 7.4, particles have been taken as perturbation of the wavefields reflected and refracted at flat interfaces and advantage has been taken of the linearity of the scattering potential to treat the problem of set of particles. But, for instance in the case of islands, upon increasing the particle coverage, multiple scattering and absorption are expected to become important, in particular when the incidence $\theta_{in}$ and exit $\theta_{sc}$ angles are close to the critical angle as the path of the beams are considerably increased inside the layer of particles. Improvement of the DWBA treatment can be obtained by using as a starting point of the perturbation $E_{ref}$ the wavefields inside the graded interface obtained after averaging along the surface the profile of electronic density that includes the particle layer itself. In this way, multiple scattering of the primarily scattered wave without change of parallel wavevector (i.e., only in the specular direction) is exactly included. The result is expected to be better than using flat interfaces because the scattering potential has a zero mean when averaged along the surface (see Sect. 4.3.4).

The method will be illustrated for islands on a flat surface [58] (see Sect. 7.3.1) by using the results of Sect. 7.3.3. A generalization to all the morphologies of Fig. 7.3 is straightforward. An analogous treatment of scattering from roughness in multi-layered systems can also be found in Chap. 6. Let us call $\widetilde{n}_0^2(z)$ the average dielectric constant perpendicular to the surface:

$$
\widetilde{n}_0^2(z) = \begin{cases} 1 & \text{if } z > t \\ \widetilde{n}_l^2(z) = \frac{n_i^2}{A} \int_A \sum_j \mathscr{S}_j(\mathbf{r}_\| - \mathbf{r}_{\|,j}, z) d\mathbf{r}_\| & \text{if } 0 < z < t \\ n_s^2 & \text{if } z < 0 \end{cases} \tag{7.96}
$$

where $t$ is the thickness of the island layer (i.e., the limit with vacuum) and $\mathscr{S}_j(\mathbf{r}_\| - \mathbf{r}_{\|,j}, z)$ the shape of the particle located at $\mathbf{r}_{\|,j}$. The $z$-dependence of the unperturbed wavefield is similar to the case of a single layer Eq. (7.14) (see Chap. 3):

$$
E_l^{PW}(\mathbf{k}_{inz,0}, z) = \begin{cases} \widetilde{A}_0^+ e^{ik_{inz,0}z} + e^{-ik_{inz,0}z} & \text{for } z > t \\ \widetilde{A}_1^+(z) e^{ik_{inz,1}(z)z} + \widetilde{A}_1^-(z) e^{-ik_{inz,1}(z)z} & \text{for } 0 < z < t \\ \widetilde{A}_2^- e^{-ik_{inz,2}z} & \text{for } z < 0 \end{cases} \tag{7.97}
$$

but with the amplitudes of the upward and downward waves $\widetilde{A}_1^\pm(z)$ that vary continuously across the layer as $k_{inz}(z) = -\sqrt{\widetilde{n}_l^2(z)k_0^2 - k_\|^2}$. These amplitudes can be calculated in a discrete way using the Abelès' matrix method developed in Sect. 3.2.1 for the reflectivity of multilayers. The perturbation potential $\delta n^2(\mathbf{r})$ measures the departure from $\widetilde{n}_0(z)$ due to either the islands or the holes in between:

$$
\delta n^2(\mathbf{r}_\|, z) = [n_i^2 - \widetilde{n}_l^2(z)] \sum_j \mathscr{S}_j(\mathbf{r}_\| - \mathbf{r}_{\|,j}, z) + [1 - \widetilde{n}_l^2(z)] \mathscr{S}_{hole}(\mathbf{r}_\|, z). \tag{7.98}
$$

The shape factor for vacuum between islands is nothing else than the inverse fingerprint of the island $\mathscr{S}_{hole}(\mathbf{r}_\|, z) = \Theta(z) - \Theta(z-t) - \sum_i \mathscr{S}_i(\mathbf{r}_\| - \mathbf{r}_{\|,i}, z)$, $\Theta(z)$ being the step function ($\Theta(z) = 0$ for $z < 0$ and $\Theta(z) = 1$ for $z > 0$). Thus

$$
\delta n^2(\mathbf{r}_\|, z) = [n_i^2 - 1] \sum_j \mathscr{S}_j(\mathbf{r}_\| - \mathbf{r}_{\|,j}, z) + [1 - \widetilde{n}_l^2(z)][\Theta(z) - \Theta(z-t)]. \tag{7.99}
$$

The second term can be dropped out as, once inserted in the DWBA cross section due to the lack of $\mathbf{r}_\|$ dependence, it gives a singular term at origin $\delta(\mathbf{q}_\|)$; in fact, only the contrast between island and vacuum scatters. Moreover, the additivity in the perturbation induced by the islands used in Sect. 7.4 can still be used. After having put the Fresnel wavefield of the graded interface (Eq. (7.97)) in the far-field Green function, one ends up with an expression analogous to Eq. (7.8) or Eq. (7.18) but with a more complex effective form factor:

$$
\begin{aligned}
\mathscr{F}(\mathbf{q}_{\|}, k_{iz}, k_{fz}) = & \int d\mathbf{r}_{\|} e^{i\mathbf{r}_{\|} \cdot \mathbf{q}_{\|}} \int dz \mathscr{S}(\mathbf{r}_{\|}, z) \\
& \times \Big\{ \widetilde{A}_1^- [k_{inz,1}(z)] \widetilde{A}_1^- [-k_{scz,1}(z)] e^{i[+k_{scz,1}(z) - k_{inz,1}(z)]z} \\
& + \widetilde{A}_1^+ [k_{inz,1}(z)] \widetilde{A}_1^- [-k_{scz,1}(z)] e^{i[+k_{scz,1}(z) + k_{inz,1}(z)]z} \\
& + \widetilde{A}_1^- [k_{inz,1}(z)] \widetilde{A}_1^+ [-k_{scz,1}(z)] e^{i[-k_{scz,1}(z) - k_{inz,1}(z)]z} \\
& + \widetilde{A}_1^+ [k_{inz,1},(z)] \widetilde{A}_1^+ [-k_{scz,1}(z)] e^{i[-k_{scz,1}(z) + k_{inz,1}(z)]z} \Big\}.
\end{aligned}
\tag{7.100}
$$

The above form factor includes in a continuous way the propagation effects inside the embedding profile of refraction index $\widetilde{n}_0(z)$. In each particle slice, four scattering events from and to upward or downward propagating waves take place. For isolated particles, $\widetilde{n}_0(z) \simeq n_0(z)$, $k_{z,0} \simeq k_{z,1}$, $\widetilde{A}_1^+ \simeq \widetilde{A}_0^+$ and $\widetilde{A}_1^- \simeq \widetilde{A}_0^- \simeq r_{01}$: Eq. (7.9) is recovered.

The form factor of a layer of monodisperse spheres on a substrate is given in Fig. 7.19 as function of coverage $\Theta = \pi R^2 n_s$ and $\theta_{in}$. For a full sphere, the profile of refraction index is parabolic. Compared to the cross section of an isolated particle

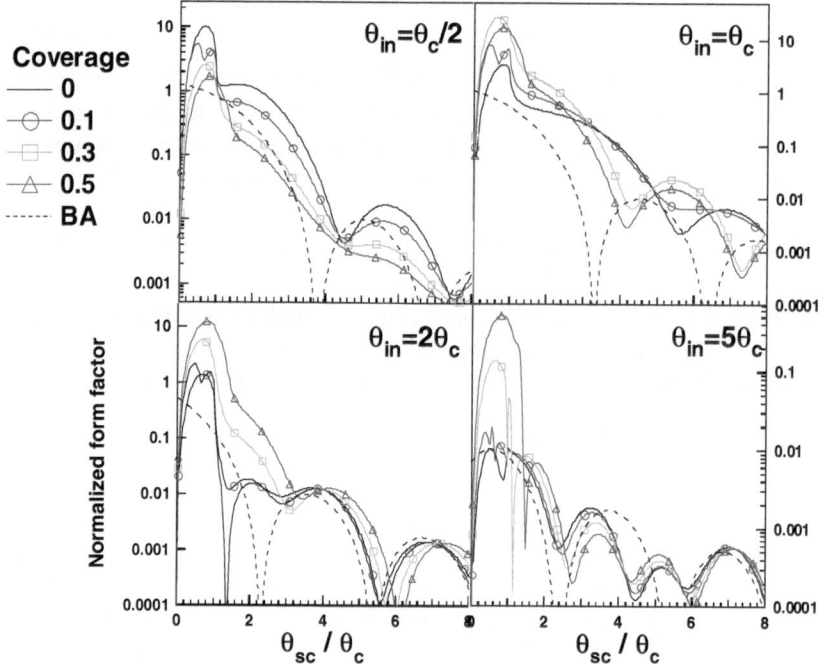

**Fig. 7.19** Form factor $\mathscr{F}(\mathbf{q}_{\|} = 0, k_{iz} = -k_0 \sin(\theta_{in}), k_{fz} = k_0 \sin(\theta_{sc}))$ Eq. (7.100) of a layer of full spheres on a substrate for increasing surface coverage $\Theta = \pi R^2 n_s$ and various incidence angle $\theta_{in}$. The form factor includes the effects of the parabolic profile of the refraction index. The intensity has been normalized by the particle volume and the scattering angle $\theta_{sc}$ by the substrate critical angle $\theta_c$. The BA scattering has been added for comparison. Adapted from [58]

$\Theta = 0$ coverage, i.e., Eq. (7.9), two main conclusions can be drawn from the form factor calculated using the graded interface Eq. (7.100). Firstly, the Yoneda peak shape is highly sensitive to the total coverage because the evanescent waves and the absorption are affected by the particle layer. Secondly, the location and sharpness of the minima of the interference fringes depend in tremendous way on the coverage, or more generally on the embedding profile of refraction index.

## 7.5 Experimental Considerations in GISAXS

Even though the pioneer work of Levine [3,4] was performed using a rotating anode, most of the GISAXS set-ups have been developed on various synchrotron sources to take advantage of the source brilliance, of the low divergence and of the multi-wavelength availability. Typically, the X-ray beam delivered by a wiggler, an undulator or a bending magnet is low band pass energy filtered by a mirror before being monochromatized and focused on the sample. Sets of horizontal and vertical slits are used to define the narrowest beam. As the scattered intensity decreases rapidly upon moving away from the origin of reciprocal space, making measurements with the highest dynamical range and the lowest background is mandatory to deduce correctly morphological parameters from correlated systems of particles. Scattering and absorbtion from air in particular for low-energy X-rays is avoided by using evacuated pipes along the X-ray path or pipes filled with helium. Great care should be taken to avoid any parasitic low-angle scattering by the beamline components. Thus, to remove slit scattering, each set of slits is associated with guard slits with an aperture slightly higher than the beam size.

The sample is mounted on a goniometric head specially designed to align the sample with its surface quasi-parallel to the X-ray beam. In particular for bending magnets, a lower beam divergence is achieved in the vertical plane. Thus, the surface is set vertical or horizontal accordingly to the desired best resolution in $q_y$ or $q_z$ that fits the beam divergence. However, for liquid sample, the beam has to be deflected by a mirror on the horizontal sample. Owing to the grazing incidence geometry, the illuminated area, the so-called footprint, is a stripe with a length equal to the sample size and a width given by the beam size, which is typically a few hundreds of microns. Special set-up design allowed to increase the lateral resolution by around two orders of magnitude in order to probe inhomogeneous sample with micro-focused beams [32,33]. Footprint down to $5 \times 300\ \mu m^2$ in the $(y - x)$ direction were achieved using beam defined by a 5 $\mu m$ pinhole. In the grazing geometry, the beam coherence length is considerably increased up to a few microns by the projection effect along the surface. The in-plane symmetry is studied by rotating the sample around its normal (azimuthal rotation $\omega$) while keeping constant the incident angle $\theta_{in}$. Thus, as $q_z \neq 0$, the full symmetry of 3D objects can be assessed in contrast to conventional SAXS where $q_z = 0$ [48].

The scattered beam is collected either (i) along $\psi$ or $\theta_{sc}$ on a position sensitive detector or (ii) on a bidimensional detector as an image plate or a CCD camera which is placed a few meters downstream. With 2D detector, several corrections have to applied before data analysis: (i) background subtraction, (ii) flat field (i.e., pixel response linearity), (iii) reference if any and (iv) camera distortion. As several decades of intensity separate the diffuse scattering from the direct beam and the specular rod, a motorized beam-stop allows to keep the detector in its linear range by suppressing the direct and reflected beam. The out-of-plane $\theta_{sc}$ and in-plane angle $\psi$ are limited to a few degrees (i.e., the small angle range) leading to a smallest accessible dimension $d_\parallel \sim \lambda / \sin(\Psi)$ or $d_\perp \sim \lambda / \sin(\theta_{sc})$ of around a nanometer. Going above becomes the field of grazing incidence diffraction from atomic scale arrangements with more specialized diffractometer design. The highest accessible parallel dimension is dictated by the X-ray wavelength, the lateral beam size, the divergence of the primary beam and the size and location of the beam-stop. Typically, $\Psi_{min} \sim 0.05°$ and $d_\parallel^{max} \sim 100\,\text{nm}$. Experimental set-ups derived from bulk ultra-small-angle scattering with large sample-detector distance (several meters) in addition to an increased collimation by entrance slits allowed to extend the $q_y$-resolution of the in-plane characteristic length up to several microns [31]. Nearly, two orders of magnitudes [72] can be gained compared to standard GISAXS. Compared to USAXS, the GIUSAXS advantage is that in reflection geometry, the direct and the reflected beams are shifted allowing to avoid, in some cases, the use of beam-stops. Moreover, using grazing incidence increases the in-plane coherence length of the beam due the projection onto the sample surface. However, in terms of resolution with conventional GISAXS set-ups, it has been shown that higher distances (few hundred of nanometers) on nanostructured surfaces can be probed by taking advantage of the forward wavevector transfer $q_x$ [73].

To perform GISAXS during sample preparation, for instance the growth of aggregates on a surface, the sample environment has to be made compatible with the constraints of low background scattering. An all in-vacuum set-up connected to a molecular beam epitaxy chamber with all the surface preparation and deposition facilities has been thoroughly described in [34]. In the same spirit, a reactor for in operando studies [35] of catalysts has been developed.

## 7.6 Examples of GISAXS Experiments in Hard Condensed Matter: Islands on Surfaces

### 7.6.1 Metal/Oxide Island Growth: The Pd/MgO(001) Case

The crystalline growth on surfaces is a field of physics and chemistry and has long attracted attention not only on fundamental point of view but also for the numerous potential applications of thin films. One challenge is the in situ monitoring of the surface morphology upon growth conditions (temperature, flux of deposition, substrate state, gaseous environment). Apart from a better control of the elaboration process, this can give some clues about the microscopic mechanisms involved during the growth (adsorption, diffusion, coalescence, etc.) through a comparison with suitable theoretical models. Even if microscopy techniques give straightforward information, the lengthy acquisition time, the growth interruption, the local view, the insulating character and the preparation of the sample set some limits to the surveys. This is particularly true in the case of metal deposited on insulating substrates where the growth proceeds through 3D islands. This is known as the Volmer–Weber mechanism. The capability of GISAXS to tackle the growth problem is obvious if the technique is applied in situ.

Pg/MgO(100) model catalyst growth has been thoroughly studied with GISAXS in situ during vapor deposition in ultra-high vacuum from the very beginning of the growth to the coalescence at various temperatures. A dedicated experimental set-up has been developed for such experiments [34] at ESRF (European Synchrotron Radiation Facility). Large Z-elements on low-density substrate are very good candidates for such GISAXS studies. This is illustrated in Fig. 7.20 where a set of GISAXS patterns are shown at various coverage during growth on a substrate kept at 650 K. Upon deposition, two scattering lobes separated by the beam-stop shrinks toward the origin of reciprocal space; in direct space, this means that all the characteristics distances increase that is to say the islands get bigger and their spacing $D$ increases. Around 0.5 nm of equivalent thickness, interference fringes (up to third order) along the perpendicular direction $\theta_{sc} \sim q_z$ show up indicating flat top islands. At the same time, a scattering rod tilted by 54.7° from the normal when the beam is

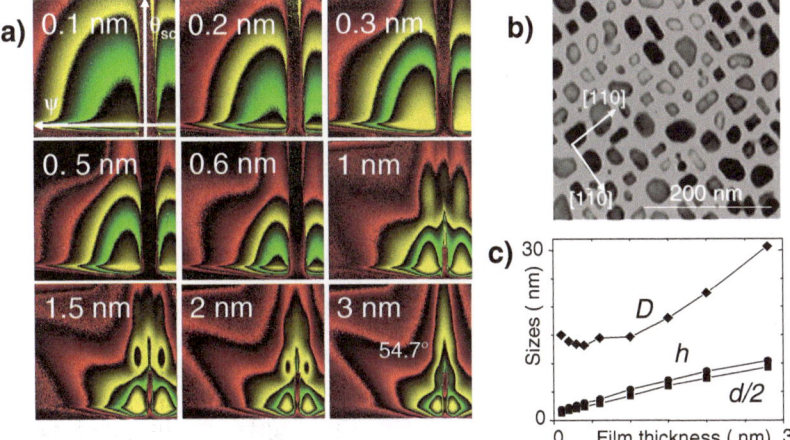

**Fig. 7.20** (a) Experimental GISAXS patterns of Pd nanoislands grown by vapor deposition on MgO(100) kept at 650 K. The equivalent deposited film thickness is given in figure. The scattering patterns that extend up to 3° in $\psi$ and $\theta_{sc}$ are displayed on a logarithmic color scale. The x-ray beam was oriented along the $[110]_{MgO}$ direction. Note the rise up of a scattering rod from the island facets above 0.5 nm. (b) Electron microscopy image ($250 \times 250$ nm) after carbon replica of the last deposit 2.8 nm. (c) Evolution of the mean island morphological parameters with the amount of deposited material as deduced from GISAXS analysis: D (*diamond*) island spacing, d (*square*) island diameter, h (*circle*) island height. From [30]

aligned along the $[110]_{MgO}$ direction demonstrates that the islands display (111) lateral facets. The facetting is driven by the cube on cube epitaxy $((100)_{Pd} \parallel (100)_{MgO}$ with $[100]_{Pd} \parallel [100]_{MgO})$. An analysis of the scattering pattern was undertaken with several shapes compatible with the island facetting. The best results were obtained with truncated cubooctahedron (top panel of Fig. 7.21), the sensitivity to the shape coming mainly from the high $q_\parallel$ range.

This finding is compatible with the ex situ transmission electron microscopy (TEM) plane view of the final deposit (see Fig. 7.20-b). An example of analysis is shown in Fig. 7.21 for the 1.5 nm thick deposit. Four intensity cross sections (two for each main azimuths with one along the Yoneda peak the other along the perpendicular direction at the position of the correlation peak) were simultaneously

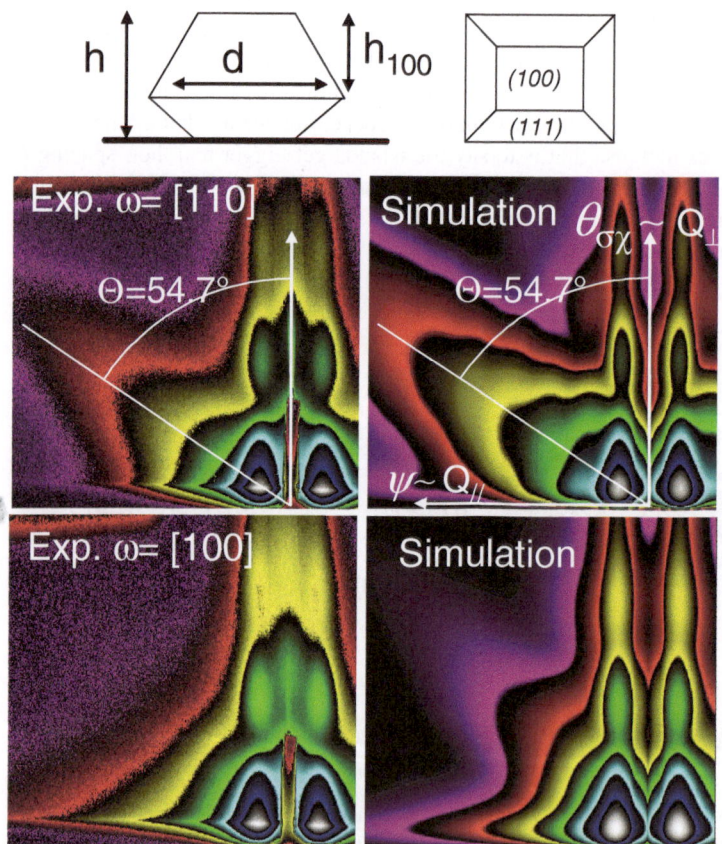

**Fig. 7.21** Experimental (*left column*) and simulated (*right column*) GISAXS pattern for a 1.5 nm thick Pd/MgO(100) deposit for two azimuthal sample orientations (*top images*: beam along $[110]_{MgO}$; *down images*: beam along $[100]_{MgO}$). The particles were modeled by truncated cubooctahedron. The scattering rod tilted by 54.7° is due to $(111)_{Pd}$ side facets; it shows up only along the $[110]_{MgO}$ azimuth because of the cube on cube epitaxy $(100)_{MgO} \parallel (100)_{Pd}$ with $[100]_{MgO} \parallel [100]_{Pd}$. From [30]

fitted. The simulation was performed using DWBA for islands with size dispersed truncated cubooctahedron and with an interference function adjusted on the TEM micrographs. LMA instead of DA was used. The size distributions were log-normal for the in-plane radius $R$ while the height $H$ was normally distributed. The obtained parameters $2R = 12.6 \, \text{nm}$, $\sigma_R(FWHM) = 6 \, \text{nm}$, $D = 21 \, \text{nm}$, $h_{001} = 5.8 \, \text{nm}$ and $h = 7.9 \, \text{nm}$ allow to reproduce the main features of the GISAXS patterns, i.e., (i) the four order of magnitude in the scattered intensity, (ii) the facet scattering rod for the beam along $[110]_{MgO}$ direction only and (iii) the perpendicular interference fringes. It is worth noticing that the specular rod was not included in the simulation. Such fit applied all along the growth gives the evolution of the mean morphological parameters as shown in Fig. 7.20-c. Three regimes in the island spacing can be distinguished: nucleation when the island density increases, particle growth at constant island density and coalescence. It is interesting to notice that the particle aspect ratio $d/2R \sim 0.62$ (height over lateral size) keeps constant until the coalescence regime; at this growth temperature, the islands are close to the equilibrium shape given by the Wulff–Kaishew construction. The adhesion energy $\beta = 1.12 \, \text{J/m}^2$ between the metal and the oxide can be deduced in one shot from the GISAXS morphological parameters; otherwise, such a determination would need lengthy TEM plane view and cross section micrographs.

Another example of analysis from [64] is shown in Fig. 7.22. The same interface is under concern but elaborated at a lower temperature $T = 550 \, \text{K}$; surface diffusion is slowed down and the islands are less faceted than at $T = 650 \, \text{K}$ as seen by microscopy (Fig. 7.22-a). As the GISAXS 2D patterns keep constant upon substrate azimuthal rotation, the chosen shape for the analysis was a truncated sphere; other shapes were unable to reproduce the intensity decrease in the high-$q$ range. The use of LMA was mandatory to reproduce the exact shape of scattering in the parallel direction. DA produced a two-intense intensity below the correlation peak that is not observed in the experiments. A close agreement between the GISAXS results and ex situ TEM was obtained ensuring the capability of the scattering to get accurate morphological parameters.

## 7.6.2 Self-Organized Growth of Aggregates: The Co/Au(111) Case

The field of research on organized nanostructures obtained via natural self-organization or surface nanopatterning is very active as the potential applications are numerous to replace top-down techniques of integration by bottom-up ones in the field of microelectronic. As a scattering technique, GISAXS is very sensitive to the early beginnings of spatial organization of nanostructures; the sharpening of the interference function peaks can be used as a monitoring tool to achieve the best organization while scanning the growth parameters.

The archetype of self-organized growth Co/Au(111) was studied in situ with GISAXS [74]. The (111) face of gold displays a complex surface reconstruction

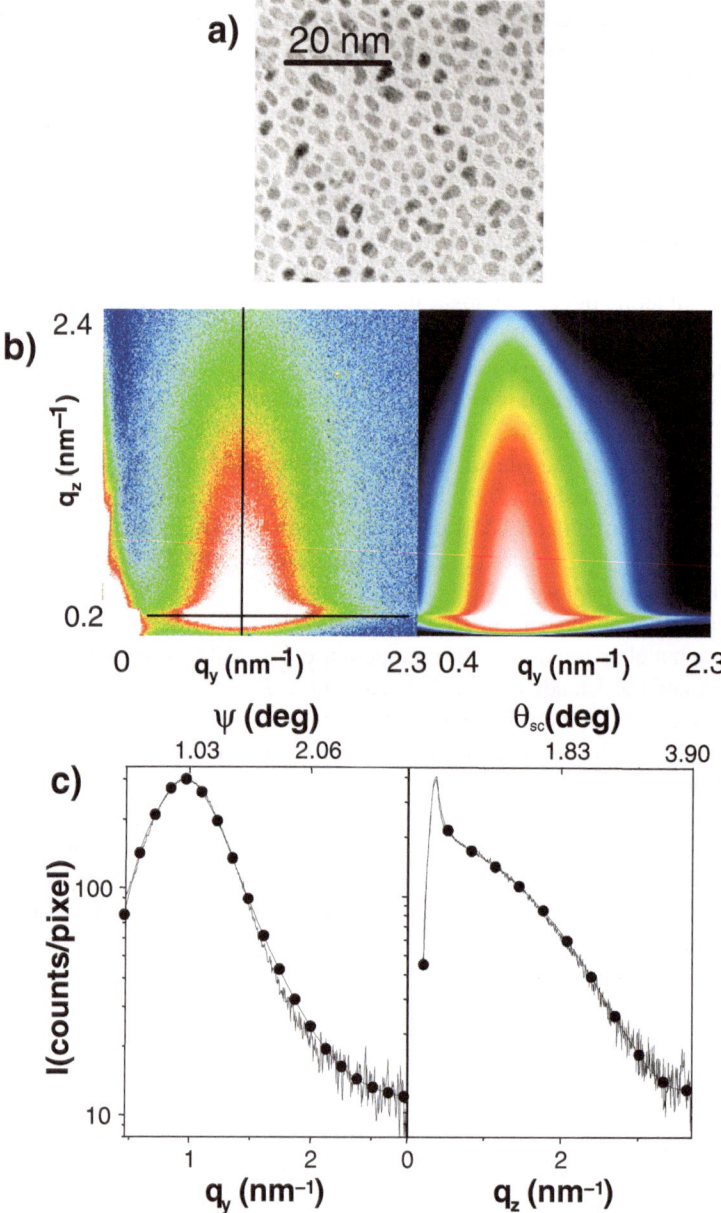

**Fig. 7.22** (a) Transmission electron microcopy image of a 0.9 nm thick Pg/MgO(100) deposit grown at 550 K. (b) Experimental and simulated GISAXS patterns on logarithmic color scale. (c) Cuts of intensity along the $q_y \sim \psi$ (*left*) and $q_z \sim \theta_{sc}$ (*right*) as indicated on the 2D experimental patterns. The symbols correspond to the simulated signal with truncated spheres while the continuous line corresponds to the experiment. The found parameters were $\langle R \rangle = 1.66$ nm, $D = 6.17$ nm, $\langle H \rangle = 2.06$ nm, $\sigma_R(FWHM) = 1.3$ nm as compared to TEM results: $R_{TEM} = 2 \pm 0.4$ nm, $D_{TEM} = 7.9 \pm 1$ nm, $\sigma_R(FWHM) = 1.3$. From [64]

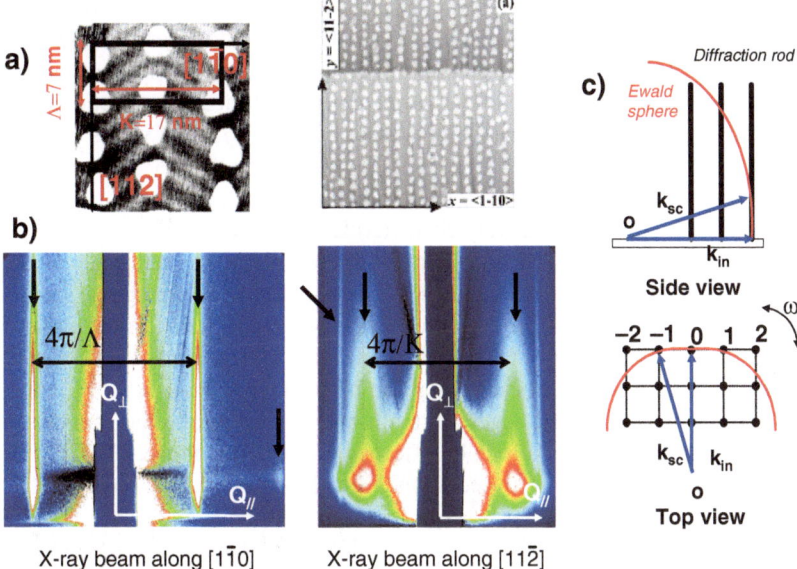

**Fig. 7.23** (**a**) Scanning tunneling microscopy images of 0.4 ML of Co on the herringbone re-construction of Au(111). Rows of cobalt dots appears on the (150 × 150 nm) large-scale image. The rectangular super-cell ($\Lambda \sim 7$ nm, $\kappa \sim 17$ nm) shown on the left image contains two dots and is aligned along the $[11\bar{2}]_{Au}$ and $[1\bar{1}0]_{Au}$. (**b**) GISAXS pattern with the beam aligned along two orthogonal directions. Sharp scattering rods (*arrows on figure*) up to second order appears when the intra-row order is probed. On the contrary, the cumulative inter-rows order leads to broader peaks. Note the appearing of a sharp rod (*tilted arrow*) from an other variant on the right image. (**c**) Representation of the reciprocal space of a 2D lattice. From [74]

known as the herringbone reconstruction (see Fig. 7.23-a). On the growth point of view, the nucleation of Co dots takes place at the elbow of the reconstruction; a network of nanoparticles is obtained with a nanometer scale rectangular unit cell oriented along the $[1\bar{1}0]_{Au}$ and $[1\bar{1}2]_{Au}$ directions. The symmetry 2 of the recon-struction combined with the threefold symmetry of the substrate leads to three vari-ants rotated by 120° depending on the terraces. The typical GISAXS patterns of Fig. 7.23-b were acquired after the growth of 0.4 monolayers (ML) of Co. Com-pared to standard nucleation and growth (see Sect. 7.6.1), sharp scattering rod are visible when the beam is aligned along one of the super-cell edge. Even second-order diffraction is visible when the beam is along $[1\bar{1}0]_{Au}$ . The reciprocal space is a 2D lattice of rods that intersects the Ewald sphere (see Fig. 7.23-c). Indeed as the Co dots are very flat (two atomic layers), the form factor decreases very slowly along the perpendicular direction. It is worth noticing that the curvature of the Ewald sphere or, in other words, the forward wavevector transfer $q_x$ cannot be neglected at all; by rotating the sample, it is possible to see the rods intersecting the Ewald sphere in an out-of-plane ($\theta_{sc} \neq 0$) location while in the perfect alignment condi-tions, the rods are tangent to the Ewald sphere. However, the rod shape depends on the azimuth. For a beam aligned along $[1\bar{1}2]_{Au}$, the narrow diffraction rods convey a long range order with a domain size (evaluated from the inverse of the peak width) of around 300 nm. This intra-dot-row order contrasts with the inter-row spacing fluctuations linked to variation of the intrinsic period of the reconstruction. This cumulative or liquid-like disorder (see microscopy on Fig. 7.23-a) yields broader peaks.

By rotating the sample, a full X-ray super-cell crystallography at the nanometer scale was undertaken to quantitatively characterize the degree of order in this system [36]. In agreement with scanning tunneling microscopy, the unit cell chosen for the simulations was rectangular with a pattern made of two triangular islands rotated by 180° (see Fig. 7.24-c). The three variants were included in the simulation. The particle radius was normally distributed while its height was kept constant around two atomic layers. The centering of the unit cell was fitted as its varies all along the growth. The unit cell edges were allowed to fluctuate in the framework of the perfect 2D paracrystal. Refraction effects were included in the DWBA. The fit results for the previously introduced images are shown in Fig. 7.24-a as well as the simulated GISAXS patterns for two special azimuths. The main result is that the degree of position disorder remains constant until the coalescence; it is in fact fully determined by the substrate itself.

The last feature is the reminiscence of a sharp scattering rods well above the static coalescence of the dots (up to nine monolayers) while STM topography tends to

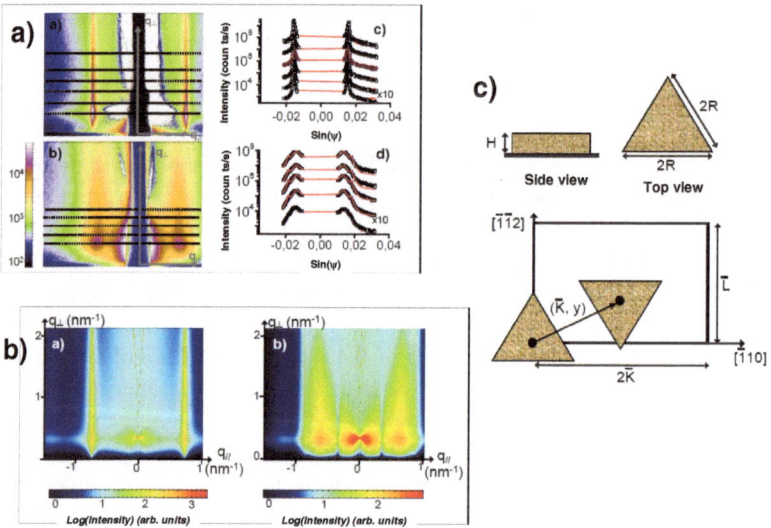

**Fig. 7.24** (**a**) Fits within the DWBA along the intensity cuts on the two previously shown images for a 0.4 ML Co deposit on Au(111). (**b**) Simulated GISAXS images with the found morphological parameters. (**c**) Unit cell used for the GISAXS fits filled with two triangular islands rotated by 180°. The used lattice is a 2D paracrystal with three domains rotated by 120° to account for the three variants. From [36] with permission

smooth out [74]. This demonstrates that the GISAXS signal is sensitive to a nanos-tructuration buried in the Co layer and inaccessible to STM measurements. Both periodic strain field inside the gold substrate and a periodic array of Co grain bound-aries are possible tracks of explanations. An anomalous GISAXS experiment at the K-edge of Co [36] has demonstrated that, within the error bars, the signal comes from the cobalt layer, thus corroborating the model of scattering from periodic grain boundaries.

## 7.7 Soft Condensed Matter GISAXS Studies: A Nanometer Scale Crystallographic Study of Self-Organization in Templated Silica Thin Films

Since the pioneering work of Mobil researchers, surfactants are used as supra-molecular templates to self-assemble inorganic precursors such as silica. The ap-proach takes benefit from the tendency of amphilic organic molecules (surfactants) to self-organize in liquid media into complex 1D, 2D or 3D supramolecular aggre-gates in the 1–50 nm range as function of the concentration. The synthesis of a great variety of new nanomaterials through condensation of the precursors in the organic template has been extended to the fabrication of mesoporous thin films. Various morphologies with a good long range order have been obtained (lamella, hexago-nal, cubic and 3D hexagonal structures). In the case of thin films, the procedure of evaporation-induced self-assembly (EISA) during dip-coating or spin-coating is often used. But, despite the potential utility of such films in various applications (sensors, membranes, catalysts, etc.), little is known on the kinetics of the organi-zation and the degree of ordering inside the film during the sol-condensation and the interplay with external parameters such as the evaporation rate or the external humidity. Microscopy techniques are only able to characterize the final product of-ten after the removal of the organic materiel (such as after a thermal treatment of calcination). GISAXS appears as an ideal tool to probe such an organization pro-cess in thin films owing to its sensitivity to buried interfaces and to ordering at the nanometer scale.

The role of humidity (RH) during the formation of templated silica thin films have been monitored in real time [28] on a film ($\sim 100$ nm) formed by evapora-tion of a sol containing surfactant and silica precursor on a silicon substrate. The sample was kept in a x-ray compatible cell in which the water partial pressure is controlled by flowing humid or dry nitrogen. The EISA process and the effects of RH cycles were followed by acquiring sequences of GISAXS patterns on a CCD detector (Fig. 7.25) located at 0.735 m from the sample. The measurements were performed on a liquid spectrometer at NSLS (National Synchrotron Light Source, USA); the sample is kept horizontal and the beam is deflected by Ge monochroma-tor and impinges at an incident angle close to the critical angle of the substrate. Pattern Fig. 7.25-a displays only the specular rod hidden by the beam-stop and corresponds to a disordered dilute phase. After complete solvent evaporation, the

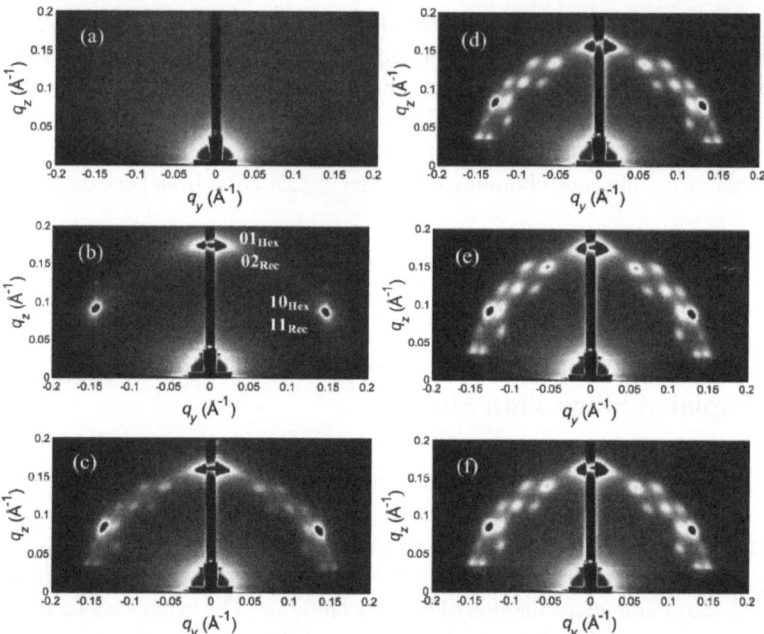

**Fig. 7.25** Evolution of GISAXS patterns during the transformation of the sol liquid film to the structured mesophases. (**a**) Liquid film, (**b**) 2D hexagonal phase (RH = 0.4), (**c**) 2D hexagonal phase and onset of the cubic phase, (**d**) 2D hexagonal + cubic phases at RH = 0.8, (**e**) 2D hexagonal + cubic phases at RH = 0.3, (**f**) 2D hexagonal + cubic phases at RH = 0.8. The wavevectors transfers are given on figure. From [28]

pattern is typical of a 2D hexagonal phase [25, 27] of $p6m$ symmetry made of cylindrical micelles aligned parallel to the substrate; of course, the sample is in-plane polycrystalline. However, at this stage, the film is still modulable and sensitive to a water uptake induced by the relative humidity. Raising RH from 0.4 to 0.8 transforms the 2D hexagonal phase to a cubic $Pm3n$ phase. Indeed, the penetration of water molecules between surfactant headgroups transform cylindrical micelles into spherical ones. The cubic scattering patterns consists of three concentric circles on which the $(200)_{cub}$, $(210)_{cub}$ and $(211)_{cub}$ Bragg reflections are aligned. This cubic phase is stable upon reduction of RH up to 0.3 with average cubic parameters $a = b = c = 8.96$ nm. This phase is located on top of the film as demonstrated by varying the incident angle and thus the probed depth. At variance, the hexagonal one located at the bottom region is much more sensitive to RH and distorts. The hexagonal phase distortion can be quantified by following the $(10)_{hex}$ and $(01)_{hex}$ Bragg reflections shown in Fig. 7.25. While at the beginning the hexagonal phase is characterized by a unique lattice parameter $a_{hex} = 8.29$ nm, it is more convenient to introduce a face-centered rectangular unit cell (see Fig. 7.26) defined by parameters $b$ and $c$ such that the reflections $(11)_{rec}$ is equivalent to $(10)_{hex}$ and $(02)_{rec}$ to $(01)_{hex}$. These parameters $b, c$ and the distortion $\eta = (\sqrt{3} - c/b)/\sqrt{3}$ defined

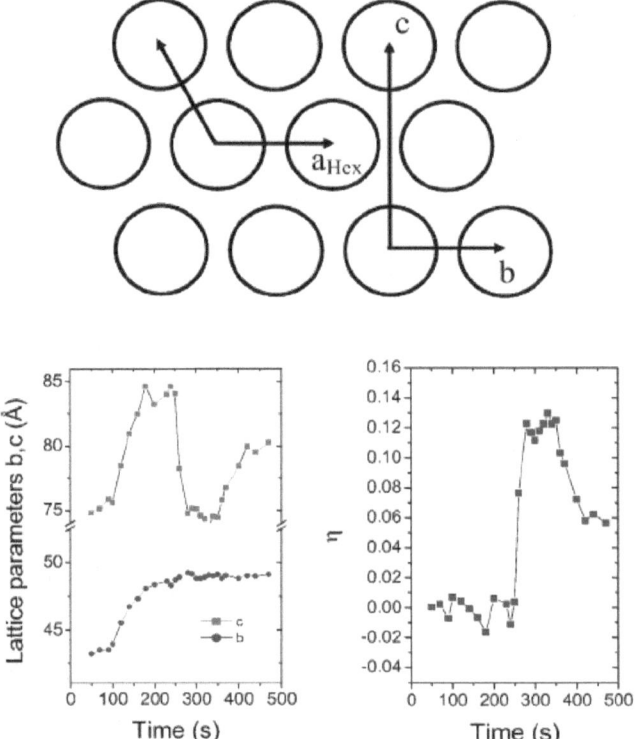

**Fig. 7.26** Drawing of the hexagonal and rectangular face-centered unit cells. Evolution of the lattice parameters $b, c$ as function of time and cycles of RH. From [28]

as the departure of $c/b$ from its $\sqrt{3}$ hexagonal value as function of RH are given in Fig. 7.26. A special care has to be taken to unfold the refraction effect in order to extract correctly the $c$ parameter. An irreversible behavior on the $b$ parameter is observed after the first humidity cycle (up to 250 s in Fig. 7.26). Both parameters $b$ and $c$ increase (from 7.4 to 8.5 nm for $c$, from 4.2 to 4.8 nm for $b$) in such a way that there is no overall distortion of the unit cell ($\eta \sim 0$). This correspond roughly to an uptake of two monolayers of water molecules per micelle. After this transient state with an exponential temporal behavior, the $b$ parameter is pinned and the film can only swell in the perpendicular direction but with a reduced flexibility due to aging of the silica network. Thickness measurements showed that it is the entire film that evolves demonstrating a quick intrusion of water molecules through the porous channels or through the grain boundaries. As the condensation of the silica network should be isotropic, this anisotropic behavior on the lattice parameters is related to patchwork of randomly oriented domains made of cylindrical micelle domains parallel to the substrate. The in-plane mosaicity hinders the dilatation along the substrate while the film/vapor interface introduces a degree of freedom in the perpendicular direction.

# References

1. Guinier, A., Fournet, G.: Small-Angle Scattering of X-Rays, Jonh Wiley & Sons, New York, (1955)
2. Glatter, G., Kratky, O.: Small Angle X-Ray Scattering, Academic Press, (1982)
3. Levine, J.R., Cohen, J.B., Chung, Y.W., Georgopoulos, P.: J. Appl. Cryst. **22**, 528 (1989)
4. Levine, J.R., Cohen, J.B., Chung, Y.W: Surf. Sci. **248**, 215 (1991)
5. Marra, W., Eisenberger, P., Cho, A.: J. Appl. Phys. **50**, 6927 (1979)
6. Robinson, I.K., Surface Crystallography, vol. 3 Elsevier, Amsterdam & New York, (1991)
7. Naudon, A., Slimani, T., Goudeau, P.: J. Appl. Cryst. **24**, 501 (1991)
8. Naudon, A., Thiaudière, D.: J. Appl. Cryst. **30**, 822 (1997)
9. Robach, O., Renaud, G., Barbier, A.: Phys. Rev. B **60**, 5858 (1999)
10. Metzger, T.H., Kegel, I., Paniago, R., Peisl, J.: J. Phys. D: Appl. Phys. **32**, A202 (1999)
11. Zhang, K., Heyn, C., Hansen, W., Schmidt, T., Falta, J.: Appl. Phys. Lett. **76**, 2229 (2000)
12. Schmidbauer, M., Wiebach, T., Raidt, H., Hanke, M., Köhler, R., Wawra, H.: Phys. Rev. B **58**, 10523 (1998)
13. Kegel, I., Metzger, T.H., Peisl, J.: Appl. Phys. Lett. **74**, 2978 (1999)
14. Stangl, J., Holý, V., Roch, T., Daniel, A., Bauer, G., Zhu, J., Brunner, K., Abstreiter, G.: Phys. Rev. B **62**, 7229 (2000)
15. Holý, V., Roch, T., Stangl, J., Daniel, A., Bauer, G., Metzger, T.H., Zhu, Y.H., Brunner, K., Abstreiter, G.: Phys. Rev. B **63**, 205318 (2001)
16. Stangl, J., Holý, V., Mikulik, P., Bauer, G., Kegel, I., Metzger, T.H., Schmidt, O.G., Lange, C., Eberl, K.: Appl. Phys. Lett. **74**, 3785 (1999)
17. Babonneau, D., Petroff, F., Maurice, J., Fettar, F., Vaurès, A., Naudon, A.: Appl. Phys. Lett. **76**, 2892 (2000)
18. Babonneau, D., Videnović, I.R., Garnier, M.G., Oelhafen, P.: Phys. Rev. B **63**, 195401 (2001)
19. Babonneau, D., Pailloux, F., Eymery, J.-P., Denanot, M.-F., Guérin, P., Fonda, E., Lyon, O.: Phys. Rev. B **71**, 035430 (2005)
20. Babonneau, D., Naudon, A., Thiaudière, D.: J. Appl. Cryst. **32**, 226 (1999)
21. Gibaud, A., Hazra, S., Sella, C., Laffez, P., Désert, A., Naudon, A., Van Tendeloo, G., Phys. Rev. B **63**, 193407 (2001)
22. Müller-Buschbaum, P., Vanhoorne, P., Scheumann, V., Stamm, M.: Europhys. Lett. **40**, 655 (1997)
23. Müller-Buschbaum, P., Gutmann, J., Cubitt, R., Stamm, M.: Colloid Polym Sci. **277**, 1193 (1999)
24. Vignaud, G., Gibaud, A., Wang, J., Sinha, S., Daillant, J., Grüber, G., Gallot, Y.: J. Phys. Condens. Matter. **9**, L125 (1997)
25. Gibaud, A., Grosso, D., Smarsly, B., Baptiste, A., Brdeau, J., Babonneau, F., Doshi, D., Chen, Z., Jeffrey Brinker, C., Sanchez, C.: J. Phys. Chem. B **107**, 6114 (2003)
26. Doshi, D., Gibaud, A., Goletto, V., Lu, M., Gerung, H., Ocko, B., Han, S., Brinker, C.J.: J. Am. Chem. Soc. **125**, 11646 (2003)
27. Gibaud, A., Baptiste, A., Doshi, D., Brinker, C.J., Yang, L., Ocko, B.: Europhys. Lett. **63**, 833 (2003)
28. Gibaud, A., Dourdain, S., Gang, O., Ocko, B.: Phys. Rev. B **70**, 161403 (2004)
29. Lairson, B.M., Payne, A.P., Brennan, S., Rensing, N.M., Daniels, B.J., Clemens, B.M.: J. Appl. Phys. **78**, 4449 (1995)
30. Renaud, G., Lazzari, R., Revenant, C., Barbier, A., Noblet, M., Ulrich, O., Leroy, F., Jupille, J., Borenstzein, Y., Henry, C.R., et al.: Science **300**, 1416 (2003)
31. Müller-Buschbaum, P., Casangrande, M., Gutmann, J., Kuhlmann, T., Stamm, M., Von Krosigk, G., Lode, U., Cunis, S., Gehrke, R.: Europhys. Lett. **42**, 517 (1998)
32. Roth, S., Burghammer, M., Riekel, C., Müller-Buschbaum, P., Diethert, A., Panagiotou, P., Walter, H.: Appl. Phys. Lett. **82**, 1935 (2003)
33. Roth, S., Müller-Buschbaum, P., Burghammer, M., Walter, H., Panagiotou, P., Diethert, A., Riekel, C.: Spectrochimica Acta Part B **59**, 1765 (2004)

34. Renaud, G., Ducruet, M., Ulrich, O., Lazzari, R.: Nucl. Inst. Meth. B **222**, 667 (2004)
35. Lager, M.C., Bailly, A., Dolle, P., Baudoing-Savois, R., Taunier, P., Garaudée, S., Cuccaro, S., Douillet, S., Geaymond, O., Perroux, G., Tissot, O.: Rev. Sci. Inst. **78**, 083902 (2007)
36. Leroy, F.: Ph.D. thesis, Université Joseph Fourier, Grenoble, France (2004)
37. Messiah, A., Quamtum mechanics, vol. 1–2 Dunod, Paris, (1964)
38. Vineyard, G., Phys. Rev. B **26**, 4146 (1982)
39. Sinha, S.K., Sirota, E.B., Garoff, S., Stanley, H.B.: Phys. Rev. B **38**, 2297 (1988)
40. Holý, V., Kubuena, J., Ohlídal, I., Lischka, K., Plotz, W.: Phys. Rev. B **47**, 15896 (1993)
41. Holý, V., Baumbach, T.: Phys. Rev. B **49**, 10668 (1994)
42. Mikulík, P., Baumbach, T.: Phys. Rev. B **59**, 7632 (1999)
43. de Boer, D., Phys. Rev. B **44**, 498 (1991)
44. de Boer, D., Phys. Rev. B **53**, 6048 (1996)
45. Lee, D., Sinha, S., Haskel, D., Choi, Y., Lang, J., Stepanov, S., Srajer, G.: Phys. Rev. B **68**, 224409 (2003)
46. Lee, D., Sinha, S., Nelson, C., Lang, J., Venkataraman, C., Srajer, G., Osgood, R.: Phys. Rev. B **68**, 224410 (2003)
47. Rauscher, M., Salditt, T., Spohn, H.: Phys. Rev. B **52**, 16855 (1995)
48. Rauscher, M., Paniago, R., Metzger, H., Kovats, Z., Domke, J., Pfannes, H.D., Schulze, J., Eisele, I.: J. Appl. Phys. **86**, 6763 (1999)
49. Weber, W., Lengeler, B.: Phys. Rev. B **46**, 7953 (1992)
50. Lazzari, R., J. Appl. Cryst. **35**, 406 (2002)
51. Salditt, T., Lott, D., Metzger, T.H., Peisl, J., Vignaud, G., Høghøj, P., Schärpf, O., Hinze, P., Lauer, R.: Phys. Rev. B **54**, 5860 (1996)
52. Yoneda, Y.: Phys. Rev. **131**, 2010 (1963)
53. Morse, P.M., Feshbach, H.: Methods of Theoretical Physics, vol. Part 1 and 2 New-York, (1953)
54. James, R.W.: The Optical Principles of the Diffraction of X-Rays, Cornell University Press, New-York, (1965)
55. Guinier, A.: Théorie et technique de la radiocristallographie, Dunod, Paris, (1956)
56. Warren, B.E.: X-Ray Diffraction, Dover Publication, Inc, New York, (1969)
57. Waseda, Y., The Structure of Non-Crystalline Materials, Mc. Graw-Hill, New-York, (1980)
58. Lazzari, R., Leroy, F., Renaud, G.: Phys. Rev. B **76**, 125411 (2007)
59. Parratt, L., Phys. Rev. **95**, 359 (1954)
60. Kotlarchyk, M., Chen, S.H.: J. Chem. Phys. **79**, 2461 (1983)
61. Pedersen, J.S.: Phys. Rev. B **47**, 657 (1993)
62. Pedersen, J.S.: J. Appl. Cryst. **27**, 595 (1994)
63. Leroy, F., Lazzari, R., Renaud, G.: Act. Cryts. A **60**, 565 (2004)
64. Revenant, C., Leroy, F., Lazzari, R., Renaud, G., Henry, C.R.: Phys. Rev. B **69**, 035411 (2004)
65. Robach, O.: Ph.D. thesis, Université Joseph Fourier Grenoble I, France, (1997)
66. Thiaudière, D.: Ph.D. thesis, Université de Poitiers, Poiters (1996)
67. Balescu, R.: Equilibrium and Nonequilibrium Statistical mechanics, Jonh Wiley & Sons, New York London Sidney Toronto, (1975)
68. Hosemann, R., Acta. Cryst. **4**, 520 (1951)
69. Hosemann, R., Bagchi, S.N.: Direct Analysis of Diffraction by Matter, North-Holland Publishing Company, Amsterdam, (1962)
70. Mu, X.-Q.: Acta. Cryst. A **54**, 606 (1998)
71. Lazzari, R., Leroy, F., Renaud, G., Jupille, J.: Phys. Rev. B **76**, 125412 (2007)
72. Müller-Buschbaum, P., Bauer, E., Maurer, E.: Schlögl, K., Appl. Phys. Lett. **88**, 083114 (2006)
73. Leroy, F., Eymery, J., Buttard, D., Renaud, G., Lazzari, R., Fournel, F.: Appl. Phys. Lett. **82**, 2598 (2003)
74. Fruchart, O., Renaud, G., Barbier, A., Noblet, M., ULrich, O., Deville, J.-P., Scheurer, F., Mane-Mane, J., Repain, V., Baudot, G., et al.: Europhys. Lett. **63**, 275 (2003)
75. Busson, B., Doucet, J.: Acta. Cryst. A **56**, 68 (2000)
76. Gazzillo, D., Giacometti, A., Carsughi, F.: Phys. Rev. E **60**, 6722 (1999)

77. Gazzillo, D., Giacometti, A., Guido Della Valle, R., Carsughi, F.: J. Chem. Phys. **111**, 7636 (1999)
78. Vrij, A.: J. Chem. Phys. **69**, 1742 (1978)
79. Vrij, A.: J. Chem. Phys. **71**, 3267 (1979)
80. Leroy, F., Eymery, J., Buttard, D., Renaud, G., Lazzari, R.: J. of. Cryst. Growth **275**, e2195 (2005)

# Main Notation Used in This Book

| | |
|---|---|
| $z$ | Direction normal to the surface |
| $x, y$ | Directions in the plane of the surface |
| $\parallel$ | Used to describe a component parallel to the interface plane |
| $xOz$ | Plane of incidence |
| $j$ | Label of layer. Numbering of layers goes from 0 (upper medium) to $N$ the last layer. $s$ is the substrate |
| $Z_j$ | Average location of the $j-1, j$ interface |
| $z_j(x, y)$ | Fluctuations of the interface location around $Z_j$ |
| $\mathbf{k}$ | Wave-vector |
| $\mathbf{k}_{in}, \mathbf{k}_r, \mathbf{k}_{tr}, \mathbf{k}_{sc}$ | Incident, reflected, transmitted and scattered wave vectors |
| $k_{in\,z,j}$ | $z$ component of the incident wavevector in the $j$th layer |
| $k_{z,j}$ | when unambiguous |
| $\mathbf{q}$ | Wave vector transfer |
| $q$ | Modulus of the wave vector transfer |
| $q_x, q_\parallel, q_z$ | Components of the wave vector |
| $\mathbf{u}$ | Scattering direction |
| $r, t$ | Reflection and transmission coefficients in amplitude |
| $R, T$ | Intensity reflection and transmission coefficients |
| $r_{j-1,j}$ | Reflection coefficient in amplitude when passing from medium $j-1$ to medium $j$ |
| $t_{j-1,j}$ | Transmission coefficient in amplitude when passing from medium $j-1$ to medium $j$ |
| $\mathbf{E}$ | Electric field |
| $\hat{\mathbf{e}}_{in}, \hat{\mathbf{e}}_{sc}$ | Polarisation vectors of the incident and scattered fields |
| $\mathbf{B}$ | Magnetic field |
| $\mathbf{j}$ | Current density |
| $\mathbf{P}$ | Electric polarisation |
| $\mathbf{A}$ | Vector potential |
| $\mathbf{S}$ | Poynting's vector |

| | |
|---|---|
| $A_j^{\pm}$ | Amplitude of the upwards and downwards propagating electric fields in layer $j$ |
| $U(\pm k_{\text{in } z,j}, z)$ | $A_j^{\pm} e^{\pm k_{\text{in } z,j} z}$ |
| $\mathcal{M}$ | Transfer matrix |
| $p_n$ | $n$-point probability distribution |
| $\sigma$ | rms roughness. $\sigma^2 = \langle z^2 \rangle$ |
| $C_{zz}(x_1, x_2, y_1, y_2)$ | Height–height correlation function. Also denoted $\langle z(x_1, y_1) z(x_2, y_2) \rangle$ |
| $g(r)$ | $2\sigma^2 - 2C_{zz}(x_1, x_2, y_1, y_2)$ |
| $G$ | Green function |
| $\overline{\overline{\mathcal{G}}}$ | Green tensor (electromagnetic case) |

$e^{i(\omega t - \mathbf{k} \cdot \mathbf{r})}$ waves are used except in Chap. 5 devoted to neutron reflectivity (see Sect. 1.2.1 for details related to the conventions used in this book, and Sect. 5.1 for the notation used in Chap. 5).

**Table 7.1** Typical length scales for x-ray reflectivity experiments

| | Definition | Value |
|---|---|---|
| Wavelength $\lambda$ | | $1\,\text{Å}$ |
| Scattering length | $b$ | $r_e = 2.818 \times 10^{-15}\,\text{m}$ for 1 electron |
| Extinction length | $L_e = \frac{\lambda}{2\pi|n-1|}$ | $1\,\mu\text{m}$ |
| Longitudinal coherence length | $\lambda^2/\delta\lambda$ | $1\,\mu\text{m}$ |
| Incidence slit opening | | $0.1\,\text{mm}$ |
| Detector slit opening normal to the plane of incidence ($y$) | $h_y$ | $10\,\text{mm}$ |
| Detector slit opening in the plane of incidence ($x$) | $h_x$ | $0.1\text{–}1\,\text{mm}$ |
| Sample-to-detector distance | $L$ | $1\,\text{m}$ |
| Transverse coherence length normal to the plane of incidence ($y$) (when fixed by the detector) | $\lambda/\Delta\theta_y$ with $\Delta\theta_y = h_y/L$ | $10\,\text{nm}$ |
| Transverse coherence length in the plane of incidence projected on the surface ($x$) (when fixed by the detector) | $\lambda/(\theta\Delta\theta)$ with $\Delta\theta_x = h_x/L$ | $100\,\mu\text{m}$ for $\theta = 10\,\text{mrad}$ |
| Illuminated area (length × width) | | $(0.1\,\text{mm}/\theta) \times$ $(1\text{–}10\,\text{mm})$ |
| Absorption length | $\mu = \lambda/4\pi\beta$ | $0.1\text{–}1\,\text{mm}$ for $\beta = 10^{-7}\text{–}10^{-8}$ |

# Index